GEORGE ALBERT BOULENGER

CATALOGUE
OF THE
LIZARDS
IN THE
BRITISH MUSEUM
(NATURAL HISTORY)

VOLUME I
GECKONIDÆ, EUBLEPHARIDÆ, UROPLATIDÆ,
PYGOPODIDÆ, AGAMIDÆ

Elibron Classics
www.elibron.com

Elibron Classics series.

© 2005 Adamant Media Corporation.

ISBN 1-4021-7015-7 (paperback)
ISBN 1-4021-0201-1 (hardcover)

This Elibron Classics Replica Edition is an unabridged facsimile
of the edition published in 1885 by the British Museum,
London.

Elibron and Elibron Classics are trademarks of
Adamant Media Corporation. All rights reserved.

This book is an accurate reproduction of the original. Any marks, names, colophons, imprints, logos or other symbols or identifiers that appear on or in this book, except for those of Adamant Media Corporation and BookSurge, LLC, are used only for historical reference and accuracy and are not meant to designate origin or imply any sponsorship by or license from any third party.

CATALOGUE

OF THE

LIZARDS

IN THE

BRITISH MUSEUM
(NATURAL HISTORY).

SECOND EDITION.

BY

GEORGE ALBERT BOULENGER.

VOLUME I.

GECKONIDÆ, EUBLEPHARIDÆ, UROPLATIDÆ,
PYGOPODIDÆ, AGAMIDÆ.

LONDON:
PRINTED BY ORDER OF THE TRUSTEES.
1885.

PRINTED BY TAYLOR AND FRANCIS,
RED LION COURT, FLEET STREET.

PREFACE.

No other order of Reptiles required so thorough an examination and rearrangement as that of Lizards. The descriptions of nearly two thirds of the species known at present were scattered over the wide range of the literature of the last forty years, and, in consequence, except to some very few zoologists, the exact determination of specimens of Lizards had become an impossible task, or, at least, one to which a great risk of failure was attached. By the student of physiogeography the absence of a critical general account of so important a type of Reptiles was still more seriously felt.

The first edition of the 'Catalogue of Lizards,' published in the year 1845, was based on a collection containing only one eighth of the number of specimens at present in the Natural History Museum, and, therefore, had long ceased to fulfil its primary purpose, viz. to serve as a guide to the collection.

Like all the other volumes of the new series of descriptive Catalogues of the Zoological Collections, the present work contains descriptions of, or references to, all the species introduced into the literature. It will consist of three volumes, and may be expected to be completed in 1886, the manuscript of the second volume being far advanced.

ALBERT GÜNTHER,
Keeper of the Department of Zoology.

British Museum, N. H.,
January 8, 1885.

INTRODUCTION.

This volume contains an account of all the species of Lizards belonging to the families *Geckonidæ, Eublepharidæ, Uroplatidæ, Pygopodidæ,* and *Agamidæ,* of which descriptions have been published. Over three fourths of the species described have been examined by myself.

The following Table will show the great progress made in our knowledge of species since the publication of the last general works on the subject, viz. Duméril and Bibron's 'Erpétologie Générale,' vols. iii.-v., 1836-1839, and Gray's first edition of this Catalogue in 1845:—

	Number of Species characterized		
Families.	by Dum. & Bibr.	by Gray.	in present volume*.
Geckonidæ	53	97	270
Eublepharidæ	—	1	7
Uroplatidæ	2	2	3
Pygopodidæ	2	7	8
Agamidæ	50	79	202
Total..	107	186	490

A comparison of the numbers of species and specimens in the National Collection in 1845 and at the present date gives the following result:—

---

* Only those species to which I have appended an ordinal number are included in this estimate, without those which are doubtful and merely referred to in footnotes or otherwise.

	1845.		1885.	
	Species.	Specimens.	Species.	Specimens.
Geckonidæ	78	166	199	1773
Eublepharidæ	1	1	6	19
Uroplatidæ	1	2	1	5
Pygopodidæ	7	20	5	93
Agamidæ	65	239	159	1265
Total	152	428	370	3155

An outline of the classification followed in this work I have recently published in Ann. & Mag. Nat. Hist. xiv. 1884, p. 117.

I have given the principal measurements of most of the species, taken from the largest or most perfect specimen in the collection. The "length of the head" is measured to the occipital condyle, and the "length of the body" signifies the distance between the latter point and the anal cleft.

The affixes to the names of donors &c., in the third column of the list of specimens, may be explained as follows:—"[P.]" signifies "Presented by;" "[C.]"="Collected by;" "[E.]"="Obtained by exchange." Where none of these signs are employed, the specimens were purchased.

G. A. BOULENGER.

British Museum, N. H.,
January 8, 1885.

SYSTEMATIC INDEX.

LACERTILIA.

Subord. I. LACERTILIA VERA.

Fam. 1. GECKONIDÆ.

	Page
1. Nephrurus, *Gthr.*	9
1. asper, *Gthr.*	9
2. Chondrodactylus, *Ptrs.*	10
1. angulifer, *Ptrs.*	11
3. Rhynchœdura, *Gthr.*	11
1. ornata, *Gthr.*	12
4. Teratoscincus, *Strauch*	12
1. scincus, *Schleg.*	12
5. Ceramodactylus, *Blanf.*	13
1. doriæ, *Blanf.*	13
2. affinis, *Murray*	14
6. Ptenopus, *Gray*	15
1. garrulus, *Smith*	15
7. Stenodactylus, *Fitz.*	16
1. orientalis, *Blanf.*	16
2. guttatus, *Cuv.*	17
3. wilkinsonii, *Gray*	18
4. petersii, *Blgr.*	18
5. tripolitanus, *Ptrs.*	19
8. Alsophylax, *Fitz.*	19
1. pipiens, *Pall.*	19
2. tuberculatus, *Blanf.*	20
9. Homonota, *Gray*	21
1. darwinii, *Blgr.*	21
2. whitii, *Blgr.*	22
10. Gymnodactylus, *Spix*	22
1. caspius, *Eichw.*	26
2. scaber, *Rüpp.*	27
3. brevipes, *Blanf.*	28
4. kotschyi, *Stdchr.*	29
5. kachhensis, *Stol.*	29
6. heterocercus, *Blanf.*	30
7. elongatus, *Blanf.*	30
8. fasciatus, *D. & B.*	31
9. stoliczkæ, *Stdchr.*	31
10. lawderanus, *Stol.*	32
11. dorbignyi, *D. & B.*	33
12. mauritanicus, *D. & B.*	33
13. trachyblepharus, *Boettg.*	34
14. steudneri, *Ptrs.*	34
15. nebulosus, *Bedd.*	34
16. jeyporensis, *Bedd.*	36
17. deccanensis, *Gthr.*	36
18. albofasciatus, *Blgr.*	37
19. oldhami, *Theob.*	38
20. triedrus, *Gthr.*	38
21. arnouxii, *A. Dum.*	39
22. geckoides, *Spix*	39
23. pelagicus, *Gir.*	40
24. heteronotus, *Blgr.*	41
25. cheverti, *Blgr.*	41
26. affinis, *Stol.*	42
27. frenatus, *Gthr.*	42
28. variegatus, *Blyth*	43
29. fasciolatus, *Blyth*	44
30. khasiensis, *Jerd.*	44
31. marmoratus, *Kuhl*	44
32. rubidus, *Blyth*	45
33. philippinicus, *Stdchr.*	46
34. pulchellus, *Gray*	46
35. consobrinus, *Ptrs.*	47
36. miliusii, *Bory*	48
37. platurus, *White*	49
atropunctatus, *Licht.*	22
tenuis, *Hall.*	22
11. Agamura, *Blanf.*	50
1. cruralis, *Blanf.*	50
2. persica, *A. Dum.*	51
12. Pristurus, *Rüpp.*	52
1. flavipunctatus, *Rüpp.*	52
2. rupestris, *Blanf.*	53
3. insignis, *Blanf.*	54
4. crucifer, *Val.*	55
5. collaris, *Stdchr.*	55
6. carteri, *Gray*	55
13. Gonatodes, *Fitz.*	56
1. albogularis, *D. & B.*	59
2. vittatus, *Licht.*	60
3. ocellatus, *Gray*	60
4. caudiscutatus, *Gthr.*	61
5. concinnatus, *O'Sh.*	61
6. humeralis, *Guich.*	62
gaudichaudii, *D. & B.*	63
7. timorensis, *D. & B.*	63
8. kendallii, *Gray*	63
9. indicus, *Gray*	64
10. wynadensis, *Bedd.*	65
11. sisparensis, *Theob.*	66
12. ornatus, *Bedd.*	66
13. marmoratus, *Bedd.*	67
14. mysoriensis, *Jerd.*	68

	Page		Page
15. kandianus, *Kel.*	68	1. marmorata, *Gray*	104
16. gracilis, *Bedd.*	70	2. ocellata, *Blgr.*	105
17. jerdonii, *Theob.*	71	3. robusta, *Blgr.*	106
18. littoralis, *Jerd.*	71	4. lesueurii, *D. & B.*	107
boiei, *Gray*	72	5. rhombifera, *Gray*	107
australis, *Gray*	72	6. ? verrillii, *Cope*	108
ferrugineus, *Cope*	56	20. Calodactylus, *Bedd.*	108
14. Ælurosaurus, *Blgr.*	73	1. aureus, *Bedd.*	108
1. felinus, *Gthr.*	73	21. Ptyodactylus, *Cuv.*	109
2. dorsalis, *Ptrs.*	74	1. lobatus, *Geoffr.*	110
3. ? brunneus, *Cope*	74	2. homolepis, *Blanf.*	111
15. Heteronota, *Gray*	74	22. Thecadactylus, *Cuv.*	111
1. binoei, *Gray*	74	1. rapicaudus, *Houtt.*	111
2. derbiana, *Gray*	75	2. australis, *Gthr.*	112
3. ? eboracensis, *Macleay.*	76	23. Hemidactylus, *Cuv.*	113
16. Phyllodactylus, *Gray*	76	1. homœolepis, *Blanf.*	117
1. tuberculosus, *Wiegm.*	79	2. bouvieri, *Bocourt*	118
2. ventralis, *O'Sh.*	80	3. reticulatus, *Bedd.*	118
3. reissii, *Ptrs.*	80	4. gracilis, *Blanf.*	119
4. pulcher, *Gray*	80	5. frenatus, *D. & B.*	120
5. spatulatus, *Cope*	81	6. mabouia, *Mor.*	122
6. galapagoensis, *Ptrs.*	82	7. muriceus, *Ptrs.*	123
7. nigrofasciatus, *Cope*	82	8. echinus, *O'Sh.*	123
8. inæqualis, *Cope*	83	9. fasciatus, *Gray*	124
9. microphyllus, *Cope*	84	10. bocagii, *Blgr.*	125
10. phacophorus, *Tsch.*	84	11. sinaitus, *Blgr.*	126
11. oviceps, *Boettg.*	85	12. turcicus, *L.*	126
12. sancti-johannis, *Gthr.*	86	13. brookii, *Gray*	128
13. stumpffi, *Boettg.*	86	14. gleadovii, *Murray*	129
14. porphyreus, *Daud.*	87	15. stellatus, *Blgr.*	130
15. marmoratus, *Gray*	88	16. guineensis, *Ptrs.*	131
16. macrodactylus, *Blgr.*	89	17. persicus, *And.*	131
17. affinis, *Blgr.*	89	18. maculatus, *D. & B.*	132
18. guentheri, *Blgr.*	90	19. triedrus, *Daud.*	133
19. europæus, *Gené*	90	20. subtriedrus, *Jerd.*	134
20. pictus, *Ptrs.*	91	21. depressus, *Gray*	134
21. lineatus, *Gray*	92	22. kushmorensis, *Murray*	135
22. ocellatus, *Gray*	93	23. leschenaultii, *D. & B.*	136
23. unctus, *Cope*	94	24. coctæi, *D. & B.*	137
24. riebeckii, *Ptrs.*	94	25. giganteus, *Stol.*	138
25. gerrhopygus, *Wiegm.*	95	26. bowringii, *Gray*	139
androyensis, *Grand.*	76	27. karenorum, *Theob.*	140
17. Ebenavia, *Boettg.*	96	28. blanfordii, *Blgr.*	141
1. inunguis, *Boettg.*	96	29. peruvianus, *Wiegm.*	141
2. boettgeri, *Blgr.*	96	30. garnotii, *D. & B.*	141
18. Diplodactylus, *Gray*	97	31. richardsonii, *Gray*	143
1. ciliaris, *Blgr.*	98	32. platyurus, *Schn.*	143
2. spinigerus, *Gray*	99	flaviviridis, *Rüpp.*	113
3. strophurus, *D. & B.*	100	angulatus, *Hall.*	113
4. vittatus, *Gray*	100	marmoratus, *Hall.*	113
5. polyophthalmus, *Gthr.*	101	mortoni, *Theob.*	113
6. steindachneri, *Blgr.*	102	sakalava, *Grand.*	113
7. pulcher, *Stdchr.*	102	tolampyæ, *Grand.*	113
8. tessellatus, *Gthr*	103	tristis, *Sauvg.*	113
annulatus, *Macleay*	97	caudiverbera, *Wagl.*	113
19. Œdura, *Gray*	104	24. Teratolepis, *Gthr.*	144

SYSTEMATIC INDEX.

	Page
1. fasciata, *Blyth*	145
25. Phyllopezus, *Ptrs.*	145
1. goyazensis, *Ptrs.*	145
26. Aristelliger, *Cope*	146
1. præsignis, *Hall.*	146
2. lar, *Cope*	147
27. Gehyra, *Gray*	147
1. mutilata, *Wiegm.*	148
2. baliola, *A. Dum.*	150
3. brevipalmata. *Ptrs.*	150
4. ? neglecta, *Gir.*	150
5. insulensis, *Gir.*	150
6. variegata, *D. & B.*	151
7. australis, *Gray*	152
8. oceanica, *Less.*	152
9. vorax, *Gir.*	153
papuensis, *Macleay*	147
ornata, *Macleay*	147
longicauda, *Macleay*	147
dubia, *Macleay*	147
marmorata, *Macleay*	147
brevicaudis, *Macleay*	147
28. Perochirus, *Blgr.*	154
1. ateles, *A. Dum.*	154
2. guentheri, *Blgr.*	155
3. depressus, *Fisch.*	155
4. scutellatus, *Fisch.*	156
5. articulatus, *Fisch.*	156
29. Spathoscalabotes, *Blgr.*	156
1. mutilatus, *Gthr.*	157
30. Microscalabotes, *Blgr.*	157
1. cowani, *Blgr.*	158
31. Lygodactylus, *Gray*	158
1. capensis, *Smith*	160
2. madagascariensis, *Boettg.*	160
3. thomensis, *Ptrs.*	161
4. gutturalis, *Bocage*	161
5. picturatus, *Ptrs.*	161
pictus, *Ptrs.*	158
bivittis, *Ptrs.*	159
hildebrandti, *Ptrs.*	159
32. Lepidodactylus, *Fitz.*	162
1. crepuscularis, *Bavay*	163
2. ceylonensis, *Blgr.*	164
3. aurantiacus, *Bedd.*	164
4. lugubris, *D. & B.*	165
5. labialis, *Ptrs.*	166
6. pulcher, *Blgr.*	166
7. guppyi, *Blgr.*	166
8. pusillus, *Cope*	167
9. cyclurus, *Gthr.*	167
10. sauvagii, *Blgr.*	168
roseus, *Cope*	162
33. Naultinus, *Gray*	168
1. elegans, *Gray*	168
2. rudis, *Fisch.*	170
34. Hoplodactylus, *Fitz.*	171

	Page
1. maculatus, *Gray*	171
2. duvaucelii, *D. & B.*	172
3. pacificus, *Gray*	173
4. granulatus, *Gray*	174
5. anamallensis, *Gthr.*	175
35. Rhacodactylus, *Fitz.*	176
1. leachianus, *Cuv.*	176
2. aubryanus, *Bocage*	177
3. chahoua, *Bavay*	177
4. trachyrhynchus, *Bocage*	178
5. auriculatus, *Bocage*	179
6. ciliatus, *Guich.*	180
36. Luperosaurus, *Gray*	181
1. cumingii, *Gray*	181
37. Gecko, *Laur.*	182
1. verticillatus, *Laur.*	183
2. stentor, *Cant.*	184
3. vittatus, *Houtt.*	185
4. monarchus, *D. & B.*	187
5. japonicus, *D. & B.*	188
6. swinhonis, *Gthr.*	189
7. subpalmatus, *Gthr.*	189
38. Ptychozoon, *Kuhl*	189
1. homalocephalum, *Crev.*	190
39. Homopholis, *Blgr.*	191
1. wahlbergii, *Smith*	171
40. Geckolepis, *Grand.*	192
1. maculata, *Ptrs.*	192
2. typica, *Grand.*	192
41. Eurydactylus, *Sauvg.*	192
1. vieillardi, *Bavay*	192
42. Æluronyx, *Fitz.*	193
1. seychellensis, *D. & B.*	193
2. trachygaster, *A. Dum.*	194
43. Tarentola, *Gray*	195
1. mauritanica, *L.*	196
2. annularis, *Geoffr.*	197
2 a. senegalensis, *Blgr.*	414
3. ephippiata, *O'Sh.*	198
4. delalandii, *D. & B.*	199
5. gigas, *Bocage*	200
americana, *D. & B.*	195
cubana, *Ptrs.*	195
clypeata, *Gray*	195
44. Pachydactylus, *Wiegm.*	200
1. bibronii, *Smith*	201
2. capensis, *Smith*	202
3. formosus, *Smith*	203
4. rugosus, *Smith*	204
5. oshaughnessyi, *Blgr.*	204
6. ocellatus, *Cuv.*	205
7. punctatus, *Ptrs.*	206
8. maculatus, *Gray*	206
9. mentomarginatus, *Smith*	207
10. mariquensis, *Smith*	207
tristis, *Hall.*	200
45. Colopus, *Ptrs.*	208

SYSTEMATIC INDEX.

1. wahlbergii, *Ptrs.* 208
46. Dactychilikion, *Thomin.* ... 209
   1. braconnieri, *Thomin.* ... 209
47. Phelsuma, *Gray* 209
   1. cepedianum, *Merr.* 211
   2. trilineatum, *Gray* 212
   3. andamanense, *Blyth* 212
   4. newtonii, *Blgr.* 212
   5. guentheri, *Blgr.* 213
   6. madagascariense, *Gray* . 214
       dubius, *Boettg.* 215
   7. laticauda, *Boettg.* 215
   8. lineatum, *Gray* 216
48. Rhoptropus, *Ptrs.* 217
   1. afer, *Ptrs.* 217
49. Sphærodactylus, *Wagl.* ... 217
   1. sputator, *Sparrm.* 219
   2. elegans, *R. & L.* 220
   3. punctatissimus, *D. & B.* 220
   4. nigropunctatus, *Gray* .. 220
   5. glaucus, *Cope* 221
   6. lineolatus, *Licht.* 221
   7. casicolus, *Cope* 222
   8. alopex, *Cope* 222
   9. oxyrrhinus, *Gosse* 222
   10. argus, *Gosse* 223
   11. fantasticus, *D. & B.* .. 223
   12. microlepis, *R. & L.* 224
   13. copii, *Stdchr.* 225
   14. anthracinus, *Cope* 225
   15. macrolepis, *Gthr.* 226
   16. notatus, *Baird* 226
   17. gilvitorques, *Cope.* 227
   18. richardsonii, *Gray* 227
Phyria, *Gray* 228
   punctulata, *Gray* 228
Gecko newtonii, *Gthr.* 228

Fam. 2. EUBLEPHARIDÆ.

1. Psilodactylus, *Gray* 229
   1. caudicinctus, *A. Dum.* . 230
2. Eublepharis, *Gray* 230
   1. hardwickii, *Gray.* 231
   2. macularius, *Blyth* 232
   3. dovii, *Blgr.* 233
   4. variegatus, *Baird.* 233
   5. fasciatus, *Blgr.* 234
3. Coleonyx, *Gray* 234
   1. elegans, *Gray* 235

Fam. 3. UROPLATIDÆ.

1. Uroplates, *Gray* 236
   1. fimbriatus, *Schn.* 237
   2. lineatus, *D. & B.* 238
   3. ebenaui, *Boettg.* 238
       caudiverbera, *L.* 236

Fam. 4. PYGOPODIDÆ.

1. Pygopus, *Merr.* 240
   1. lepidopus, *Lacép.* 240
2. Cryptodelma, *Fisch.* 242
   1. nigriceps, *Fisch.* 242
   2. orientalis, *Gthr.* 242
3. Delma, *Gray* 243
   1. fraseri, *Gray* 243
   2. impar, *Fisch.* 244
4. Pletholax, *Cope* 245
   1. gracilis, *Cope* 245
5. Aprasia, *Gray* 245
   1. pulchella, *Gray* 246
6. Lialis, *Gray* 246
   1. burtonii, *Gray* 247

Fam. 5. AGAMIDÆ.

1. Draco, *L.* 253
   1. volans, *L.* 256
   2. reticulatus, *Gthr.* 257
   3. guentheri, *Blgr.* 257
   4. everetti, *Blgr.* 258
   5. cornutus, *Gthr.* 258
   6. ornatus, *Gray* 259
   7. spilopterus, *Wiegm.* 260
   8. rostratus, *Gthr.* 261
   9. timorensis, *Kuhl* 261
   10. maculatus, *Gray* 262
   11. bimaculatus, *Gthr.* 263
   12. lineatus, *Daud.* 264
   13. beccarii, *Ptrs. & Doria* . 264
   14. spilonotus, *Gthr.* 265
   15. fimbriatus, *Kuhl* 265
   16. cristatellus, *Gthr.* 266
   17. hæmatopogon, *Gray* .. 267
   18. blanfordii, *Blgr.* 267
   19. dussumieri, *D. & B.* .. 268
   20. tæniopterus, *Gthr.* 269
   21. quinquefasciatus, *Gray* . 269
2. Sitana, *Cuv.* 270
   1. ponticeriana, *Cuv.* 270
3. Otocryptis, *Wiegm.* 271
   1. bivittata, *Wiegm.* 271
   2. beddomii, *Blgr.* 272
4. Ptyctolæmus, *Ptrs.* 273
   1. gularis, *Ptrs.* 273
5. Aphaniotis, *Ptrs.* 274
   1. fusca, *Ptrs.* 274
6. Lophocalotes, *Gthr.* 274
   1. interruptus, *Gthr.* 275
7. Oophotis, *Ptrs.* 275
   1. ceylanica, *Ptrs.* 275
   2. sumatrana, *Hubr.* 276
8. Ceratophora, *Gray* 277
   1. stoddartii, *Gray* 277

SYSTEMATIC INDEX.

2. tennentii, *Gthr.* 278
3. aspera, *Gthr.* 278
9. Harpesaurus, *Blgr.* 279
   1. tricinctus, *A. Dum.* 279
10. Phoxophrys, *Hubr.* 280
    1. tuberculata, *Hubr.* 280
11. Lyriocephalus, *Merr.* 281
    1. scutatus, *L.* 281
12. Gonyocephalus, *Kaup* 282
    1. doriæ, *Ptrs.* 284
    2. chamæleontinus, *Laur.* . 285
    3. kuhlii, *Schley.* 286
    4. sumatranus, *Schleg.* 286
    5. liogaster, *Gthr.* 286
    6. miotympanum, *Gthr.* ... 287
    7. borneensis, *Schleg.* 288
    8. bellii, *D. & B.* 288
    9. sophiæ, *Gray* 288
    10. semperi, *Ptrs.* 289
    11. interruptus, *Blgr.* 290
    12. dilophus, *D. & B.* 290
    13. tuberculatus, *Gthr.* ... 291
    14. spinipes, *A. Dum.* 292
    15. subcristatus, *Blyth* ... 292
    16. humii, *Stol.* 293
    17. modestus, *Meyer* 294
    18. geelvinkianus, *Ptrs. & Doria* 294
    19. auritus, *Meyer* 295
    20. bruijnii, *Ptrs. & Doria* . 295
    21. binotatus, *Meyer* 295
    22. godeffroyi, *Ptrs.* 295
        nigrigularis, *Meyer* ... 296
    23. papuensis, *Macleay* ... 297
    24. boydii, *Macleay* 297
    25. grandis, *Gray* 298
13. Acanthosaura, *Gray* 299
    1. capra, *Gthr.* 300
    2. armata, *Gray* 301
    3. crucigera, *Blgr.* 302
    4. lamnidentata, *Blgr.* 302
    5. coronata, *Gthr.* 303
    6. minor, *Gray* 304
    7. kakhienensis, *And.* 305
    8. major, *Jerd.* 306
    9. tricarinata, *Blyth* 306
14. Japalura, *Gray* 307
    1. variegata, *Gray* 308
    2. swinhonis, *Gthr.* 309
    3. polygonata, *Hall.* 310
    4. yunnanensis, *And.* 310
    5. planidorsata, *Jerd.* 311
    6. nigrilabris, *Ptrs.* 311
15. Salea, *Gray* 312
    1. horsfieldii, *Gray* 312
    2. anamallayana, *Bedd.* .. 313

16. Calotes, *Cuv.* 314
    1. cristatellus, *Kuhl* 316
    2. celebensis, *Gray* 318
    3. marmoratus, *Gray* 318
    4. jubatus, *D. & B.* 318
    5. smaragdinus, *Gthr.* 319
    6. tympanistriga, *Gray* .. 320
    7. versicolor, *Daud.* 321
    8. maria, *Gray* 322
    9. jerdonii, *Gthr.* 323
    10. emma, *Gray* 324
    11. mystaceus, *D. & B.* 325
    12. grandisquamis, *Gthr.* ... 325
    13. nemoricola, *Jerd.* 326
    14. liolepis, *Blgr.* 326
    15. ophiomachus, *Merr.* .. 327
    16. nigrilabris, *Ptrs.* 328
    17. liocephalus, *Gthr.* 329
    18. rouxii, *D. & B.* 330
    19. elliotti, *Gthr.* 330
        gularis, *Blyth* 314
17. Chelosania, *Gray* 331
    1. brunnea, *Gray* 331
18. Charasia, *Gray* 332
    1. dorsalis, *Gray* 332
    2. blanfordiana, *Stol.* 333
    3. ornata, *Blyth* 334
19. Agama, *Daud.* 334
    1. mutabilis, *Merr.* 338
    2. sinaita, *Heyd.* 339
    3. hartmanni, *Ptrs.* 340
    4. tournevillii, *Lataste.* 340
    5. agilis, *Oliv.* 341
    6. isolepis, *Blgr.* 342
    7. sanguinolenta, *Pall.* 343
    8. latastii, *Blgr.* 344
    9. inermis, *Reuss.* 344
    10. persica, *Blanf.* 345
    11. leucostigma, *Reuss.* 346
    12. rubrigularis, *Blanf.* 346
    13. megalonyx. *Gthr.* 347
    14. ruderata, *Oliv.* 348
    15. pallida, *Reuss* 348
    16. hispida, *L.* 349
    17. brachyura, *Blgr.* 350
    18. aculeata, *Merr.* 351
    19. armata, *Ptrs.* 352
    20. atra, *Daud.* 352
    21. mossambica, *Ptrs.* 353
    22. kirkii, *Blgr.* 354
    23. spinosa, *Gray* 355
    24. rueppellii, *Vaill.* 355
    25. colonorum, *Daud.* 356
    26. bibronii, *A. Dum.* 357
    27. planiceps, *Ptrs.* 358
    28. atricollis, *Smith* 358

SYSTEMATIC INDEX.

29. cyanogaster, *Rüpp.* 359
30. annectens, *Blanf.* 360
31. stoliczkana, *Blanf.* 360
32. tuberculata, *Gray* 361
33. dayana, *Stol.* 362
34. himalayana, *Stdchr.* 362
35. agrorensis, *Stol.* 363
36. melanura, *Blyth* 363
37. lirata, *Blanf.* 364
38. nupta, *De Fil.* 365
39. microlepis, *Blanf.* 366
40. caucasica, *Eichw.* 367
41. stellio, *L.* 368
20. Phrynocephalus, *Kaup* .. 369
    1. olivieri, *D. & B.* 370
    2. helioscopus, *Pall.* 371
    3. vlangalii, *Strauch* 372
    4. theobaldi, *Blyth* 373
    5. versicolor, *Strauch* 374
    6. frontalis, *Strauch* 375
    7. caudivolvulus, *Pall.* 375
    8. przewalskii, *Strauch* .. 377
    9. affinis, *Strauch* 377
    10. maculatus, *And.* 377
    11. axillaris, *Blanf.* 378
    12. interscapularis, *Licht.* ... 378
    13. mystaceus, *Pall.* 379
        varius, *Eichw.* 369
        melanurus, *Eichw.* 369
        nigricans, *Eichw.* 369
21. Amphibolurus, *Wagl.* 380
    1. maculatus, *Gray* 381
    2. imbricatus, *Ptrs.* 382
    3. ornatus, *Gray* 382
    4. cristatus, *Gray* 383
    5. caudicinctus, *Gthr.* 384
    6. decresii, *D. & B.* 385
    7. pictus, *Ptrs.* 385
    8. reticulatus, *Gray* 386
    9. adelaidensis, *Gray* 387
    10. pulcherrimus, *Blgr.* 388
    11. pallidus, *Blgr.* 388

    12. angulifer, *Gray* 389
    13. muricatus, *White* 390
    14. barbatus, *Cuv.* 391
        jugularis, *Macleay* 380
22. Tympanocryptis, *Ptrs.* ... 392
    1. lineata, *Ptrs.* 392
    2. cephalus, *Gthr.* 393
23. Diporophora, *Gray* 393
    1. bilineata, *Gray* 394
    2. australis, *Stdchr.* 394
    3. bennettii, *Gray* 395
24. Physignathus, *Cuv.* 395
    1. gilberti, *Gray* 396
    2. longirostris, *Blgr.* 397
    3. temporalis, *Gthr.* 397
    4. maculilabris, *Blgr.* 398
    5. lesueurii, *Gray* 398
    6. cochinchinensis, *Cuv.* .. 399
    7. mentager, *Gthr.* 400
        lateralis, *Macleay* 395
25. Chlamydosaurus, *Gray* .. 401
    1. kingii, *Gray* 401
26. Lophura, *Gray* 402
    1. amboinensis, *Schloss.* .. 402
27. Liolepis, *Cuv.* 403
    1. bellii, *Gray* 403
28. Uromastix, *Merr.* 405
    1. ornatus, *Rüpp.* 406
    2. acanthinurus, *Bell* 406
    3. spinipes, *Daud.* 407
    4. microlepis, *Blanf.* 407
    5. hardwickii, *Gray* 408
    6. asmussii, *Strauch* 409
    7. loricatus, *Blanf.* 409
29. Aporoscelis, *Blgr.* 410
    1. princeps, *O'Sh.* 410
    2. batilliferus, *Vaill.* 411
30. Moloch, *Gray* 411
    1. horridus, *Gray* 411

    Oreodeira, *Gir.* 412
    gracilipes, *Gir.* 413

CATALOGUE

OF

LIZARDS.

Order LACERTILIA.

"Quadrate bone articulated to the skull; parts of the ali- and orbito-sphenoid regions fibro-cartilaginous; rami of the mandible united by suture; temporal region without or with only one horizontal bar. Anal cleft transverse. Copulatory organs present, paired."—*Günther, Phil. Trans.* clviii. 1867, p. 625.

Suborder I. LACERTILIA VERA.

Nasal bones entering the border of the nasal apertures; pterygoid in contact with quadrate. Clavicle present whenever the limbs are developed. Tongue flattened.

A. Tongue smooth, or with villose papillæ; clavicle dilated, loop-shaped proximally; no postorbital or postfronto-squamosal arches.

Fam. 1. **Geckonidæ.** Vertebræ amphicœlian; parietal bones distinct.

Fam. 2. **Eublepharidæ.** Vertebræ procœlian; parietal single.

B. Tongue smooth or with villose papillæ; clavicle not dilated proximally.

Fam. 3. **Uroplatidæ.** Vertebræ amphicœlian; interclavicle minute; no postorbital or postfronto-squamosal arches.

Fam. 4. **Pygopodidæ.** No postorbital or postfronto-squamosal arches; pre- and postfrontal bones in contact, separating the frontal from the orbit.

Fam. 5. **Agamidæ.** Postorbital and postfronto-squamosal arches present; supratemporal fossa not roofed over by bone; tongue thick; acrodont.

Fam. 6. **Iguanidæ.** Postorbital and postfronto-squamosal arches present; supratemporal fossa not roofed over by bone; tongue thick; pleurodont.

Fam. 7. **Xenosauridæ.** Postorbital and postfronto-squamosal arches present; supratemporal fossa not roofed over; anterior portion of tongue retractile.

Fam. 8. **Zonuridæ.** Postorbital and postfronto-squamosal arches complete; supratemporal fossa roofed over; tongue simple.

Fam. 9. **Anguidæ.** Postorbital and postfronto-squamosal arches present; supratemporal fossa roofed over; body with osteodermal plates, each provided with a system of irregularly arranged arborescent or radiating tubules; anterior portion of tongue retractile.

Fam. 10. **Aniellidæ.** No interorbital septum, no columella cranii, no arches.

Fam. 11. **Helodermatidæ.** Postorbital arch present, postfronto-squamosal arch absent; pre- and postfrontals in contact, separating the frontal from the orbit.

Fam. 12. **Varanidæ.** Postorbital arch incomplete; postfronto-squamosal arch present; supratemporal fossa not roofed over; nasal bone single; tongue deeply bifid, sheathed posteriorly.

C. Tongue covered with imbricate scale-like papillæ or with oblique plicæ; clavicle dilated proximally, frequently loop-shaped.

Fam. 13. **Xantusiidæ.** Parietals distinct; postorbital and post-fronto-squamosal arches present; supratemporal fossa roofed over.

Fam. 14. **Teiidæ.** Postorbital and postfronto-squamosal arches present; supratemporal fossa not roofed over; no osteodermal plates.

Fam. 15. **Amphisbænidæ.** No interorbital septum; no columella cranii; no arches; premaxillary single.

Fam. 16. **Lacertidæ.** Arches present; supratemporal fossa roofed over; premaxillary single; no osteodermal plates on the body.

Fam. 17. **Gerrhosauridæ.** Arches present; supratemporal fossa roofed over; premaxillary single; body with osteodermal plates, each provided with a regular system of tubules (a transverse one anastomosing with longitudinal ones).

Fam. 18. **Scincidæ.** Arches present; premaxillary double; body with osteodermal plates as in the preceding.

Fam. 19. **Anelytropidæ.** Premaxillary single; no arches; no osteodermal plates.

Fam. 20. **Dibamidæ.** Premaxillary double; no interorbital septum; no columella cranii; no arches; no osteodermal plates.

Suborder II. RHIPTOGLOSSA.

Nasal bones not bounding nasal apertures; pterygoid not reaching quadrate. Clavicle absent, limbs well developed. Tongue vermiform, projectile.

Fam. 21. **Chamæleontidæ.**

Suborder I. LACERTILIA VERA.

Leptoglossi, Pachyglossi, Annulati, *Wiegmann, Herp. Mex.* 1834.
Cyclosaura, Geissosaura, Nyctisaura, Strobilosaura, *Gray, Cat. Liz.* 1845.
Amphisbænia, *Gray, Cat. Tort., Croc., Amph.* 1842.
Amphisbænoidea, Kionocrania, *Stannius, Zoot. Amph.* 1856.
Acrodonta Pachyglossa, Nyctisaura, Pleurodonta, Ophiosauri, *Cope, Proc. Acad. Philad.* 1864.
Amphisbænoidea, Cionocrania, Nyctisaura, *Günther, Phil. Trans.* clviii. 1867.
Pachyglossa, Nyctisaura, Pleurodonta, Ophiosauri, *Cope, Proc. Amer. Assoc. Adv. Sc.* xix. 1871.
Lacertilia vera, *Boulenger, Ann. & Mag. N. H.* (5) xiv. 1884.

Fam. I. GECKONIDÆ.

Geckotiens, part., *Cuvier, Règne Anim.* ii. 1817.
Ascalabotæ, part., *Merrem, Tent. Syst. Amph.* 1820.
Geckotidæ, part., *Gray, Ann. Phil.* (2) x. 1825.
Ascalabotoidea, part., *Fitzinger, Neue Classif. Rept.* 1826.
Platyglossæ, part., *Wagler, Syst. Amph.* 1830.
Ascalabotæ, part., *Wiegmann, Herp. Mex.* 1834.
Geckotiens ou Ascalabotes, part., *Duméril & Bibron, Erp. Gén.* iii. 1836.
Ascalabotæ, part., *Fitzinger, Syst. Rept.* 1843.
Geckotidæ, part., *Gray, Cat. Liz.* 1845.
Gecconidæ, part., *Cope, Proc. Am. Assoc. Adv. Sc.* xix. 1871.
Geckonidæ, *Boulenger, Ann. & Mag. N. H.* (5) xiv. 1884.

The skull is generally much depressed, and its bones are thin. The nasals remain distinct; the frontal is either single or with distinct suture; the jugal is rudimentary, the orbit not being bound posteriorly by a bony arch; the postfronto-squamosal arch is likewise absent; the pterygoids are widely separated and devoid of teeth; a columella cranii is present. The mandible contains only five bones, the angular and articular having coalesced; the dentition belongs to the pleurodont type; the teeth are small, numerous, closely set, with long, slender, cylindrical shaft and obtuse point; the new teeth hollow out the base of the old ones. Only in a few instances does the derm of the head coalesce with the skull, and a supraorbital bone is present only in a few species of *Tarentola*. Both pairs of limbs are constantly well developed and pentadactyle. The clavicle is dilated, and perforated proximally; and the interclavicle is subrhomboidal or of a shape intermediate between that and the cruciform. The vertebræ are biconcave; the ribs are long, and so prolonged as to form more or less ossified hoops across the whole of the abdominal region.

The digits vary considerably, and afford excellent characters for systematic arrangement. Some Geckos (living in barren regions) have the digits similar to those of many Agamoids, *i. e.* they are subcylindrical or feebly depressed, and frequently keeled inferiorly or denticulated laterally; other forms with non-dilated digits have them angularly bent at the articulations and provided with strong claws; but the greater number have the whole or part of the digits dilated into adhesive organs with symmetrical plates or lamellæ inferiorly, the arrangement of which varies considerably. Then also the claw may be retractile, either between some of the lamellæ or into a special sheath. Membranes may unite the digits, but the web serves only for the purpose of obtaining a greater adhesive surface, and never for swimming, none of the Geckos entering the water.

The body is generally more or less depressed, and may be bordered by cutaneous expansions, the object of which appears to be, in most cases, the same as that of the interdigital membrane; but in the curious genus *Ptychozoon*, in which the lateral membranes attain the greatest development, they act as a parachute. The tail presents almost every possible shape, from the leaf-like tail of *Gymnodactylus platurus* and the grotesque rudimentary tail of *Nephrurus* to the slender rat-like tail of *Agamura* and the compressed crested tail of *Pristurus*. This organ is, except in *Agamura*, extremely fragile and rapidly reproduced, in which case, however, it generally assumes an abnormal shape and lepidosis. In some forms the tail proves to be prehensile, a faculty which is possessed by few Lizards other than the Chameleons; and I am induced to believe that a careful examination of the Geckos, when alive, will show this character to be not unfrequent.

The eye is generally large and with vertical pupil, which, when strongly contracted, is frequently denticulated or assumes the shape of two superposed rhombs; some diurnal forms have the eye smaller and the pupil circular. The eye is exposed, as in Snakes, covered by a transparent lid under which it moves freely, the valvular lids being rudimentary; in *Ælurosaurus*, however, there are connivent movable lids, and in *Ptenopus* the upper lid is sufficiently developed to cover the eye nearly completely.

The tympanum is more or less exposed, except in *Teratolepis*, in which genus it appears to be completely concealed under the scales. The tongue is fleshy, moderately elongate, very feebly incised anteriorly, and capable of protrusion out of the mouth.

The teguments are nearly always soft, and consist generally of granules or tubercles on the dorsal surface, of small imbricated cycloid or hexagonal scales on the ventral surface. Some Geckos are entirely covered with scales of the latter description, which attain their highest development in *Teratoscincus*, *Teratolepis*, and *Geckolepis*.

The habits of the Geckos are highly interesting and deserve special attention, as but few observations have been made on them. Some inhabit arid regions, sometimes burrowing in the sand; others are arboreal, living on shrubs or in woods, concealing themselves under stones or under the bark of trees during the daytime; others live on rocks; others have become the commensals of man, and they again may be divided into two groups—those living inside, those living outside houses. Most are nocturnal, but some are diurnal. Col. Tytler, in a very interesting paper on the habits of Geckos[*], observes that "although several species of Geckos may inhabit the same locality, yet, as a general rule, they keep separate and aloof from each other; for instance, in a house the dark cellars may be the resort of one species, the roof of another, and crevices in the walls may be exclusively occupied by a third species. However, at night they issue forth in quest of insects, and may be found mixed up together in the same spot; but on the slightest disturbance, or when they have done feeding, they return hurriedly to their particular hiding-places."

Many Geckos utter sounds, probably produced chiefly by a movement of the tongue against the palate, and in which *yecko*, *chucko*, *tockee*, or something similar, is distinctly audible. A. Smith says that a South-African Sand-Gecko (*Ptenopus garrulus*) utters during the day a sharp sound somewhat like *chick*, *chick*; and he adds that the number thus occupied is at times so great, and the noise so disagreeable, as to cause the traveller to change his quarters.

The eggs are round and with a hard shell. Ovoviviparism has not been observed in this family. Males are generally distinguished from females by a larger size, the swelling of the base of the tail, and the presence of femoral or præanal pores, which are constantly absent in the latter.

The *Geckonidæ* are represented in the hotter parts of all the regions of the world. They are most numerous in the Indian and Australian regions.

---

[*] Journ. As. Soc. Beng. xxxiii. 1864, pp. 535-548.

Synopsis of the Genera.

I. Digits short, cylindrical, the skin swollen on the palmar surface and under the articulations.

Digits clawed; tail extremely short, terminating in a globular knob.
 1. **Nephrurus**, p. 9.

No claws .... 2. **Chondrodactylus**, p. 10.

II. Digits straight, not dilated, clawed, without pads.

Digits granular inferiorly, not fringed laterally; rostral and mental plates projecting, nail-like ...... 3. **Rhynchœdura**, p. 11.

Digits granular inferiorly, strongly fringed laterally; dorsal scales large, imbricate .... 4. **Teratoscincus**, p. 12.

Digits covered inferiorly with small imbricate pointed scales; dorsal scales small .... 5. **Ceramodactylus**, p. 13.

Digits inferiorly with a series of narrow transverse plates; toes strongly fringed laterally; fingers not fringed.
 6. **Ptenopus**, p. 15.

Digits inferiorly with a series of narrow transverse plates, fringed or denticulated laterally .... 7. **Stenodactylus**, p. 16.

Digits inferiorly with a series of narrow transverse plates, not fringed nor denticulated laterally; dorsal scales juxtaposed; male with a series of præanal pores.
 8. **Alsophylax**, p. 19.

Digits inferiorly with a series of narrow transverse plates, not fringed nor denticulated laterally; dorsal scales imbricate; no præanal pores .... 9. **Homonota**, p. 21.

III. Digits not or but slightly dilated at the base, the two or three distal joints more or less compressed and angularly bent, inferiorly with a series of transverse plates; all the digits clawed.

A. Claw between two scales, a smaller superior and a large latero-inferior.

Pupil vertical; tail fragile .... 10. **Gymnodactylus**, p. 22.

Pupil vertical; tail very slender, not fragile.
 11. **Agamura**, p. 50.

Pupil round; body not depressed; tail compressed.
 12. **Pristurus**, p. 52.

Pupil round; body more or less depressed; tail not compressed.
 13. **Gonatodes**, p. 56.

B. Claw between three scales, a smaller superior and two large latero-inferior.

Upper and lower eyelids well developed, connivent; ungual scales forming a large compressed sheath. 14. **Ælurosaurus**, p. 73.

No compressed ungual sheath ...... 15. **Heteronota**, p. 74.

IV. Digits dilated at the apex, which is furnished inferiorly with two plates separated by a longitudinal groove.

Digits not dilated at the base, clawed, the distal expansion covered above with scales strongly differentiated from those of the basal part .... 16. **Phyllodactylus**, p. 76.

No claws .... 17. **Ebenavia**, p. 96.

Digits not dilated at the base, clawed, the distal expansion covered above with small tubercular scales similar to those on the basal part .... 18. **Diplodactylus**, p. 97.

Digits dilated at the base, the basal expansion anteriorly with paired oblique lamellæ .... 19. **Œdura**, p. 104.

The penultimate joint with an expansion bearing two plates exactly similar to the distal .... 20. **Calodactylus**, p. 108.

V. Digits dilated at the apex, which is furnished inferiorly with two diverging series of lamellæ; digits clawed, the claw sessile and retractile in the anterior notch of the distal expansion .... 21. **Ptyodactylus**, p. 109.

VI. Digits entirely dilated, with a double series of lamellæ inferiorly, clawed, the claw sessile and retractile in the median groove .... 22. **Thecadactylus**, p. 111.

VII. Digits dilated, the distal phalanges compressed.

A. The distal joint long, free, rising from within the extremity of the digital expansion.

Infradigital plates in a double series; inner digit with compressed clawed phalanx; dorsal lepidosis composed of small scales or tubercles .... 23. **Hemidactylus**, p. 113.

Infradigital plates double; dorsal scales large, imbricate.
24. **Teratolepis**, p. 144.

Infradigital plates in a simple series; inner digit with compressed clawed phalanx, similar to the other digits.
25. **Phyllopezus**, p. 145.

Infradigital plates in a simple series; inner digit clawed, the claw retractile laterally, inferiorly with a circular plate.
26. **Aristelliger**, p. 146.

Infradigital plates in a simple or double series; inner digit clawless.
27. **Gehyra**, p. 147.

Infradigital plates in a simple series; inner digit rudimentary, of fore limb clawless, of hind limb clawed.
28. **Perochirus**, p. 154.

B. The free distal joint at the extremity of the digital expansion; a double series of infradigital lamellæ.

Pupil vertical; digits narrow at the base, the dilatation strong and discoid, the distal joint long and slender; inner digit rudimental.
29. **Spathoscalabotes**, p. 156.

Pupil round; eyelid distinct all round the eye; digits narrow at the base, the dilatation strong and discoid, the distal joint free; inner digit rudimental, with strong, very distinct claw.
30. **Microscalabotes**, p. 157.

Pupil round; eyelid distinct all round the eye; digits narrow at the base, the dilatation strong and discoid, the distal joint strongly curved, the claw retractile between the anterior lamellæ; inner digit rudimental, with very small, frequently indistinct claw.
31. **Lygodactylus**, p. 158.

Pupil vertical; distal joint of digits short; thumb clawless.
32. **Lepidodactylus**, p. 162.

C. The free distal joint at the extremity of the digital expansion; a single series of infradigital lamellæ.

1. The distal joint long.

Digits narrowing gradually towards the end, the narrower portion not forming an angle with the dilated basal part; eyelid distinct all round the eye.............. 33. **Naultinus**, p. 168.

The slender distal portion of the digit forming an angle with the dilated basal portion .... 34. **Hoplodactylus**, p. 171.

2. The distal joint very short.

Digits more or less webbed, inner clawed.
35. **Rhacodactylus**, p. 176.

Digits half-webbed, inner clawless.. 36. **Luperosaurus**, p. 181.

Digits free or slightly webbed, inner clawless.
37. **Gecko**, p. 182.

Digits entirely webbed, inner clawless. 38. **Ptychozoon**, p. 189.

VIII. Digits entirely dilated, clawed, without compressed ungual phalanx, inferiorly with a single series of lamellæ.

Body covered with small imbricate scales.
39. **Homopholis**, p. 191.

Body covered with large imbricate scales.
40. **Geckolepis**, p. 192.

Body covered above with large juxtaposed scales, largest and sub-symmetrical on the head........ 41. **Eurydactylus**, p. 192.

Body covered above with granular scales; the claws of the three inner digits turned inwards, those of the two outer turned outwards .... 42. **Æluronyx**, p. 193.

  IX. Digits dilated, only the third and fourth clawed.
     43. **Tarentola**, p. 194.

  X. Digits (the fingers at any rate) more or less dilated, clawless.

Pupil vertical; tips of toes dilated, with simple transverse lamellæ inferiorly................... 44. **Pachydactylus**, p. 200.

Pupil vertical; tips of toes rather narrowed, with only two small lamellæ inferiorly .... 45. **Colopus**, p. 208.

Digits dilated at the apex only, inferiorly with transverse lamellæ furnished on their hinder edge with fine fringes.
     46. **Dactychilikion**, p. 209.

Pupil circular; eyelid distinct all round the eye.
     47. **Phelsuma**, p. 209.

  XI. Digits dilated at the apex only, with very small sheathed claw, the sheath opening laterally.

Digital expansion with transverse lamellæ inferiorly.
     48. **Rhoptropus**, p. 217.

Digital expansion with a circular plate inferiorly.
     49. **Sphærodactylus**, p. 217.

1. NEPHRURUS.

Nephrurus, *Günth. Journ. Mus. Godeffr.* xii. 1876, p. 46

Digits short, cylindrical, clawed, outer opposed to inner, inferiorly with uniform small spinose tubercles; the extremity of the digits with verticils of keeled scales; the skin swollen on the palmar surface and under the articulations of the digits, simulating pads. Derm of head completely involved in cranial ossification. Body covered with spinose tubercles. Tail extremely short and small, terminating in a globular knob. Pupil vertical. No præanal nor femoral pores.

Australia.

1. Nephrurus asper. (PLATE I.)

Nephrurus asper, *Günth. l. c.*

Head large, subtriangular, not much depressed, very distinct from neck; snout a little longer than the diameter of the orbit, a little shorter than the distance between the latter and the ear-opening; eye very large; nostril rather large, transversely elliptical, directed backwards; loreal region and forehead concave; ear-opening a long vertical slit, measuring three fifths the diameter of the orbit, with the tympanum deeply sunk. Body short, not much depressed. Limbs long, slender. Head above with irregular small rough tubercles confluent with the cranial ossification, intermixed with conical ones on the temples and upper eyelids, larger and keeled on the internasal region and in front of the orbit, smallest in the frenal and frontal concavities; a series of larger tubercles bordering the orbit superiorly. Rostral broad and very low; no nasals, the nostril being surrounded by minute granules and widely separated from the rostral and labial plates; latter small, twelve upper and thirteen lower; mental like the rostral, but narrower; no chin-shields. Body and limbs finely granular, above and on the sides with round groups of conical spinose tubercles, the one in the centre being the largest; gular region similarly tuberculate, but the spinose tubercles smaller; the granules round the mandible larger than the others and keeled. The extremely small tail is swollen in its proximal, and thin and tapering in its distal half, and scaled like the body; the globular knob terminating it covered with small keeled granules; this knob is emarginate inferiorly in front, being thus kidney-shaped. Brownish above, with many of the tubercles white; faint indications of whitish transverse lines on the back; head with a wide-meshed network of blackish lines, simulating symmetrical plates; lower parts whitish.

Total length	114	millim.
Head	27	,,
Width of head	25	,,
Body	65	,,
Fore limb	42	,,
Hind limb	48	,,
Tail	22	,,

Eastern Australia.

a. ♂. Peak Downs. Museum Godeffroy. (Type.)

2. CHONDRODACTYLUS.

Chondrodactylus, *Peters, Mon. Berl. Ac.* 1870, p. 110.

Digits very short, cylindrical, clawless, with obtuse tips bent downwards, inferiorly with uniform minute spinose granules; the skin swollen on the palmar surface and under the articulations of the digits, simulating pads. Body covered above with flat granules intermixed with larger tubercles, below with imbricate scales. Pupil vertical. No præanal nor femoral pores.

South Africa.

1. Chondrodactylus angulifer. (PLATE II. fig. 5.*)

Chondrodactylus angulifer, *Peters, l. c.* p. 111, pl.—. fig. 1.

Head large, swollen; snout very short and convex, as long as the diameter of the orbit or the distance between the latter and the ear-opening; interorbital space narrow, a little concave; ear-opening an oblique slit, measuring half the diameter of the orbit. Body short, not depressed. Limbs rather slender. Snout and crown covered with polygonal scales, largest round the orbit; temples and occiput with small granules intermixed with keeled tubercles; rostral small, pentagonal; nostril pierced between three nasals, the anterior being the largest and in contact with its fellow; nine or ten upper and ten or eleven lower labials; mental narrow, a little longer than the adjacent labials; no chin-shields. Upper surface of body covered with irregular flat granules intermixed with round keeled tubercles. Throat granulate; abdominal region with small smooth subhexagonal imbricate scales. Tail cylindrical, tapering, covered with irregular flat scales, intermixed above with conical tubercles arranged in transverse series. Light grey-brown above, with five more or less marked blackish angular transverse bands on the back; sometimes round white spots on the sides of the back; a median longitudinal dark streak on the nape; an oblique dark band from the eye towards the latter; tail with dark annuli above; lower surfaces whitish.

Total length	126	millim.
Head	22	,,
Width of head	19	,,
Body	52	,,
Fore limb	27	,,
Hind limb	32	,,
Tail	52	,,

South Africa.

a–b. ♂.	Karroo.	The Trustees of the South-African Museum [P.].
c. Hgr.	S. Africa.	

3. RHYNCHŒDURA.

Rhynchœdura, *Günth. Ann. & Mag. N. H.* (3) xx. 1867, p. 50.

Digits cylindrical, slightly compressed, clawed, covered with uniform small granules. Scales uniformly granular, subimbricate on ventral region. Rostral and mental plates projecting, nail-like. Pupil vertical.

Australia.

---

* Lower surface of hand, × 2; *a.* Lower surface of fourth finger, × 4.

1. Rhynchœdura ornata. (Plate II. fig. 1.)

Rhynchœdura ornata, *Günth. l. c.* p. 51.

Head resembling that of a young bird, high, the snout pointed and compressed, beak-like; the length of the snout equals the diameter of the orbit, and is much more than the distance between the latter and the very small, roundish ear-opening; the upper eyelid is broad, and may cover a good part of the eye. Body rather elongate, scarcely depressed. Limbs moderately elongate; the digits are rather slender, and the outer toe inserted far down. Head and body covered with small granules, from which the labials can hardly be distinguished, smallest on the throat, largest, flattened and subimbricate on the belly; enlarged scales before and behind the vent. Nostrils large, pierced between granules; rostral and mental small, nail-like, prominent, forming two small hooks. Tail short, rounded, swollen, covered with rings of smooth square scales. Light greyish-brown above, with round, faint, whitish spots; each side with confluent blackish-brown half rings; a blackish-brown band across the occiput; lower parts white.

Total length	72	millim.
Head	11	,,
Width of head	7	,,
Body	36	,,
Fore limb	16	,,
Hind limb	19	,,
Tail	25	,,

North-western Australia.

a. ♀? Nicol Bay. Mr. Duboulay [C.]. (Type.)

4. TERATOSCINCUS.

Teratoscincus, *Strauch, Bull. Ac. St. Pétersb.* vi. 1863, p. 480, *and Mél. Biol.* vi. 1867, p. 553.

Digits not dilated, furnished with a long claw, depressed, with a lateral fringe of long pointed scales, covered inferiorly with minute granular scales. Body covered with uniform large cycloid imbricated scales. Pupil vertical. No præanal nor femoral pores.

Persia; Turkestan.

1. Teratoscincus scincus. (Plate II. fig. 3.*)

Stenodactylus scincus, *Schleg. Handl. Dierk.* ii. p. 16.
Teratoscincus keyserlingii, *Strauch, l. c.*; *Blanford, 2nd Yark. Miss., Rept.* p. 11.

Head large, high; snout obtuse, slightly longer than the diameter of the orbit or the distance between the eye and the ear-opening; eye large; ear-opening large, elliptic, oblique, three fifths the

---

* Lower surface of fourth toe, ×4.

diameter of the eye. Body depressed. Limbs moderate; toes rather long. Head covered with small granules, largest on the snout; rostral quadrangular, broader than high, with median cleft above; nostril pierced between the rostral and three nasals; nine or ten upper and as many lower labials; mental quadrangular, a little longer than the adjacent labials; no regular chin-shields. Scales of body large, cycloid, smooth, imbricate, largest on the abdomen, of limbs smaller. Tail round at the base, compressed in its posterior half, covered inferiorly and laterally with scales similar to those of the body, above with a series of large, transverse, nail-like plates. Cream-coloured above, with traces of the brown transverse bands which are distinct in the young; lower surfaces white.

Total length	126	millim.
Head	22	,,
Width of head	19	,,
Body	57	,,
Fore limb	30	,,
Hind limb	42	,,
Tail	47	,,

Persia; Turkestan.

a. ♀.　　　　Eastern Turkestan.　　　　Dr. J. Scully [P.].

5. CERAMODACTYLUS.

Ceramodactylus, *Blanford, Ann. & Mag. N. H.* (4) xiii. 1874, p. 454.

Digits not dilated, furnished with a long claw, depressed, with a lateral fringe of long pointed scales, covered inferiorly with small imbricate pointed scales. Body covered with small flat scales. Pupil vertical. Males with præanal pores.

Persia; Arabia.

1. Ceramodactylus doriæ. (PLATE II. fig. 4*.)

Ceramodactylus doriæ, *Blanf. l. c.*, and *Zool. E. Persia*, p. 353, pl. xxiii. fig. 2.

Head large, oviform; snout obtusely pointed, as long as the diameter of the orbit or the distance between the eye and the ear-opening; eye large; ear-opening small, elliptic, vertical. Body little depressed. Limbs long, slender; if stretched forward the fore limb reaches beyond the tip of the snout, the hind limb a little beyond the axil; toes rather long. Head covered with granular scales, which are slightly keeled on the snout; rostral pentagonal, broader than high, with median cleft above; nostril pierced in the centre of a slight swelling followed by a concavity, between the rostral, the first labial, and three nasals; fourteen or fifteen upper

---

* Lower surface of fourth toe, ×5.

and as many lower labials; mental rather large, squarish, broader than high; no chin-shields. Body covered with small flat, slightly keeled, polygonal scales of uniform size. Male with two widely separated præanal pores. Tail rather slender, cylindrical, gradually tapering to a very fine point, covered with uniform, small, feebly keeled scales. Pale buff above, with brown network enclosing round whitish spots; tail above annulate brown and white; lower surfaces white.

Total length	111 millim.
Head	17 ,,
Width of head	13 ,,
Body	49 ,,
Fore limb	26 ,,
Hind limb	34 ,,
Tail	45 ,,

Persia; Arabia.

a. ♂.	Arabia.	Capt. Burton [P.].
b. ♂.	Sinaitic Peninsula.	H. C. Hart, Esq. [C.].

2. Ceramodactylus affinis.

Ceramodactylus affinis, *Murray, Ann. & Mag. N. H.* (5) xiv. 1884, p. 103.

Distinguished from the preceding by a stouter habit. The head is more convex, and the limbs and digits shorter; if stretched forwards the fore limb does not extend beyond the tip of the snout, the hind limb does not reach the axil. The tail is shorter and ends more abruptly. Ten or eleven upper and as many lower labials; mental as long as broad, its posterior edge curved. Sandy grey above, with darker specks, and four dark cross bands, the anterior on the occiput, the posterior on the loins; these bands curved, with the concavity in front; lips with dark vertical bars; lower surfaces whitish.

Total length	85 millim.
Head	16 ,,
Width of head	12 ,,
Body	37 ,,
Fore limb	21 ,,
Hind limb	28 ,,
Tail	32 ,,

Persia.

a. ♀.   Tanjistan.   J. A. Murray, Esq. [P.]. (One of the types.)

6. PTENOPUS.

Ptenopus, *Gray, Proc. Zool. Soc.* 1865, p. 640; *Cope, Proc. Ac. Philad.* 1868, p. 321.

Digits not dilated, furnished with a long claw, inferiorly with a series of narrow transverse plates. Fingers subcylindrical. Toes depressed, with a lateral fringe of long pointed scales. Body covered with granular scales. Pupil vertical. Lower eyelid distinct, upper much developed. No præanal nor femoral pores.

South Africa.

1. Ptenopus garrulus. (PLATE II. fig. 2.)

Stenodactylus garrulus, *Smith, Ill. S. Afr., Rept., App.* p. 6; *A. Dum. Cat. Méth. Rept.* p. 47, *and Arch. Mus.* viii. p. 488; *Günth. Zool. Rec.* ii. p. 149.
Ptenopus maculatus, *Gray, l. c.*; *Cope, l. c.*

Head short, swollen; snout rounded, as long as the diameter of the orbit, shorter than the distance between the eye and the ear-opening; upper eyelid well developed, meeting almost the rudimentary lower lid; ear-opening a small oblique slit. Body moderately elongate, depressed; limbs moderate; a more or less distinct transverse gular fold. Scales uniform small granules, a little larger on the belly. Rostral small, six-sided, nearly as broad as high; nostril pierced between two nasals, the anterior being the largest and in contact with the rostral and first labial; seven or eight upper, and seven to nine lower labials; mental small, a little longer than the adjacent labials; no chin-shields. Tail short, cylindrical, tapering, covered with uniform, small, squarish flat scales. Pale buff or cream-coloured above, with dark reddish-brown reticulations forming spots; lower surfaces white.

Total length	80	millim.
Head	12	,,
Width of head	11	,,
Body	37	,,
Fore limb	19	,,
Hind limb	25	,,
Tail	31	,,

South Africa.

a. Several specs.: ♂, ♀, S. Africa. Sir A. Smith [P.].
  and hgr. (Types.)
b–c. ♂ ♀. Damaraland. (Types of *Ptenopus maculatus.*)

7. STENODACTYLUS.

Stenodactylus, *Fitzing. N. Classif. Rept.* p. 13, *and Syst. Rept.*
 p. 89; *Cuv. Règne Anim.* 2nd ed. ii. p. 58; *Dum. & Bibr.* iii.
 p. 433; *Gray, Cat.* p. 177.
Ascalabotes, *Wagl. Syst. Amph.* p. 143.
Gymnodactylus, part., *Wiegm. Herp. Mex.* p. 19.
Tolarenta, *Gray, Zool. Misc.* p. 58.
Tropiocolotes, *Peters, Mon. Berl. Ac.* 1880, p. 306.

Digits not dilated, furnished with a long claw, and a lateral fringe or denticulation of pointed scales; inferiorly with a series of keeled scales. Body covered with juxtaposed or imbricated scales. Pupil vertical. No præanal nor femoral pores.

North Africa; South-western Asia; Sind.

Synopsis of the Species.

I. No chin-shields; dorsal scales small.

Dorsal lepidosis formed of granules intermixed with scattered tubercles; digits
 strongly fringed .... 1. *orientalis*, p. 16.
Dorsal scales uniform, juxtaposed; rostral
 entering the nostril .... 2. *guttatus*, p. 17.
Dorsal scales uniform, subimbricate; rostral
 not entering the nostril .... 3. *wilkinsonii*, p. 18.

II. Two pairs of chin-shields; dorsal scales large, imbricate.

Dorsal scales feebly keeled, abdominal scales
 smooth .... 4. *petersii*, p. 18.
Dorsal and ventral scales strongly keeled.. 5. *tripolitanus*, p. 19.

1. Stenodactylus orientalis. (PLATE III. fig. 1.)

Stenodactylus orientalis, *Blanf. Journ. As. Soc. Beng.* xlv. 1876, p. 21.
—— dunstervillei, *Murray, Zool. Sind,* p. 363, *& erratum.*

Head rather depressed, regularly oviform; snout slightly longer than the diameter of the orbit, as long as the distance between the latter and the ear-opening; forehead very slightly concave; ear-opening an oblique slit measuring two thirds the diameter of the eye. Body moderate, depressed. Limbs moderate; digits elongate, depressed, with well-developed lateral fringe; the transverse inferior lamellæ quinquecarinate. Head covered with small granules, which are feebly keeled on the snout and between the eyes; rostral quadrangular, broader than high, with median cleft above; nostril pierced between the rostral, the first labial, and three nasals; twelve upper and thirteen lower labials; mental broader than high, subpentagonal, the lower angle rounded; no chin-shields. Body covered with small granules, keeled on the ventral region, intermixed with small, keeled, irregularly scattered roundish tubercles on the back. Tail cylindrical, tapering to a fine point, covered with small keeled scales arranged in rings. Colour pale sandy, with indistinct darker transverse bands; a darker line from the eye down each side; dorsal tubercles darker; lower surface whitish.

Total length	84	millim.
Head	14	,,
Width of head	10	,,
Body	35	,,
Fore limb	19	,,
Hind limb	26	,,
Tail	35	,,

Sind.

a. ♀. Near Rohri. W. T. Blanford, Esq. [P.]. (Type.)

2. Stenodactylus guttatus. (PLATE III. fig. 2.*)

Geoffr. Descr. Egypte, pl. v. fig. 2.
Ascalabotes sthenodactylus, *Licht. Verz. Doubl. Mus. Berl.* p. 102.
Stenodactylus guttatus, *Cuv. R. A.* 2nd ed. ii. p. 58; *Dum. & Bibr.* iii. p. 434; *Strauch, Erp. Alg.* p. 24; *Peters, Mon. Berl. Ac.* 1880, p. 306.
Trapelus savignyi, *Aud. Descr. Egypte, Rept., Suppl.* p. 167, pl. i. figs. 3 & 4.
Stenodactylus elegans, *Fitz. Syst. Rept.* p. 89.
—— mauritanicus, *Guich. Expl. Sc. Alg., Rept.* p. 5, pl. i. fig. 1; *A. Dum. Cat. Méth. Rept.* p. 47, and *Arch. Mus.* viii. p. 87; *Strauch, l. c.* p. 25.

Head very variable in shape, more or less depressed; snout rounded or more or less pointed, as long as the diameter of the orbit, longer than the distance between the eye and the ear-opening; eye very large; ear-opening oval, oblique, not half the diameter of the eye. Body moderately elongate, depressed. Limbs long and slender; digits elongate, scarcely depressed, feebly denticulated laterally; the transverse inferior lamellæ tricarinate. Head covered with small keeled granular scales, hexagonal on the snout; rostral broader than high, with median cleft above; nostril pierced in the middle of a slight swelling, between the rostral, the first labial, and three nasals; eleven to thirteen upper, and ten to twelve lower labials; mental quadrangular; no chin-shields. Body covered with small flat juxtaposed scales, which may be slightly keeled on the back and are always so on the belly. Tail cylindrical, tapering, probably prehensile, covered with small juxtaposed keeled scales. Light buff or sandy above, with round whitish spots inclosed in the meshes of a brown network; tail with brown annuli; back sometimes with rather indistinct dark cross bands; lower surfaces white.

---

* Lower surface of fourth toe, × 4.

Total length	98 millim.
Head	17 ,,
Width of head	12·5 ,,
Body	41 ,,
Fore limb	23 ,,
Hind limb	30 ,,
Tail	40 ,,

North Africa; South-western Asia.

a. ♂.	Egypt.	
b–c. ♂ ♀.	Cairo.	Sir R. Owen [P.].
d. ♂.	Sinaitic Peninsula.	H. C. Hart, Esq. [C.].
e. ♀.	Mount Sinai.	
f. ♂.	Dead Sea.	Rev. H. B. Tristram [C.].
g. ♂.	Jaffa.	

3. Stenodactylus wilkinsonii. (PLATE III. fig. 3.*)

Stenodactylus guttatus (non Cuv.), *Gray, Cat.* p. 177.
Tolarenta wilkinsonii, *Gray, Zool. Misc.* p. 58.

Snout acutely pointed. Limbs very slender and elongate; digits slender, very feebly denticulated laterally. Scales flat, small, slightly keeled, subimbricate. Nostril pierced in the centre of a very strong swelling between the first labial and three nasals; fourteen upper and twelve lower labials. Otherwise as in *S. guttatus*. Brown above, white-spotted.

Total length	94 millim.
Head	15 ,,
Width of head	9 ,,
Body	38 ,,
Fore limb	23 ,,
Hind limb	30 ,,
Tail	41 ,,

Egypt.

a. Bad state. Egypt. Sir J. G. Wilkinson [P.]. (Type.)

4. Stenodactylus petersii. (PLATE III. fig. 4.)

Head moderately depressed; snout pointed, as long as the diameter of the orbit, or the distance between the eye and the ear-opening; latter small, round. Body and limbs rather slender; digits elongate, scarcely depressed, feebly denticulated laterally; the transverse inferior lamellæ tricarinate. Head covered with large convex granules; rostral as high as broad, with median cleft above; nostril pierced between the rostral, the first labial, and two nasals; nine upper and eight lower labials; mental large, triangular; two pairs of chin-shields. Body covered with rather large cycloid imbricated scales, which are feebly keeled on the back and

---

* Profile of head, × 2.

smooth on the belly. Tail cylindrical, tapering, covered with uniform keeled scales forming rings. Colour above sandy, with three series of brown spots on the back; a brown streak on the side of the head, passing through the eye; lower surfaces white.

Total length	51 millim.
Head	7·5 ,,
Width of head	4·5 ,,
Body	20·5 ,,
Fore limb	10 ,,
Hind limb	14 ,,
Tail (reproduced)	23 ,,

Egypt.

a. ♀?   Egypt.

5. Stenodactylus tripolitanus.

Tropiocolotes tripolitanus, *Peters, Mon. Berl. Ac.* 1880, p. 306, fig. 1.

This species appears to agree in every point with the preceding, except that all the scales of the body are strongly keeled, and there are only seven upper and six lower labials. The tail is considerably longer than head and body.

Tripoli.

8. ALSOPHYLAX.

Alsophylax, *Fitzing. Syst. Rept.* p. 90.
Bunopus, *Blanf. Ann. & Mag. N. H.* (4) xiii. 1874, p. 454, *and Zool. E. Persia,* p. 348.

Digits not dilated, clawed, not denticulated laterally, inferiorly with a series of lamellæ. Body covered above with juxtaposed scales intermixed with enlarged tubercles, inferiorly with imbricate scales. Pupil vertical. Males with præanal pores.

Turkestan; Persia; Baluchistan.

1. Alsophylax pipiens. (PLATE III. fig. 5.*)

Lacerta pipiens, *Pallas, Zoogr. Ross.-As.* iii. p. 27.
Ascalabotes pipiens, *Lichtenst. in Eversm. Reise,* p. 145, *and Verz. Doubl. Mus. Berl.* p. 103.
Gymnodactylus pipiens, *Eichw. Zool. Spec.* iii. p. 181.
Stenodactylus pipiens, *Fitzing. N. Classif. Rept.* p. 47.
—— eversmanni, *Fitzing. Syst. Rept.* p. 90.
Eublepharis (Alsophylax) pipiens, *Fitzing. l. c.*
Gymnodactylus microtis, *Blanf. Journ. As. Soc. Beng.* xliv. 1875, p. 193, *and 2nd Yark. Miss., Rept.* p. 15, pl. ii. fig. 1.

Head rather small, not much depressed; snout obtuse, a little longer than the diameter of the orbit or the distance between the

---

* Lower surface of fourth toe, × 4.

eye and the ear-opening; latter very small. Body moderately depressed. Limbs moderate; digits slender, inferiorly with simple transverse lamellæ. Head covered with large convex granules; rostral rather large, pentagonal, with median cleft above; nostril pierced between the rostral, the first labial, and one nasal; seven or eight upper and five or six lower labials, anterior very large; mental very large, trapezoid, in contact with two small chin-shields, followed by others passing gradually into the rather large flat gular granules. Back covered with irregular flat juxtaposed scales, intermixed with small, very irregularly scattered keeled tubercles. Abdominal scales large, hexagonal, imbricate. Male with an angular series of seven or nine præanal pores. Tail cylindrical, tapering, covered with smooth imbricated scales, inferiorly with a median series of enlarged scales. Colour sandy above, with rather indistinct darker cross bands; lips brown-spotted; lower surfaces white.

Total length	82 millim.
Head	10 ,,
Width of head	6·5 ,,
Body	27 ,,
Fore limb	14 ,,
Hind limb	19 ,,
Tail	45 ,,

Turkestan.

a–c. ♂ ♀.   Gt. Mt. Bogdo.   St. Petersburg Museum [E.].

2. Alsophylax tuberculatus.

Bunopus tuberculatus, *Blanf. Ann. & Mag. N. H.* (4) xiii. p. 454, and *Zool. E. Persia,* p. 348, pl. xxii. fig. 4.

Head moderate; snout obtuse, a little longer than the diameter of the orbit or the distance between the eye and the ear-opening; forehead very slightly concave; ear-opening elliptic, vertical, nearly half the diameter of the eye. Body moderately depressed. Limbs moderate; digits slender, the inferior lamellæ furnished with projecting tubercles. Head covered with large granules, smaller and intermixed with round tubercles on the temporal and occipital regions; rostral subquadrangular, not much broader than high, with median cleft above; nostril pierced between the rostral, the first labial, and three nasals; about ten to twelve upper and eight to ten lower labials; mental subtrapezoid, broader than long; no chin-shields; gular granules minute. Back covered with small irregular flat granules, intermixed with large trihedral tubercles, forming about fourteen irregular longitudinal series. Abdominal scales rather small, subhexagonal, imbricate. Males with seven or eight præanal pores, forming a slightly angular series. Tail cylindrical, slightly depressed, verticillate, with rings of keeled tubercles; no enlarged scales inferiorly. Colour sandy, darker-spotted; a darker streak on the side of the head, passing through the eye, sometimes meeting its fellow on the occiput, sometimes extending to the side of the body; lower surfaces white.

Total length	104	millim.
Head	13	,,
Width of head	9	,,
Body	40	,,
Fore limb	17	,,
Hind limb	22	,,
Tail	51	,,

Baluchistan; Southern Persia.

a–b. ♂ ♀.	Bahu Kalat, Baluchistan.	W. T. Blanford,
c–d. ♀ & hgr.	Mand, Baluchistan.	Esq. [C.].
e. ♂.	Near Bampur, Baluchistan.	(Types.)

9. HOMONOTA.

Homonota, *Gray, Cat. Liz.* p. 171.

Digits not dilated, clawed, inferiorly with a series of lamellæ, not denticulated laterally. Body covered with uniform imbricate scales. Pupil vertical. No præanal nor femoral pores.

South-eastern South America.

1. Homonota darwinii. (PLATE III. fig. 7.)

Homonota gaudichaudi (*non D. & B.*), *Gray, l. c.*
Gymnodactylus gaudichaudii (*non D. & B.*), *Bell, Zool. 'Beagle,' Rept.* p. 26, pl. xvi. fig. 1.

Head short, convex; snout rounded, longer than the diameter of the orbit, as long as the distance between the eye and the ear-opening; latter a small oblique slit. Body rather elongate, feebly depressed. Limbs moderate; digits short, rather thick. Head covered with convex granules, which are largest on the snout; rostral pentagonal or quadrangular, broader than high, with median cleft above; nostril pierced between the rostral, the first labial, and three nasals; six or seven upper and five or six lower labials; mental large, pentagonal subcampanuliform; no regular chin-shields, but dilated granules border the mental and infralabials, and pass gradually into the smaller granules of the throat, which are flat and become imbricate as they approach the neck. Body covered with uniform cycloid imbricated scales, which are a little larger on the belly; those on the back sometimes very indistinctly keeled. Tail cylindrical, tapering to a very fine point, sometimes rather swollen; the scales uniform, cycloid, imbricate, smooth, rather larger inferiorly. Light brown above, marbled with darker; a somewhat indistinct dark streak on the side of the head, passing through the eye; lower surfaces whitish.

Total length	97 millim.
Head	13 ,,
Width of head	10 ,,
Body	35 ,,
Fore limb	16 ,,
Hind limb	22 ,,
Tail	49 ,,

Eastern Patagonia; Buenos Ayres; Uruguay.

a–d. ♂ ♀.	Port Desire.	C. Darwin [P.]. (Typical specimens of Bell's *G. gaudichaudii*.)
e. ♂.	Montevideo.	
f. ♂.	——?	T. Bell, Esq. [P.].

2. Homonota whitii. (PLATE III. fig. 6.)

Distinguished from the preceding in having the head and body considerably depressed, the head longer, the body shorter, and the digits more slender. Rostral quadrangular, twice as broad as high; eight upper and six lower labials. A median series of enlarged infracaudals. Grey-brown, marbled with darker.

Total length	74 millim.
Head	11 ,,
Width of head	7·5 ,,
Body	26 ,,
Fore limb	12 ,,
Hind limb	16·5 ,,
Tail	37 ,,

Cordova.

a. ♂.	Cosquin.	E. W. White, Esq. [C.].

10. GYMNODACTYLUS *.

Phyllures, *Cuv. R. A.* ii. p. 50.
Gymnodactylus, *Spix, Spec. Nov. Lacert. Bras.* p. 17; *Wagl. Syst. Amph.* p. 144; *Fitzing. Syst. Rept.* p. 90; *Gray, Cat. Liz.* p. 175; *Günth. Rept. Brit. Ind.* p. 112.
Stenodactylus, *Fitzing. N. Class. Rept.* p. 13.
Phyllurus, *Fitzing. l. c., and Syst. Rept.* p. 92; *Gray, Cat.* p. 176.
Gonyodactylus, *Kuhl, Isis*, 1827, p. 290; *Wagl. l. c.*; *Fitzing. l. c.* p. 92.
Cyrtodactylus, *Gray, Phil. Mag.* (2) iii. 1827, p. 55, *and Zool. Journ.* iii. 1828, p. 224, *and Cat.* p. 173.
Gymnodactylus, part., *Wiegm. Herp. Mex.* p. 19; *Dum. & Bibr.* iii. p. 408.
Anomalurus, *Fitzing. l. c.* p. 90.

---

* *Gymnodactylus atropunctatus*, Lichtenst. Nomencl. Rept. Mus. Berol. p. 6.—Turkestan.
—— *tenuis*, Hallow. Proc. Ac. Philad. 1856, p. 149.—Manilla.

Saurodactylus, *Fitzing. l. c.* p. 91.
Dasyderma, *Fitzing. l. c.* p. 92.
Cyrtopodion, *Fitzing. l. c.* p. 93.
Cubina, *Gray, Cat.* p. 175.
Puellula, *Blyth, Journ. As. Soc. Beng.* xxix. 1860, p. 109.
Geckoella, *Gray, Proc. Zool. Soc.* 1868, p. 99.
Quedenfeldtia, *Boettg. Abh. Senck. Ges.* xiii. 1883, p. 125.

Digits not dilated, clawed, cylindrical or slightly depressed at the base; the two or three distal phalanges compressed, forming an angle with the basal portion of the digits; the claw between two enlarged scales, the inferior of which is more or less deeply notched under the claw; digits inferiorly with a row of more or less distinct transverse plates. Body variously scaled. Pupil vertical. Males with or without præanal or femoral pores.

Borders of the Mediterranean; Southern Asia; Australia; islands of the Pacific; Tropical America.

Synopsis of the Species.

I. Tail not, or but slightly, swollen.
  A. Stone- or Sand-Geckos with rows of trihedral sharply keeled tubercles on the back and tail; ventral scales not larger than the dorsal tubercles.
    1. Infracaudal scales not keeled.

Dorsal tubercles very large, not longer than broad, in 14 longitudinal series; 26 longitudinal series of scales across the middle of the belly; male with a long series of præanal and femoral pores .... 1. *caspius*, p. 26.
Dorsal tubercles large, in 12 or 14 longitudinal series; about 20 series of scales across the middle of the belly .... 2. *scaber*, p. 27.
Dorsal tubercles small, in 10 longitudinal series; about 20 series of scales across the middle of the belly; the fore limb, when stretched forwards, does not reach the tip of the snout .... 3. *brevipes*, p. 28.
Dorsal tubercles small, in 10 or 12 longitudinal series; about 30 series of scales across the middle of the belly; the hind limb stretched forwards does not extend beyond the shoulder .... 4. *kotschyi*, p. 29.
Dorsal tubercles moderate, in 12 or 14 longitudinal series; about 30 series of scales across the middle of the

belly; the hind limb stretched forwards reaches constriction of neck . . . . . . 5. *kachhensis*, p. 29.
Dorsal tubercles small, irregularly arranged; about 25 scales across the middle of the belly; no granules or small scales between the keeled scales of the upper surface of the tail .... 7. *elongatus*, p. 30.

2. Infracaudal scales keeled.

The fore limb stretched forwards reaches the tip of the snout ...... 6. *heterocercus*, p. 30.
The fore limb does not reach the tip of the snout .... 8. *fasciatus*, p. 31.

B. Stone- or Sand-Geckos with long slender digits and the dorsal scales uniform or intermixed with slightly enlarged smooth tubercles.

1. Dorsal scales juxtaposed.

    a. Enlarged tubercles on the back or on the tail.

Tail swollen, with enlarged tubercles. 9. *stoliczkæ*, p. 31.
Tail without enlarged tubercles; mental plate large .... 10. *lawderanus*, p. 32.
Tail without enlarged tubercles; mental scarcely larger than the adjacent labials...................... 11. *dorbignyi*, p. 33.

    b. No enlarged tubercles either on the back or on the tail.

Tail cylindrical .... 12. *mauritanicus*, p. 33.
Tail depressed; border of the eyelid with prominent spine-like scales .. 13. *trachyblepharus*, p. 34.

2. Dorsal scales imbricate, like the ventrals .... 14. *steudneri*, p. 34.

C. Tree-Geckos, with uniform or heterogeneous dorsal scales; clawed phalanges strong.

1. No lateral fold.

    a. Dorsal tubercles smooth or unicarinate.

        α. Ventral scales small.

Dorsal scales very small, granular, uniform or intermixed with very small tubercles..................... 15. *nebulosus*, p. 34.
Dorsal scales very large and flat, smooth, squarish, uniform; seven lower labials...................... 16. *jeyporensis*, p. 36.

Dorsal scales very large and flat, smooth or feebly keeled, uniform; body and tail with transverse white bands .. 17. *deccanensis*, p. 36.
Back with tubercles of unequal size; ventral scales tubercular, slightly keeled; body and tail with transverse white bands .... 18. *albofasciatus*, p. 37.
Back granular, with about 30 longitudinal series of tubercles contained between an area bounded by faint keels margining the belly .... 19. *oldhami*, p. 38.
Back with small granules intermixed with numerous small trihedral tubercles .... 20. *trihedrus*, p. 38.
Back with 16 regular longitudinal series of roundish tubercles ...... 21. *arnouxii*, p. 39.

β. Ventral scales very large, in 16 longitudinal series .... 22. *geckoides*, p. 39.

b. Dorsal tubercles ribbed; ventral scales keeled.

Dorsal tubercles forming very regular longitudinal series .... 23. *pelagicus*, p. 40.
Dorsal tubercles in about 10 irregular longitudinal series .... 24. *heteronotus*, p. 41.
Dorsal tubercles in more than 10 irregular longitudinal series .... 25. *cheverti*, p. 41.
Dorsal tubercles not in regular series. 26. *affinis*, p. 42.

2. A slight fold from axilla to groin. Large species, with large head.

Dorsal tubercles small, smooth, irregularly scattered; male with two pairs of præanal pores .... 27. *frenatus*, p. 42.
Dorsal tubercles large, trihedral; 26 longitudinal series of ventral scales; male with 16 præanofemoral pores on each side .... 28. *variegatus*, p. 43.
Dorsal tubercles large, trihedral; 36 longitudinal series of ventral scales; male with 5 or 6 præanal pores on each side .... 29. *fasciolatus*, p. 44.
Dorsal tubercles small, roundish, feebly keeled; infradigital lamellæ much developed; male with 10 or 13 præanal pores, forming an angle...... 30. *khasiensis*, p. 44.
Dorsal tubercles small, roundish, feebly keeled; male with 12 or

13 præanal, and, on each side, 5 or 6 femoral pores, the former in a longitudinal groove	31. *marmoratus*, p. 44.
Dorsal tubercles small, roundish, feebly keeled; no femoral pores; male with a few præanal pores in a longitudinal groove	32. *rubidus*, p. 45.
Dorsal tubercles small, subconical; infradigital lamellæ hardly distinguishable; no femoral pores; male with a few præanal pores in a longitudinal groove	33. *philippinicus*, p. 46.
Dorsal tubercles small, roundish, keeled; male with a few præanal pores in a longitudinal groove, and 14 or 15 femoral pores on each side; body with transverse brown bands, narrower than the interspaces between them	34. *pulchellus*, p. 46.
Dorsal tubercles very small; male with an angular series of 9 to 11 præanal pores, without pubic groove or femoral pores; body with transverse brown bands narrower than the interspaces between them	35. *consobrinus*, p. 47.

II. Tail at least nearly as broad as the body.

Tail carrot-shaped	36. *miliusii*, p. 48.
Tail leaf-shaped	37. *platurus*, p. 49.

1. Gymnodactylus caspius.

Gymnodactylus caspius, *Eichw. Zool. Spec. Ross. Pol.* iii. p. 181, *and Faun. Casp. Cauc.* p. 91, pl. xv.; *A. Dum. Cat. Méth. Rept.* p. 45, *and Arch. Mus.* viii. p. 482; *Stoliczka, Proc. As. Soc. Beng.* 1872, pp. 80 & 126.

Uromastix fasciatus, *Ménétr. Cat. Rais.* p. 64.

? Gymnodactylus geckoides (*non Spix*), *Blyth, Journ. As. Soc. Beng.* xxii. 1853, p. 410; *Theobald, Cat. Rept. As. Soc. Mus.* p. 31.

Head moderate, oviform; snout slightly longer than the diameter of the orbit, longer than the distance between the eye and the ear-opening; eye large; forehead very slightly concave; ear-opening rather small, elliptical, vertical, slightly oblique. Body moderate. Limbs long, the fore limb carried forwards reaching a little beyond the tip of the snout, the hind limb reaching constriction of neck; digits long and slender, cylindrical at the base, inferiorly with well-developed lamellæ. Head covered with large convex tubercles, which are slightly keeled and intermixed with small granules on the occipital and temporal regions; rostral quadrangular, broader than high, with median cleft above; nostril pierced between the rostral, the first labial, and three nasals; ten upper and eight lower labials; mental large, pentagonal, longer than broad; two pairs of chin-shields, median largest and forming a short suture behind the point of the mental. Body covered above with irregular small flat scales and large trihedral subtriangular tubercles, forming fourteen longitudinal series; these tubercles are very large, slightly broader than long, the diameter of the largest equalling three fifths the diameter of the eye; the keel is very sharp, terminating in a small spine on the largest tubercles. Abdominal scales moderate, cycloid, imbricate, smooth, in twenty-six longitudinal series in the middle of the belly. Male with a long continuous series of about twenty femoral and præanal pores. Tail rounded, tapering, slightly depressed, above with rows of large spinose trihedral tubercles, inferiorly with a median series of enlarged transverse plates. Sandy-coloured above, with rather indistinct darker transverse bands; lower surfaces white.

Total length	115	millim.
Head	18	,,
Width of head	12·5	,,
Body	42	,,
Fore limb	26	,,
Hind limb	37	,,
Tail (injured)	55	,,

Turkestan; Punjab?

a. ♂. Krasnowodsk. St. Petersburg Museum [E.].

2. Gymnodactylus scaber.

Stenodactylus scaber, *Rüpp. Atl. N. Afr., Rept.* p. 15, pl. iv. fig. 2.
Gymnodactylus scaber, part., *Dum. & Bibr.* iii. p. 421.
Gonyodactylus (Cyrtopodion) scaber, *Fitz. Syst. Rept.* p. 93.
Gymnodactylus geckoides (*non Spix*), *Schreib. Herp. Eur.* p. 482.
—— scaber, *Murray, Ann. & Mag. Nat. Hist.* (5) xiv. 1884, pp. 102, 110.

Head moderate, narrower and smaller than in the preceding; snout slightly longer than the diameter of the orbit, longer than the distance between the eye and the ear-opening; eye large, forehead very slightly concave; ear-opening rather small, elliptical, vertical. Body moderate. Limbs long, the fore limb carried forwards, reaching as far as the tip of the snout, the hind limb reaching articulation of neck; digits long and slender, cylindrical at the base, inferiorly with well-developed lamellæ. Snout covered with large convex polygonal granules, which are considerably smaller than in *G. caspius*; hinder part of head with small granules intermixed with roundish tubercles; rostral quadrangular, with median cleft above; nostril pierced between the rostral, the first labial, and three nasals; ten upper and eight or nine lower labials; mental triangular or pentagonal, not longer than broad; two pairs of chin-shields, median largest and forming a suture behind the point of the mental. Body covered above with irregular small flat scales and large trihedral subtriangular tubercles, forming twelve or fourteen longitudinal series; these tubercles are not so large as in *G. caspius*, though likewise much larger than the interspaces between them, the diameter of the largest not quite equal to half that of the eye; they are strongly keeled. Abdominal scales large, cycloid-hexagonal, imbricate, smooth, in about twenty longitudinal series in the middle of the belly. Male with five or six præanal pores. Tail feebly depressed, tapering, above with rows of large spinose trihedral tubercles, inferiorly with a median series of enlarged transverse plates. Sandy-coloured above, brown-spotted; tail with brown annuli; lower surfaces white.

Total length	105	millim.
Head	13	,,
Width of head	9	,,
Body	33	,,
Fore limb	19	,,
Hind limb	27	,,
Tail	59	,,

From Egypt southwards to Abyssinia and eastwards to Afghanistan and Sind.

a. ♂.	Egypt.	J. Doubleday, Esq. [P.].
b-c. ♂.	Fao, S. Mesopotamia.	J. A. Murray, Esq. [P.].
d. ♂.	Shiraz.	Kotschy [C.].
e. ♂.	Candahar.	Col. Swinhoe [P.].

3. Gymnodactylus brevipes.

Gymnodactylus brevipes, *Blanford, Ann. & Mag. Nat. Hist.* (4) xiii. 1874, p. 453, *and Zool. E. Persia*, p. 344, pl. xxii. fig. 2, *and Journ. As. Soc. Beng.* xlv. 1876, p. 20.

Differs from *G. scaber* in the following points:—Limbs shorter; the fore limb reaches between the eye and the tip of the snout, the hind limb to the shoulder. Dorsal tubercles small, nearly equal to the small ear-opening in size, forming ten longitudinal series. Nine upper and seven lower labials. Præanal pores four. In all these characters this species approaches *G. kotschyi*, from which it differs chiefly in the larger ventral scales, which are as in *G. scaber*. The enlarged plates on the lower surface of the tail are about equally long and broad, and many of them are divided into two. Colour grey, with three rather imperfect longitudinal dusky bands on the back, formed of arrowhead-shaped marks; a dusky line, not very strongly marked, from the eye to the shoulders. Size of *G. kotschyi*.

Baluchistan.

4. Gymnodactylus kotschyi.

Gymnodactylus geckoides (*non Spix*), *Gray, Cat.* p. 175.
Stenodactylus guttatus (*non Cuv.*), *Bibr. in Bory, Expéd. Sc. Mor.* iii. p. 69, pl. xi. fig. 3.
Gymnodactylus scaber, part., *Dum. & Bibr.* iii. p. 421.
—— kotschyi, *Steindachn. Sitzb. Ak. Wien,* lxii. i. 1870, p. 329, pl. i. fig. 1; *Schreib. Herp. Eur.* p. 481; *De Betta, Atti Ist. Venet.* (5) v. 1879, p. 382; *Boettger, Ber. Senck. Ges.* 1878–79, p. 75; *Bedriaga, Bull. Mosc.* lvi. 1882, p. 54; *Peracca, Zool. Anz.* 1884, p. 572.

Distinguished from *G. scaber* in the following characters:—Body and limbs stouter, less elongate, the fore limb reaching hardly the tip of the snout, and the hind limb the axilla or the shoulder. Ear-opening larger, suboval or roundish. Dorsal tubercles much smaller, not larger than the ear-opening, the largest much longer than broad; they form ten or twelve longitudinal series. Abdominal scales smaller, in about thirty longitudinal series in the middle of the belly. Four præanal pores. Nine upper and seven lower labials. Grey above, with angular transverse bands; lower surfaces whitish.

Total length	94 millim.
Head	13 ,,
Width of head	9 ,,
Body	29 ,,
Fore limb	17 ,,
Hind limb	22·5 ,,
Tail	52 ,,

South Italy, Greece, Turkey (?), Cyprus, Syria, Persia, Egypt, island of Goree.

a–b. ♀.	S. Europe.	
c. ♂.	Milo Island, Cyclades.	G. A. Boulenger, Esq. [P.].
d. ♂.	Mount Carmel.	Rev. H. B. Tristram [C.].
e. Yg.	Lake of Galilee.	Rev. H. B. Tristram [C.].

5. Gymnodactylus kachhensis.

Gymnodactylus kachhensis, *Stoliczka, Proc. As. Soc. Beng.* 1872, p. 79; *Blanford, Journ. As. Soc. Beng.* xlv. 1876, p. 20.
—— petrensis, *Murray, Zool. Sind,* p. 362, pl. —. fig. 1.

Allied to the preceding species. Body rather short, as in *G. kotschyi*; fore limb reaching as far as the tip of the snout or a little beyond, hind limb to constriction of neck. Dorsal tubercles in twelve or fourteen longitudinal series, intermediate in size between those of *G. scaber* and *G. kotschyi*, in shape similar to the former. Abdominal scales in about thirty longitudinal rows, as in *G. kotschyi*. No regular series of infracaudal plates. Præanal pores four to six. Ten or eleven upper and eight to ten lower labials. Ear-opening narrow, vertical. Sandy-coloured above, with rather indistinct darker spots on body and limbs and annuli on the tail.

Total length	80 millim.
Head	12 „
Width of head	8·5 „
Body	28 „
Fore limb	18 „
Hind limb	25 „
Tail (injured)	40 „

Cutch; Sind.

a–c. ♂ ♀ & hgr.	Cutch.	F. Stoliczka [C.]. (Typical specimens.)
d–e, f–i. ♂ ♀.	Sind.	J. A. Murray, Esq. [P.]. (As typical of *G. petrensis*.)

6. Gymnodactylus heterocercus.

Gymnodactylus heterocercus, *Blanford, Ann. & Mag. Nat. Hist.* (4) xiii. 1874, p. 453, *and Zool. E. Persia*, p. 345, pl. xxii. fig. 3.

Distinguished from all the preceding species, to which it is allied, in having the scales on the lower portion of the tail small, strongly keeled, sharply pointed behind, imbricate, and not arranged in regular verticils. Dorsal tubercles about equal to the small ear-opening in size, very little, if at all, longer than broad, and arranged in twelve longitudinal rows. Ventral scales in twenty-five to thirty longitudinal rows. The fore limb reaches the tip of the snout and the hind limb the shoulder. Eight to ten upper and seven or eight lower labials. Uniform grey. From snout to vent about 40 millim.

Eastern Persia.

7. Gymnodactylus elongatus.

Gymnodactylus elongatus, *Blanford, Journ. As. Soc. Beng.* xliv. 1875, p. 193, *and 2nd Yark. Miss., Rept.* p. 14, pl. ii. fig. 2.

Allied to the preceding species. Body rather elongate. Limbs long, the fore limb extending to the end of the snout, the hind limb extending some distance in front of the shoulder. Surface of the head granular, granules nearly uniform; nostril between the rostral, the first labial, and two nasals, which are rather swollen. Dorsal tubercles triangular, nearly as large as the small ear-opening; they are not arranged in regular rows, but about twelve may be counted across the back. About twenty-five larger scales across the belly. Præanal pores about six, in a V-shaped line. Tail thin, very regularly attenuate, verticillate, covered with trapezoidal or subtrapezoidal keeled scales, the posterior row of each ring larger, but without any granules or small scales between, so that there are no distinct tubercles; lower surface of the tail, except near the base, with a row of large plates about as broad as long. Colour pale grey, with darker transverse bands on the body, limbs, and tail.

Total length	127	millim.
Head	16	,,
Fore limb	26	,,
Hind limb	32	,,
Tail	68	,,

Eastern Turkestan.

8. Gymnodactylus fasciatus.

Cubina fasciata, *Gray, Cat.* p. 175.
Gymnodactylus fasciatus, *Dum. & Bibr.* iii. p. 420.

Head not much depressed, oviform; snout a little longer than the diameter of the orbit, as long as the distance between the eye and the ear-opening; forehead very slightly concave; ear-opening elliptical, oblique, about two fifths the diameter of the eye. Body and limbs moderate; digits slightly depressed at the base, with well-developed transverse plates inferiorly. Snout covered with rather large polygonal granules; hinder part of head with small granules intermixed with roundish tubercles; rostral subquadrangular, twice as broad as high, with median cleft above; nostril pierced between the rostral, the first labial, and three nasals; seven or eight upper and six or seven lower labials; mental broadly triangular; a pair of chin-shields forming a suture behind the mental, followed by a row of smaller shields. Body covered above with small granules and large keeled subtriangular tubercles arranged in about twelve pretty regular longitudinal series. Abdominal scales large, cycloid, imbricate, smooth. Tail cylindrical, with rings of large keeled tubercles. Greyish above, with darker transverse lines.

From snout to vent	51	millim.
Head	19	,,
Width of head	9·5	,,
Fore limb	18	,,
Hind limb	25	,,

Martinique.

a. ♀. Martinique.

9. Gymnodactylus stoliczkæ.

Gymnodactylus stoliczkæ, *Steindachn. Novara, Rept.* p. 15, pl. ii. fig. 2; *Blanf. Journ. As. Soc. Beng.* xliv. 1875, p. 193, *and 2nd Yark. Miss., Rept.* p. 12.
Cyrtodactylus yarkandensis, *Anders. Proc. Zool. Soc.* 1872, p. 381, fig.

Head moderate, oviform, much depressed; snout much longer than the diameter of the orbit, slightly longer than the distance between the eye and the ear-opening; latter very small, suboval. Limbs rather long; digits long, slender, slightly depressed at the base, inferiorly with well-developed lamellæ. Body much depressed. Head and body covered with rather large, flat, round granules, some of which are slightly enlarged [and generally tubercular]. Rostral subpentagonal, with median cleft above; nostril between the rostral, the first labial, and three nasals; ten upper and nine lower labials; mental large, triangular, a little longer than broad; two or three pairs of chin-shields, median largest and forming a short suture behind the mental. Abdominal scales moderate, sub-hexagonal, slightly imbricate. No femoral or præanal pores. Tail swollen, depressed, [about as long as head and body,] ringed, with three enlarged blunt tubercles at each side of the ring. Grey above, with darker wavy cross bands on the back.

From snout to vent	41	millim.
Head	14	,,
Width of head	9	,,
Fore limb	19	,,
Hind limb	26	,,

Ladak.

a. ♀ ? Yarkand (?) Indian Museum [P.].
(One of the types of *Cyrtod. yarkandensis.*)

10. Gymnodactylus lawderanus.

Gymnodactylus lawderanus, *Stoliczka, Journ. As. Soc. Beng.* xli. 1872, p. 105, pl. ii. fig. 4.

Body rather slender and elongate, depressed, covered above with numerous granules intermixed with small roundish tubercles. Upper side of head equally granular, the granular scales being somewhat larger on the snout. Rostral large, broad; nostril pierced between the rostral, the first labial, and two nasals; nine upper and eight lower labials; mental triangular, partially wedged in between two elongate chin-shields, forming a suture below it; each of the chin-shields is followed along the labials by three other somewhat rounded shields. Ear-opening small, rounded. Ventral scales small. Two pairs of præanal pores in the male, close together, and forming an angle. General colour above greyish brown, very densely marbled and spotted with dark brown, with some indistinct, undulating, whitish cross bands on the body, margined on the anterior edges with blackish brown; a somewhat indistinct dark band from the nostril through the eye to the ear; front and hind edges of the eye white; labials spotted and speckled with brown; below whitish.

Length of body nearly 50 millim.

Almorah.

11. Gymnodactylus dorbignyi.

Cubina d'orbignii, *Gray, Cat.* p. 175.
Gymnodactylus dorbignii, *Dum. & Bibr.* iii. p. 418; *Guichen. in Gay, Hist. Chile, Rept.* p. 18, and in *D'Orb. Voy. Amér. Mér.* v. *Rept.* p. 7, pl. ii. figs. 1–5.

Head rather elongate and depressed; ear-opening oval. Digits long and slender, inferiorly with a row of small squarish plates. Nostril between the rostral, the first labial, and three nasals, which are slightly swollen; eight upper and as many lower labials; mental small, hexagonal, followed by two small hexagonal chin-shields. Body covered with minute granules intermixed with small lenticular tubercles. Abdominal scales lozenge-shaped. Tail rounded, tapering, covered with uniform flat, squarish, imbricate scales. Grey above, with numerous darker dots; the border of the eyelid pure white; lower surfaces greyish white.

Total length	87 millim.
Head	13 ,,
Body	34 ,,
Fore limb	15 ,,
Hind limb	20 ,,
Tail	40 ,,

Chili.

12. Gymnodactylus mauritanicus.

Goniodactylus? mauritanicus, *Gray, Cat.* p. 172.
Gymnodactylus mauritanicus, *Dum. & Bibr.* iii. p. 414; *Boettg. Abh. Senck. Ges.* ix. 1874, p. 137.
Saurodactylus mauritanicus, *Fitz. Syst. Rept.* p. 91; *Boettg. Abh. Senck. Ges.* xiii. 1883, p. 125.

Head little depressed; snout rounded. Upper surfaces covered with uniform juxtaposed round scales. Rostral pentagonal; nostril between the rostral, the first labial, and three nasals; five upper and as many lower labials; mental very large, rhomboidal; one or two pairs of chin-shields. Digits long and very slender, cylindrical. Abdominal scales rhomboidal, larger than dorsals. No femoral nor præanal pores. Tail cylindrical, covered above with uniform rounded scales which are larger than those on the back, inferiorly with a series of enlarged transverse plates. Grey-brown above, dotted with darker and with a few white, black-edged spots; a dark streak on the side of the head; lower surfaces whitish, throat brownish.

Total length	59 millim.
Head	8 ,,
Fore limb	9 ,,
Hind limb	11 ,,
Tail	30 ,,

Algeria; Morocco.

13. Gymnodactylus trachyblepharus.

Gymnodactylus trachyblepharus, *Boettg. Abh. Senck. Ges.* ix. 1874, p. 138, pl. i. fig. 3.
Saurodactylus (Quedenfeldtia) trachyblepharus, *Boettg. Abh. Senck. Ges.* xiii. 1883, p. 126.

Head rather broad and depressed; snout rounded; ear-opening transversely oval. Limbs slender; digits long and slender. Upper surfaces covered with uniform small round scales, largest on the snout; upper eyelid with several projecting triangular scales on its free border; rostral pentagonal, nearly twice as broad as high, with median cleft above; nostril between the rostral, the first labial, and four nasals; seven upper and six lower labials; mental large, subtriangular; chin-shields very small. Abdominal scales rather large, subhexagonal. No femoral nor præanal pores. Tail slender, depressed, covered above with uniform small scales, inferiorly with a median series of enlarged transverse plates. Greyish-olive above; tail with rather indistinct yellowish cross bands.

Total length .... 97 millim.
Head .... 10·5 „
Tail.................. 57 „

Mogador.

14. Gymnodactylus steudneri.

Gymnodactylus steudneri, *Peters, Mon. Berl. Ac.* 1869, p. 788.

Digits well developed and of the same proportions as in *G. scaber.* Ear-opening very small. Head granulate, the granules a little larger and polygonal on the snout; nostril between the rostral, the first labial, and two nasals; nine upper and seven lower labials; mental wedged in between two trapezoid chin-shields, on the side of which are a larger and a smaller shield. Body and tail covered with uniform smooth imbricate scales, those on the belly being of the same size as those on the back. Bluish-white, with dark-brown markings, viz. a band from the nostril through the eye and above the ear to the middle of the neck, three rows of irregular spots on the body, a cross band on the neck, a second on scapular region, and others on the tail, and irregular spots on outer side of limbs. From snout to vent 30 millim.

Sennaar.

15. Gymnodactylus nebulosus. (PLATE IV. fig. 1.)

Gymnodactylus nebulosus, *Beddome, Madras Journ. Med. Sc.* 1870; *Günth. Proc. Zool. Soc.* 1875, p. 226.
—— speciosus, *Bedd. l. c.*
—— collegalensis, *Bedd. l. c.*
—— frenatus, juv., *Günth. Ann. & Mag. N. H.* (4) ix. 1872, p. 86.

Head rather large, oviform, generally very convex, sometimes more depressed; snout longer than the diameter of the orbit or the distance between the eye and the ear-opening; forehead slightly concave; ear-opening elliptical, oblique, one third to one half the diameter of the eye. Body rather short, not much depressed. Limbs moderate; digits short, thick, slightly depressed at the base, compressed at the end, inferiorly with enlarged plates. Head covered with granules, which are larger on the snout; rostral quadrangular, generally nearly twice as broad as high; nostril pierced between the rostral, the first labial, and several small scales; nine to eleven upper and seven to nine lower labials; mental triangular; a pair of large chin-shields forming a suture behind the point of the mental, surrounded by several smaller shields. Upper surface of body covered with small granules, which are uniform or intermixed with more or less numerous, irregularly scattered, small roundish keeled tubercles. Abdominal scales small, cycloid, imbricate, smooth. No femoral nor præanal pores. Tail cylindrical, tapering, probably prehensile, covered with small imbricate smooth scales largest inferiorly. Pale brownish above, variously ornate with brown spots or bands becoming blackish towards their borders and more or less distinctly finely margined with lighter; all the specimens have the upper surface of the head more or less marbled or elegantly marked with insuliform brown spots, and a brown band passes through the eye; the lower surfaces are whitish, the throat reticulated with brown, which reticulation has a tendency to form oblique lines.

Total length	94	millim.
Head	15·5	,,
Width of head	11	,,
Body	38·5	,,
Fore limb	18	,,
Hind limb	24	,,
Tail	40	,,

Southern India, Ceylon.

A. Three broad cross bands, broader than the interspaces between them, which are of a very light hue; the anterior band narrowest, on the nape, confluent with the band extending from the eye, the two others across the back.

a. ♀.	Erode.	Col. Beddome [C.].
		(Type of *G. speciosus*.)
b. ♀.	Ceylon.	G. H. K. Thwaites, Esq. [C.].

B. Large round or oval brown spots on the body, generally forming two regular longitudinal series.

c, d. ♀ & yg.	Balarangams.	Col. Beddome [C.].
		(Types of *G. collegalensis*.)
e–g. ♂ & hgr.	Manar.	Col. Beddome [C.].
h–n. ♂ ♀.	Foot of Nilgherries.	Col. Beddome [C.].
o–p. ♀.	High Wavy, Madura.	Col. Beddome [C.].
q–t. ♂ ♀.	S. India.	Col. Beddome [C.].

C. The spots are smaller and less strongly marked, also arranged in pairs, but more transverse and sometimes uniting into angular cross lines.

u–x. Hgr. & yg.	Golconda Hills.	Col. Beddome [C.].
		(Types of *G. nebulosus.*)
y–z. Yg.	Gorge Hill, Godavery.	Col. Beddome [C.].
a–δ. ♂ & hgr.	Nellumba.	Col. Beddome [C.].
ε–ζ. ♂ ♀.	Russelconda.	Col. Beddome [C.].

16. Gymnodactylus jeyporensis. (Plate IV. fig. 2.)

Gymnodactylus jeyporensis, *Beddome, Proc. Zool. Soc.* 1877, p. 685.

Head rather large, oviform; snout longer than the diameter of the orbit or the distance between the eye and the ear-opening; forehead and loreal region slightly concave; ear-opening an oblique slit, three fifths the diameter of the eye. Body depressed, rather short. Limbs moderate; digits rather short, cylindrical in their basal, compressed in their distal portion, inferiorly with enlarged plates. Head covered with large subequal flat granules; rostral quadrangular, twice as broad as high, with median cleft above; nostril pierced between the rostral, the first labial, and three nasals; ten upper and seven lower labials; mental triangular; a pair of large chin-shields forming a suture behind the point of the mental, surrounded by much smaller chin-shields. Body covered above with uniform juxtaposed large, squarish, hexagonal flat scales arranged regularly like the bricks of a wall. Abdominal scales smooth, round, imbricate, not half the size of the dorsals. No enlarged præanal or femoral scales nor pores. Tail cylindrical, tapering, slightly swollen, covered with uniform rather large smooth scales, which are imbricate inferiorly. Light yellowish brown above, with large reddish-brown, black-edged spots arranged in pairs; two transverse spots on the nape, the posterior crescent-shaped; head with small darker spots and a streak from eye to ear; lips and side of throat with small brown dots and marblings; lower surfaces brownish white.

Total length	92	millim.
Head	16	,,
Width of head	11	,,
Body	38	,,
Fore limb	20	,,
Hind limb	27	,,
Tail	38	,,

Jeypoor.

a. ♂.   Patinghe Hill, Jeypoor.   Col. Beddome [C.]. (Type.)

17. Gymnodactylus deccanensis.

Gymnodactylus dekkanensis, *Günth. Rept. Brit. Ind.* p. 115, pl. xii. fig. E.

Head rather large, oviform; snout longer than the diameter of the orbit or the distance between the eye and the ear-opening; forehead and loreal region concave; ear-opening suboval, vertical, about one third the diameter of the eye. Body moderately elongate, depressed. Limbs rather long; digits moderately elongate, cylindrical in the basal, compressed in the distal phalanges; the plates under the basal phalanx very small, little larger than the surrounding tubercles. Head covered with convex granules, largest on the snout and temples; rostral quadrangular, not quite twice as broad as high, with median cleft above; nostril pierced between the rostral, the first labial, and three nasals; nine to eleven upper, and as many lower labials; mental triangular or pentagonal; a pair of large chin-shields forming a long suture behind the point of the mental, in contact externally and posteriorly with two smaller pairs. Body covered above with large juxtaposed subequal tubercles arranged in more or less regular transverse series; these tubercles flat or very slightly keeled, generally with a small raised point in the centre. Abdominal scales round, smooth, subimbricate, much smaller than dorsals. Males with enlarged præanal and femoral scales, but without pores. Tail cylindrical, tapering, covered with uniform smooth scales arranged in rings. Reddish brown above, with narrow white, black-edged cross bars; the first semicircular, extending from one eye to the other across the nape; a second on scapular region, two on the body, and a fifth on sacrum; similar bands forming annuli round the tail; lower surfaces whitish.

Total length	122	millim.
Head	19	,,
Width of head	13	,,
Body	45	,,
Fore limb	24	,,
Hind limb	33	,,
Tail	58	,,

Deccan.

a. ♂. Deccan. Col. Sykes [P.]. (Type.)
b–e. ♂ ♀. Matheran. Dr. Leith [P.].

18. Gymnodactylus albofasciatus. (PLATE IV. fig. 3.)

This species resembles *G. deccanensis* strikingly in size, proportions, and colour, but the following characters will at once distinguish it. The dorsal pholidosis is not composed of uniform large tubercles, but of intermixed smaller and larger tubercles, the latter being mostly feebly keeled; the small scales on the limbs and upper part of the tail intermixed with larger keeled tubercles. The ventral scales are larger, tubercular, and feebly keeled. No chin-shields behind the median pair. The tubercular plates under the basal phalanx of all the digits much more developed. The ground colour of the upper parts is darker than in our specimens of *G. deccanensis*, chestnut-brown.

Total length	127	millim.
Head	20	,,
Width of head	14	,,
Body	47	,,
Fore limb	27	,,
Hind limb	36	,,
Tail	60	,,

South Canara.

a–c, d. ♂.   S. Canara.   Col. Beddome [C.].

19. Gymnodactylus oldhami.

Gymnodactylus oldhami, *Theobald, Cat. Rept. Brit. Ind.* p. 81.

Crown of head behind the eyes finely granular, the scales in front being larger; eleven upper and ten lower labials; three pairs of chin-shields. Back granular, with about thirty longitudinal rows of tubercles contained between an area bounded by faint keels margining the belly. Enlarged præanal and femoral scales, but no pores. Colour above vinous brown; a white semicircular line joins the superciliary ridges; a second horseshoe-shaped white line runs from the gape below the ear to the opposite side, enclosing a nuchal collar; behind the collar two closely approximated white lines run along the spine, becoming soon broken up into spots merely; on the sides are more white spots, tending to form three distinct lines, the lowest of which coincides with the ventral keel; belly whitish. South Canara.

According to Col. Beddome this species is identical with the preceding; but it will be remarked that the two descriptions by no means agree, and I must therefore hold them distinct until we obtain further information.

20. Gymnodactylus triedrus.

Gymnodactylus triedrus, *Günth. Rept. Brit. Ind.* p. 113.
Geckoella punctata, *Gray, Proc. Zool. Soc.* 1867, p. 99, pl. ix.

Head rather large, oviform; snout longer than the diameter of the eye, slightly longer than the distance between the eye and the ear-opening; forehead concave; ear-opening small, suboval, horizontal. Body and limbs moderate; digits short, cylindrical in their basal, compressed in their distal portion, with well-developed tubercle-like plates inferiorly. Head covered with small granules, largest on the snout; rostral quadrangular, not twice as broad as high; nostril pierced between the rostral, the first labial, and several small scales; ten or eleven upper and nine lower labials; mental triangular; two or three pairs of chin-shields, median largest and forming a long suture behind the point of the mental. Upper surface of body covered with small granules intermixed with numerous small trihedral tubercles. Abdominal scales rather small, cycloid, imbricate, smooth. Males with three or four præanal pores, Tail cylindrical, tapering, rather swollen, covered with roundish smooth scales, which are small and subimbricate superiorly, much larger and strongly imbricate inferiorly. Brown above, with small whitish spots; lower surfaces light brown.

Total length	115 millim.
Head	19 ,,
Width of head	13 ,,
Body	43 ,,
Fore limb	22 ,,
Hind limb	29 ,,
Tail	53 ,,

Ceylon.

a. ♂.	Ceylon.	(Type.)
b–d. ♂ ♀.	Ceylon.	(Types of *Geckoella punctata.*)
e–g. ♂ ♀.	Ceylon.	

21. Gymnodactylus arnouxii.

Gymnodactylus arnouxii, *A. Dum. Cat. Méth. Rept.* p. 44, *and Arch. Mus.* viii. p. 479, pl. xviii. fig. 5.

In habit similar to *G. pelagicus.* Sixteen longitudinal very regular series of round, convex, smooth tubercles. Abdominal scales small, smooth. Tail with uniform small smooth scales, forming rings. Nostril pierced between the rostral, the first labial, and several nasals, the antero-superior of which is enlarged; eight upper and seven lower labials; mental very large, subtriangular, extending beyond the labials; a small chin-shield on each side of the mental. Brown, lighter beneath; eight transverse dark bands between the occiput and the base of the tail.

Head and body 44 millim., tail 41 millim.

New Zealand.

22. Gymnodactylus geckoides.

Cubinia darwinii, *Gray, Cat.* p. 274.
Gymnodactylus geckoides, *Spix, Spec. Nov. Lacert. Bras.* p. 17, pl. xviii. fig. 1.
Cyrtodactylus spixii, *Gray, Griff. A. K.* ix. *Syn.* p. 52.
Gonyodactylus (Dasyderma) spinulosus, *Fitzing. Syst. Rept.* p. 92.
Gymnodactylus girardi, *Steindachn. Novara, Rept.* p. 15, pl. ii. fig. 3, *and Sitzb. Ak. Wien,* lxii. i. 1870, p. 344.

Appearance very similar to that of *G. pelagicus.* The tubercles form fourteen very regular longitudinal series on the back; these tubercles are very close together in each series; they are small, roundish, and keeled, not ribbed. The ventral scales are very large, cycloid and strongly imbricate, smooth, in sixteen longitudinal series. Otherwise as in *G. pelagicus.* Brown above, with darker spots, forming bands across the back; a ∪-shaped dark line from eye to eye over the nape. From snout to vent 37 millim.

Brazil.

a–b. Hgr.   Bahia or Rio Janeiro.   C. Darwin [P.]. (Types of *Cubina darwinii*.)

23. Gymnodactylus pelagicus.

Heteronota pelagica, *Girard, Proc. Ac. Philad.* 1857, p. 197, *and U.S. Explor. Exped., Herp.* p. 306.

Gymnodactylus multicarinatus, *Günth. Ann. & Mag. Nat. Hist.* (4) x. 1872, *and in Brenchley, Curaçoa*, p. 404, pl. xxiv. fig. A.

—— (Heteronota) arfakianus, *Meyer, Mon. Berl. Ac.* 1874, p. 129.

—— arnouxii (*non A. Dum.*), *Peters & Doria, Ann. Mus. Genov.* xiii. 1878, p. 372; *Bouleng. Proc. Zool. Soc.* 1883, p. 129, pl. xxii. fig. 8.

Head rather large, oviform, moderately depressed; snout a little longer than the distance between the eye and the ear-opening, about once and two thirds the diameter of the orbit; forehead concave; ear-opening small, roundish. Body short, depressed. Limbs moderate; digits slightly depressed at the base, compressed in the remaining portion, with well-developed transverse plates under the basal phalanx. Head covered with very small granular scales; rostral subquadrangular, not quite twice as broad as high, with median cleft above; nostril pierced between the rostral, the first upper labial, two nasals, and two or three granules; seven or eight upper and as many lower labials; mental very large, triangular, pentagonal, or trapezoid, extending considerably beyond the adjacent labials, generally with a pair of chin-shields posteriorly; no other chin-shields. Upper surfaces covered with very small granules; back with sixteen to twenty longitudinal series of round, subconical, distinctly ribbed tubercles; these series regular, equidistant, the tubercles generally very close together, forming lines; hind limbs with similar, irregularly scattered tubercles. Gular scales minute, granular; abdominal scales very small, imbricate, keeled. Males generally with a very short angular series of seven or eight præanal pores. Tail cylindrical, tapering, generally with uniform very small keeled scales, occasionally intermixed with large tubercles. Brown, lighter beneath; back and limbs with more or less distinct darker transverse markings; generally a dark streak from the eye to the ear-opening.

Total length	125	millim.
Head	18	,,
Width of head	12	,,
Body	42	,,
Fore limb	19	,,
Hind limb	29	,,
Tail	65	,,

Islands of the Pacific; New Guinea; Cape York.

a. ♂.	Murray Island.	Rev. S. Macfarlane [C.].
b-c. ♂ ♀.	Islands of Torres Straits.	Rev. S. Macfarlane [C.].
d. Hgr.	Somerset, Cape York.	H.M.S. 'Challenger.'
e-g. ♂ ♀.	Duke of York Island.	Rev. G. Brown [C.].
h-l. ♂ ♀.	Shortland Islands, Solomon Group.	H. B. Guppy, Esq. [P.].
m-p. ♂ ♀.	Api, New Hebrides.	H.M.S. 'Challenger.'
q-s. ♂ ♀.	New Hebrides.	J. Brenchley, Esq. [P.]. (Types
t. ♀.	Aneiteum.	J. Brenchley, Esq. [P.]. of G.
u-v. ♀.	Tongatabu.	J. Brenchley, Esq. [P.]. multicarinatus.)
w. ♀.	Lifu, Loyalty Islands.	Rev. S. J. Whitmee [C.].
x. Hgr.	Vanikoro, Santa Cruz.	W. W. Perry, Esq. [P.].
y. ♀.	Viti.	Godeffroy Museum.
z-a. ♂ & yg.	Kandavu, Fiji.	H.M.S. 'Challenger.'
β-δ. ♀.	Fiji.	J. McGillivray, Esq. [P.].
ε-ζ. ♂ & yg.	Fiji.	H.M.S. 'Herald.'
η. ♀.	Erromanga.	

24. Gymnodactylus heteronotus.

Heteronota fasciata, *Macleay, Proc. Linn. Soc. N. S. W.* ii. 1877, p. 100.

" Form elongate; head broader than the neck; internasal shields medium size and not contiguous, two granular scales intervening; upper rostral large, truncate behind, and depressed in the middle; under rostral very large and triangular behind, with a small subtriangular plate in the angle formed between it and the first labial; labials seven on each side above and below, the two last shields very small; eyes large, with a heavy scaly curtain above, pupils round; ear-openings distinct, round, the tympanum not deep; scales on the back granular, with about ten irregular series of roundish tubercles, each with several minute keels or striæ, those on the legs and under surface of the body small, round, convex, and bicarinate, and those on the tail larger, oval, unicarinate, and in concentric rings; legs long and slender; tail about the length of the body; colour mottled grey, with brownish cross bands of irregular form from the muzzle to the base of the tail. Length 3½ inches."

Hall Sound.

25. Gymnodactylus cheverti.

Heteronota marmorata, *Macleay, Proc. Linn. Soc. N. S. W.* ii. 1877, p. 100.

" Form moderately elongate; internasal shields very short and contiguous, chin-shields smaller and more rounded than in the last species; upper labials eight, the last two very small; lower labials six, the last one very minute; pupil elliptical; ear-openings round;

scales on the back as in the last species, but with the tubercles much more numerous; tail round, tapering, about the length of the body, very slightly ringed, and with the scales beneath smooth, but not large. Colour above yellowish brown, spotted and barred with deep brown or black, under surface dingy yellow. Length from 3 to 4 inches."

Fitzroy Island and Endeavour River.

26. Gymnodactylus affinis.

Cyrtodactylus affinis, *Stoliczka, Journ. As. Soc. Beng.* xxxix. 1870, p. 167, pl. x. fig. 1.

Body rather depressed. Digits very slender and elongate. Ear-opening moderate, vertically elongated. Head-scales small, those in front slightly enlarged and flattened; rostral very large, and grooved posteriorly; twelve upper and eleven lower labials; mental very large, subtriangular; a few of the chin-shields next to the rostral are squarish, very little larger than others, but none are elongated. Body covered with granules intermixed with small subtrihedral tubercles, each of which has from three to five grooves. Abdominal scales small, subtubercular, and carinated. No femoral or præanal pores. Tail round, with a few indistinct rings of enlarged tubercles near the base, without enlarged subcaudals. General colour above pale vinaceous ashy, finely marbled and mottled with dark, especially on the head, sides, and limbs; a V-shaped blackish mark on the nape, followed by a black spot on the neck; then follow five other angular blackish bands across the body, the first across the shoulders, the last between the hind limbs; tail with blackish broad bands; lower parts whitish with a slight purplish tinge. From snout to vent about 50 millim.

Pinang.

27. Gymnodactylus frenatus.

Gymnodactylus frænatus, *Günth. Rept. Brit. Ind.* p. 113, pl. xii. fig. D.

Head large, depressed, oviform; snout longer than the diameter of the orbit, which equals its distance from the ear-opening; forehead concave; ear-opening suboval, oblique, nearly one third the diameter of the eye. Body and limbs rather elongate. Digits strong, distinctly depressed at the base, strongly compressed in the remaining portion; the basal phalanx with well-developed transverse plates inferiorly. Head granular, the granules largest on the snout; a few scattered round tubercles on the temples; rostral subquadrangular, twice as broad as high, with median cleft above, entering considerably the nostril; latter pierced between the rostral, the first labial, and three nasals; ten to twelve upper and nine or ten lower labials: mental broadly triangular; a pair of chin-shields, forming a suture behind the mental; a few small chin-shields on each side of the median pair; gular scales minutely granular. Body covered above with flat granules intermixed with irregularly scattered small round smooth tubercles. A slight fold from axilla to groin. Abdominal scales rather small, cycloid, imbricate. Males with two pairs of præanal pores. Tail cylindrical, tapering, covered above with small flat subquadrangular scales, inferiorly with a median series of large transverse plates. Light pinkish brown above, with five pairs of angular dark-brown spots confluent into cross bands, the anterior on the nape and uniting with a dark-brown band extending to the eye, and which is continued, more or less interrupted, along the side of the body; tail with dark-brown annuli; lower surfaces whitish.

Total length	192 millim.
Head	24 ,,
Width of head	18 ,,
Body	66 ,,
Fore limb	32 ,,
Hind limb	42 ,,
Tail	102 ,,

Ceylon.

a, b, c. ♂, ♀, and hgr.　　　Ceylon.　　　(Types.)
d. ♂.　　　Ceylon.

28. Gymnodactylus variegatus.

Naultinus variegatus, *Blyth, Journ. As. Soc. Beng.* xxviii. 1859, p. 279; *Theob. Cat. Rept. As. Soc. Mus.* p. 32.
Gymnodactylus variegatus, *Günth. Rept. Brit. Ind.* p. 116; *Anders. Proc. Zool. Soc.* 1871, p. 161.

Limbs and digits slender; the basal joints not very distinct from the distal ones, which are strongly compressed, and provided with transverse imbricate plates below and a series of much smaller ones on the compressed phalanges. Rostral notched behind and grooved; nostril between the rostral, the first labial, and three nasals; ten to eleven upper and eleven lower labials; mental partially wedged in between two large chin-shields, which form a broad suture with each other. Body granular, with numerous large trihedral tubercles. Twenty-six longitudinal rows of rather elongated leaf-like scales on the middle of the belly. A fold of skin along the side, indicating where the abdominal scales terminate and the granules begin. Sixteen femoral pores on each side along the whole length of the thigh. Tail cylindrical, the granules arranged in verticils, and the tubercles in rings; a series of enlarged subcaudals. Grey above, spotted and marbled with black, set off with subdued white; a broad dark streak bordered with whitish behind each eye, and continued irregularly round the occiput; tail irregularly banded above; lower parts whitish. Total length about 150 millim., the tail measuring nearly half that length.

Moulmein.

29. Gymnodactylus fasciolatus.

Naultinus fasciolatus, *Blyth, Journ. As. Soc. Beng.* xxix. 1860, p. 114; *Theob. Cat. Rept. As. Soc. Mus.* p. 32.
Gymnodactylus fasciolatus, *Günth. Rept. Brit. Ind.* p. 116; *Anders. Proc. Zool. Soc.* 1871, p. 161.

Differs from *G. variegatus* in the larger ventral scales, which form only thirty-six longitudinal series in the middle of the belly, and in the small number of femoral (or rather præanal) pores; these are five or six on each side, and extend outwards in a line with the commencement of the thigh. A dark-brown band, edged behind with white, from the eye to the occiput, where it meets its fellow of the opposite side; a brown similarly white-edged band on the nape, with seven cross bands on the body; tail cross-barred.

Subathoo, Western Himalayas.

30. Gymnodactylus khasiensis.

Pentadactylus? khasiensis, *Jerdon, Proc. As. Soc. Beng.* 1870, p. 75.
Gymnodactylus khasiensis, *Anders. Proc. Zool. Soc.* 1871, p. 162; *Theobald, Cat. Rept. Brit. Ind.* p. 84.

Very closely allied to *G. marmoratus*, from which it differs in the following points:—Digits more elongate; the plates under the basal phalange larger, the one under the articulation being nearly as long as broad. Males without pubic groove or femoral pores, with a series of ten to thirteen præanal pores forming a very open angle. Brown above, with darker angular spots, forming more or less regular series along the back; a dark streak on the side of the head, passing through the eye; tail with dark annuli; lower surfaces brownish.

Total length	175	millim.
Head	22	,,
Width of head	15	,,
Body	56	,,
Fore limb	30	,,
Hind limb	42	,,
Tail	97	,,

Khasi Hills.

a. Many spec., ♂ ♀	Khasi Hills.	T. C. Jerdon, Esq. [P.].
b. Hgr.	Assam.	W. T. Blanford, Esq. [P.].
c. ♀.	N.E. Bengal.	T. C. Jerdon, Esq. [P.].

31. Gymnodactylus marmoratus.

Cyrtodactylus marmoratus, part., *Gray, Cat.* p. 173.
Phyllurus marmoratus, (*Kuhl*) *Fitzing. N. Class. Rept.* p. 47.
Gonyodactylus marmoratus (*Kuhl*), *Gray, Griff. A. K.* ix. *Syn.* p. 51; *Girard, U.S. Explor. Exped., Herp.* p. 304.
Gymnodactylus marmoratus, *Dum. & Bibr.* iii. p. 426, pl. xxxiv. fig. 1; *Schleg. Abbild.* p. 8, pl. ii.; *Steindachn. Novara, Rept.* p. 17.

Head rather large, depressed, oviform; snout longer than the diameter of the orbit, which equals its distance from the ear-opening; forehead concave; ear-opening suboval, oblique, not quite one third the diameter of the eye. Body and limbs rather elongate. Digits strong, slightly depressed at the base, strongly compressed in the remaining portion; the basal phalange with well-developed transverse plates inferiorly. Head granular, with very small tubercles on the occipital and temporal regions, the granules rather enlarged on the snout; rostral subquadrangular, not twice as broad as high, with median cleft above, entering considerably the nostril; latter slightly directed posteriorly, pierced between the rostral, the first labial, and several scales; ten to twelve upper and nine or ten lower labials; mental triangular; two or three pairs of chin-shields, median largest and in contact behind the point of the mental; throat minutely granulate. Body and limbs covered above with small granules intermixed with small roundish, feebly keeled, subtrihedral tubercles; a more or less defined series of tubercles from axilla to groin, limiting the abdominal region; ventral scales small, cycloid, imbricate. Males with præanal and femoral pores; the former twelve or thirteen altogether, in a ∧-shaped series, enclosing a groove; the latter widely separated from the former, four to six on each side. Tail cylindrical, tapering, covered with uniform small flat scales, above with a few scattered tubercles. Light brown above, with chestnut-brown spots, which are sometimes confluent into cross bands on the back; tail, when intact, with dark-brown annuli; a chestnut-brown streak on the side of the head, passing through the eye; lower surfaces brownish.

Total length	148	millim.
Head	21	,,
Width of head	15	,,
Body	55	,,
Fore limb	26	,,
Hind limb	34	,,
Tail	72	,,

East-Indian archipelago.

a–b. ♂.	Java.	Leyden Museum.
c. ♀.	Java.	Dr. Ploem [C.].
d. ♀.	Java.	Dr. Bleeker.
e. ♀.	Agam.	Dr. Bleeker.
f. ♀.	Sumatra.	H. O. Forbes, Esq. [C.].
g. ♀.	Matang.	
h. ♀.	N. Celebes.	Dr. A. B. Meyer [C.].

32. Gymnodactylus rubidus.

Puellula rubida, *Blyth, Journ. As. Soc. Beng.* xxix. 1860, p. 109; Günth. *Rept. Brit. Ind.* p. 118.
Gecko tigris, *Tytler, Journ. As. Soc. Beng.* xxxiii. 1864, p. 546.
Cyrtodactylus rubidus, *Stoliczka, Journ. As. Soc. Beng.* xxxix. 1870, p. 165.

Agrees in every respect with *G. marmoratus*, except that the ventral scales are a little larger, and there are no femoral pores; the præanal pores few, and in a groove, as in *G. philippinicus*.
Andaman Islands.

a. ♀. ———? W. Theobald, Esq.

33. Gymnodactylus philippinicus.

Cyrtodactylus marmoratus, part., *Gray, Cat.* p. 173.
Gymnodactylus philippinicus, *Steindachn. Novara, Rept.* p. 17, pl. ii. fig. 1.

Also closely allied to *G. marmoratus*. Head a little larger. Digits rather more elongate; the plates under the basal phalanx so slightly developed as to be hardly distinguishable from the surrounding small scales. Dorsal tubercles more conical and rather more widely separated. Abdominal scales smaller. Eleven lower labials. Male with a longitudinal groove on the pubic region, containing two parallel series of six præanal pores. Coloration as in *G. marmoratus*.

Total length	187	millim.
Head	25	,,
Width of head	19	,,
Body	65	,,
Fore limb	35	,,
Hind limb	45	,,
Tail	97	,,

Philippine Islands.

a-d. ♂, hgr., & yg. Philippines.
e. ♀. Luzon. Dr. A. B. Meyer [C.].
f. ♀. Dinagat Island. A. Everett, Esq. [C.].

34. Gymnodactylus pulchellus.

Cyrtodactylus pulchellus, *Gray, Cat.* p. 173.
Cyrtodactylus pulchellus, *Gray, Zool. Journ.* iii. 1828, p. 224, *and Illustr. Ind. Zool.*
Gonyodactylus pulchellus, *Wagl. Syst. Amph.* p. 144.
Gymnodactylus pulchellus, *Dum. & Bibr.* iii. p. 423, pl. xxxiii. fig. 7; *Cantor, Cat. Mal. Rept.* p. 25; *Günth. Rept. Brit. Ind.* p. 113; *Stoliczka, Journ. As. Soc. Beng.* xlii. 1873, p. 118.

Head large, much depressed, oviform; snout longer than the diameter of the orbit, which equals its distance from the ear-opening; forehead and loreal region concave; ear-opening suboval, vertical, slightly oblique, one third or two fifths the diameter of the eye. Body and limbs rather elongate. Digits strong, slightly depressed at the base, strongly compressed in the remaining portion; the basal phalanx with well-developed transverse plates inferiorly. Head granular, with small round tubercles on the occipital and temporal regions, the granules enlarged on the snout, except in the frontal and loreal concavities. Rostral subquadrangular, nearly twice as broad as high, with median cleft above, entering considerably the nostril; latter directed posteriorly, pierced between the rostral, the first labial, and three or four nasals; ten to thirteen upper and ten to twelve lower labials; mental triangular; two or three pairs of chin-shields, median largest and in contact behind the point of the mental; throat minutely granulate. Body and limbs above with small flat granules intermixed with small roundish, keeled, subtrihedral tubercles; a series of keeled tubercles from axilla to groin, limiting the abdominal region; ventral scales cycloid, imbricate, moderately large. Males with a longitudinal groove on the pubic region containing two parallel series of præanal pores, forming a right angle with a long series of femoral pores; altogether eighteen to twenty pores on each side, four or five of which are in the groove. Tail cylindrical, tapering, above with small flat scales and annuli of feebly keeled tubercles, inferiorly with a series of large transverse plates. Light brown above, with broad chestnut-brown, light-edged cross bands, which are narrower than the interspaces between them; the anterior horseshoe-shaped, from eye to eye over the nape; the second crescent-shaped, on scapular regions; three others on the body; tail with chestnut-brown complete annuli; lower surfaces dirty white.

Total length	231	millim.
Head	28	,,
Width of head	20	,,
Body	73	,,
Fore limb	42	,,
Hind limb	54	,,
Tail	130	,,

Malay peninsula to Bengal.

a, b. ♀ & hgr.	Singapore.	Gen. Hardwicke [P.]. (Types.)
c–d. ♂.	Singapore.	Dr. Cantor.
e. Yg.	Bengal.	W. Masters, Esq. [P.].
f. ♂.	——?	E. India Company.
g. ♀.	——?	

35. Gymnodactylus consobrinus.

Gymnodactylus consobrinus, *Peters, Mon. Berl. Ac.* 1871, p. 569, and *Ann. Mus. Genov.* iii. 1872, pl. ii. fig. 1.

Closely allied to *G. pulchellus*. Distinguished by the following characters:—Granules and tubercles of the upper surfaces smaller, the dorsal tubercles not or but very indistinctly keeled, conical; ventral scales much smaller. Ear-opening smaller. Males with an angular series of nine to eleven præanal pores; no pubic groove nor femoral pores. Grey-brown above, with darker transverse cross bands, which are much wider than the interspaces between them; there are eight of these bands, from the nape to the sacrum inclusively; tail with large dark and small light annuli; lower surfaces dirty white. In the young, the dark bands are almost black and the interspaces white.

Total length	258	millim.
Head	30	,,
Width of head	22	,,
Body	85	,,
Fore limb	44	,,
Hind limb	57	,,
Tail	143	,,

Borneo.

a–e. ♂, hgr., & yg.   Matang.

36. Gymnodactylus miliusii.

Phyllurus miliusii, *Gray, Cat.* p. 176.
Phyllurus miliusii, *Bory de St. Vinc. Dict. Hist. Nat.* vii. p. 183, pl. —. fig. 1; *Gray, Zool. Ereb. & Terror,* pl. xvii. fig. 2.
Cyrtodactylus nilii, *Gray, Griff. A. K.* ix. *Syn.* p. 52.
Gymnodactylus miliusii, *Dum. & Bibr.* iii. p. 430, pl. xxxiii. fig. 1; *Peters, Mon. Berl. Ac.* 1863, p. 229.
—— (Anomalurus) miliusii, *Fitz. Syst. Rept.* p. 90.

Head large, oviform; snout a little longer than the diameter of the orbit, as long as the distance between the eye and the ear-opening; forehead and loreal region concave; ear-opening elliptical, vertical, about three fifths the diameter of the eye. Body moderate. Limbs long, slender; digits rather short, subcylindrical. Snout covered with granules of unequal size; hinder part of head with minute granules intermixed with round tubercles; rostral subquadrangular, three times as broad as high; nostril directed posteriorly, separated from the rostral and first labial by two nasals; labials small, eleven to fourteen upper and ten to twelve lower; mental broadly trapezoid; no regular chin-shields; gular granules minute. Body and limbs covered above with small granules intermixed with small round conical tubercles; belly covered with flat granules. Tail short, thick, swollen, and nearly as broad as the body in its anterior half, depressed, tapering to a fine point posteriorly; it is covered with small granules, and, on the upper surface, small conical tubercles arranged in transverse series. Chestnut-brown above, with white cross bands on the back and tail; head and limbs white-spotted; lower surfaces white.

Total length	135	millim.
Head	25	,,
Width of head	19	,,
Body	65	,,

```
Fore limb ............  36 millim.
Hind limb ............  43   ,,
Tail .................  55   ,,
```
Australia.

a–b, c. ♂, ♀, & hgr.	Australia.	
d. ♂.	Australia.	G. Krefft, Esq.
e. Hgr.	Sydney.	
f. Hgr.	Champion Bay, N.W. Australia.	Mr. Duboulay [C.].
g. ♂, skeleton.	——?	

37. Gymnodactylus platurus.

Phyllurus platurus, *Gray, Cat.* p. 176.
Phyllurus inermis, *Gray, l. c.*
Lacerta platura, *White, Journ. N. S. Wales,* p. 246, pl. —. fig. 2; *Shaw, Gen. Zool.* iii. p. 247.
Stellio phyllurus, *Schneid. Amph. Phys.* ii. p. 31.
—— platurus, *Daud. Rept.* iv. p. 24.
Agama platyura, *Merr. Syst. Amph.* p. 51.
—— discosura, *Merr. l. c.*
Phyllurus platurus, *Fitzing. N. Class. Rept.* p. 47; *Guérin, Icon. R. A., Rept.* pl. xiv. fig. 1; *Gray, Zool. Ereb. & Terror,* pl. xvii. fig. 3; *Girard, U.S. Explor. Exped., Herp.* p. 303.
—— cuvieri, *Bory de St. Vinc. Dict. Hist. Nat.* vii. p. 183, pl. —. fig. 2.
Gymnodactylus platyurus, *Wagl. Syst. Amph.* p. 144.
Cyrtodactylus platura, *Gray, Griff. A. K.* ix. *Syn.* p. 52.
Gymnodactylus phyllurus, *Dum. & Bibr.* iii. p. 428.
Gonyodactylus (Phyllurus) platurus, *Fitz. Syst. Rept.* p. 92.
Phyllurus inermis, *Gray, Zool. Ereb. & Terror,* pl. xvii. fig. 1.

Head large, much depressed, elongate triangular, very distinct from neck; derm of the head more or less confluent with cranial ossification; snout measuring once and two thirds to once and three fourths the diameter of the orbit, much longer than the distance between the eye and the ear-opening; forehead and loreal region slightly concave; ear-opening elliptical, vertical, not quite half the diameter of the eye. Body moderate. Limbs long; digits strong, subcylindrical at the base, compressed in their distal portion. Head covered with small granules, intermixed with conical, spinose tubercles; rostral subquadrangular, three times as broad as high, generally with median cleft above; nostril directed posteriorly, nearly always separated from the rostral and first labial; labials small, thirteen to sixteen upper and eleven to thirteen lower; mental broadly trapezoid; no chin-shields. Body and limbs covered above with small granules, intermixed with conical more or less spinose tubercles; belly covered with flat granules. Tail short, depressed, very broad, leaf-like, contracted at the base, attenuated at the tip; when intact, this organ is studded above with spine-like tubercles, but when reproduced it is covered with uniform granular scales. Brown above, marbled with darker, or with broad angular dark cross bands on the back; lower surfaces light brown, sometimes dotted with darker.

Total length	183	millim.
Head	34	,,
Width of head	27	,,
Body	76	,,
Fore limb	53	,,
Hind limb	70	,,
Tail	73	,,

Australia.

a. Ad., dry.	Australia.	
b. Ad.	Australia.	Allan Cunningham, Esq. [P.].
c, d. Ad.	Australia.	(Types of *Phyllurus inermis*.)
e–i. Ad.	Australia.	J. Gould, Esq. [C.].
k–l. Ad.	Australia.	J. B. Jukes, Esq. [P.].
m, n. Ad.	Australia.	G. Krefft, Esq.
o. Ad.	Australia.	
p–r. Ad. & hgr.	Queensland.	H.M.S. 'Challenger.'
s. Ad.	Pt. Curtis, Queensland.	Dr. Coppinger [C.].
t. Ad.	Sydney.	
u. Ad.	Pt. Macquarie.	G. Krefft, Esq.

11. AGAMURA.

Agamura, *Blanford, Ann. & Mag. N. H.* (4) xiii. 1874, p. 455.

Digits slender, clawed, cylindrical at the base; the distal phalanges compressed, forming an angle with the basal portion of the digits; the claw between two enlarged scales; digits inferiorly with a row of plates. Body covered with small granules intermixed with enlarged tubercles. Tail very slender, not fragile. Pupil vertical. Males with or without præanal pores.

Persia and Baluchistan.

1. Agamura cruralis.

Agamura cruralis, *Blanf. l. c., and Zool. E. Persia*, p. 356, pl. xxiii. fig. 3.

Head short, feebly depressed; snout as long as, or slightly longer than, the diameter of the orbit, or the distance between the eye and the ear-opening; eye large; forehead not concave; ear-opening subelliptical, vertical, one third or two fifths the diameter of the eye. Body feebly depressed, rather short. Limbs very long and slender; the hind limb if carried forward reaches the eye; the fore limb being stretched forward, the wrist attains the tip of the snout or slightly beyond; digits long and slender, unequal. Head covered with flat granules, largest on the snout; rostral broader than high, with median cleft above; nostril pierced between the rostral, the first labial, and three rather swollen nasals; thirteen or fourteen upper and ten or eleven lower labials; mental elongate trapezoid, twice as long as broad; no chin-shields. Body covered above with small, rather irregular, flat granules, intermixed with rather numerous scattered round flat tubercles. Ventral scales small, roundish-hexagonal, juxtaposed, or subimbricate, smooth. [Male with two præanal pores.] Tail very thin and rounded, of nearly the same thickness to the end, covered above with smooth imbricate elongate squarish scales, inferiorly with a row of larger plates. Greyish-brown or sandy above, more or less dotted with darker, especially on the sides of the head and neck; usually a dark cross band on the back of the neck and four [or five] others on the back. Lower surfaces white; throat more or less brown-dotted.

Total length	135 millim.
Head	19 ,,
Width of head	14 ,,
Body	49 ,,
Fore limb	41 ,,
Hind limb	57 ,,
Tail	67 ,,

Baluchistan.

a–c. ♀ & hgr. Bahu Kalat & Askan.   W. T. Blanford, Esq. [C.].
(Types.)

2. Agamura persica.

Gymnodactylus persicus, *A. Dum. Arch. Mus.* viii. p. 481.
Agamura persica, *Blanf. E. Persia,* p. 358, pl. xxiii. fig. 4.

Differs from the preceding in the following characters:—Head more depressed. Limbs shorter, the hind limb not reaching the eye, and the wrist reaching half-way between the eye and the tip of the snout. The granules on the snout are quite as small as those on the occiput, where there are numerous enlarged tubercles; the rostral is more than twice as broad as high, and completely divided into two; the mental is not twice as long as broad. The dorsal tubercles are larger, becoming subconical on the hinder part of the back; numerous enlarged tubercles on the hind limbs. No præanal pores in the male.

Total length	98 millim.
Head	16 ,,
Width of head	11 ,,
Body	37 ,,
Fore limb	29 ,,

Hind limb .... 37 millim.
Tail 45   ,,

Persia.

a. ♂.   Rayin, S.E. of Karman (8000 ft.).   W. T. Blanford, Esq. [P.].

12. PRISTURUS.

Pristurus, *Rüppell, N. Wirbelth. Faun. Abyss., Rept.* p. 16; *Fitzing. Syst. Rept.* p. 91; *Gray, Cat.* p. 171.
Gymnodactylus, part., *Dum. & Bibr.* iii. p. 408.
Spatalura, *Gray, Proc. Zool. Soc.* 1863, p. 236.

Digits slender, clawed, cylindrical at the base; the distal phalanges compressed, forming an angle with the basal portion of the digits, the lower surface of which has a row of plates. Body not depressed, covered with uniform granules. Tail compressed, keeled. Pupil circular; eyelid distinct all round the eye. No præanal nor femoral pores.

North-east Africa; South-west Asia.

Synopsis of the Species.

I. Rostral plate entering the nostril.

Hind limb, when stretched forwards, reaching axilla; upper caudal keel denticulate...................... 1. *flavipunctatus*, p. 52.
Hind limb reaching the ear-opening or between the shoulder and the ear-opening; upper caudal keel denticulate 2. *rupestris*, p. 53.
Hind limb reaching to between the ear-opening and the eye; upper caudal keel not denticulate .... 3. *insignis*, p. 54.

II. Nostril separated from the rostral.

Tail at least as long as the body, crestless; abdominal scales much larger than dorsals .... 4. *crucifer*, p. 55.
Tail at least as long as the body, with strong upper and lower crest ...... 5. *collaris*, p. 55.
Tail much shorter than the body, with strong upper and lower crest ...... 6. *carteri*, p. 55.

1. Pristurus flavipunctatus.

Pristurus flavipunctatus, *Gray, Cat.* p. 171.
Pristurus flavipunctatus, *Rüpp. N. Wirbelth. Faun. Abyss., Rept.* p. 17, pl. vi. fig. 3.
Gymnodactylus flavipunctatus, *Dum. & Bibr.* iii. p. 417.
Saurodactylus (Pristiurus) flavipunctatus, *Fitz. Syst. Rept.* p. 91.

Head short and high; snout subacuminate, longer than the distance between the eye and the ear-opening, once and one third the diameter of the orbit; forehead very slightly concave; ear-opening oval, oblique, hardly one third the diameter of the orbit. Limbs long, the hind limb when stretched forwards reaching the axilla; digits long and slender. Head, body, and tail covered with granular scales, largest and of about equal size on the snout and belly. Rostral subquadrangular, more than twice as broad as high, with median cleft above; nostril pierced between the rostral and three nasals; seven or eight upper and five or six lower labials; mental very large, truncate posteriorly; no regular chin-shields. Tail longer than head and body, slender, strongly compressed, keeled above and beneath; the two keels denticulate, the upper most strongly, especially in males, in which it forms a regular crest; in males the dorsal keel forming also a denticulation. Light greyish-brown above, with more or less distinct darker transverse bars; a darker streak passing through the eye; lower surfaces whitish; throat sometimes white-dotted.

Total length	79	millim.
Head	9·5	,,
Width of head	6·5	,,
Body	29·5	,,
Fore limb	17	,,
Hind limb	21	,,
Tail (injured)	40	,,

Abyssinia.

a–b. ♀.	Abyssinia.	Frankfort Museum.
c–d. ♂.	Syria (?).	
e. ♀.	——— ?	

2. Pristurus rupestris.

Pristurus rupestris, *Blanford, Ann. & Mag. N. H.* (4) xiii. 1874, p. 454; *id. Zool. E. Persia*, p. 350, pl. xxiii. fig. 1; *id. Proc. Zool. Soc.* 1881, p. 465; *Murray, Zool. Sind*, p. 365, pl. —. fig. 1.

Head short and high; snout acuminate, once and two fifths to once and a half the diameter of the orbit; forehead scarcely concave; ear-opening oval, oblique, about one third the diameter of the orbit. Limbs long, the hind limb when stretched forwards reaching the ear-opening or between the shoulder and the ear-opening; digits long and slender. Snout covered with polygonal convex scales; the remainder of the head, the body, and the limbs covered with small granules; those on the belly larger, though smaller than the scales on the snout. Rostral more than twice as broad as high, with median cleft above; nostril pierced between the rostral and two or three nasals; seven or eight upper and five or six lower labials; mental very large, truncate posteriorly; no regular chin-shields. Tail longer than head and body, slender, strongly compressed, keeled above and beneath; the two keels denticulate, the upper most strongly, especially in males, in which it forms a regular crest, which, however, never extends to the body.

Grey-brown or olive above, clouded with darker, and frequently with round lighter spots; a dark streak passing through the eye; small bright red dots may be present on the sides of the body; sometimes a light reddish vertebral band.

Total length	85 millim.
Head	9 ,,
Width of head	5·5 ,,
Body	23 ,,
Fore limb	16 ,,
Hind limb	21 ,,
Tail	53 ,,

Arabia and Socotra to Sind.

a–b. ♀.	Muscat, and Island of Karrack, near Busheer, Persian Gulf.	W. T. Blanford, Esq. [P.]. (Types.)
c. Several spec., ♂ ♀.	Socotra.	Prof. I. B. Balfour [C.].

3. Pristurus insignis.

Pristurus insignis, Blanford, *Proc. Zool. Soc.* 1881, p. 466, pl. xlii. fig. 1.

Head short and high, *Anolis*-like; snout subacuminate, longer than the distance between the eye and the ear-opening, once and a half the diameter of the orbit; forehead feebly concave; ear-opening large, oval, vertical, nearly half the diameter of the orbit. Limbs very long, the hind limb when stretched forwards reaching to between the eye and the ear-opening; digits very long and slender. Snout covered with polygonal convex scales; the remaining portion of the head, as well as the upper parts of the body, limbs, and tail, covered with minute granules. Rostral subquadrangular, more than twice as broad as high, with median cleft above; nostril pierced between the rostral and three nasals; seven to nine upper and five or six lower labials; mental extremely large, truncate posteriorly, followed by three small chin-shields. Abdominal scales granular, a little larger than the dorsals. Tail much longer than head and body, slender, strongly compressed and keeled above, crestless. Earthy brown above, with rather darker but indistinct cross bands and numerous red spots of irregular shape scattered over the back and sides; lower parts paler; chin, throat, and breast mottled with brown.

Total length	153 millim.
Head	15 ,,
Width of head	9·5 ,,
Body	41 ,,
Fore limb	30·5 ,,
Hind limb	40 ,,
Tail	97 ,,

Socotra.

a–b. Ad.	Socotra.	Prof. I. B. Balfour [C.]. (Types.)

4. Pristurus crucifer.

Gymnodactylus crucifer, *Val. C. R. Ac. Paris,* lii. 1861, p. 433; *Vaill. Miss. Révoil aux Pays Somalis, Rept.* p. 17, pl. iii. fig. 1.
Pristurus longipes, *Peters, Mon. Berl. Ac.* 1871, p. 566.

Head short and high; snout subacuminate, longer than the distance between the eye and the ear-opening, once and one third the diameter of the orbit; forehead not concave; ear-opening small, oval, oblique, not one third the diameter of the orbit. Limbs very long, the hind limb when stretched forwards reaching to between the ear-opening and the eye; digits very long and slender, with extremely long claws. Snout covered with flat polygonal scales; upper parts of body and limbs with much smaller flat granules; abdominal scales flat, hexagonal, subimbricate, a little larger than the scales on the snout. Rostral large, subpentagonal, with truncate posterior angle, more than twice as broad as high, with median cleft above; nostril separated from the rostral and labials, pierced between three rather swollen nasals; six upper and five lower labials; mental very large, truncate posteriorly; no chin-shields. Tail feebly compressed, not keeled (?). Grey-brown above, with darker markings and white spots; a white vertebral band; on each side a series of almost confluent white spots; white inferiorly, throat with a few grey spots.

Total length	61 millim.
Head	8 ,,
Width of head	5 ,,
Body	24 ,,
Fore limb	16 ,,
Hind limb	24 ,,
Tail	29 ,,

Arabian Gulf.

a. ♀.   Kursi, near Aden.   Marquis G. Doria [P.].
(One of the types of *Pristurus longipes*.)

5. Pristurus collaris.

Spatalura collaris, *Steindachn. Novara, Rept.* p. 20.

Closely allied to *P. carteri,* from which it differs in having the tail longer, as long as or a little longer than the body, and the upper caudal crest extending on the back. A deep black band across the throat, extending on the sides of the neck.

Habitat unknown.

6. Pristurus carteri.

Spatalura carteri, *Gray, Proc. Zool. Soc.* 1863, p. 236, pl. xx. fig. 2; *Carter, op. cit.* 1864, p. 135.

Head short and very high; snout triangular, acutely pointed, longer than the distance between the eye and the ear-opening, slightly longer than the diameter of the orbit; forehead not concave; ear-opening rather large, oval, vertical, the upper part concealed under a fold of the skin. Six larger incisor teeth. Limbs very long, the hind limb when stretched forwards reaching the eye; digits very long and slender, with very long claws. Upper surfaces covered with small flat granules, largest on the snout. Abdominal scales smaller still, convex, those on the median line pointed and erect. Rostral transversely suboval, not twice as broad as high, with median cleft; nostril pierced between two large and one very small nasal, separated from rostral and labials; eight upper and seven lower labials; mental large, subtrapezoid; no chin-shields. Tail shorter than the body, strongly compressed, in profile rounded at the extremity, which is a little higher than the base, furnished above with a much-developed crest of closely set slender linear scales. The life-coloration is as follows, according to Carter:—"Ground cinereous, six pairs of white spots between the back of the head and root of tail, symmetrically placed; six to eight lines of red spots on each side, broken and terminating in small points towards the belly; buff-coloured irregular spots on the sides among the red lines; belly bright yellow, passing into cinereous towards the roots of the posterior and anterior extremities; legs and tail spotted with red towards their proximal ends, with white spots towards their extremities; head irregularly marked with red and white spots having a transverse direction."

Total length	94	millim.
Head	17	„
Width of head	13	„
Body	48	„
Fore limb	35	„
Hind limb	53	„
Tail	29	„

South-east coast of Arabia.

a–b. Ad., bad state.   Makulla.   H. Carter, Esq. [P.]. (Types.)

13. GONATODES.*

Gymnodactylus, part., *Dum. & Bibr.* iii. p. 408; *Günth. Rept. Brit. Ind.* p. 112.
Gonatodes, *Fitzing. Syst. Rept.* p. 91.
Goniodactylus (*non Kuhl*), *Gray, Cat.* p. 171.
Heteronota, part., *Gray, l. c.* p. 174.

Digits slender, clawed, cylindrical or depressed at the base (in one species dilated); the distal phalanges compressed, forming an angle with the basal portion of the digits, the lower surface of which has

---

* *Gonatodes ferrugineus,* Cope, Proc. Ac. Philad. 1862, p. 102.—Trinidad.

a row of plates. Body more or less depressed, granular or tubercular above. Tail not compressed. Pupil circular; eyelid distinct all round the eye. Males with or without præanal or femoral pores.

Tropical America; East Indies.

Synopsis of the Species.

I. AMERICAN SPECIES.—Males more brilliantly coloured than the females, without femoral or præanal pores.

   A. Basal phalanges of digits cylindrical.

Snout obtusely pointed, little longer than the diameter of the orbit; male with a broad bluish-black band on each side of the belly .... 1. *albogularis*, p. 59.
Snout longer than the diameter of the orbit, pointed; a broad whitish, black-edged vertebral band, and a broad bluish-black band on each side of the belly in the male .... 2. *vittatus*, p. 60.
Snout acutely pointed, once and a half the diameter of the orbit; head white, black-marbled; one or two large ocelli on each side of the back .... 3. *ocellatus*, p. 60.
Snout not once and a half the diameter of the orbit; head white, with reticulated black lines; a more or less distinct ocellus above axilla .... 4. *caudiscutatus*, p. 61.

   B. Basal phalanges of digits distinctly depressed.

Chin-shields scarcely enlarged; male with the head and neck uniform yellowish, the rest of the body grey-blue with black vermiculations .... 5. *concinnatus*, p. 61.
Dorsal granules exceedingly small; relatively large plates under the basal phalanges; male brown, finely vermiculated with darker above .... 6. *humeralis*, p. 62.

II. MALAYAN SPECIES.—Males coloured like the females, without femoral or præanal pores.

Snout short and pointed; upper surfaces covered with uniform small granules .. 7. *timorensis*, p. 63.
Snout long and broad, with strong canthal ridges; body above with small granules intermixed with keeled tubercles .... 8. *kendallii*, p. 63.

III. INDIAN SPECIES.—Males coloured like the females, with femoral or præanal pores.

A. Digits not dilated.
1. Flanks without spine-like projecting tubercles.
   a. Males with femoral pores, without præanal pores.

Back with uniform keeled granules; 4 or 5 pores on each side .... 9. *indicus*, p. 64.
Back with uniform or heterogeneous round granules, each with a raised central point or short keel; 4 to 6 pores on each side .... 10. *wynadensis*, p. 65.
Back with uniform round granules; 8 pores on each side .... 11. *sisparensis*, p. 66.

b. Males without femoral pores, with præanal pores.

Back with small smooth granules irregularly intermixed with small smooth or slightly keeled tubercles; ventral scales smooth; 6 to 9 pores .... 12. *ornatus*, p. 66.
Back with small granules irregularly intermixed with slightly larger ones, all with a central raised point or short keel; ventral scales keeled; 6 to 8 pores.... 13. *marmoratus*, p. 67.

c. Males with præanal and femoral pores.

Back with small strongly keeled tubercles, and scattered enlarged ones on the flanks; 2 or 3 præanal, and on each side 3 or 4 femoral pores .... 14. *mysoriensis*, p. 68.

2. Flanks with small spine-like projecting tubercles.

Scales under the neck keeled; tail with rings of spine-like tubercles; 3 or 4 præanal, and on each side 3 to 5 femoral pores........................... 15. *kandianus*, p. 68.
Scales under the neck smooth; tail with rings of spine-like tubercles; 2 to 4 præanal, and on each side 3 to 5 femoral pores........................... 16. *gracilis*, p. 70.
No enlarged tubercles on the back; no rings of spines on the tail; no præanal pores; 5 to 12 femoral pores on each side .... 17. *jerdonii*, p. 71.

B. Digits dilated at the base, with large plates inferiorly .... 18. *littoralis*, p. 71.

1. Gonatodes albogularis.

Goniodactylus albogularis, *Gray, Cat.* p. 172.
Gymnodactylus albogularis, *Dum. & Bibr.* iii. p. 415; *Cocteau, in R. de la Sagra, Hist. Cuba, Rept.* p. 174, pl. xix.; *Steindachn. Novara, Rept.* p. 16.
Gonatodes albigularis, *Fitzing. Syst. Rept.* p. 91.
? Gymnodactylus notatus, *Reinh. & Lütk. Vidensk. Meddel.* 1862, p. 280.
Gymnodactylus maculatus, *Steindachn. l. c.* pl. i. fig. 4.

Head small, high; snout obtusely pointed, short, measuring the distance between the eye and the ear-opening, little longer than the diameter of the orbit; forehead not concave: ear-opening small, oval. Body and limbs moderate; digits slender, narrow at the base, the scales under the basal joint small, subequal. Upper parts covered with small granules, largest on the snout. Rostral pentagonal, nearly twice as broad as high, with median cleft above: nostril pierced between the rostral and three or four nasals; five or six upper and four or five lower labials; mental very large, truncate behind, in contact with two or three small chin-shields followed by others passing gradually into the minute granules of the gular region. Abdominal scales rather large, flat, hexagonal, imbricate. Tail cylindrical, tapering, covered above with uniform small round slightly imbricate scales, inferiorly with larger scales, those of the median series transversely dilated. Males grey-brown above, unspotted; throat pure white, bordered on each side by a broad dark blue band; a white or pale blue, black-edged, vertical line in front of the fore limb; a broad black (steel-blue?) band of each side of the belly, the lower part of which, as well as the lower surface of the thighs, is bright yellow. Females grey-brown above, with small darker spots; the white line in front of the shoulder more or less distinct; frequently a lighter vertebral band; lower surfaces whitish. Young white-spotted.

Total length	76 millim.
Head	9 ,,
Width of head	6 ,,
Body	23 ,,
Fore limb	13 ,,
Hind limb	16 ,,
Tail	44 ,,

West Indies.

a. Hgr.	Rio S. Juan, Cuba.
b–c. ♀.	Jamaica.
d–i. ♂ ♀.	—— ?
k–l. Yg.	—— ?

Var. fuscus.

Stenodactylus fuscus, *Hallow. Journ. Ac. Philad.* (2) iii. 1855, p. 33.
Gymnodactylus fuscus, *A. Dum. Arch. Mus.* viii. p. 477; *Bocourt, Miss. Sc. Mex., Rept.* p. 48, pl. x. fig. 5.

Gymnodactylus albogularis, var., *A. Dum. l. c.* p. 472.
Goniodactylus braconnieri, *O'Shaughn. Ann. & Mag. N. H.* (4) xvi. 1875, p. 265.

The gular region is light brown (orange?) with a median whitish line, bifurcating on the chin; no blue band on the side of the neck, but a few pale blue spots on the lips; back finely vermiculated with blackish; end of tail yellow. Otherwise like the preceding. Female more vermiculated.

Colombia; Central America.

a–b, c–d. ♂ ♀.	Baranquilla.	Mr. Rippon. (Types of *G. braconnieri.*)
e. ♂.	——?	
f. ♀.	Panama.	Mdme. Pfeiffer [C.].
g–h. ♀.	Panama.	

In one of these two specimens the chin-shields in contact with the mental are united into one, as in the *G. fuscus* figured by Bocourt.

2. Gonatodes vittatus.

Gymnodactylus vittatus (*Wiegm.*), *Lichtenst. Nomencl. Rept. Mus. Berol.* p. 6; *Reinh. & Lütk. Vidensk. Meddel.* 1863, p. 283.
Gonatodes gilli, *Cope, Proc. Ac. Philad.* 1863, p. 102, & 1868, p. 97.

Very closely allied to *G. albogularis*. The snout is a little longer and more pointed. Grey above; a broad white (yellow) vertebral band commencing from the tip of the snout, bordered with black; sides of belly steel-blue; the lower belly yellow in the middle; throat white, uniform or with deep black variegations. Female unknown.

West Indies; Northern South America.

a. ♂.	Dominica.
b–c. ♂.	Island of Granada.
d. ♂.	Island of Curaçao.
e. ♂.	——?

3. Gonatodes ocellatus. (PLATE V. fig. 1.)

Goniodactylus? ocellatus, *Gray, Cat.* p. 172.
Cyrtodactylus ocellatus, *Gray, Zool. Misc.* p. 59.

Differs from *G. albogularis* in the following points :—Snout acutely pointed, a little longer than the distance between the eye and the ear-opening, once and a half the diameter of the orbit. Scales on the throat quite as large as those on the middle of the back. Light brown; head and neck above and beneath with large dark brown marblings; a light blue, black-edged transverse streak on scapular region, extending across base of humerus; one or two large light blue, black-edged ocelli on the side of the back; belly dark.

Total length	86 millim.
Head	13 ,,
Width of head	8 ,,
Body	33 ,,
Fore limb	17 ,,
Hind limb	22 ,,
Tail	40 ,,

Tobago.

a. ♂. Tobago. Zoological Society. (Type.)

4. Gonatodes caudiscutatus. (PLATE V. fig. 2.)

Gymnodactylus caudiscutatus, *Günth. Proc. Zool. Soc.* 1859, p. 410.
Goniodactylus caudiscutatus, *O'Shaughn. Ann. & Mag. N. H.* (4) xvi. 1875, p. 265.

Head considerably more depressed than in *G. albogularis*; snout a little longer. In the males the supraocular spine-like scales are much developed, and the subcaudal shields very broad. Males dark grey on the back and limbs, with light blue, black-edged spots; a more or less distinct larger ocellus above axilla; head white above, with reticulated black lines, one from the eye towards the snout being very constant; chin, throat, and breast white, uniform or with a few black specks; belly grey or blackish. Female grey-brown above, with darker spots symmetrically arranged in pairs on the back and tail: lower surfaces a little lighter, the throat with brown reticulation.

Total length	79 millim.
Head	12 ,,
Width of head	7 ,,
Body	30 ,,
Fore limb	16 ,,
Hind limb	20 ,,
Tail	37 ,,

Ecuador and Colombia.

a–d. ♂ ♀.   W. Ecuador.   Mr. Fraser [C.]. (Types.)
e–f. ♂.   Panama.   Mdme. Pfeiffer [C.].

5. Gonatodes concinnatus.

Goniodactylus concinnatus, *O'Shaughn. Proc. Zool. Soc.* 1881, p. 237, pl. xxxiii. fig. 2.
—— buckleyi, *O'Shaughn. l. c.* p. 238, pl. xxiii. fig. 3.

The snout is a little longer and more pointed than in *G. albogularis*; the digits are slightly depressed at the base, as in *G. humeralis*; the anterior chin-shields are very small, and can hardly be termed such. Males: head and fore part of body above and below as far as the shoulder, and including the fore limb, pale brown or yellowish, abruptly terminated by two vertical humeral bands, sometimes meeting above and forming a regular collar of pure white with black borders; the rest of the body, with the hind limb, blue, with black vermiculations elaborately interwoven; tail darker, with the variegations continued; inferior surface from chest blue, paler again at the hind limb and anal region. Females: ground-colour greyish brown; head variegated with black; back with two parallel longitudinal rows of black blotches, pointed in front and separated by the median line; a narrow white vertical streak on the shoulder; gular region, from the chin to the chest, with alternating black and white oblique stripes converging behind, and making a triangular pattern.

Total length	96	millim.
Head	13	,,
Width of head	9	,,
Body	38	,,
Fore limb	19	,,
Hind limb	25	,,
Tail	45	,,

Ecuador.

a–c. ♂.	Canelos.	Mr. Buckley [C.].	(Types.)
d–e. ♀.	Canelos.	Mr. Buckley [C.].	(Types of *Gonio-*
f. ♀.	Pallatanga.	Mr. Buckley [C.].	*dactylus buckleyi*.)

6. Gonatodes humeralis. (PLATE V. fig. 3.*)

Gymnodactylus humeralis, *Guichen. in Casteln. Voy. Amér. Mér., Rept.* p. 6, pl. iii. fig. 1; *A. Dum. Arch. Mus.* viii. p. 474.
—— incertus, *Peters, Mon. Berl. Ac.* 1871, p. 397.
Goniodactylus sulcatus, *O'Shaughn. Ann. & Mag. N. H.* (4) xvi. 1875, p. 265.

Snout pointed, longer than the distance between the eye and the ear-opening; all the scales smaller than in the preceding species; six or seven upper and five or six lower labials; the basal joint of the fingers and toes slightly though distinctly depressed, with relatively large plates; the basal joint of the fourth toe longer than the remaining part of the digit. Male brown, above finely vermiculated with darker; a narrow, dark-edged whitish vertical streak in front of and above the fore limb; tail with alternating darker and lighter cross bars; a few dark spots on the side of the back. Female brown, lighter beneath, with black spots symmetrically arranged in pairs on the back and tail; a narrow white ante-humeral line; head, limbs, and throat vermiculated or clouded with brown.

---

* Lower surface of fourth toe, ×4.

13. GONATODES.

Total length	73 millim.
Head	10 „
Width of head	6·5 „
Body	27 „
Fore limb	13 „
Hind limb	17 „
Tail	36 „

Peru.

a. ♂.	[Cuba.] —— ?	(Type of *Goniodactylus sulcatus*.)
b–d. ♂ ♀.	Yurimaguas, Huallaga River.	Dr. Hahnel [C.].
e. ♀.	Cayaria.	Messrs. Veitch [P.].
f. ♂ ♀.	Puerto del Mairo.	Messrs. Veitch [P.].
g. ♂.	Santarem.	Mr. Wickham [C.].

[*Gymnodactylus gaudichaudii*, Dum. & Bibr. iii. p. 413, from Coquimbo, Chili, belongs doubtless to this genus, but it is not possible from the description to say in what points it differs from its American congeners, to which it seems to be very closely related.]

7. Gonatodes timorensis.

Gymnodactylus timoriensis, *Dum. & Bibr.* iii. p. 411.

Snout short, pointed; rostral pentagonal, a little broader than high; five pairs of labials; mental very large, lozenge-shaped; a very small chin-shield on each side of the mental. Digits long and slender. Upper surfaces covered with uniform small granules. Abdominal scales hexagonal, imbricate. No præanal or femoral pores (?). Head and back reddish, sides grey-brown; brown spots on each side of the vertebral line; a black interrupted band along the flanks; reddish-grey inferiorly.

Head	11 millim.
Body	24 „
Fore limb	12 „
Hind limb	14 „

Timor.

8. Gonatodes kendallii. (PLATE V. fig. 4.)

Heteronota kendallii, *Gray, Cat.* p. 174.

Habit very slender. Head oval; snout long and broad, rounded, depressed, with strong canthal ridges, much longer than the distance between the eye and the ear-opening, once and a half the diameter of the orbit; eyes large; ear-opening vertically oval, two fifths the diameter of the eye. Limbs long; digits long and slender, compressed, inferiorly with small lamellæ and a large oval plate at the articulation of the basal and proximal phalanges. Upper surfaces covered with minute granules, intermixed on the body with irregularly arranged small keeled tubercles. Rostral large, quadrangular, not twice as broad as high, with median cleft; nostril between the rostral and several granules; eleven to thirteen upper and ten to twelve lower labials; mental very large, subtriangular; two large chin-shields, in one specimen fused with the mental. Abdominal scales very small, juxtaposed, convex, keeled. No præanal or femoral pores. Tail cylindrical, slender, with small keeled scales and large pointed tubercles, of which there is a median series on the lower surface. Brown above, more or less distinctly clouded with darker; lower surfaces lighter.

Total length	118	millim.
Head	15	,,
Width of head	9	,,
Body	41	,,
Fore limb	28	,,
Hind limb	39	,,
Tail	62	,,

Borneo.

a–b. ♂ ♀.   Borneo.   Sir E. Belcher [P.]. (Types.)
c. ♂.   Matang.

9. Gonatodes indicus. (PLATE VI. fig. 1.)

Goniodactylus indicus, *Gray, Ann. & Mag. N. H.* xviii. 1846, p. 429.
Gymnodactylus indicus, *Jerdon, Journ. As. Soc. Beng.* xxii. 1853, p. 469; *Günth. Rept. Brit. Ind.* p. 115.

Head short; snout obtusely pointed, slightly longer than the distance between the eye and the ear-opening, once and two thirds the diameter of the orbit; forehead not concave; ear-opening very small, round. Body and limbs moderate; the hind limb reaching hardly the axilla. Digits moderately slender; the basal joint not dilated, scarcely wider than the distal joints, inferiorly with larger, subequal plates. Upper surfaces covered with uniform strongly keeled granules, which are larger on the back than on the snout. Rostral subquadrangular, twice as broad as high, with median cleft above; nostril pierced between the rostral and three nasals; seven or eight upper and five or six lower labials; mental large, broad, subtriangular or pentagonal, with truncate posterior angle; small chin-shields passing gradually into the gular granules. Ventral scales hexagonal, imbricate, smooth or feebly keeled. Males with four or five femoral pores on each side. Tail cylindrical, tapering, covered above with uniform small keeled scales, inferiorly with larger scales, those of the median series being slightly enlarged. Brown above, marbled with darker and lighter; sometimes a light vertebral line; brownish-white beneath, the throat frequently brown, or brown-marbled.

Total length	83	millim.
Head	10	,,
Width of head	8	,,
Body	30	,,
Fore limb	13·5	,,
Hind limb	18	,,
Tail	43	,,

Southern India.

a-c. ♂ ♀.	Madras Presidency.	T. C. Jerdon, Esq. [P.]. (Types.)
d. Many spec.: ♂, ♀, & hgr.	Nilgherries.	Col. Beddome [C.].

10. Gonatodes wynadensis. (PLATE VI. fig. 2.)

Gymnodactylus wynaadensis, *Beddome, Madras Journ. Med. Sc.* 1870.
Goniodactylus wynadensis, *Günth. Proc. Zool. Soc.* 1875, p. 226.

Habit of *G. indicus.* The basal part of the digits inferiorly with very small scales, the two distal ones only being a little enlarged and plate-like. Head covered with very small granules, which are keeled on the snout. Upper surface of body with large round granules, each with a raised central point or short keel, largest on the sides; these tubercles are either homogeneous, or intermixed with much smaller ones, and as there are specimens which are intermediate in this respect, no great importance can be attached to this difference. Abdominal scales smooth. Six upper and seven or eight lower labials. Male with four or five femoral pores on each side. Tail generally with a median series of transversely dilated plates inferiorly. Brown above, marbled with darker and lighter; the median dorsal zone sometimes lighter, dark-bordered; brownish inferiorly, the throat brown-marbled; tail inferiorly dark-brown, generally lighter-spotted.

Total length	87	millim.
Head	12	,,
Width of head	8·5	,,
Body	29	,,
Fore limb	14	,,
Hind limb	20	,,
Tail	46	,,

Southern India.

a. Many spec.: ♂, ♀, & hgr.	Wynaad.	Col. Beddome [C.]. (Types.)
b. ♂.	Anamallays.	Col. Beddome [C.].
c-i. ♂, ♀, & hgr.	Bolumputta Hills.	Col. Beddome [C.].
k-l. ♀ & yg.	Tinnevelly.	Col. Beddome [C.].
m-n. Yg.	Nellicottah.	Col. Beddome [C.].

11. Gonatodes sisparensis.

Gymnodactylus maculatus (*non* Steind.), *Beddome, Madras Journ. Med. Sc.* 1870.
—— sisparensis, *Theobald, Cat. Rept. Brit. Ind.* p. 86.

Closely allied to *G. wynadensis*, but the digits much more elongate. Eight femoral pores on each side. Dorsal tubercles homogeneous. Brown, with regular transverse dark bands across the body and tail. From snout to vent 62 millim.

Southern India.

a. ♂, very bad state.	Sholakal, Sispara Ghat.	Col. Beddome [C.].	(Type.)
b. ♂.	S.W. India.	Col. Beddome [C.].	

12. Gonatodes ornatus. (Plate VI. fig. 3.)

Gymnodactylus ornatus, *Beddome, Madras Journ. Med. Sc.* 1870.

Head rather elongate; snout acuminate, longer than the distance between the eye and the ear-opening, once and a half to once and two thirds the diameter of the orbit; forehead feebly concave; ear-opening small, vertically oval. Body moderate, limbs rather slender; the hind limb reaching the shoulder or a little beyond. Digits elongate, slender; the basal part not dilated, scarcely wider than the distal, inferiorly with very small plates and a large discoid one under the articulation. Snout covered with keeled granules; hinder part of head with uniform minute granules. Rostral four-sided, twice as broad as high, with median cleft above; nostril pierced between the rostral and three nasals; six to nine upper and seven or eight lower labials; mental large, triangular or pentagonal, its posterior angle truncate and in contact with a small median chin-shield; two pairs of larger chin-shields. Upper surface of body covered with small smooth granules irregularly intermixed with small round convex, sometimes slightly keeled, tubercles. Abdominal scales small, round, imbricate, smooth. Males with six to nine præanal pores forming an angular line. Tail cylindrical, becoming slender in its distal half, covered above with small smooth scales, in its anterior portion with semiannuli of pointed keeled tubercles; inferiorly with large smooth scales, the median series being a little enlarged. Brown above, with blackish and whitish markings; head generally with black and white angular lines; generally a black and white band across the neck and a white, black-edged ocellus in front of the arm; tail with complete black and white annuli; lower surfaces brownish, the lower lip brown-edged.

Total length .... 117 millim.
Head .... 16 ,,
Width of head .... 10·5 ,,

Body	39 millim.
Fore limb	24 "
Hind limb	32 "
Tail	62 .,

Southern India.

a. Many specimens: ♂, ♀, & yg.	Tinnevelly.	Col. Beddome [C.]. (Types.)
b–f. ♂, ♀, & hgr.	Tinnevelly.	Col. Beddome [C.].
g. ♂.	Travancore.	Col. Beddome [C.].
h. ♂.	S. India.	Col. Beddome [C.].

13. Gonatodes marmoratus. (PLATE VI. fig. 4.)

Gymnodactylus marmoratus (*non* D. & B.), *Beddome, Madras Journ. Med. Sc.* 1870.
—— *beddomei, Theobald, Cat. Rept. Brit. Ind.* p. 88.

Head rather elongate; snout acuminate, longer than the distance between the eye and the ear-opening, once and half to once and two thirds the diameter of the orbit; forehead feebly concave; ear-opening small, vertically oval. Body moderate, limbs rather slender; the hind limb reaching the shoulder or a little beyond. Digits elongate, slender; the basal part not dilated, scarcely wider than the distal, inferiorly with regular plates, that under the articulation being large and discoid. Head covered with small granules, largest and keeled on the snout. Rostral four-sided, twice as broad as high, with median cleft above; nostril pierced between the rostral and three nasals; six to eight upper and as many lower labials; mental large, triangular or pentagonal, its posterior angle truncate; chin-shields very small. Upper surface of body covered with small granules irregularly intermixed with slightly larger ones; all these tubercles with a central raised point or a short keel. Abdominal scales very small, juxtaposed, convex, keeled. Males with six to eight præanal pores. Tail cylindrical, tapering, covered above with uniform small feebly keeled scales, inferiorly with larger smooth scales, those of the median series sometimes a little enlarged. Brown, above clouded with darker and lighter markings; tail with ill-defined darker and lighter annuli; the lower lip edged with dark brown; another dark-brown streak, parallel to the latter, on each side of the throat.

Total length	100 millim.
Head	15 "
Width of head	10 "
Body	36 "
Fore limb	28 "
Hind limb	28 "
Tail	49 "

Southern India.

a. Several spec.: ♂, ♀, & yg.	Travancore.	Col. Beddome [C.]. (Types.)
b. Several spec.: ♂, ♀, & yg.	Wynaad.	T. C. Jerdon, Esq. [P.].
c–d. ♂ & hgr.	Tinnevelly.	Col. Beddome [C.].

14. Gonatodes mysoriensis.

Gymnodactylus mysoriensis, *Jerdon, Journ. As. Soc. Beng.* xxii. 1853, p. 469; *Günth. Rept. Brit. Ind.* p. 114.

Head short; snout obtusely pointed, longer than the distance between the eye and the ear-opening, once and a half the diameter of the orbit; forehead not concave; ear-opening very small, oval. Body and limbs moderate; the hind limb not reaching beyond axilla. Digits slender but rather short, the basal part not dilated, and with small plates inferiorly. Upper surfaces covered with small strongly keeled tubercles, smallest on the hinder part of the head, largest on the sides of the body, where they are intermixed with irregularly scattered tubercles, differing only in size from the granules of the back. Rostral quadrangular, nearly twice as broad as high, with median cleft and emargination above; nostril pierced between the rostral and three nasals; six or seven upper and seven or eight lower labials; mental large, broadly triangular, its posterior angle truncate; small chin-shields passing gradually into the rather large, flat, gular granules. Abdominal scales moderate, hexagonal, imbricate, smooth. Males with two or three præanal, and on each side three or four femoral pores. Tail cylindrical, tapering, covered above with small keeled scales and large pointed keeled tubercles in six longitudinal series, inferiorly with larger smooth scales, those of the median series slightly enlarged. Brown above, frequently with a light vertebral band and a series of dark-brown marks on the back; digits very conspicuously dark-barred; lower surfaces whitish, throat sometimes brown-marbled.

Total length	64 millim.
Head	8 ,,
Width of head	5·5 ,,
Body	19 ,,
Fore limb	10 ,,
Hind limb	13 ,,
Tail	37 ,,

Southern India.

a, b. Many spec.: ♂, ♀, & hgr..	Shevaroys.	Col. Beddome [C.].
c–d. Hgr.	Malabar.	Col. Beddome [C.].

15. Gonatodes kandianus.

Gymnodactylus kandianus, *Kelaart, Prodr. Faun. Zeyl.* p. 186; *Günth. Rept. Brit. Ind.* p. 114.

?Gymnodactylus wicksii, *Stoliczka, Journ. As. Soc. Beng.* xlii. 1873, p. 165.
?Gymnodactylus humei, *Theob. Cat. Rept. Brit. Ind.* p. 89.

Head rather elongate; snout obtusely pointed, longer than the distance between the eye and the ear-opening, once and a half the diameter of the orbit; forehead not concave; ear-opening small, oval. Body and limbs rather slender; the hind limb reaching the axilla or the shoulder. Digits slender, the basal part not dilated, scarcely wider than the distal, and with enlarged plates inferiorly. Snout covered with suboval keeled granules; the rest of the head minutely granulate; rostral twice as broad as high, with median emargination and cleft above; nostril pierced between the rostral and three or four nasals; seven or eight upper and as many lower labials; mental large, triangular, with truncate posterior angle; numerous small chin-shields passing gradually into the gular granules, which are feebly keeled. Upper surface of body covered with small, more or less distinctly keeled granules, intermixed with irregularly scattered small keeled tubercles; flanks with small, widely separated, spine-like tubercles. Scales on the limbs keeled. Ventral scales cycloid, imbricate, those under the neck keeled, the others smooth. Males with three or four præanal, and on each side three to five femoral pores. Tail cylindrical, tapering, above with very small keeled scales and annuli of spine-like tubercles, inferiorly with larger scales, but no transversely dilated median plates. Brown above, variegated with darker and lighter, these variegations generally forming transverse markings on the back and tail; sometimes a light vertebral band; the spine-like tubercles on the flanks white; lower surfaces light brown or dirty white, the throat sometimes blackish.

Total length	75 millim.
Head	9·5 ,,
Width of head	5·5 ,,
Body	27·5 ,,
Fore limb	13 ,,
Hind limb	18 ,,
Tail	38 ,,

Ceylon; S. India.

a–c, d–h, i. ♂ ♀.	Ceylon.	Dr. Kelaart. (Types.)
k–m. ♂ ♀.	Ceylon.	G. H. K. Thwaites, Esq. [P.].
n, o. Several spec.: ♂, ♀, & yg.	S. Canara.	Col. Beddome [C.].
p–s. ♂ ♀.	Anamallays.	Col. Beddome [C.].
t–u. ♂ ♀.	Tinnevelly.	Col. Beddome [C.].
v–w. ♂ ♀.	Sevagherry Hills.	Col. Beddome [C.].
x. ♂.	Malabar.	Col. Beddome [C.].
y–z. ♂ ♀.	S. India.	T. C. Jerdon, Esq. [P.].

The following specimens agree so well in structure and in their different modes of coloration with *G. kandianus* that I cannot consider them specifically distinct from the latter. However, as their ventral scales are keeled, they should be kept distinct from the typical form. They may be called the *tropidogaster* form.

a. ♂.	Ceylon.	G. H. K. Thwaites, Esq. [P.].
β. ♀.	Ceylon.	Col. Beddome [C.].
γ–θ, ι. ♂, ♀, & hgr.	Tinnevelly.	Col. Beddome [C.].
κ. ♀.	Nilgherries.	Col. Beddome [C.].
λ–ν. ♀, hgr., & yg.	Wynaad.	Col. Beddome [C.].

16. Gonatodes gracilis. (Plate VI. fig. 5.)

? Gymnodactylus malabaricus, *Jerdon, Journ. As. Soc. Beng.* xxii. 1853, p. 469.
Gymnodactylus gracilis, *Beddome, Madras Journ. Med. Sc.* 1870; *Günth. Proc. Zool. Soc.* 1875, p. 226.

Closely allied to *G. kandianus*, from which it differs in the following points:—The gular granules are larger, flat, smooth; so are also all the ventral scales; the mental is very seldom truncated posteriorly, and the median chin-shields form a suture behind its point. Males with two or four præanal, and on each side three to five femoral pores. Grey-brown above, with darker and lighter spots; generally a median row of light blotches on the centre of the back; temples and sides of neck and throat with oblique dark brown lines; tail with blackish semiannuli; digits conspicuously blackish-barred; whitish beneath.

Total length	76	millim.
Head	9·5	,,
Width of head	6	,,
Body	23·5	,,
Fore limb	13	,,
Hind limb	17	,,
Tail	43	,,

Southern India; Ceylon.

a. Several spec.: ♂, ♀, & yg.	Palghat Hills.	Col. Beddome [C.]. (Types.)
b–d. ♀ & yg.	Palghat Hills.	Col. Beddome [C.].
e. Several spec.: ♂, ♀, & yg.	Sevagherry Hills.	Col. Beddome [C.].
f–k, l. ♂, ♀, & hgr.	Wynaad.	Col. Beddome [C.].
m. ♀.	Salem.	Col. Beddome [C.].
n. Several spec.: ♂ ♀.	Malabar.	Col. Beddome [C.].
o–r. ♂, ♀, & yg.	Forests of S. India.	T. C. Jerdon, Esq. [P.]. (As *Gymnodactylus malabaricus*.)
s–v. ♂ ♀.	Forests of S. India.	T. C. Jerdon, Esq. [P.].
w. Several spec.: ♂, ♀, & yg.	Ceylon.	Col. Beddome [C.].

17. Gonatodes jerdonii.

Gymnodactylus jerdoni, *Theobald, Cat. Rept. As. Soc. Mus.* p. 31, and *Cat. Rept. Brit. Ind.* p. 89.

In habit similar to *G. kandianus* and *gracilis*. Digits not elongate, but with rather large plates under the basal part, the distal of these plates being the largest and longitudinally oval in shape. Upper surfaces covered with uniform small granules, smooth on the back, a little larger and keeled on the snout; a few erect spine-like tubercles on the flanks. Rostral four-sided, nearly twice as broad as high, with median cleft above; nostril pierced between the rostral and three nasals; eight to ten upper and seven or eight lower labials; mental large, triangular or pentagonal; small chinshields passing gradually into the granules of the throat, which are rather large, flat, and smooth. Ventral scales hexagonal, imbricate, smooth. Males with five to twelve femoral pores on each side; no præanal pores. Tail cylindrical, tapering, covered with small smooth scales, in its basal half with a few scattered larger tubercles; median series of subcaudals enlarged. Grey-brown above, clouded with darker; the small lateral spines white; sometimes a black cervical spot; whitish beneath, the throat sometimes brown-dotted.

Total length	72 millim.
Head	11·5 ,,
Width of head	8 ,,
Body	31·5 ,,
Fore limb	18 ,,
Hind limb	23 ,,
Tail (injured)	29 ,,

Ceylon; Southern India.

a, b. ♂ ♀.	Ceylon.	Dr. Kelaart.
c–e. ♂ ♀.	Ceylon.	G. H. K. Thwaites, Esq. [P.].
f. ♀.	Ceylon.	
g. ♂.	S. India.	Col. Beddome [C.].
h–i. ♂.	Lamparis Peak (5000 feet).	Col. Beddome [C.].

18. Gonatodes littoralis. (PLATE VI. fig. 6.*)

Gymnodactylus littoralis, *Jerdon, Journ. As. Soc. Beng.* xxii. 1853, p. 469, *and Proc. As. Soc. Beng.* 1870, p. 75; *Günth. Proc. Zool. Soc.* 1875, p. 226.

—— planipes, *Beddome, Madras Journ. Med. Sc.* 1870.

Habit very slender, *Anolis*-like. Head long; snout pointed, much longer than the distance between the eye and the ear-opening, once and two thirds to once and three fourths the diameter of the orbit; forehead very slightly concave; ear-opening small, oval, vertical. The hind limb does not reach beyond axilla. Digits

---

* Lower surface of fourth toe, × 4.

strongly dilated at the base, with large plates inferiorly; the distal of these plates is the largest, longer than broad, truncate anteriorly. Upper surfaces covered with equal smooth granules, minute on the back, much larger on the snout; a few very small subconical tubercles are scattered on the flanks. Rostral large, quadrangular, twice as broad as high, with median cleft above; nostril pierced between the rostral and three nasals; eight to ten upper and six to eight lower labials; mental large, broadly triangular, its posterior angle truncate; small chin-shields passing gradually into the small smooth gular granules. Ventral scales hexagonal, imbricate, smooth. Males with sixteen to eighteen femoral pores on each side; no præanal pores. Tail cylindrical, tapering, above with uniform small smooth scales, inferiorly with a median series of transversely dilated plates. Grey-brown above, generally with a row of lighter dark-edged spots along the back, and a black spot on the nape; nearly always a black line bordering the lower lip, and extending as far as the ear; lower surfaces whitish.

Total length	68 millim.
Head	10 ,,
Width of head	5 ,,
Body	23 ,,
Fore limb	14 ,,
Hind limb	16 ,,
Tail	35 ,,

Southern India.

a. Many spec.: ♂, ♀, & yg.   Nellicottah.   Col. Beddome [C.]. (Types of *G. planipes*.)
b, c–d. Several spec.: ♂, ♀, & yg.   Nilambar.   Col. Beddome [C.].

Gonatodes boiei.

Goniodactylus boiei, *Gray, Zool. Misc.* p. 58.
—— timorensis (*non D. & B.* ?), part., *Gray, Cat.* p. 172.

"Blackish (discoloured? in spirits); back with close, squarish, minute, uniform, rather prominent scales; tail slender, longer than the body; lower labial shields unequal, three front large, the second largest, rest small; the chin-scales moderate, triangular behind, with two small polygonal shields behind them; rest of throat-scales small, polygonal."

a–d.   India.   Gen. Hardwicke [P.]. (Type.)

In too bad state to be described; doubtless a *Gonatodes* distinct from the other Indian species, and apparently allied to *G. jerdonii*.

The following specimens, types of *Goniodactylus australis*, Gray, belong probably to this genus, but they are in too bad a state to be determinable:—

Goniodactylus australis, Gray, Cat. p. 172.—" Pale brown, darker-marbled; scales very minute, of head rather larger; lower labial plates 10-1-10; the lower rostral small, trigonal, the other plates very gradually smaller; chin granular; those in the centre of the front rather larger, 6-sided; tail ——?".

a–b.   Australia.   Mr. Buchanan. (Types.)

14. ÆLUROSAURUS.

Pentadactylus, *Günther, Rept. Brit. Ind.* p. 117 (*nec Gray*).

Digits short, cylindrical at the base, and with transverse lamellæ inferiorly, compressed in the distal phalanges, which are raised and furnished with a claw rectractile between two large plates forming a compressed sheath; a narrow plate covering the upper suture of the two distal plates. Body covered with small juxtaposed flat scales. Upper and lower eyelids well developed, connivent. Pupil vertical. Males with præanal pores.

East Indies; Australia?

1. Ælurosaurus felinus. (PLATE III. fig. 8.*)

Pentadactylus felinus, *Günth. l. c.* pl. xii. fig. F.
—— borneensis, *Günth. l. c.*

Habit slender. Head depressed, elongate, very distinct from neck; snout pointed, longer than the distance between the eye and the ear-opening, once and one third the diameter of the orbit; ear-opening small, suboval, oblique. Body long, not depressed. Limbs long and thin. All the scales uniform, small, flat granules. Rostral broad and low, separated from the nostril, which is pierced between two larger anterior nasals, the first labial, and small granules posteriorly; two large internasals followed by three other plates; thirteen upper and as many lower labials, the latter in contact with another series of equally large plates; mental small, very low. Tail short, rounded, vertically oval when intact, ending in a very obtuse point, probably prehensile. Male with twenty-one præanal pores in an angular series. Brown above, with two dorsal series of more or less distinct lighter, dark-edged roundish spots; sometimes scattered white dots on the sides of the body, forming spots on the tail (spec. b); upper lip whitish; lower surfaces whitish, brown-dotted.

Total length	157	millim.
Head	23	,,
Width of head	14	,,
Body	71	,,
Fore limb	29	,,
Hind limb	38	,,
Tail	63	,,

Singapore; Borneo.

---

* Lower and lateral view of fourth toe, × 4.

a. ♂.   Singapore.   (Type.)
b. ♀.   Borneo.   A. R. Wallace, Esq. [C.]. (Type of *Pentadactylus borneensis*.)

2. Ælurosaurus dorsalis.

Pentadactylus dorsalis, *Peters, Mon. Berl. Ac.* 1871, p. 569, *and Ann. Mus. Genov.* iii. 1872, pl. ii. fig. 2.

"Brown with a yellow median dorsal band, which is bordered on each side by a row of small flat tubercles well distinguished from the surrounding fine granulation. On each side of the body a row of scattered yellow dots. Otherwise quite similar to *Æ. felinus*."
Sarawak.

3. Ælurosaurus? brunneus.

Pentadactylus brunneus, *Cope, Proc. Ac. Philad.* 1868, p. 320.

"Nostrils surrounded by four small shields and the first labial, the rostral being excluded. The anterior of these scales separated from its fellow by a polygonal scale, which is not included in a notch of the rostral. Rostral fissured above. Superior labials eleven, last two minute; two or three of them longer than high. Distinguishable inferior labials nine; two first much deeper than long. Infralabials not marked, forming some four or five rows of small ovate scales. Scaling of the body coarse. No superciliary spine; no præanal pores. Free joints of the toes, especially of the thumbs, thick. Tail with whorls of flat hexagonal scales, abruptly separated from those of the sacrum above. Colour above brown, with seven irregular undulate, transverse bars of very deep brown, between rump and nape. Below pale."
Australia.

15. HETERONOTA.

Heteronota, part., *Gray, Cat.* p. 174.

Digits not dilated, clawed, inferiorly with a row of plates; claw between three enlarged scales, the latero-inferior pair forming a longitudinal suture. Body covered above with granules and tubercles, inferiorly with imbricate scales. Pupil vertical. Males with præanal pores.
Australia.

1. Heteronota binoei.

Heteronota binoei, *Gray, Cat.* p. 174.
Heteronota binoei, part., *Günth. Ann. & Mag. N. H.* (3) xx. 1867, p. 50.

Head oviform, moderately depressed, large, measuring a little more than one third the distance between end of snout and vent; snout a little longer than the diameter of the orbit, as long as the distance between the eye and the ear-opening; forehead concave; ear-opening oval, oblique. Body short, moderately depressed. Limbs moderate; digits rather long and slender. Snout covered with large keeled granules; hinder part of head with minute granules intermixed with round tubercles; rostral square, twice as broad as high, with median cleft above; nostril pierced between the rostral, the first labial, and three or four nasals; seven or eight upper and six or seven lower labials; mental large, broadly pentagonal; two large chin-shields forming a suture behind the mental. Body covered above with small keeled granules, and large subtriangular strongly keeled tubercles; the latter are placed close together in twelve or fourteen very regular longitudinal series. Abdominal scales moderately large, cycloid, imbricate, smooth. Male with five præanal pores. Tail cylindrical, tapering, verticillate, above with keeled scales and tubercles, inferiorly with a median series of transversely enlarged plates. Light brown above; seven dark transverse bands, broader than the interspaces between them, from nape to base of tail; a U-shaped dark streak from eye to eye, over the nape; lower surfaces dirty white.

Total length	80	millim.
Head	10·5	,,
Width of head	6	,,
Body	30·5	,,
Fore limb	11·5	,,
Hind limb	16	,,
Tail	39	,,

Western Australia.

a. ♂.	Houtman's Abrolhos.	Mr. Gilbert [C.]. (Type.)
b. ♂.	Champion Bay.	Mr. Duboulay [C.].

2. Heteronota derbiana.

Eublepharis derbianus, *Gray, Cat.* p. 274.
Hoplodactylus (Pentadactylus) australis, *Steindachn. Novara, Rept.* p. 18, pl. i. fig. 2.
Phyllodactylus anomalus, *Peters, Mon. Berl. Ac.* 1867, p. 14.
Heteronota binoei, part., *Günth. Ann. & Mag. N. H.* (3) xx. 1867, p. 50.

Differs from the preceding as follows:—Head not three times in the distance from tip of snout to vent. Dorsal tubercles generally less regularly arranged, more distant from one another, those of one series being separated by interspaces at least equal to their length. Præanal pores four or six. Light brown above, spotted or marbled with darker; some of the tubercles sometimes whitish; a dark temporal streak; lower surfaces whitish.

Total length	102	millim.
Head	14	,,
Width of head	9	,,

Body	32 millim.
Fore limb	15 ,,
Hind limb	22 ,,
Tail	56 ,,

Northern and Eastern Australia.

a. ♀.	Port Essington.	Lord Derby [P.]. (Type.)
b-c. ♀.	N. Australia.	J. Elsey, Esq. [P.].
d. ♂.	Peak Downs.	Godeffroy Museum.
e. ♀.	Rockhampton.	
f. ♂.	Queensland.	
g. ♂.	Australia.	Sir J. Richardson [P.].

3. Heteronota? eboracensis.

Heteronota eboracensis, *Macleay, Proc. Linn. Soc. N. S. W.* ii. 1877, p. 101.

" Form rather robust; head broad; tail much shorter than the body; internasal shields rather short and not contiguous; labials nine upper and eight lower, the last two of each very small; scales mostly tricarinate; tubercles on the back numerous and nearly smooth; scales on the tail all smooth; colour, above speckled grey and black; legs and under surface pale with numerous minute black dots. Length 3 inches."

Cape York.

16. PHYLLODACTYLUS *.

Phyllodactylus, *Gray, Spicil. Zool.* 1830, p. 3; *Fitz. Syst. Rept.* p. 94; *Gray, Cat.* p. 150.
Sphærodactylus, part., *Wagler, Syst. Amph.* p. 143.
Phyllodactylus, part., *Wiegm. Herp. Mex.* p. 20; *Dum. & Bibr.* iii. p. 388.
Euleptes, *Fitz. l. c.* p. 95.
Discodactylus, *Fitz. l. c.* p. 95.
Diplodactylus, part., *Wiegm. l.c.*; *Fitz. l. c.* p. 94; *Gray, Cat. Liz.* p. 148.
Parœdura, *Günth. Ann. & Mag. N. H.* (5) iii. 1879, p. 218.

Digits more or less slender, free, all clawed, with transverse lamellæ or tubercles inferiorly, the extremity more or less dilated, with two large plates inferiorly, separated by a longitudinal groove in which the claw is retractile; the distal expansion covered above with scales strongly differentiated from those on the basal portion of the digit. Upper surfaces covered with juxtaposed scales, uniform or intermixed with larger tubercles; abdominal scales generally imbricate. Pupil vertical. Males without præanal nor femoral pores.

Tropical America; Australia; Africa; islands of the Mediterranean.

---

* *Phyllodactylus androyensis*, Grandid. Rev. Mag. Zool. xix. 1867, p. 233.— S.W. Madagascar.

16. PHYLLODACTYLUS.

Synopsis of the Species.

I. Back with unequal lepidosis.
   A. Skin of crown free from cranial ossification.
      1. Dorsal tubercles keeled.
         a. Digital expansion much wider than the rest of the digit, trapezoid.

Dorsal tubercles in 14 more or less irregular longitudinal series; ventral scales in 25 to 30 longitudinal and 65 transverse series .... 1. *tuberculosus*, p. 79.
Dorsal tubercles in 14 longitudinal series; ventral scales in 18 to 20 longitudinal, and 45 to 50 transverse series ...... 2. *ventralis*, p. 80.
Hinder part of head with uniform minute granular scales .... 3. *reissii*, p. 80.
Dorsal tubercles in 20 to 22 longitudinal series; ventral scales in 22 longitudinal and 53 transverse series ...... 4. *pulcher*, p. 80.
Dorsal tubercles in 20 longitudinal series; ventral scales in about 25 longitudinal series; labials $\frac{6}{5}$ .... 5. *spatulatus*, p. 81.

   b. Digital expansion not much wider than the rest of the digit, rounded.

Dorsal tubercles small, in 12 longitudinal series .... 6. *galapagoensis*, p. 82.
Dorsal tubercles large, in 16 longitudinal series; digits slender, the digital expansions very small .... 10. *phacophorus*, p. 84.

      2. Dorsal tubercles not keeled.
         a. Digital expansion well developed.

Dorsal tubercles very distinct from the granular scales; mental very large .. 7. *nigrofasciatus*, p. 82.
Dorsal scales of unequal sizes, the tubercles differing only in size from the other dorsal scales .... 8. *inæqualis*, p. 83.
Dorsal tubercles very distinct from the granular scales; two elongate chin-shields; head large; limbs long .... 11. *oviceps*, p. 85.

         b. Digital expansion very small.

Dorsal tubercles a little smaller than the ventral scales.................... 9. *microphyllus*, p. 84.

   B. Skin of head confluent with cranial ossification.

Infradigital lamellæ divided medially .. 12. *sancti-johannis*, p. 86.
Infradigital lamellæ entire .... 13. *stumpffi*, p. 86.

II. Back with uniform lepidosis.
 A. Digital expansions well developed.
  1. Dorsal scales not larger than those on the snout.
   a. Dorsal scales not keeled.
    α. Digital expansion considerably wider than the digit.
     * Mental not twice as broad as long.

Rostral four-sided, entering considerably the nostril; first labial not entering the nostril .... 14. *porphyreus*, p. 87.
Rostral pentagonal or hexagonal, the latero-superior border entering the nostril; first labial entering the nostril; digital expansion much smaller than the eye; 7 or 8 entire lamellæ under the fourth toe .... 15. *marmoratus*, p. 88.
Digital expansion large, that of the fourth toe measuring nearly the diameter of the eye; ear-opening minute ...... 16. *macrodactylus*, p. 89.
Rostral separated from the nostril; 10 entire lamellæ under the fourth toe .. 17. *affinis*, p. 89.
10 or 11 entire lamellæ under the fourth toe; ear-opening large, oval, measuring two thirds or three fourths the diameter of the eye .... 18. *guentheri*, p. 90.

 ** Mental twice as broad as long.

Scales on the snout not larger than those on the back .... 19. *europæus*, p. 90.

  β. Digital expansion scarcely wider than the digit.

The scales on the upper surface of the head much larger than the granules on the back .... 20. *pictus*, p. 91.
Dorsal scales flat, as large as those on the snout; two median larger chin-shields .... 21. *lineatus*, p. 92.

   b. Dorsal scales feebly keeled.

The scales on the back about as large as those on the snout................ 22. *ocellatus*, p. 93.

  2. Dorsal scales larger than those on the snout.

7 upper and 6 lower labials .... 23. *unctus*, p. 94.
11 upper and 10 lower labials .... 24. *riebeckii*, p. 94.

 B. Digital expansion extremely small.
Digits slender .... 25. *gerrhopygus*, 95.

1. Phyllodactylus tuberculosus.

Phyllodactylus tuberculatus, *Gray, Cat.* p. 150.
Phyllodactylus tuberculosus, *Wiegm. Nova Acta Ac. Leop.-Carol.* xvii. p. 241, pl. xviii. fig. 2; *Dum. & Bibr.* iii. p. 396; *Baird, U.S. Mex. Bound. Surv.* ii. *Herp.* pl. xxiii. figs. 1–8; *Bocourt, Miss. Sc. Mex., Rept.* p. 43, pl. x. fig. 3; *O'Shaughn. Ann. & Mag. N. H.* (4) xvi. 1875, p. 262.
Discodactylus tuberculosus, *Fitzing. Syst. Rept.* p. 96.
Phyllodactylus xanti, *Cope, Proc. Ac. Philad.* 1863, p. 102.

Head oviform, much longer than broad; snout rounded, longer than the distance between the eye and the ear-opening, once and one third the diameter of the orbit; forehead concave; ear-opening rather large, narrow, oblique. Body and limbs moderate. Digits slender, the distal dilatation large, truncate, its diameter one half or two fifths that of the eye; the slender part of the digit with narrow transverse lamellæ, about ten under the fourth toe, the distal one divided. Snout covered with equal granules; hinder part of head and temples with minute granules intermixed with large ones; the scales on the border of the eyelid forming a rather strong denticulation. Rostral four-sided, twice as broad as high, with trace of cleft above; nostril pierced between the rostral, the first labial, and three or four nasals, the anterior, or upper, large and in contact with its fellow; six or seven upper and as many lower labials; mental large, pentagonal, in contact with two chin-shields, which are followed by extremely small polygonal shields passing rapidly into the minute granules of the gular region. Back and limbs above with small unequal granules intermixed with large trihedral keeled tubercles; on the back these tubercles are arranged in seven more or less irregular longitudinal series on each side of the vertebral line; these tubercles somewhat variable in size, sometimes nearly flat. Lower surfaces with small imbricate smooth scales, in 25 to 30 longitudinal series, and about 65 transverse series between constriction of neck and vent. Tail cylindrical, tapering, covered with small imbricate smooth scales; larger, pointed, feebly keeled scales forming verticils on the upper part of the tail; a median series of transversely dilated scales inferiorly. Light grey-brown above, with dark-brown spots; a dark-brown streak on the side of the head, passing through the eye; tail with dark annuli; lower surfaces white.

Total length	119	millim.
Head	18	,,
Width of head	13	,,
Body	47	,,
Fore limb	21	,,
Hind limb	27	,,
Tail	54	,,

California; Central America.

a. Ad.	California.	Capt. Belcher [P.].
b–e. Ad. & hgr.	California.	J. O. W. Fabert, Esq. [P.].
f–h. Ad. & hgr.	Ventanas, Mexico.	Hr. A. Forrer [C.].
i–l, m. Ad. & hgr.	Presidio, W. Mexico.	Hr. A. Forrer [C.].
n–o. Ad.	Tres Marias.	Hr. A. Forrer [C.].
p–q. Ad.	Rio Montagua.	O. Salvin, Esq. [C.].
r. Ad.	San Geronimo.	O. Salvin, Esq. [C.].
s. Ad.	——?	

2. Phyllodactylus ventralis.

Phyllodactylus ventralis, *O'Shaughn. Ann. & Mag. N. H.* (4) xvi. 1875, p. 263.

This species differs from *P. tuberculosus* in the following points:— Head narrower. Ventral shields much larger, in 18 to 20 longitudinal and 45 to 50 transverse rows. Mental narrower, and more produced posteriorly; chin-shields on second row larger.

Total length	102	millim.
Head	15	,,
Width of head	9	,,
Body	37	,,
Fore limb	16	,,
Hind limb	20	,,
Tail	50	,,

Jamaica?; Nicaragua.

a. Ad.	Jamaica (?).	(Type.)
b. Ad.	Granada.	

3. Phyllodactylus reissii.

Phyllodactylus reissii, *Peters, Mon. Berl. Ac.* 1862, p. 626.

Agrees in almost every point with *P. tuberculosus*. The differences are that the scales on the hinder part of the head are uniform minute granules, and that the mental is almost entirely between the front pair of infralabials, which are hardly smaller than that plate.

Guayaquil; Peru.

4. Phyllodactylus pulcher.

Phyllodactylus pulcher, *Gray, Cat.* p. 150.
Gecko triedrus (non *Daud.*), *Wolf, Abbild. u. Beschr. merkw. nat. Gegenst.* ii. p. 73, pl. xx. fig. 1.
Phyllodactylus pulcher, *Gray, Spicil. Zool.* i. p. 3, pl. iii. fig. 1; *Dum. & Bibr.* iii. p. 397.
Discodactylus pulcher, *Fitzing. Syst. Rept.* p. 95.

Head oval, much longer than broad; snout rounded, longer than the distance between the eye and the ear-opening, once and a half the diameter of the orbit; forehead slightly concave; ear-opening rather large, narrow, oblique. Body and limbs moderate. Digits slender, the distal dilatation large, triangular, the front margin being straight; the diameter of the dilatation of the fourth toe equals two thirds the diameter of the eye; the slender portion of the digits with transverse lamellæ inferiorly, about ten under the fourth toe, the distal one divided. Snout covered with equal large granules; hinder part of head and temples with minute granules intermixed with large ones; the scales on the border of the eyelid forming a rather strong denticulation. Rostral four-sided, twice as broad as high, with trace of cleft above; nostril pierced between the rostral, the first labial, and three nasals, the anterior, or upper, large and in contact with its fellow; seven upper and seven lower labials; mental large, pentagonal, in contact with two chin-shields, which are followed by others passing gradually into the granules of the gular region. Back and limbs above with minute granules intermixed with large trihedral keeled tubercles; on the back these tubercles are arranged into pretty regular longitudinal series, ten or eleven on each side of the vertebral line; on the flanks these tubercles are so close to each other as nearly to overlap. Lower surfaces with moderate-sized imbricate smooth scales, in 22 longitudinal series, and 53 transverse series between constriction of neck and vent. Tail with small imbricate smooth scales intermixed with larger slightly keeled ones forming rather indistinct verticils; inferiorly with a median series of transversely dilated scales.

Head	18	millim.
Width of head	12	,,
Body	48	,,
Fore limb	23	,,
Hind limb	28	,,

Tropical America.

a. Ad.　　　　　　Tropical America.　　　　　　　　(Type.)

5. Phyllodactylus spatulatus.

Phyllodactylus spatulatus, *Cope, Proc. Ac. Philad.* 1862, p. 176.

"Muzzle elongate, depressed, extending anterior to the orbit once and one third the diameter of the latter. Frontal and nasal regions closely squamulose tuberculous, each tubercle as large as those that are scattered upon the occiput. Superior labials six, the last minute; inferior labials five. Symphyseal elongate campanuliform in outline, succeeded by three or four transverse series of mental plates. The anterior is composed of three (median smallest), which are much longer than broad; the posterior are hexagonal. About twenty-five rows of abdominal plates, and twenty rows of elongate trihedral dorsal tubercles. Extremities coarsely tuber-

culous. Length of head to angle of mandible 8 lines; from this point to vent 1 inch 9 lines; of hinder extremity 1 inch; tail? Above pale yellowish; a dark brown line from orbit to shoulder; dark brown longitudinal lines, which inosculate on the nape and anterior dorsal region; on the posterior dorsal and sacral they form cross bands. Extremities banded. Beneath immaculate."
Barbadoes.

6. Phyllodactylus galapagoensis.

Phyllodactylus galapagensis, *Peters, Mon. Berl. Ac.* 1869, p. 720.

Head oval, much longer than broad; snout rounded, a little longer than the distance between the eye and the ear-opening, once and one third the diameter of the ear-opening; forehead slightly concave; ear-opening narrow, vertical. Body and limbs moderate. Basal part of the digits moderately depressed, with transverse lamellæ inferiorly; digital expansions not much wider, rounded. Snout covered with large granular scales; hinder part of head with minute granules, among which larger ones are scattered. Rostral subpentagonal, twice as broad as high, with median cleft above; nostril pierced between the rostral, the first labial, and three nasals; seven upper and six lower labials; mental large, pentagonal, in contact with two chin-shields, which are followed by smaller ones passing gradually into the minute granules of the throat. Back covered with minute granules; small keeled tubercles form six very regular longitudinal series on each side of the vertebral line, this arrangement resembling that seen in *Gymnodactylus pelagicus*. Belly with small imbricate smooth scales. Tail cylindrical, tapering, covered with smooth imbricate scales, larger inferiorly but not forming a regular median series. Grey-brown above, with darker markings; a rather indistinct dark streak on the side of the head, passing through the eye; tail with dark annuli above; lower surfaces brownish white.

Total length	74	millim.
Head	10	,,
Width of head	6	,,
Body	24	,,
Fore limb	11	,,
Hind limb	13	,,
Tail	40	,,

Galapagos Islands.

a–b. Ad.? Charles Island. Commander Cookson [P.].

7. Phyllodactylus nigrofasciatus.

Phyllodactylus nigrofasciatus, *Cope, Proc. Amer. Philos. Soc.* xvii. 1877, p. 36.

"The large dermal tubercles are not prominent nor angular, nor arranged in regular longitudinal rows. They are round, and very distinct from the small round scales between them, and not almost assimilated to them as in *P. inæqualis*. There are eight superior labials to below the pupil of the eye. The mental scutum is very large and urceolate; it has two lateral and a short posterior median facet, each one corresponding to a scutum; the anterior of these is the first labial, which is about twice as large as the scutum that follows it. Behind these is a transverse row of five subround scales, of which the median is in contact with the mental. The next row embraces eight, arranged in an undulating manner. The scales diminish but slowly to the size of the gulars. The toes are slender as in *P. microphyllus*, but the expansions are large as in *P. inæqualis*. When the limbs are appressed to the side, the elbow reaches the base of the toes in this species, but only to their tips in *P. inæqualis*; the length of the toes in *P. microphyllus* is intermediate. The ground-colour is very light, brilliantly white on the inferior surfaces. Between the axilla and groin the back is crossed above by six narrow black cross bands. These bifurcate or break up on the sides; the axillar band breaks up on the back, and two anterior to it are represented by spots. A broad dark band passes from the nostril through the eye and breaks up on the sides of the neck. Limbs indistinctly cross-barred."

Length to vent	43 millim.
Length to ear..........	13 ,,
Width of head at ears ..	7 ,,
Length of fore limb	14 ,,
Length of hind limb	21 ,,

Chimbote Valley, Peru.

8. Phyllodactylus inæqualis.

Phyllodactylus inæqualis, *Cope, Journ. Ac. Philad.* (2) viii. 1876, p. 174.

" Scales of back and sides subequal, the former of unequal sizes but without elevated or keeled tubercles. Ventral scales larger, subround, smooth, those bordering the vent in front smaller. Muzzle with convex scales larger than those on the occiput. Labials to below the pupil, six above; below five, followed by three others; the inferior first three are the larger. Mental scutum longer than wide, angulate behind, with an oval scute on each side of the angle, which meet at an angle, each on the middle line. Behind these are round scales from which others graduate into the granules of the throat. Meatus auditorius a very small slit. Eye contained nearly twice in length of muzzle. The fore limbs extended reach the front of the orbit; the hind limbs extend to the appressed elbow. Scales of the normal tail square and flat above, rounded and a little larger below. Above yellowish, with seven blackish cross bands from nape to groin, somewhat connected by oblique and longitudinal lines on the sides. A dark band from nostril, through eye, to shoulder. Limbs and tail cross-banded. Head with coarse blackish reticulation above. A brown spot on each labial. Below straw-colour, immaculate."

Length to vent	38	millim.
Length to ear	10	,,
Length of hind limb	70	,,
Width of head	60	,,

Pacasmayo, Peru.

9. Phyllodactylus microphyllus.

Phyllodactylus microphyllus, *Cope, Journ. Ac. Philad.* (2) viii. 1876, p. 175.

"Small scales of the back and sides with larger ones scattered irregularly among them; these are not very much larger, not keeled, but smoothly convex, and a little smaller than the smooth flat belly-scales. Gular scales granular; labials to pupil $\frac{7}{7}$. Mental large, convex behind, bounded by four round scales of small size. Scales of top of muzzle twice as large as those of vertex. Tail-scales uniform. Transverse series of inferior side of the digits rather short; large plates of the end of the toes remarkably small, permitting the ungual phalange to project very freely. A row of prominent scales behind the thighs on each side of the base of the tail. Colour very pale, with a few very indistinct transverse shades; in the young these shades are cross bars."

Total length	101	millim.
Length to vent	48	,,
Length to ear	12	,,
Width of head	10	,,
Length of hind limb	20	,,

Valley of Jequetepeque, Peru.

10. Phyllodactylus phacophorus.

Discodactylus phacophorus, *Tschudi, Faun. Per., Herp.* p. 38; *Girard, U.S. Explor. Exped., Herp.* p. 300, pl. xxv. figs. 25–32.

Head elongate, oviform; snout as long as the distance between the eye and the ear-opening, once and one third the diameter of the orbit; forehead slightly concave; ear-opening narrow, vertical. Body and limbs moderate. Digits slender, scarcely depressed, with a series of transverse lamellæ inferiorly, thirteen to fifteen under the fourth toe; digital expansions very small, longitudinally suboval, scarcely wider than the slender digits. Snout covered with small polygonal convex scales; hinder part of head with minute granules intermixed with round tubercles. Rostral four-sided, not twice as broad as high, with median cleft above; nostril pierced between the rostral, the first labial, and three nasals, the anterior largest and in contact with its fellow; seven upper and as many lower labials; mental large, bell-shaped, much longer than broad; very small irregular chin-shields. Back covered with very small granules intermixed with large suboval, trihedral, strongly keeled tubercles arranged in sixteen longitudinal series. Limbs with imbricate scales and large tubercles. Abdominal scales rather large, roundish, smooth, imbricate. Tail cylindrical, tapering, with small imbricated smooth scales, larger inferiorly, above with rings of large, pointed, keeled tubercles. Brown above; a dark brown, whitish-edged streak on the side of the head, passing through the eye; back with broad transverse dark-brown bands separated by series of white tubercles; lower surfaces whitish.

Total length	99	millim.
Head	14	,,
Width of head	9	,,
Body	37	,,
Fore limb	17	,,
Hind limb	22	,,
Tail	48	,,

Peru.

a. Ad.	Lima.	Prof. W. Nation [P.].
b. Ad.	Lima.	J. M. Cowper, Esq. [P.].

11. Phyllodactylus oviceps.

Phyllodactylus oviceps, *Boettger, Zool. Anz.* 1881, p. 359, *and Abh. Senck. Ges.* xii. 1882, p. 475, pl. iii. fig. 10.

Head rather large, oviform; snout acuminate; forehead concave; eyes large, prominent. Body moderate. Limbs long, slender. Digits long, slender, with transverse lamellæ inferiorly; digital expansions moderate, trapezoid. Rostral trapezoid; fourteen upper and twelve lower labials; mental triangular, in contact with two elongate chin-shields, on each side of which are two smaller ones. Back covered with granules intermixed with roundish oval tubercles arranged irregularly. Belly with rather small roundish smooth scales. Tail tapering, covered with small scales and annuli of small spinose tubercles. Grey above, variegated with black; back with four **Λ**- or **V**-shaped markings interrupted on the median line; tail with black annuli.

Total length	76	millim.
Head	15·5	,,
Width of head	9	,,
Body	23·5	,,
Fore limb	17	,,
Hind limb	23	,,
Tail	37	,,

Nossi Bé, Madagascar.

12. Phyllodactylus sancti-johannis. (Plate VII. fig. 1.)

Parœdura sancti johannis, *Günth. Ann. & Mag. Nat. Hist.* (5) iii. 1879, p. 218.

Head large, oviform, very distinct from neck; the skin confluent with the cranial ossification; forehead with a deep longitudinal concavity; eye large, very prominent; snout longer than the distance between the eye and the ear-opening, once and one third the diameter of the orbit; ear-opening narrow, vertical, half the diameter of the eye. Body moderate. Limbs very long and slender. Digits slender, depressed at the base, with two rows of lamellæ inferiorly, which are broken up into small tubercles some distance before the distal expansion; latter subcordiform, moderately large, measuring about one third the diameter of the eye. Head covered with polygonal scales, elongate and feebly keeled on the snout and interorbital space; temple with a row of four larger keeled tubercles. Rostral four-sided, twice as broad as high; nostril pierced between the rostral, the first labial, and five nasals; ten upper and as many lower labials; mental small, triangular, in contact with two large elongate hexagonal chin-shields; a smaller outer pair of similar shields and a second row of small polygonal chin-shields. Upper surface of body with irregular small flat scales intermixed with larger, oval, keeled tubercles; these are arranged in about eight irregular series along the back. Abdominal scales smooth, roundish, subimbricate. Limbs above with rather large, imbricate, keeled scales. Tail cylindrical, tapering, rather swollen in its anterior half, covered with irregular smooth scales intermixed with large keeled tubercles arranged in transverse series on the dorsal side; towards the end of the tail all the tubercles are spine-like, closely set. Brownish above; a black spot on the occiput; back with three broad transverse dark bands; tail with dark annuli above; lower surfaces dirty white.

Total length	117	millim.
Head	19	,,
Width of head	13	,,
Body	48	,,
Fore limb	24	,,
Hind limb	33	,,
Tail	50	,,

Comoro Islands.

a. Ad. Johanna. C. E. Bewsher, Esq. [C.]. (Type.)

13. Phyllodactylus stumpffi.

Phyllodactylus stumpffi, *Boettger, Ber. Senck. Ges.* 1878–79, p. 85, *and Abh. Senck. Ges.* xi. 1878, p. 18, and xii. 1882, p. 472, pl. ii. fig. 9.

Head large, oviform, very distinct from neck; the skin confluent with the cranial ossification; upper surface of head concave, the borders forming a ridge; eyes large, very prominent. Body moderate; limbs rather long. Digits moderate, with a series of transverse lamellæ inferiorly; the digital expansions well developed, roundish subtrapezoid. Upper surface of head covered with rather elongate slightly keeled scales; the scales on the supraorbital region roundish, flat. Rostral trapezoid; rostral and first labial touching the nostril; twelve upper and as many lower labials; mental triangular, a little broader than long, in contact with two much elongate hexagonal chin-shields; two other chin-shields on each side, and also several smaller ones behind, central largest. Upper surface of head and limbs covered with small granules intermixed with large trihedral tubercles, becoming almost spine-like on the hinder part of the back; these tubercles are arranged in six regular longitudinal series on the back, and there are besides many more, but irregularly disposed, down the sides; abdominal scales rather large, roundish, smooth. Tail cylindrical, tapering, with rings of large spinose tubercles. Dark grey above; a ⊔-shaped white line enclosing the hinder part of the head; a light vertebral line, black-edged and crossed by four large black-edged light bands; tail with dark annuli; dirty white beneath; a violet-grey spot in front of the vent.

Total length .... 143 millim.
Head .... 23·5 ,,
Width of head .... 17 ,,
Body .... 46·5 ,,
Tail................. 73 ,,

Nossi Bé, Madagascar.

14. Phyllodactylus porphyreus. (Plate VII. fig. 5.*)

Gecko porphyreus, *Daud. Rept.* iv. p. 130; *Merrem, Tent.* p. 43.
Sphærodactylus porphyreus, *Wagl. Syst. Amph.* p. 143.
Phyllodactylus porphyreus, part., *Dum. & Bibr.* iii. p. 393.

Head oviform, much longer than broad; snout rounded, a little longer than the distance between the eye and the ear-opening, once and two fifths the diameter of the orbit; forehead very slightly concave; ear-opening small, roundish or oval, its diameter one third to one half that of the eye. Body rather elongate; limbs moderate. Digits not much depressed; digital expansions moderate, rounded, subtrapezoid; the diameter of the disk of the fourth toe equals two thirds the diameter of the eye; the slender part of the digit with regular transverse lamellæ inferiorly, which are broken up into small scales a short distance in advance of the distal expansion; seven or eight entire lamellæ under the fourth toe. Upper surfaces covered with uniform small granules, largest on the snout, smallest on the hinder part of the head. Rostral four-sided, twice as broad

---

* End of snout, × 4.

as high, without cleft above; the nostril is pierced just above the suture of the rostral and first labial, between the former and three small nasals, the rostral entering considerably and the first labial being separated by the narrow infraposterior nasal; eight or nine upper and as many lower labials; mental trapezoid or pentagonal, not larger than the adjacent labials; no regular chin-shields, but small polygonal scales, passing gradually into the minute granules of the gular region. Abdominal scales moderate, smooth, subhexagonal, slightly imbricate. Tail long, cylindrical, tapering in its posterior half, covered with uniform small smooth scales, rather larger inferiorly, arranged in rings. Greyish or reddish brown above, variegated with dark brown; whitish inferiorly.

Total length	97	millim.
Head	12	,,
Width of head	9	,,
Body	35	,,
Fore limb	15	,,
Hind limb	18	,,
Tail	50	,,

South Africa; Madagascar.

a. Ad.	Damara Land.	
b–d. Ad. & yg.	Cape of Good Hope.	Dr. Jones [P.].
e–f. Ad.	Cape of Good Hope.	Prof. Busk [P.].
g. Ad.	Cape of Good Hope.	Voy. of the 'Herald.'
h–k. Ad.	Cape of Good Hope.	Rev. C. Searle [P.].
l. Ad.	Cape of Good Hope.	
m. Ad.	Cape Town.	Rev. G. H. R. Fisk [P.].
n. Ad.	Madagascar.	

15. Phyllodactylus marmoratus. (PLATE VII. fig. 6.*)

Diplodactylus marmoratus, *Gray, Cat.* p. 149.
Diplodactylus marmoratus, *Gray, Zool. Erebus & Terror*, pl. xv. fig. 6.
Phyllodactylus porphyreus, part., *Dum. & Bibr.* iii. p. 393.
—— peronii, *Fitz. Syst. Rept.* p. 95.

The rostral is pentagonal or hexagonal, the posterior angle being truncate, the latero-superior angles touching the nostril; the latter is pierced posteriorly to the suture of the rostral and first labial, and between the latter and three nasals. This is the only constant difference I can detect between this form and the preceding; but it appears to me of sufficient importance to separate specifically the Australian from the African form. I might add that the scales are generally rather smaller and the mental broader in *S. marmoratus* than in *S. porphyreus*, but these differences are not absolutely constant.

---

* End of snout, × 4.

Total length	119 millim.
Head	13 ,,
Width of head	10 ,,
Body	38 ,,
Fore limb	17 ,,
Hind limb	21 ,,
Tail	68 ,,

Australia.

a. Ad.	Australia.	(Type.)
b. Ad.	Australia.	G. Krefft, Esq.
c–d, e. Ad.	Australia.	J. S. Bowerbank, Esq.
f, g. Ad.	Freemantle, N.Australia.	J. S. Bowerbank, Esq.
h–n. Ad.	W. Australia.	
o–r. Ad. & yg.	Houtman's Abrolhos.	Mr. Gilbert [C.]. (Types.)
s–t. Ad.	Swan River.	
u. Ad.	Kangaroo Island.	

16. Phyllodactylus macrodactylus. (PLATE VII. fig. 2.)

Distinguished from the preceding in the thicker digits with much larger discoidal dilatations; that of the fourth toe measures nearly the diameter of the eye. The ear-opening is smaller. Upper parts light reddish brown with darker variegations.

Total length	116 millim.
Head	13 ,,
Width of head	10 ,,
Body	44 ,,
Fore limb	17 ,,
Hind limb	23 ,,
Tail	59 ,,

Australia.

a–b. Ad.	——?	G. Krefft.

17. Phyllodactylus affinis. (PLATE VII. fig. 4.*)

Differs from *P. marmoratus* in the following points:—Rostral separated from the nostril by the anterior nasal. Digits longer; as many as ten entire lamellæ under the fourth toe. Dorsal granules larger, flatter. Grey above, marbled with blackish.

Total length (tail mutilated)	70 millim.
Head	14 ,,
Width of head	10 ,,
Body	43 ,,
Fore limb	19 ,,
Hind limb	23 ,,

New Hebrides.

a. Ad.	Aneiteum.	J. MacGillivray, Esq. [C.].

---

* End of snout and lower surface of fourth toe, × 4.

18. Phyllodactylus guentheri. (PLATE VII. fig. 3.)

Head oviform, much longer than broad; snout rounded, as long as the distance between the eye and the ear-opening, once and a half the diameter of the orbit; forehead slightly concave; ear-opening large, oval, oblique, measuring two thirds or three fourths the diameter of the eye. Body and limbs moderate. Digits rather elongate; the claw more raised above the digit and more exposed than in the other species of the genus, resembling the structure in *Gecko*; the slender part of the digits slightly dilated, with regular transverse lamellæ inferiorly, which are broken up into small scales some distance in front of the distal expansion; ten or eleven entire lamellæ under the fourth toe; digital expansion small, roundish, measuring about three fifths the diameter of the eye. Upper surfaces covered with uniform granular scales, extremely minute on the back, larger on the snout and limbs. Rostral four-sided, twice as broad as high, without cleft above; nostril pierced just above the suture of the rostral and first labial, between both the latter and four small nasals; seven or eight upper and as many lower labials; mental trapezoid, not larger than the adjacent labials; no regular chin-shields, but small polygonal scales passing gradually into the minute granules of the gular region. Abdominal scales small, smooth, roundish hexagonal, slightly imbricate. Tail cylindrical, tapering, covered with uniform small smooth scales, a little larger inferiorly. Light reddish brown above; rather indistinct transverse dark markings on the back, interrupted on the vertebral line, which is sometimes lighter; lower surfaces brownish white, speckled with brown.

Total length (tail mutilated)	110	millim.
Head	23	,,
Width of head	18	,,
Body	66	,,
Fore limb	31	,,
Hind limb	38	,,

Australia.

a–c. Ad.	Norfolk Island.	F. M. Rayner, Esq. [C.].
d. Hgr.	Lord Howe's Island.	Voy. of the 'Herald.'
e. Ad.	Champion Bay, N.W. coast of Australia.	Mr. Duboulay [C.].

19. Phyllodactylus europæus.

Phyllodactylus europæus, *Gené, Mem. Ac. Tor.* (2) 1839, i. p. 263, pl. i. fig. 1; *Bonap. Faun. Ital., and Amph. Eur.* p. 29; *A. Dum. Cat. Méth. Rept.* p. 41, *and Arch. Mus.* viii. p. 455; *De Betta, Faun. Ital., Rett. Anf.* p. 21; *Schreiber, Herp. Eur.* p. 485; *Lataste,*

Bull. Soc. Zool. France, 1877, p. 467; *Camerano, Atti Ac. Tor.* xiv. 1878, p. 219; *De Betta, Atti Ist. Venet.* (5) v. 1879, p. 380.
Phyllodactylus (Euleptes) wagleri, *Fitz. Syst. Rept.* p. 95.
—— doriæ, *Lataste, l. c.* p. 467.

Head oviform, much longer than broad; snout rounded, as long as the distance between the eye and the ear-opening, once and one third the diameter of the orbit; forehead concave; ear-opening small, round. Body short; limbs rather strong. Digits relatively much depressed in their basal part, inferiorly with transverse lamellæ, the greater number of which are generally broken up into small scales forming three longitudinal series; digital expansion moderately large, roundish, its diameter half or three fifths that of the eye. Upper surfaces covered with equal small smooth granules. Rostral pentagonal, the posterior angle frequently truncate; nostril pierced between the rostral, the first labial, and three nasals, the anterior largest; nine or ten upper and as many lower labials; mental broadly trapezoid; no regular chin-shields, but very small polygonal scales passing gradually into the minute granules of the throat. Abdominal scales small, smooth, imbricate. Tail cylindrical, slightly depressed, tapering, prehensile, covered with equal small squarish scales arranged in verticils. Grey-brown above, marbled with darker and dotted with lighter; a more or less distinctly marked dark streak on the side of the head, passing through the eye; lower surfaces whitish.

Total length	70	millim.
Head	11	,,
Width of head	8	,,
Body	29	,,
Fore limb	13	,,
Hind limb	17	,,
Tail	30	,,

Islands of the Mediterranean west of Italy.

a–f. Ad. & yg.   Tinetto.   Marquis G. Doria [P.].

20. Phyllodactylus pictus.

Diplodactylus pictus, *Peters, Mon. Berl. Ac.* 1854, p. 615.
Phyllodactylus pictus, *Peters, Reise n. Mossamb.* iii. p. 29, pl. v. fig. 1.

Head oviform, much longer than broad; snout rounded, as long as the diameter of the orbit; ear-opening narrow, vertical. Body short. Fingers and toes depressed in their basal part, with three series of scales inferiorly; digital expansion scarcely wider, roundish. Upper surface of head covered with small polygonal scales much larger than the dorsal granules. Rostral hexagonal, twice as broad as high; nostril pierced between several small scales, the anterior of which is the largest; nine upper labials; mental rather large, rounded; a row of four small chin-shields, followed by smaller ones passing gradually into the minute granules of the throat. Temples and upper surface of body, limbs, and tail covered with small granular scales intermixed with scattered, roundish, triangular, keeled tubercles, which form eighteen to twenty-two longitudinal series on the middle of the body. Lower surfaces covered with small imbricated smooth scales. Tail cylindrical, tapering. Upper surfaces with alternate yellowish-green and brown cross bands; lower surfaces dirty white.

Total length	62	millim.
Head	10	,,
Width of head	8	,,
Body	26	,,
Fore limb	15	,,
Hind limb	17	,,
Tail	26	,,

Western Madagascar.

21. Phyllodactylus lineatus.

Diplodactylus lineatus, *Gray, Cat.* p. 150.
Phyllodactylus lineatus, *Smith, Ill. S. Afr., Rept., App.* p. 6.

Head oviform, longer than broad, very convex; snout rounded, a little shorter than the distance between the eye and the ear-opening, slightly longer than the diameter of the orbit; forehead not concave; ear-opening small, roundish. Body short. Limbs moderate. Digits depressed in their basal part, with three series of scales inferiorly, the central series of which is slightly enlarged transversely; digital expansion rounded, scarcely wider than the basal part, measuring about half the diameter of the eye. Upper surface of head covered with small convex granules, largest on the snout; rostral subpentagonal, nearly twice as broad as high, with median cleft above; nostril pierced between the rostral, the first labial, and three small nasals, which are generally not distinguishable from the surrounding granules; six or seven upper and as many lower labials; mental pentagonal; a row of small chin-shields, the two median largest, and in contact with the mental; behind these, convex granules becoming gradually smaller. Upper surface of body and limbs covered with flat, smooth, subimbricate roundish scales, as large as the granules on the snout; abdominal scales larger, hexagonal, imbricate, smooth. Tail cylindrical, tapering, covered with uniform, strongly imbricate smooth scales. Upper surfaces greyish brown, generally with four or six dark-brown longitudinal lines on the back and tail, sometimes replaced by transverse undulated brown lines; head generally variegated with dark brown; a dark line passing through the eye; lower surfaces brownish white.

Total length	59 millim.
Head	8 ,,
Width of head	6 ,,
Body	22 ,,
Fore limb	9·5 ,,
Hind limb	12 ,,
Tail	29 ,,

South Africa.

a–b. Ad.	Cape of Good Hope.	Lord Derby [P.].	(Types.)
c–i. Ad.	Cape of Good Hope.	Sir A. Smith [P.].	

22. Phyllodactylus ocellatus.

Diplodactylus ocellatus, *Gray, Cat.* p. 149.
Diplodactylus bilineatus, *Gray, l. c.*
Diplodactylus ocellatus, *Gray, Zool. Erebus and Terror,* pl. xv. fig. 4; *Günth. Ann. & Mag. N. H.* (3) xx. 1867, p. 49.
—— bilineatus, *Gray, l. c.* pl. xv. fig. 3.

Head oviform, much longer than broad; snout rounded, slightly longer than the distance between the eye and the ear-opening, about once and a half the diameter of the orbit; forehead not concave; ear-opening small, roundish or suboval. Body and limbs moderate. Digits short, depressed; digital expansions moderately large, rounded, measuring about three fifths the diameter of the eye; three series of tubercles under the digits, central more dilated transversely. Head covered with granular scales, a little larger on the snout; rostral four-sided, twice as broad as high, with trace of median cleft above; nostril pierced between the rostral, the first labial, and four nasals, the anterior dilated transversely, the others very small; seven or eight upper and as many lower labials; mental small, trapezoid, narrower than the adjacent labials; no chin-shields. Back covered with uniform, slightly keeled, granular scales, about the size of the granules on the snout; abdominal scales moderate, roundish, smooth, imbricate. Tail cylindrical, prehensile, covered with irregular smooth scales arranged in rings. Grey-brown above, variegated with darker and with more or less distinct light ocelli, arranged in two rows on the back; sometimes a light streak on each side of the head and back; lower surfaces dirty white, more or less scantily brown-dotted.

Total length	53 millim.
Head	9 ,,
Width of head	6·5 ,,
Body	23 ,,
Fore limb	9 ,,
Hind limb	11 ,,
Tail	21 ,,

West Australia.

a. Ad.	Australia.	(Type).
b, c. Ad.	Houtman's Abrolhos.	Mr. Gilbert [C.]. (Types of *P. bilineatus*.)
d–f. Ad.	Champion Bay, N.W. coast of Australia.	Mr. Duboulay [C.].
g. Ad.	North-west coast of Australia.	Haslar Collection.

23. Phyllodactylus unctus.

Diplodactylus unctus, *Cope, Proc. Ac. Philad.* 1863, p. 102.

"Scales rounded, flat, in about thirty-six rows on the dorsal region; those of the muzzle smaller than dorsal, little larger than those on the occiput. Internasals in contact. Seven superior labials, including that under the pupil; six inferior, to the same point, the last two very small, the first larger than the second, extensively in contact with the first pair of mental plates, which bound the symphyseal posteriorly. Each of the former is bounded by three smaller, and these are succeeded by a few rows which diminish in size. Terminal disks with straight outlines. Tail without tubercles. Anal scales similar to the abdominal. Auricular opening as long as pupil. Premaxillary teeth five. Above grey, shining, with five broad, blackish, centrally-pale cross bands, from base of tail to interscapular region; a dark band from the muzzle through the eye, a cross band on occiput, and various irregular spots on the top of head and labial regions. Length from end of muzzle to auricular meatus 12''', from the same point to vent 4·5'''."

Cape St. Lucas, Lower California.

24. Phyllodactylus riebeckii.

Diplodactylus riebeckii, *Peters, Sitzb. Nat. Freunde Berlin*, 1882, p. 43.

Digital expansion with straight anterior border; seven transverse lamellæ under the middle toe. Ear-opening a crescent-shaped slit, the convexity postero-inferior. Head covered with fine granules much larger than the dorsal scales. Rostral four-sided, almost twice as broad as high; nostril pierced between the rostral, the first labial, and three nasals; eleven upper and ten lower labials; mental once and a half the width of the first infralabial, its posterior obtuse angle between two elongate pentagonal chin-shields, on each side of which there are three others rapidly decreasing in size. Dorsal scales flat, slightly larger than the ventrals. The fore limb reaches the end of the snout; the hind limb reaches the shoulder. Grey above, with blackish-brown spots arranged in cross bands on the back, in longitudinal lines on the nape; broad cross bands on the tail; yellowish grey inferiorly, the throat spotted and marbled with black.

Total length	290 millim.
Head	33 ,,
Width of head	28 ,,
Body	107 ,.
Fore limb	52 ,,
Hind limb	70 ,,
Tail	150 ,,

Socotra.

25. Phyllodactylus gerrhopygus.

Diplodactylus gerrhopygus, *Gray, Cat.* p. 150.
Diplodactylus gerrhopygus, *Wiegm. Nova Acta Ac. Leop.-Carol.* xvii. p. 242, pl. xviii. fig. 3; *Peters, Mon. Berl. Ac.* 1862, p. 627.
Phyllodactylus gymnopygus, *Dum. & Bibr.* iii. p. 394.
—— gerrhopygus, *Dum. & Bibr.* iii. p. 399.
Diplodactylus lepidopygus, *Tschudi, Faun. Per., Herp.* p. 38.

Head oviform; snout rounded-acuminate, longer than the distance between the eye and the ear-opening, once and one fourth the diameter of the orbit; forehead not concave; ear-opening small, oval, oblique. Body moderate. Limbs rather slender and elongate. Digits slender, with regular transverse lamellæ inferiorly (thirteen under the fourth toe); digital expansion very small, scarcely wider than the narrow basal part, truncate. Upper surfaces covered with uniform, rather large granules. Rostral six-sided, with longitudinal cleft above; nostril pierced between the rostral, the first labial, and three swollen nasals; nine upper and eight lower labials; mental four-sided, narrow, much elongate, extending considerably beyond the adjacent labials; no chin-shields. Gular region covered with minute granules, belly with moderate subhexagonal, subimbricate, smooth scales; a large semicircular plate in front of the vent. Tail cylindrical, tapering, rather slender, covered with equal squarish smooth scales, forming rings. Light grey above; head, sides, and limbs variegated with dark brown; five dark-brown bands across the back; tail with dark-brown annuli above; lower surfaces white.

Total length	94 millim.
Head	12 ,,
Width of head	8 ,,
Body	38 ,,
Fore limb	16 ,,
Hind limb	19 ,,
Tail	44 ,,

Peru; Chili.

a. Ad. Chili.

17. EBENAVIA.

Ebenavia, *Boettger, Abh. Senck. Ges.* xi. 1878, p. 276.

Differs from *Phyllodactylus* in having all the digits destitute of claws.

Madagascar.

1. Ebenavia inunguis.

Ebenavia inunguis, *Boettg. l. c.* pl. i. fig. 3.

Snout pointed. Body and limbs moderate. Digits with regular transverse lamellæ inferiorly and large truncate trapezoid digital expansion. Head covered with small granules; a series of conical tubercles from the orbit to the back; rostral narrow; ten upper and nine lower labials; mental triangular; no regular chin-shields. Upper surface of body covered with small granules intermixed with tubercles which are not much more than twice larger, trihedral, and somewhat pointed; they form eight to ten irregular series along the back. Abdominal scales small, granular. Tail cylindrical, with annuli of large spinose tubercles. Olive-brown above; a broad blackish-brown lateral band from the nostril through the eye to the base of the tail; the series of tubercles behind the eye yellowish white; lower surfaces greyish, darker-dotted.

Total length (tail injured)	53·5 millim.
Head	10·5 ,,
Width of head	6 ,,
Body	22·5 ,,

Nossi Bé.

2. Ebenavia boettgeri. (PLATE VIII. fig. 1.)

Head very long, pyramidal, depressed; snout pointed, much longer than the distance between the eye and the ear-opening, once and two thirds the diameter of the orbit; ear-opening small, oval, horizontal. Body elongate; limbs short. Digital expansions well developed, trapezoid, separated from the infradigital lamellæ by three rows of granules; lamellæ, ten under the fourth toe. Front part of head covered with small polygonal, elongate, tricarinate scales; hinder part with small granules; rostral four-sided, very low, four times as broad as high; nostril pierced between the first labial and four nasals, the anterior of which is very large and in contact with the rostral, the following smaller, and the two hinder ones not differentiated from the adjoining small scales; eight upper and nine lower labials; mental campanuliform, small, smaller than the adjacent labials; no chin-shields. Body covered above with small tricarinate granules intermixed with larger, roundish, slightly keeled tubercles; these tubercles are about twice as large as the scales on the snout, and form about twelve very irregular longitudinal series. Abdominal scales small, larger than the dorsal granules, hexagonal, juxtaposed, tricarinate. Tail cylindrical, reproduced in the unique specimen. Reddish brown above, with darker dots and four rather indistinct dark lines along the back; a dark band on each side of the head, passing through the eye; lower surfaces brownish grey, with scattered brown dots.

Total length	61 millim.
Head	9·5 ,,
Width of head	5·5 ,,
Body	24·5 ,,
Fore limb	9 ,,
Hind limb	12 ,,
Tail	27 ,,

Madagascar.

a. Ad.   Madagascar.

18. DIPLODACTYLUS *.

Diplodactylus, *Gray, Proc. Zool. Soc.* 1832, p. 40; *Fitz. Syst. Rept.* p. 94.
Phyllodactylus, part., *Dum. & Bibr.* iii. p. 388.
Strophurus, *Fitz. l. c.* p. 96.
Strophura, *Gray, Cat. Liz.* p. 148.
Diplodactylus, part., *Gray, l. c.*
Stenodactylopsis, *Steindachn. Sitz. Ak. Wien,* lxii. i. 1870, p. 343.

Digits free, not dilated at the base, slightly at the apex, all clawed, the claw retractile between two plates under the extremity of the digits; the basal portion of the digits inferiorly with transverse lamellæ or tubercles; the upper surface of the digits covered with uniform small tubercular scales. Upper surfaces covered with juxtaposed scales, uniform or intermixed with larger tubercles; abdominal scales juxtaposed. Pupil vertical.

Australia.

Synopsis of the Species.

I. Back covered with granular scales intermixed with larger tubercles; rostral divided; males with præanal pores.

Supraciliary border and tail with long
  spines .... 1. *ciliaris*, p. 98.

---

* *Diplodactylus annulatus,* Macleay, Proc. Linn. Soc. N. S.W. ii. 1877, p. 97.—Palm Islands.—Nothing is said of the structure of the digits, and, from the other characters given, I doubt whether the species has been referred to the proper genus.

Tail with long spines; no supraciliary spines	2. *spinigerus*,	p. 99.
Tail without spines	3. *strophurus*,	p. 100.

II. Back covered with uniform granular scales; males without præanal pores.

A. Digits with large transversely dilated tubercles inferiorly.

Snout as long as the distance between the eye and the ear-opening	4. *vittatus*,	p. 100.
Snout longer than the distance between the eye and the ear-opening ..	5. *polyophthalmus*,	p. 101.

B. Digits with small round tubercles inferiorly.

Dorsal scales small, granular; rostral and first labial entering the nostril.	6. *steindachneri*,	p. 102.
Dorsal scales small, granular; rostral and first labial not entering the nostril; snout acutely pointed	7. *pulcher*,	p. 102.
Dorsal scales large, flat; rostral and first labial entering the nostril	8. *tessellatus*,	p. 103.

1. Diplodactylus ciliaris. (Plate VIII. fig. 2.)

Head oviform, convex; snout rounded, longer than the distance between the eye and the ear-opening and than the orbit; eye large; ear-opening small, narrow, oblique. Body and limbs moderate. Digits much depressed, with large transverse lamellæ inferiorly, five under the fourth toe, the distal incised, cordiform; the plates under the apex of the digit large, together cordiform. Upper surfaces covered with rather large granules, intermixed on the back with enlarged conical tubercles, forming two irregular longitudinal series. Rostral subquadrangular, completely divided medially; nostril pierced between the rostral, the first labial, and three nasals; twelve upper and as many lower labials; mental trapezoid, scarcely larger than the adjacent labials; no chin-shields; supraciliary border with four spine-like scales, the second the largest, measuring half or two thirds the diameter of the eye. Lower surfaces covered with uniform granules, which are largest on the belly. Tail short, cyclo-tetragonal, prehensile (?), covered with granular scales; on each side of its upper surface a series of long curved spines. Yellowish brown above, indistinctly marbled with darker; some markings on the head and some of the enlarged tubercles and spines black, the other caudal spines reddish brown; lower surfaces yellowish, with brown dots.

Total length	125	millim.
Head	22	,,
Width of head	16	,,
Body	56	,,

Fore limb .... 28 millim.
Hind limb .... 35 ,,
Tail .... 47 ,,

A larger specimen, with injured tail, measures 85 millim. from snout to vent.

North Australia.

a–b. ♀.   Port Darwin.   R. G. S. Buckland, Esq. [C.].

2. Diplodactylus spinigerus.

Strophura spinigera, *Gray, Cat.* p. 148.
Diplodactylus spinigerus, *Gray, Zool. Misc.* p. 53.
Phyllodactylus spinigerus, *A. Dum. Cat. Méth. Rept.* p. 41, *and Arch. Mus.* viii. p. 467.
Strophura spinigera, *Gray, Zool. Erebus & Terror,* pl. xvi. fig. 5.

Head oviform, very convex; snout rounded, longer than the distance between the eye and the ear-opening, a little longer than the diameter of the orbit; eye large; ear-opening small, roundish. Body and limbs moderate. Digits much depressed, with large transverse lamellæ inferiorly, about seven under the fourth toe, the middle ones chevron-shaped, the distal one heart-shaped, the basal ones divided into two rounded plates; the plates under the apex of the digit large, together cordiform. Upper surfaces covered with minute granules, intermixed with few irregularly scattered conical spine-like tubercles. Rostral pentagonal, completely divided medially; nostril pierced between the rostral, the first labial, and three or four nasals; thirteen to fifteen upper and as many lower labials; mental small, trapezoid, not larger than the adjacent labials; no chin-shields. Lower surfaces covered with small juxtaposed granules, largest on the belly. Males with a doubly arched series of eleven to fourteen præanal pores, and three or four large conical tubercles on each side of the base of tail. Tail short, subcylindrical, prehensile, covered with small granules; on each side of its upper surface a series of long black spines. Upper surfaces olive-grey, speckled with black; the spinose tubercles black; sometimes a rather indistinct broad zigzag band along the back; lower surfaces dirty white, uniform or black-speckled.

Total length .... 114 millim.
Head .... 17 ,,
Width of head .... 12 ,,
Body .... 52 ,,
Fore limb .... 23 ,,
Hind limb .... 30 ,,
Tail .... 45 ,,

West and North Australia.

a, b. ♂ & hgr.	Houtman's Abrolhos.	Mr. Gilbert [C.]. (Types.)
c. ♀.	West Australia.	Mr. Duboulay [C.].
d. ♂.	Champion Bay, North-west Australia.	Mr. Duboulay [C.].
e–f. ♂ ♀.	Freemantle, North Australia.	J. S. Bowerbank, Esq.
g–m, n. ♂ ♀.	Australia.	J. S. Bowerbank, Esq.
o. ♀.	Australia.	

3. Diplodactylus strophurus.

Phyllodactylus strophurus, *Dum. & Bibr.* iii. p. 397, pl. xxxii. fig. 1.
Discodactylus (Strophurus) dumerilii, *Fitz. Syst. Rept.* p. 96.

Differs from the preceding in the absence of caudal spines and the presence on the back of large, round, obtusely conical tubercles forming two irregular longitudinal series. The ear-opening is still smaller than in *D. spinigerus*; and there are only ten to twelve labials. The head is not black-speckled, but has dark undulated longitudinal lines, a lower passing through the eye, and an upper meeting its fellow on the snout.

South-east Australia.

a–b. ♀.	Sydney.	G. Krefft, Esq. [P.].

4. Diplodactylus vittatus. (PLATE VIII. fig. 3.*)

Diplodactylus vittatus, *Gray, Cat.* p. 148.
Diplodactylus ornatus, *Gray, l. c.* p. 149.
Diplodactylus vittatus, *Gray, Proc. Zool. Soc.* 1832, p. 40, *and Zool. Erebus & Terror*, pl. xvi. fig. 3.
Phyllodactylus vittatus, *Dum. & Bibr.* iii. p. 400.
Diplodactylus furcosus, *Peters, Mon. Berl. Ac.* 1863, p. 229.
—— ornatus, *Gray, Zool. Erebus & Terror*, pl. xvi. fig. 2.

Head short, very convex; snout rounded-acuminate, measuring the diameter of the orbit or the distance between the eye and the ear-opening; latter rather small, round. Body short; limbs moderate. Digits short, depressed, with small apical dilatation, inferiorly with a series of large transversely oval tubercles, some of them breaking up into two rounded tubercles; the extremity of the digit is raised and bears inferiorly two roundish plates separated from the large tubercles of the basal part by three or four rows of small granules. Upper surfaces covered with uniform small granular scales. Rostral four-sided, twice as broad as high, with median cleft above; nostril pierced between the rostral, the first labial, and five or six nasals, the anterior or upper largest and generally in contact with its fellow, the others granular; ten or eleven upper and as many lower labials; mental trapezoid, a little larger than the adjacent labials; no chin-shields. Abdominal scales granular, scarcely larger than those on the upper surfaces. Tail short, swollen, root-shaped, with rings of uniform small squarish

---

* Lower surface of foot, ×2.

scales. Male with a small group of conical tubercles on each side the base of the tail. Brown above; a light dark-edged festooned vertebral band, bifurcating on the nape, sometimes broken up into angular spots; sides and limbs with light spots; lower surfaces dirty white.

Total length	88 millim.
Head	15 ,,
Width of head	12 ,,
Body	40 ,,
Fore limb	19 ,,
Hind limb	24 ,,
Tail	33 ,,

Australia.

a. ♀.	Australia.	A. Cunningham, Esq. [P.]. (Type.)
b. ♂.	Houtman's Abrolhos.	Mr. Gilbert [C.]. (Type of *D. ornatus*.)
c. ♀.	W. Australia.	G. F. Moore, Esq. [P.].
d. ♀.	Champion Bay, N.W. Australia.	Mr. Duboulay [C.].
e. ♂.	Sydney.	
f–g, h. ♂, ♀, & yg.	Australia.	G. Krefft, Esq.
i. ♂.	Australia.	Godeffroy Museum.
k. ♀.	Australia.	

5. Diplodactylus polyophthalmus. (Plate VIII. fig. 4.*)

Diplodactylus polyophthalmus, *Günth. Ann. & Mag. N. H.* (3) xx. 1867, p. 49.

Œdura marmorata, jun.?, *Gray, Zool. Erebus & Terror,* pl. xvi. fig. 1.

Very closely allied to the preceding, from which it differs in the following points:—Snout a little longer; digits narrower and longer; the infradigital tubercles smaller, rounded or shortly oval. Brown above, reticulated with darker, and with more or less regular roundish light spots; lower surfaces white.

Total length	73 millim.
Head	17 ,,
Width of head	10 ,,
Body	30 ,,
Fore limb	16 ,,
Hind limb	20 ,,
Tail	26 ,,

North-west Australia.

a. ♀.	Champion Bay.	Mr. Duboulay [C.]. } (Types.)
b. Yg.	Nicol Bay	Mr. Duboulay [C.].
c. ♂.	W. Australia.	Mr. Duboulay [C.].
d. ♂.	N. Australia.	
e. ♀.	Australia.	J. S. Bowerbank, Esq.

---

* Lower surface of foot, ×3.

6. Diplodactylus steindachneri. (PLATE VIII. fig. 5.)

Head short, very convex; snout rounded-acuminate, longer than the distance between the eye and the ear-opening, scarcely longer than the diameter of the orbit; ear-opening small, round. Body short; limbs moderate. Digits rather short, not much depressed, inferiorly with small, irregularly arranged tubercles; apical dilatation small, the inferior plates rounded. Upper surfaces covered with minute granular scales, a little larger on the snout. Rostral twice as broad as high, without median cleft, emarginate posteriorly to receive a small lozenge-shaped plate bounded posteriorly by the large supero-anterior nasals; latter in contact; nostril pierced between the rostral, the first labial, and five nasals; eleven upper and as many lower labials; mental trapezoid, moderate; no chin-shields. Border of the upper eyelid with a fringe of conical scales. Abdominal scales very small, granular, considerably larger than the dorsal granules. Male with two conical scales on each side the base of the tail. Light brown above: a broad lighter, black-edged vertebral line, bifurcating on the neck, where it encloses a large rhomboidal black-edged spot; three black-edged ocelli on the light vertebral band, two on the back, one on the base of the tail; a whitish, dark-edged streak from axilla to groin; sides with small light spots; lower surfaces white.

Total length	60	millim.
Head	13	„
Width of head	9	„
Body	30	„
Fore limb	16	„
Hind limb	21·5	„
Tail (reproduced)	17	„

New South Wales.

c. ♂.   Sydney.   Museum Godeffroy.

7. Diplodactylus pulcher.

Stenodactylopsis pulcher, *Steindachn. Sitzb. Ak. Wien*, lxii. i. 1870, p. 343, pl. ii. figs. 3–5.

Head short, very convex; snout acutely pointed, longer than the distance between the eye and the ear-opening, once and one third the diameter of the orbit; ear-opening minute. Body moderate; limbs rather long, slender. Digits not much depressed, inferiorly with small round tubercles; the apex not dilated, with two oval plates inferiorly. Upper surfaces covered with small granular scales, largest on the middle of the back. Rostral large, high, subpentagonal, with emarginate posterior border, and trace of median cleft; nostril pierced between five or six nasals, the anterior and the inferior larger; a polygonal plate behind the rostral between the nasals; eleven or twelve upper and as many lower labials, the anterior larger, the others very small; mental subpyramidal, twice as long as the adjacent labials; no chin-shields. Abdominal scales very small, granular, smaller than the granules on the middle of the back. Tail short, swollen, root-shaped, with rings of small subquadrangular convex scales largest above. Male with a group of conical scales on each side the base of the tail. Reddish brown above, with irregular dark-edged light transverse spots on the back and tail; sides dotted with whitish; lower surfaces white.

Total length	70 millim.
Head	11 ,,
Width of head	8 ,,
Body	33 ,,
Fore limb	17 ,,
Hind limb	20 ,,
Tail	26 ,,

West Australia.

a. ♂. N.W. Australia.

8. Diplodactylus tessellatus. (Plate VIII. fig. 6.)

Stenodactylopsis tessellatus, *Günth. Zool. Erebus & Terror*, p. 16.

Head large, oviform, very convex; snout rounded, as long as the diameter of the orbit or the distance between the eye and the ear-opening; latter small, round. Body short; limbs long, slender. Digits rather long, slender, feebly depressed, not dilated at the end, inferiorly with small granules; apical plates small, oval. Head with small granular scales; rostral four-sided, emarginate above, more than twice as broad as high, with trace of median cleft; nostril pierced between the rostral, the first labial, and six nasals; latter, anterior larger, posterior very small granules; nine upper and ten lower labials; mental elongate, not larger than the adjacent labials; no chin-shields. Back covered with flat tessellated juxtaposed scales, much larger on the middle of the back. Abdominal scales flat, subimbricate, not half the size of the larger dorsal scales. Male with a group of conical tubercles on each side the base of the tail. Greyish white above, with faint irregular brownish variegation; white beneath.

Total length	69 millim.
Head	14 ,,
Width of head	9 ,,
Body	34 ,,
Fore limb	19 ,,
Hind limb	23 ,,
Tail (reproduced)	21 ,,

Australia.

a. ♂. Australia. G. Krefft, Esq. (Type.)

19. ŒDURA.

Œdura, *Gray*, Zool. Misc. 1842, p. 52, *and Cat. Liz.* p. 147.
Phyllodactylus, part., *Dum. & Bibr.* iii. p. 388.
Pachyurus, *Fitzing. Syst. Rept.* p. 94.

Digits free, all clawed, dilated at the base, with raised distal joint bearing a discoid dilatation; latter with two large plates inferiorly, separated by a longitudinal groove in which the claw is retractile; basal expansion inferiorly with paired lamellæ. Upper surfaces covered with homogeneous juxtaposed scales; abdominal scales juxtaposed or slightly imbricate. Pupil vertical.

Australia.

Synopsis of the Species.

I. Dorsal scales flat, about as large as ventrals.

Tail much depressed, short, at least as broad as the body; first infralabials in contact behind the mental................ 1. *marmorata*, p. 104.
Tail slightly depressed, much narrower than the body; first infralabials separated.... 2. *ocellata*, p. 105.

II. Dorsal scales small, convex, granular.

Four divided lamellæ under the median toes; tail much depressed; males with præanal pores............................. 3. *robusta*, p. 106.
Two or three divided lamellæ under the median toes; tail depressed; no præanal pores............................. 4. *lesueurii*, p. 107.
Two or three divided lamellæ under the median toes; tail cylindrical; body much elongate; males with præanal pores .... 5. *rhombifer*, p. 107.

III. Dorsal scales flat, smaller than ventrals.

Only one divided lamella at the end of the antepenultimate phalange .... 6. *verrillii*, p. 108.

1. Œdura marmorata. (PLATE IX. fig. 2.*)

Œdura marmorata, *Gray, Cat.* p. 147.
Œdura marmorata, *Gray, Zool. Misc.* p. 52, *and Zool. Erebus & Terror*, pl. xvi. fig. 4.

Head large, much depressed, oviform; snout as long as the distance between the eye and the ear-opening, once and a half the diameter of the orbit; ear-opening oblique, half the diameter of the eye. Body moderately elongate, much depressed. Limbs moderate. Digits strongly dilated, as broad as the apical expansion; latter large,

---

* Chin.

rounded, broader than long; four or five pairs of broad infradigital plates under the median toes, followed by three or four undivided ones. Head covered with uniform, roundish, small flat scales; rostral four-sided, twice as broad as high, with median cleft above; nostril pierced between the rostral, the first labial, and four nasals, the two upper large, the anterior in contact with its fellow on the other side; ten or eleven upper and nine or ten lower labials; mental triangular, shorter than the adjacent labials, which are in contact behind it; a small median chin-shield. Back covered with juxtaposed flat round scales, larger than those on the head, arranged in regular transverse series; these scales become much smaller on the flanks; abdominal scales juxtaposed, flat, hexagonal, about the same size as the dorsal scales. Males with seventeen præanal pores, forming a gently curved series interrupted in the middle. Tail thick, short, much depressed, oval, the end tapering into a point; its width is contained twice or twice and a half in its length, and equals at least the width of the body; it is covered with equal square scales, arranged like the bricks of a wall; males with two or three large tubercles on each side the base of the tail. Dark brown above, with whitish spots and cross bands; latter not half as wide as the interspace between them, five on body, anterior on neck, posterior on sacrum; lower surfaces white.

Total length	143	millim.
Head	27	,,
Width of head	20	,,
Body	73	,,
Fore limb	30	,,
Hind limb	38	,,
Tail	43	,,

North Australia.

a–f. ♂ ♀.	Port Essington.	Mr. Gilbert [C.]. (Types.)
g. ♀.	Port Essington.	J. B. Jukes, Esq. [P.].
h. ♀.	Australia.	Lord Derby [P.].
i–l. ♂ ♀.	Australia.	Haslar Collection.
m. Hgr.	Australia.	

2. Œdura ocellata. (PLATE IX. fig. 1.)

Closely allied to the preceding; distinguished by the following characters:—Anterior infralabials not meeting behind the mental, which is truncate posteriorly and in contact with a small chin-shield hardly distinguishable from the neighbouring scales. Tail cylindrical, slightly depressed, tapering, its width contained five or six times in its length, much narrower than the body. Upper surfaces reticulated with brown and whitish, and with more or less distinct light, dark-edged ocelli.

The male has nineteen præanal pores.

Total length	143	millim.
Head	22	,,
Width of head	16·5	,,
Body	64	,,
Fore limb	25	,,
Hind limb	32	,,
Tail	57	,,

Australia.

a. ♂.	Australia.	
b, c, d, e, f. ♀.	Australia.	G. Krefft, Esq.

3. Œdura robusta. (Plate X. fig. 1.)

Head oviform, much depressed; snout longer than the distance between the eye and the ear-opening, once and a half the diameter of the orbit; ear-opening suboval, oblique, half the diameter of the eye. Body moderately elongate, depressed. Limbs strong. Digits as in *Œ. marmorata*. Upper surfaces covered with small granular scales, a little larger on the snout. Rostral four-sided, twice as broad as high, with median cleft above; nostril pierced between the rostral, the first labial, and four nasals; eleven or twelve upper and as many lower labials; mental small, triangular; no regular chin-shields, but small polygonal scales passing gradually into the minute granules of the gular region. Abdominal scales hexagonal, smooth, subimbricate, much larger than the dorsal scales. Males with fifteen or seventeen præanal pores arranged in a transverse series interrupted in the middle. Tail much depressed, tapering to a point, when intact about five times as long as broad, narrower than the body; it is covered with uniform small flat hexagonal scales arranged regularly; three or four large tubercles on each side the base of the tail in the males. Brownish white above, with more or less distinctly ✕-shaped dark-brown markings or cross bands on the back, confluent on the sides, and enclosing large spots of the light ground-colour; a dark band on each side of the head, passing through the eye; lower surfaces whitish.

The largest specimen measures 80 millim. from snout to vent, but the tail is reproduced. The following are the measurements of a specimen with intact tail:—

Total length	127	millim.
Head	19	,,
Width of head	14·5	,,
Body	54	,,
Fore limb	22	,,
Hind limb	28	,,
Tail	54	,,

Australia.

a, b. ♂ ♀.	Australia.	
c, d. ♂ ♀.	Australia.	G. Krefft, Esq.

4. Œdura lesueurii. (Plate X. fig. 2.*)

Œdura rhombifer, part., *Gray, Cat.* p. 147.
Phyllodactylus lesueurii, *Dum. & Bibr.* iii. p. 392.
Diplodactylus (Pachyurus) lesueurii, *Fitz. Syst. Rept.* p. 94.

Head oviform, depressed; snout longer than the distance between the eye and the ear-opening, once and a half the diameter of the orbit; ear-opening round or suboval, two fifths the diameter of the eye. Body and limbs moderate. Digits less dilated than in the preceding species, the basal dilatation not being quite as wide as the apical; the lamellæ also fewer, only the two or three distal ones divided. Upper surfaces covered with minute granular scales, larger on the snout. Rostral four-sided or subpentagonal, about twice as broad as high, with median cleft above; nostril pierced between the rostral, the first labial, and four or five nasals; ten or eleven upper and as many lower labials; mental small, triangular; no regular chin-shields, but small polygonal scales passing gradually into the minute granules of the gular region. Abdominal scales hexagonal, smooth, subimbricate. No præanal pores. Tail depressed, oval in section, tapering; when intact its width is contained at least six times in its length; it is covered with uniform, small, flat hexagonal scales; four or five large tubercles on each side the base of the tail in the males. Light brown above, variegated with darker; two zigzag dark-brown lines on the back, enclosing rhomboidal spots of the lighter ground-colour; sides and limbs with small light spots; lower surfaces whitish.

Total length	98	millim.
Head	15	,,
Width of head	10	,,
Body	38	,,
Fore limb	17	,,
Hind limb	21	,,
Tail	45	,,

Australia.

a–b. ♂ & hgr.	Sydney.	
c–e, f, g. ♂ ♀.	Australia.	G. Krefft, Esq.
h–i. ♂.	Australia.	Museum Godeffroy.
k. ♀.	Australia.	(One of the types of *Œ. rhombifer*.)
l. ♀.	Australia.	

5. Œdura rhombifera.

Œdura rhombifer, part., *Gray, Cat.* p. 147.
Œdura rhombifer, *Gray, Zool. Erebus & Terror*, pl. xvi. fig. 6.

Very closely allied to the preceding. Distinguished by the smaller head, the rather shorter snout, the more elongate body, the perfectly cylindrical tail, and the presence of præanal pores in the male; latter ten to twenty. Pale brown, middle of back and tail paler, with a zigzag brown line on each side.

---

* Lower surface of fourth toe, ×4.

Total length	90 millim.
Head	11 ,,
Width of head	7·5 ,,
Body	36 ,,
Fore limb	13 ,,
Hind limb	17 ,,
Tail	43 ,,

Australia.

a-b. ♂ ♀.	Australia.	(Types.)
c-d. ♂ ♀.	Australia.	
e. ♂.	Islands of Torres Straits.	Rev. S. McFarlane [C.].

6. Œdura ? verrillii.

Œdura verrillii, *Cope, Proc. Ac. Philad.* 1868, p. 318.

"The femoral pores in a series arched angularly forwards, and not extending on the femora. The plates of the under surfaces of the toes are, besides the terminal disks, one pair only, as large as the terminal, and at the end of the antepenultimate phalange. Labials regular, 8-7 to below pupil; two rows infralabials across chin. Rostral undivided. Gular scales granular; thoracic and ventral flat, larger than the flat dorsals. Muzzle-scales tubercular. A tubercle on each side vent. Head as broad as from end of muzzle to halfway between orbit and ear. Colour very pale above, with six very deep brown cross bands from nape to sacrum, which are more or less connected on the sides. A brown band through orbit, and one behind, crossing the occiput.

"Muzzle to ear 12″, to axilla 20″·5 ; axilla to groin 25″ ; tail lost. Fore limb 13″ ; hind limb 18″."

Australia.

20. CALODACTYLUS.

Calodactylus, *Beddome, Madras Med. Journ.* 1870, p. 30.

Digits slender at the base, free, with squarish scales inferiorly, with large trapezoid distal and penultimate expansions, the lower surface of which is covered by two large plates separated by a longitudinal groove; all the digits clawed, the claw retractile between the distal plates; in the inner digit, the penultimate expansion is absent. Body covered above with small granular scales, intermixed with larger tubercles; abdominal scales juxtaposed. Pupil vertical. Males without præanal nor femoral pores.

Southern India.

1. Calodactylus aureus. (PLATE X. fig. 3.*)

Calodactylus aureus, *Bedd. l. c.* pl. ii.

Head large, oviform, very distinct from neck; a strong rounded

---

* Upper view of head; *a*, lower surface of foot, ×2.

supraorbital and canthal ridge; five deep concavities, viz. a frontal, two postnasals, and two loreals; snout longer than the distance between the eye and the ear-opening, once and one third the diameter of the orbit; ear-opening vertical, measuring half the diameter of the eye. Body not much depressed. Limbs long, slender. The width of the digital expansion measures about half the diameter of the eye. Head covered with very small granules, largest on the canthal ridges; rostral four-sided, twice as broad as high, its posterior border concave; nostril pierced between the rostral, the first labial, and three nasals, the anterior large and in contact with its fellow; twelve or thirteen upper and as many lower labials; mental as large as, or smaller than, the adjacent labials; no regular chin-shields, but small polygonal scutes passing gradually into the granules which cover the gular region. Upper surfaces covered with minute granules; back with scattered, scarcely prominent, smooth, round, larger tubercles, hardly as large as the ventral scales; latter flat, smooth, squarish, juxtaposed, arranged like the bricks of a wall. Tail long, cylindrical, remarkably slender, covered with squarish scales which are much larger inferiorly. Brownish white above (golden during life), dotted or vermiculated with brown; lower surfaces whitish.

Total length	169	millim.
Head	23	,,
Width of head	18	,,
Body	66	,,
Fore limb	35	,,
Hind limb	47	,,
Tail	80	,,

Southern India.

a–b. Ad. Eastern Ghats. Col. Beddome [P.]. (Types.)
c. Several spec. N. Arcot. Col. Beddome [C.].

21. PTYODACTYLUS.

Ptyodactyles, part., *Cuv. R. A.* ii. p. 49.
Ptyodactylus, *Gray, Ann. Phil.* (2) x. 1825, p. 498; *Fitzing. N. Class. Rept.* p. 13; *Wagl. Syst. Amph.* p. 143; *Wiegm. Herp. Mex.* p. 20; *Fitzing. Syst. Rept.* p. 96; *Gray, Cat.* p. 151.
Ptyodactylus, part., *Dum. & Bibr.* iii. p. 375.

Digits slender, free, with a series of transverse plates inferiorly, the extremity strongly dilated, with two diverging series of lamellæ inferiorly; every digit armed with a claw retractile in the anterior notch of the distal expansion. Body covered above with small granular scales, uniform or intermixed with enlarged tubercles; abdominal scales scarcely imbricate. Pupil vertical. No præanal nor femoral pores.

North Africa; South-western Asia; Sind.

1. Ptyodactylus lobatus.

Ptyodactylus gecko, *Gray, Cat.* p. 151.
Hasselquist, Reise n. Paläst. p. 356.
Stellio gecko, part., *Schneid. Amph. Phys.* ii. p. 12.
Gecko lobatus, *Geoffr. Descr. Egypte, Rept.* p. 130, pl. v. fig. 5; *Aud. eod. loc., Suppl.* p. 166, pl. i. fig. 2.
—— ascalabotes, *Merr. Tent.* p. 40.
Ptyodactylus lobatus, *Gray, Ann. Phil.* (2) x. p. 498.
—— guttatus, *Rüpp. Atl. N. Afr., Rept.* p. 13, pl. iv.
—— hasselquistii, *Dum. & Bibr.* iii. p. 378, pl. xxxiii. fig. 3; *Boettg. Ber. Senck. Ges.* 1879-80, p. 194.
—— oudrii, *Lataste, Le Natur.* 1880, p. 299.

General proportions varying considerably. Head large, oviform; snout a little longer than the distance between the eye and the ear-opening, once and one third the diameter of the orbit; forehead concave; ear-opening subcrescentic, vertical. Body rather short, depressed; limbs long and slender. Head covered with granular scales, which are much larger on the snout. Nostril pierced in the centre of a more or less marked globular swelling; rostral and first labial generally entering the nostril. Rostral generally quadrangular, not twice as broad as high; twelve or thirteen upper and ten to twelve lower labials; a row of chin-shields. Upper surfaces covered with granules intermixed with small keeled tubercles; lower surfaces with flat, smooth, slightly imbricate, small subhexagonal scales, which are enlarged on the femoral and inter-femoral regions. Tail slender, tapering, rounded, covered above with very small scales intermixed with keeled tubercles, inferiorly with large imbricated scales. Greyish or yellowish brown above, spotted with darker and with yellowish white; lower surfaces uniform white.

Total length	135	millim.
Head	24	,,
Width of head	18	,,
Body	53	,,
Fore limb	37	,,
Hind limb	49	,,
Tail	58	,,

Algeria, Egypt, Nubia, Abyssinia, Arabia, Syria.

a. ♂.	Egypt.	J. Burton, Esq. [P.].
b-c. ♂ & hgr.	Egypt.	Sir J. Wilkinson [P.].
d. ♀.	Egypt.	A. Christy, Esq. [P.].
e. ♀.	Edfou.	Dr. Anderson [P.].
f. ♂.	Between Khan Tubb Tusef and Ain Mellaha.	Dr. Anderson [P.].
g-h. ♂ ♀.	Sinaitic Peninsula.	H. C. Hart, Esq. [C.].
i. ♀.	Mt. Sinai.	
k. ♂.	Jerusalem.	Dr. Anderson [P.].
l. ♂.	Jerusalem.	Rev. H. B. Tristram [C.].
m. ♂.	Mt. Hermon.	Rev. H. B. Tristram [C.].
n. ♀.	Dead Sea.	Rev. H. B. Tristram [C.].

o. ♀.	Galilee.	Dr. Anderson [P.].
v-s. ♂ & hgr.	Bou Saada, Algerian Sahara.	M. F. Lataste [P.]. (As typical of *P. oudru.*)

2. Ptyodactylus homolepis.

Ptyodactylus homolepis, *Blanford, Proc. As. Soc. Beng.* 1876, p. 19, pl. ii.

General characters of *P. lobatus*. Upper surfaces not intermixed with tubercles. Nostril pierced between three swollen nasals, neither the rostral nor the first labial entering the nostril; fourteen or fifteen upper and as many lower labials. Light brownish grey above, with broad transverse wavy bands of lighter and darker shades; white beneath.

Total length	190 millim.
Head	29 ,,
Width of head	23 ,,
Body	76 ,,
Fore limb	48 ,,
Hind limb	60 ,,
Tail	85 ,,

Sind.

a-b. Ad. & hgr. Khirthar Range. W. T. Blanford, Esq. [P.]. (Types.)

22. THECADACTYLUS.

Thecadactyles, *Cuv. R. A.* ii. p. 48.
Thecadactylus, *Gray, Ann. Phil.* (2) x. 1825, p. 198; *Fitzing. N. Classif. Rept.* p. 13; *Gray, Cat.* p. 146.
Thecodactylus, *Gray, Griff. A. K.* ix. *Syn.* p. 50; *Wagler, Syst. Amph.* p. 142; *Wiegm. Herp. Mex.* p. 20; *Fitzing. Syst. Rept.* p. 98.
Platydactylus, part., *Dum. & Bibr. Erp.* iii. p. 290.

Digits strongly dilated, more or less webbed, inferiorly with two series of regular transverse lamellæ divided by a median groove, with a retractile claw sheathed in a groove between the anterior lamellæ; the claw of the inner digit very indistinct or absent. Upper surfaces covered with juxtaposed scales; belly with imbricate scales. Pupil vertical.

Tropical America; islands of Torres Straits.

1. Thecadactylus rapicaudus.

Thecadactylus rapicaudus, *Gray, Cat.* p. 146.
Gecko rapicauda, *Houttuyn, Verh. Genotsch. Vlissing.* ix. p. 322, pl. iii. fig. 1; *Daud. Rept.* iv. p. 141, pl. li.
Lacerta rapicauda, *Gmel. S. N.* i. p. 1068.
Stellio perfoliatus, *Schneid. Amph. Phys.* ii. p. 26.
Gecko lævis, *Daud. l. c.* p. 112.
— — surinamensis, *Daud. l. c.* p. 126.
Thecadactylus lævis, *Gray, Ann. Phil.* (2) x. p. 198.
Platydactylus theconyx, *Dum. & Bibr.* iii. p. 306, pl. xxxiii. fig. 2.

Head large, longer than broad; snout subtriangular, slightly longer than the distance between the eye and the ear-opening, once and a half the diameter of the orbit; forehead concave; ear-opening small, oval, horizontal. Body and limbs moderate; digits subequal, half-webbed, the web being as much developed between the two outer; a rudimentary claw generally distinguishable in the inner digit. Head covered with very small granular scales; rostral quadrangular, twice as broad as high, with trace of median cleft above; nostril pierced between the rostral, the first upper labial, a large supero-nasal, which is in contact with its fellow, and a few granules; nine to twelve upper and eight to eleven lower labials; mental small, pentagonal; a row of small chin-shields, median pair elongate. Back, limbs, and throat covered with very small granular scales; belly covered with small imbricate cycloid scales. No præanal nor femoral pores. Tail, when intact, cylindrical, tapering, covered with very small imbricated scales, largest inferiorly; generally strongly swollen at the base when reproduced. Brown or greyish-brown above, variegated with darker and lighter; lower surfaces whitish, immaculate.

Total length	195	millim.
Head	24	,,
Width of head	20	,,
Body	81	,,
Fore limb	30	,,
Hind limb	36	,,
Tail	90	,,

South and Central America; West Indies.

a–b. ♂ ♀.	Yucatan.	
c. ♀.	Granada.	
d–g. ♂ ♀.	St. Thomas, W. Indies.	Mr. Riise [C.].
h. ♂.	St. Thomas, W. Indies.	Capt. Sawyer [P.].
i–l. ♂, ♀, & hgr.	Sta. Cruz, W. Indies.	A. Newton, Esq. [P.].
m. ♀.	Nevis.	T. Cottle, Esq. [P.].
n. ♂.	Antigua.	Zool. Soc.
o–p. ♀ & hgr.	Anguilla, a rock near Trinidad.	W. J. Cooper, Esq. [P.].
q. ♂.	Caracas.	
r. ♀.	British Guiana.	
s. ♂.	Surinam.	Hr. Kappler [C.].
t. ♀.	Pebas.	H. W. Bates, Esq. [C.].
u–v. ♂ ♀.	Canelos, Ecuador.	Mr. Buckley [C.].
w–x. ♂.	Sarayacu, Peru.	W. Davis [C.].
y, z, α. ♂ ♀.	S. America.	

2. Thecadactylus australis. (PLATE XI. fig. 1.)

Thecadactylus australis, *Günth. Ann. & Mag. Nat. Hist.* (4) xix. 1877, p. 414.

Head large, oviform elongate; snout as long as the distance between the eye and the ear-opening, once and two thirds the diameter of the orbit; forehead slightly concave; ear-opening small, oval, oblique. Body and limbs moderate; digits webbed at the base, the web absent between the two outer toes; inner digit clawless. Snout and forehead with large, rough, vermiculated tubercles confluent with the cranial ossification; hinder part of head with juxtaposed flat tubercles; rostral subquadrangular, twice as broad as high, with trace of median cleft above; nostril pierced between the first upper labials and several tubercles; eleven upper and ten lower labials; mental large, subtrapezoid, separating two pairs of small chin-shields. Upper surfaces with small, flat, juxtaposed, granular scales; throat granulate; abdominal scales not much larger than dorsals, subhexagonal, scarcely imbricate. A subtriangular patch of eighteen præanal pores. A globular swelling at the base of the tail, behind the vent, covered with large hexagonal scales. Tail cylindrical, covered with small flat square scales arranged in verticils. Upper parts brownish violet, marbled with reddish; lower surfaces brownish white.

Total length	167 millim.
Head	27 ,,
Width of head	20 ,,
Body	82 ,,
Fore limb	32 ,,
Hind limb	40 ,,
Tail	58 ,,

Islands of Torres Straits.

a. ♂. Islands of Torres Straits. Rev. S. Macfarlane [C.]. (Type.)

23. HEMIDACTYLUS*.

Hemidactyles, *Cuv. R. A.* ii. p. 47.
Hemidactylus, *Gray, Ann. Phil.* (2) x. 1825, p. 199; *Fitzinger, N. Classif. Rept.* p. 13; *Wagler, Syst. Amph.* p. 142; *Wiegm. Herp. Mex.* p. 20; *Fitzing. Syst. Rept.* p. 104; *Gray, Cat.* p. 152; *Günth. Rept. Brit. Ind.* p. 100.
Crossurus, *Wagl. l. c.* p. 141; *Fitzing. Syst. Rept.* p. 106; *Gray, l. c.* p. 158.
Hemidactylus, part., *Dum. & Bibr.* iii. p. 344.
Boltalia, *Gray, Zool. Misc.* p. 58, and *Cat.* p. 158.
Hoplopodion, *Fitzing. Syst. Rept.* p. 103.

---

* *H. flaviviridis*, Rüppell, N. Wirbelth. Faun. Abyss. p. 18, pl. vi. fig. 2.—Massoa.
H. angulatus, Hallow. Proc. Ac. Philad. 1852, p. 63.—W. Africa.
H. marmoratus, Hallow. eod. loc. 1860, p. 491.—Loo Choo.
H. mortoni, Theobald, Journ. Linn. Soc. x. p. 32.—Pegu.
H. sakalava, Grandid. Rev. Mag. Zool. 1867, p. 233.—Madagascar.
H. tolampyæ, Grandid. Ann. Sc. Nat. xv. 1872, art. 20, p. 8.—Madagascar.
H. tristis, Sauvage, Bull. Soc. Philom. (7) iii. 1878, p. 49.—New Guinea.
Mythical species:—*Crossurus caudiverbera*, Wagler, Syst. Rept. p. 141. Gray, Cat. p. 158. *Hemidactylus sebæ*, Dum. & Bibr. iii, p. 373.

Cosymbotus, *Fitzing. l. c.* p. 104.
Microdactylus, *Fitzing. l. c.*
Onychopus, *Fitzing. l. c.*
Tachybates, *Fitzing. l. c.* p. 105.
Pnoëpus, *Fitzing. l. c.*
Velernesia, *Gray, Cat.* p. 156.
Doryura, *Gray, l. c.*
Platyurus, *Gray, l. c.* p. 157.
Leiurus, *Gray. l. c.*
Nycteridium, *Günth. Rept. Brit. Ind.* p. 111.

Digits free or more or less webbed, dilated, inferiorly with two rows of lamellæ; all the digits provided with slender distal clawed joints angularly bent and rising from within the extremity of the dilated portion. Dorsal lepidosis uniform or heterogeneous. Pupil vertical. Males with præanal or femoral pores.

Southern Europe; Southern Asia; Africa; Tropical America; Polynesia.

Synopsis of the Species.

I. Digits quite free.

    A. Back with uniform flat subimbricate scales .... 1. *homœolepis*, p. 117.

    B. Free distal joints of all the digits remarkably short.

3 lamellæ* under the inner, and 4 or 5 under the median toe; dorsal scales uniformly granular .... 2. *bouvieri*, p. 118.
5 or 6 lamellæ under the inner, and 8 or 9 under the median toe; dorsal granules intermixed with small round keeled tubercles .... 3. *reticulatus*, p. 118.
5 lamellæ under the inner, and 9 under the median toe; dorsal tubercles oval, subtrihedral, strongly keeled........ 4. *gracilis*, p. 119.

    C. Free distal joints of digits long.

        1. Dorsal tubercles, if present, small, smooth, conical, or very feebly keeled.

            a. Less than 8 lamellæ under the inner toe.

                a. Tail with rows of enlarged tubercles.

Inner digit very short, with sessile claw; no tubercles on hinder part of head .. 5. *frenatus*, p. 120.
Dorsal tubercles small, convex or very feebly keeled; 5 or 6 lamellæ under the inner, and 7 to 9 under the median

---

* The infradigital lamellæ are counted longitudinally, those which are divided on the median line being reckoned as one.

toe; male with 15 to 30 femoral pores on each side .... 6. *mabouia*, p. 122.

Dorsal tubercles pointed, conical; on each side a feeble longitudinal fold on which the lower tubercles are inserted .... 7. *muriceus*, p. 123.

Dorsal tubercles round, smooth; 7 lamellæ under the median toe; male with 10 or 12 femoral pores on each side .... 22. *kuchmorensis*, p. 135.

Dorsal tubercles, if present, quite smooth; 6 or 7 lamellæ under the inner, and 9 to 11 under the median toe; male with 12 to 16 femoral pores on each side.......................... 23. *leschenaultii*, p. 136.

β. Tail with uniform small scales on its upper surface.

No enlarged dorsal tubercles; tail oval in section .... 26. *bowringii*, p. 139.

Back with numerous small convex tubercles; tail with sharpish denticulated lateral edge .... 27. *karenorum*, p. 140.

No enlarged dorsal tubercles; tail with sharpish denticulated lateral edge.... 28. *blanfordii*, p. 141.

No enlarged dorsal tubercles; tail with sharpish denticulated lateral edge; digits very slightly dilated .... 29. *peruvianus*, p. 141.

b. 8 or more lamellæ under the inner toe.

8 to 10 lamellæ under the inner, and 11 to 13 under the median toe; first labial entering the nostril .... 24. *coctæi*, p. 137.

11 or 12 lamellæ under the inner, and 13 to 15 under the median toe; first labial not entering the nostril ...... 25. *giganteus*, p. 138.

2. Dorsal tubercles strongly keeled.

a. Not more than 8 lamellæ under the inner toe and than 11 under the median toe.

5 or 6 lamellæ under the inner, and 8 to 10 under the median toe; tubercles small; no enlarged subcaudals; snout considerably longer than the distance between the eye and the ear-opening; male with præanal pores only ...... 10. *bocagii*, p. 125.

5 lamellæ under the inner, and 8 under the median toe; tubercles large; no enlarged subcaudals; snout not longer than the distance between the eye and the ear-opening; male with præanal pores only .... 11. *sinaitus*, p. 126.

6 to 8 lamellæ under the inner, and 9 to 11 under the median toe; tubercles in 14 or 16 longitudinal series; male with præanal pores only .... 12. *turcicus*, p. 126.

4 to 6 lamellæ under the inner, and 7 or 8 under the median toe; tubercles in 16 to 20 longitudinal series; male with 10 to 20 femoral pores on each side (African).................. 13. *brookii*, p. 128.

4 to 6 lamellæ under the inner, and 7 or 8 under the median toe; tubercles in 16 to 20 longitudinal series; male with 8 to 12 femoral pores on each side (Indian) .... 14. *gleadovii*, p. 129.

4 or 5 lamellæ under the inner, and 7 or 8 under the median toe, the basal ones tubercle-like; male with 8 femoral pores on each side; back with pure white spots..................... 15. *stellatus*, p. 130.

2 or 3 lamellæ under the inner, and 4 or 5 under the median toe; tubercles in 20 or 22 longitudinal series .... 16. *guineensis*, p. 131.

6 or 7 lamellæ under the inner, and 8 to 10 under the median toe; dorsal tubercles large, in 16 to 20 longitudinal series; male with 12 to 16 præanal pores; back with dark transverse bands and rows of pure white tubercles 19. *triedrus*, p. 133.

b. More than 8 lamellæ under the inner toe, and than 11 under the median toe.

Infradigital lamellæ obliquely curved; dorsal tubercles in 14 to 16 longitudinal series; male with a few præanal pores .... 17. *persicus*, p. 131.

Infradigital lamellæ straight; dorsal tubercles in about 20 longitudinal series; male with 19 to 25 femoral pores on each side .... 18. *maculatus*, p. 132.

Infradigital lamellæ slightly oblique; dorsal tubercles large .... 20. *subtriedrus*, p. 134.

II. A more or less developed interdigital membrane.

A. A more or less marked fold on the side of the body, but no cutaneous expansion.

1. Tail cylindrical or quadrangular.

A series of long, spine-like scales on each side of the belly; tail quadrangular.. 8. *echinus*, p. 123.

Dorsal tubercles small, round, convex;
tail cylindrical .... 9. *fasciatus*, p. 124.

 2. Tail depressed, flat inferiorly, with sharp denticulated lateral edge.

Back with enlarged trihedral tubercles.. 21. *depressus*, p. 134.
Back uniformly granular; digits very slightly webbed................... 30. *garnoti*, p. 141.
Back granular, some of the granules slightly enlarged; digits one third webbed .... 31. *richardsonii*, p. 143.

 B. A much-developed cutaneous expansion from axilla to groin.

Digits nearly half-webbed............ 32. *platyurus*, p. 143.

1. Hemidactylus homœolepis.

Hemidactylus (Liurus) homœolepis, *Blanford, Proc. Zool. Soc.* 1881, p. 464. pl. xlii. fig. 2.

Snout longer than the distance between the eye and the ear-opening, once and one third the diameter of the orbit; forehead scarcely concave; ear-opening small, oval, oblique. Body and limbs moderate. Digits moderately dilated, free, with rather short distal joints; six lamellæ under the thumb, eight under the fourth finger, five under the inner toe, and nine or ten under the fourth toe. Head covered with small convex granules, largest on the snout; rostral subquadrangular, not twice as broad as high, with median cleft above; nostril pierced between the rostral, the first labial, and three nasals; nine upper and seven lower labials; mental large, triangular, more than twice as long as the adjacent labials; four chin-shields, median pair largest and in contact behind the mental. Back covered with flat, subimbricate, smooth round scales, largest on the sides. Abdominal scales small, cycloid, imbricate, scarcely larger than the dorsals. Male with four præanal pores. Tail cylindrical, tapering, covered above with uniform small, smooth, subimbricate, flat scales, inferiorly with a median series of transversely dilated plates, commencing some distance behind the vent. Grey above, spotted with darker; a dark streak passing through the eye; tail with black annuli; lower surfaces dirty white.

Total length	77	millim.
Head	10	,,
Width of head	6	,,
Body	27	,,
Fore limb	10·5	,,
Hind limb	13·5	,,
Tail	40	,,

The larger, tail-less specimen measures 40 millim. from snout to vent.

Socotra.

a–b. ♂ ♀.   Socotra.   Prof. I. B. Balfour [C.]. (Types.)

2. Hemidactylus bouvieri.

Emydactylus bouvieri, *Bocourt, Arch. Mus.* vi. 1870, *Bull.* p. 17.
Hemidactylus cessacii, *Bocage, Jorn. Sc. Lisb.* iv. 1873, p. 211.

Snout acutely pointed, longer than the distance between the eye and the ear-opening, once and a half the diameter of the orbit; forehead scarcely concave; ear-opening very small, round. Body and limbs short. Digits moderately dilated, free, with short distal joints; three lamellæ under the inner digits, and four or five under the median digits. Upper surfaces covered with uniform rather large granules, smallest on the hinder part of the head. Rostral four-sided, not twice as broad as high, with median cleft above; nostril pierced between the rostral, the first labial, and three or four nasals; seven or eight upper and six or seven lower labials; mental large, triangular, nearly twice as long as the adjacent labials; two rather small chin-shields, followed by smaller ones passing gradually into the minute granules of the gular region. Abdominal scales large, cycloid, imbricate, smooth. Males with two præanal pores. Tail thick, cylindrical, tapering, covered above with uniform small squarish smooth scales, inferiorly with a median series of transversely dilated plates. Light brown above, with transverse darker bands on the back and tail; upper lip white; a dark streak on the side of the head, passing through the eye; lower surfaces white.

Total length	67	millim.
Head	10·5	,,
Width of head	7	,,
Body	25·5	,,
Fore limb	10	,,
Hind limb	13	,,
Tail	31	,,

Cape Verde Islands.

a. ♂.   San Jago.   Prof. B. du Bocage [P.].
  (As typical of *H. cessacii.*)
b–c. ♂ ♀.   S. Vicente.   Rev. R. T. Lowe [P.].

3. Hemidactylus reticulatus. (PLATE XI. fig. 2.)

Hemidactylus reticulatus, *Beddome, Madr. Journ. Med. Sc.* 1870.

Head short, oviform, very convex; snout a little longer than the distance between the eye and the ear-opening and than the diameter of the orbit; forehead not concave; ear-opening small, roundish. Body and limbs short. Digits short, free, with very short distal joint, moderately dilated; five or six lamellæ under the inner digits, seven or eight under the fourth finger, and eight or nine under the fourth toe. Snout covered with keeled granules; the rest of the head with smaller granules intermixed with round tubercles. Rostral four-sided, not twice as broad as high, with median cleft above; nostril pierced between the rostral, the first labial, and three or four nasals; nine or ten upper and seven to nine lower labials; mental large, triangular or pentagonal; four chin-shields, inner pair largest and in contact behind the point of the mental. Body covered above with coarse granules intermixed with numerous irregularly arranged, small, round, keeled tubercles. Abdominal scales rather small, cycloid, imbricate, smooth. Male with six to nine præanal pores forming an angular series. Tail cylindrical, tapering, covered above with small granular scales and rings of six or eight large conical tubercles, inferiorly with uniform small imbricated scales. Brown above, with a network of darker lines; many of the tubercles whitish; lower surfaces whitish, the throat sometimes vermiculated with brown.

Total length	64	millim.
Head	10·5	,,
Width of head	8·5	,,
Body	29·5	,,
Fore limb	12	,,
Hind limb	14·5	,,
Tail (reproduced)	24	,,

South India.

a-b. ♂ ♀.	Colegal.	Col. Beddome [C.]. (Types.)
c-d, e-g. ♂, ♀, & hgr.	Shevaroys.	Col. Beddome [C.].

4. Hemidactylus gracilis.

Hemidactylus gracilis, *Blanford, Journ. As. Soc.* xxxix. 1870, p. 362, pl. xvi. figs. 4-6.

Head narrow, elongate; snout a little longer than the distance between the eye and the ear-opening, once and one third or once and two fifths the diameter of the orbit; forehead not concave; ear-opening small, roundish. Body and limbs slender, feebly depressed. Digits free, very feebly dilated, inner well developed; the distal joint short; five lamellæ under the inner digits, seven under the fourth finger, and nine under the fourth toe. Snout with rather large polygonal rugose scales; hinder part of head with small granules. Rostral four-sided, not twice as broad as high, with median cleft above; nostril pierced between the rostral and four nasals; nine upper and seven lower labials; mental large, triangular, more than twice as long as the adjacent labials; four large chin-shields, inner pair in contact behind the point of the mental. Body covered above with coarse granules intermixed with oval, subtrihedral, strongly keeled tubercles arranged in about twelve irregular longitudinal series. Abdominal scales large, cycloid, imbricate, smooth. Male with six præanal pores forming an angular series. [Tail round, but slightly depressed at the base, and not at all further back, tapering, without any enlarged or spinose tubercles whatever.] Grey above, with subquadrangular black spots arranged in longitudinal series, confluent into lines on the rachis and sides; a black, above white-edged streak on the side of the head, passing through the eye; whitish beneath, specimen *a* with longitudinal grey lines.

From snout to vent	36	millim.
Head	10	,,
Width of head	5·5	,,
Fore limb	11	,,
Hind limb	14	,,

South-east Berar and Rajpoor.

a. ♂.	S.E. Berar.	W. T. Blanford, Esq. [C.]. (One of the types.)
b. ♂.	Rajpoor.	W. T. Blanford, Esq. [P.].

5. Hemidactylus frenatus.

Hemidactylus frenatus, *Gray, Cat.* p. 155.
Hemidactylus vittatus, *Gray, l. c.*
Hemidactylus frenatus (*Schleg.*), *Dum. & Bibr.* iii. p. 366; *Kelaart, Prodr. Faun. Zeyl.* p. 161; *Günth. Rept. Brit. Ind.* p. 108; *Stoliczka, Journ. As. Soc. Beng.* xxxix. 1870, p. 164, & xli. 1872, p. 96; *Anders. Zool. W. Yunnan,* p. 801.
—— (Pnoepus) javanicus, *Fitzing. Syst. Rept.* p. 106.
? Hemidactylus punctatus, *Jerdon, Journ. As. Soc. Beng.* xxii. 1854, p. 467.
Hemidactylus inornatus, *Hallow. Proc. Ac. Philad.* 1860, p. 492.
? Hemidactylus pumilus, *Hallow. l. c.* p. 502.
Gecko chaus, *Tytler, Journ. As. Soc. Beng.* xxxiii. 1864, p. 547.
—— caracal, *Tytler, l. c.*
Hemidactylus vittatus, *Gray, Zool. Erebus & Terror,* pl. xv. fig. 5.
—— longiceps, *Cope, Proc. Ac. Philad.* 1868, p. 320.

Snout longer than the distance between the eye and the ear-opening, once and one third to once and a half the diameter of the orbit; forehead concave; ear-opening small, roundish. Body and limbs moderate. Digits moderately dilated, free, inner with sessile claw; four or five lamellæ under the inner digits, seven or eight (seldom nine) under the fourth finger, and nine or ten under the fourth toe. Upper surfaces covered with small granules, largest on the snout; on the back these granules are intermixed with more or less numerous, sometimes almost entirely absent, irregularly scattered round convex tubercies always much smaller than the ear-opening. Rostral four-sided, nearly twice as broad as high, with median cleft above; nostril pierced between the rostral, the first labial, and three nasals; ten to twelve upper and eight to ten lower labials; mental large, triangular or pentagonal; two or three pairs of chin-shields, the median in contact behind the point of the mental. Abdominal scales moderate, cycloid, imbricate. Male with a series of thirty to thirty-six femoral pores, not interrupted on the præanal region. Tail rounded, feebly depressed, covered above with very small smooth scales and six longitudinal series of keeled tubercles, inferiorly with a median series of transversely dilated plates. Greyish or pinkish brown above, uniform or more or less distinctly marbled with darker; head generally variegated with brown; a more or less defined brown, above light-edged streak on the side of the head, passing through the eye, extending sometimes along the side of the body; lower surfaces whitish.

Total length	113	millim.
Head	16	,,
Width of head	11·5	,,
Body	39	,,
Fore limb	18	,,
Hind limb	21	,,
Tail	58	,,

China, Indo-China, Malay peninsula, Southern India, islands of the Western Pacific and Indian Oceans, St. Helena.

a. ♀.	Corea.	Sir E. Belcher [P.].
b-c. ♀.	Hainan.	R. Swinhoe, Esq. [C.].
d-f. ♀ & hgr.	Taiwanfoo, S. Formosa.	
g-h. ♂.	Nilgherries.	F. Day, Esq. [P.].
i. ♀.	Salem.	Col. Beddome [C.].
k. ♂	Calcad Hills.	Col. Beddome [C.].
l-o. ♂ ♀.	Anamallays.	Col. Beddome [C.].
p. ♂.	S. India.	Col. Beddome [C.].
q-r, s. ♂, ♀, & hgr.	Ceylon.	
t. Dried.	Ceylon.	Dr. Kelaart.
u-v. ♂ ♀.	Ceylon.	H. Barnes, Esq. [P.].
w-y. ♂ & hgr.	Birma.	F. Day, Esq. [P.].
z-a. ♂.	Gamboja.	M. Mouhot [C.].
β. ♀.	Siam.	M. Mouhot [C.].
γ-ε. ♂.	Pinang.	Dr. Cantor.
ζ-θ. ♂ ♀.	Andamans.	Col. Tytler [P.].
ι. ♂.	Nias.	Hr. H. Sandemann [C.].
κ. ♂.	Java.	
λ. ♂.	Batavia.	
μ-ν. ♀.	Borneo.	Sir E. Belcher [P.].
		(Types of *H. vittatus*.)
ξ-o. ♂.	Luzon.	Dr. A. B. Meyer [C.].
π. ♂.	Puerto Princesa.	A. Everett, Esq. [C.].
ρ. Several spec.: ♂, ♀, & hgr.	Zebu, Philippines.	H.M.S. 'Challenger.'
σ. Several spec., ♂ ♀.	Manado, Celebes.	Dr. A. B. Meyer [C.].
τ. ♀.	Murray Island.	Rev. S. McFarlane [C.].
υ. ♂.	Port Essington.	(Type of *H. vittatus*.)
φ. ♀.	Ke Daulau, Ki Islands.	H.M.S. 'Challenger.'
χ. Several spec.: ♂, ♀, yg., & eggs.	Eagle Island, Amirantes.	H.M.S. 'Alert.'
ψ. ♂.	Mauritius.	
ω-cc. ♂ ♀.	St. Helena.	J. C. Melliss, Esq. [P.].
dd, ee. ♂.	—— ?	

6. Hemidactylus mabouia.

Hemidactylus mabouia, *Gray, Cat.* p. 154.
Hemidactylus mercatorius, *Gray, l. c.* p. 155.
Gecko mabouia, *Moreau de Jonnès, Bull. Soc. Philom.* 1818, p. 138; *Cuv. R. A.* ii. p. 54.
—— tuberculosus, *Raddi, Mem. Soc. Modena,* xix. 1823, p. 63.
—— aculeatus, *Spix, Spec. Nov. Lacert.* p. 16, pl. xviii. fig. 3.
—— cruciger, *Spix, l. c.* p. 16.
Thecadactylus pollicaris, *Spix, l. c.* p. 17, pl. xviii. fig. 2.
Gecko incanescens, *Wied, Beitr. Naturg. Bras.* p. 102.
—— armatus, *Wied, l. c.* p. 104.
Hemidactylus mabouia, *Dum. & Bibr.* iii. p. 362; *Guichen. in Casteln. Amer. Sud, Rept.* p. 12; *Bianconi, Spec. Zool. Mossamb.* p. 21, pl. i. f. 1; *Girard, U.S. Explor. Exped., Herp.* p. 283, pl. xxv. figs. 9–16; *Boettger, Abh. Senck. Ges.* xi. 1879, p. 478, & xii. 1881, p. 467; *Peters, Reise n. Mossamb.* iii. p. 27. pl. v. fig. 3.
—— mercatorius, *Gray, Zool. Misc.* p. 58; *Boettger, Abh. Senck. Ges.* xi. 1877, p. 23, pl. i. fig. 4.
—— (Tachybates) mabuya, *Fitzing. Syst. Rept.* p. 105.
—— (——) armatus, *Fitzing. l. c.*
—— platycephalus, *Peters, Mon. Berl. Ac.* 1854, p. 615; *Bocage, Jorn. Sc. Lisb.* iv. 1873, p. 209.
—— hexaspis, *Cope, Proc. Ac. Philad.* 1868, p. 320.
—— tuberculosus, *Peters, eod. loc.* 1877, p. 414.
—— frenatus (*non D. & B.*), *Boettg. eod. loc.* 1878, p. 275, pl. i. fig. 1.

Snout considerably longer than the distance between the eye and the ear-opening, once and a half to once and two thirds the diameter of the orbit; forehead concave; ear-opening oval, oblique, half or two fifths the diameter of the eye. Body and limbs moderate. Digits free, moderately dilated, inner well developed; five or six lamellæ under the inner digits, and seven to nine under the median ones. Snout covered with small convex granules; hinder part of head with minute granules intermixed with round tubercles; rostral four-sided, nearly twice as broad as high, with median cleft above; nostril pierced between the rostral, the first labial, and three or four nasals; ten to fourteen upper and nine or ten lower labials; mental large, triangular; two or three pairs of chin-shields, median largest and in contact behind the point of the mental. Upper surface of body covered with small granules intermixed with small, irregularly scattered, convex or subtrihedral tubercles Abdominal scales small, cycloid, imbricate. Male with a long uninterrupted series of femoral pores, fifteen to thirty on each side. Tail cylindrical, feebly depressed, covered above with very small scales intermixed with large conical tubercles in six longitudinal series; inferiorly with a median series of transversely dilated plates. Grey or light brown above, with dark spots or undulated cross bands; white beneath.

Total length	166	millim.
Head	22	,,
Width of head	17	,,
Body	59	,,

```
Fore limb  ............  27 millim.
Hind limb ............  36   ,,
Tail  ...............   85   ,,
```

South America; West Indies; Madagascar; South Africa.

a-d, e. ♂, ♀, & hgr.	Brazil.	
f. ♂.	Brazil.	Lord Stuart de Rothsay [P.].
g. ♂.	Rio Janeiro.	
h, i. Yg.	Bahia.	
k. Hgr.	Pernambuco.	Mrs. J. P. G. Smith [P.].
l, m. ♂ & hgr.	Para.	
n. ♂.	Santarem.	H. W. Bates, Esq. [C.].
o. ♀.	Vera Cruz.	
p. ♂.	Mexico (?).	
q. ♂.	S. Domingo.	
r. ♀.	S. Domingo.	M. Sallé [C.].
s. ♂.	Dominica.	G. F. Angas, Esq. [P.].
t. ♂.	Madagascar.	Dr. J. E. Gray [P.]. (Type of *H. mercatorius.*)
u-v. ♂.	Madagascar.	Mr. Bartlett [C.].
w-z. ♂ ♀.	E. Betsileo.	Rev. W. Deans Cowan [P.].
α. ♂.	Antananarivo.	Rev. J. Wills [C.].
β. Several spec., ♂ ♀.	Johanna, Comoro Islands.	C. E. Bewsher, Esq. [P.].
γ. ♂.	Zambezi.	Dr. Livingstone [C.].
δ. ♂.	Shiré Valley, Zambezi.	
ε-θ. ♂, ♀, & hgr	Zanzibar.	Sir J. Kirk [C.].
ι-λ. ♀.	Carangigo.	Dr. Welwitsch [C.].
μ-ν, ξ. Hgr. & yg.	—— ?	C. Darwin, Esq. [C.].

7. Hemidactylus muriceus.

Hemidactylus muriceus, *Peters, Mon. Berl. Ac.* 1870, p. 641.

" Numerous small pointed conical tubercles irregularly scattered among the fine dorsal granulation. Tail with similar longer tubercles. Snout with larger convex scales, eight or nine infralabials. A larger chin-shield, followed by smaller ones, on each side behind the pointed mental. Ear-opening almost vertical, rather narrow. Chin and throat finely granulate. Ventral scales in the middle in thirty-three longitudinal rows; on each side a feeble longitudinal fold on which the lower tubercles are inserted. Grey-brown, with narrow **M**- or **V**-shaped dark irregular transverse bands; similar cross bands on the tail and limbs. Lower surfaces lighter with small darker dots, more numerous and confluent on the chin. Total length 86 millim.; head 13 millim.; tail 46 millim."

Keta, Guinea.

8. Hemidactylus echinus. (PLATE XI. fig. 3.)

Hemidactylus echinus, *O'Shaughn. Ann. & Mag. N. H.* (4) xvi. 1875, p. 264.

Snout rather pointed, longer than the distance between the eye and the ear-opening, once and one third the diameter of the orbit;

forehead concave; ear-opening very small, oblique. Body and limbs moderate. Digits webbed at the base, moderately dilated, inner well developed; infradigital lamellæ very numerous, oblique; ten under the inner digits, eleven under the fourth finger, and thirteen under the fourth toe. Head covered with minute granules, largest on the snout, intermixed on the supraorbital, occipital, and temporal regions with larger round tubercles; rostral four-sided, nearly twice as broad as high, with median cleft above; nostril pierced between the rostral, the first labial, a large superonasal, and several minute granules; twelve or thirteen upper and ten lower labials; mental large, triangular, its posterior angle truncate and in contact with a small median chin-shield; four other small chin-shields. Upper surface of body minutely granular, with numerous small convex tubercles scattered irregularly; on each side of the belly a series of long, spine-like scales. Abdominal scales very small, scarcely imbricate. Male with a curved series of eight præanal pores. Tail quadrangular, minutely scaled, the ridges with projecting spinous tubercles. Brown, lighter inferiorly; a dark spot on the centre of the nape; a light streak on the side of the head, passing through the eye; some rather indistinct lighter mottlings on the back; a round yellow spot on the hind part of the thigh, close to the root of the tail; latter with darker and lighter annuli.

Total length	114 millim.
Head	16 ,,
Width of head	11·5 ,,
Body	42 ,,
Fore limb	20 ,,
Hind limb	25 ,,
Tail	56 ,,

Gaboon.

a. ♂. Gaboon. (Type.)

9. Hemidactylus fasciatus. (PLATE XI. fig. 4.)

Hemidactylus fasciatus, *Gray, Cat.* p. 154.
Leiurus ornatus, *Gray, l. c.* p. 157.
Hemidactylus formosus, *Hallow. Proc. Ac. Philad.* 1856, p. 148.

Snout slightly longer than the distance between the eye and the ear-opening, once and one fourth the diameter of the orbit; forehead concave; ear-opening suboval, oblique, half or three fifths the diameter of the eye. Body and limbs moderate. Digits distinctly webbed at the base, moderately dilated, inner well developed; six to eight lamellæ under the inner digits, and nine or ten under the median ones. Snout covered with small convex granules; hinder part of head with minute granules intermixed with small round tubercles; rostral four-sided, not twice as broad as high, with median cleft above; nostril pierced between the rostral, the first labial, and three or four nasals; nine to eleven upper and eight to ten lower labials; mental large, triangular or pentagonal; two chin-shields, in contact behind the mental. Upper surface of body covered with small granules intermixed with small round convex tubercles, forming about twenty more or less regular longitudinal series. Abdominal scales rather small, cycloid, imbricate. Males with a long series of femoral pores, interrupted in the middle, composed of eighteen to twenty pores on each side. Tail cylindrical, tapering, covered above with very small, uniform, smooth, imbricate scales, inferiorly with a median series of large transverse plates. Pale chocolate-brown above; a large U-shaped band from the eyes across the nape, two other cross bars on the back and one on the base of the tail, and complete broad rings round the tail, dark brown, light-edged; lower surfaces dirty white.

Total length	172	millim.
Head	21	,,
Width of head	17	,,
Body	59	,,
Fore limb	26	,,
Hind limb	32	,,
Tail	92	,,

West Africa.

a. ♂.	——?	(Type.)
b. Yg.	W. Africa.	(Type of *Leiurus ornatus*.)
c. ♂.	W. Africa.	Dr. A. Günther [P.].
d. ♂.	Ashantee.	
e. ♀.	Camaroons.	
f–g, h. ♂, ♀. & hgr.	Fernando Po.	Mr. Fraser [C.].
i. ♀.	Gaboon.	

10. Hemidactylus bocagii.

Hemidactylus longicephalus, *Bocage, Jorn. Sc. Lisb.* iv. 1873, p. 210.
—— longiceps, (non *Cope*) *O'Shaughn. Zool. Rec.* x. p. 89.

Head elongate oviform; snout much longer than the distance between the eye and the ear-opening, about once and two fifths the diameter of the orbit; forehead feebly concave; ear-opening suboval or subtriangular, about one third the diameter of the eye. Body rather elongate; limbs moderate. Digits moderate, free, inner well developed; five or six lamellæ under the inner digits, seven or eight under the fourth finger, and eight to ten under the fourth toe. Snout covered with large keeled granules; hinder part of head with minute granules intermixed with small tubercles. Rostral four-sided, nearly twice as broad as high, with median cleft above; nostril pierced between the rostral and three or four nasals, the rostral rarely entering the nostril; nine or ten upper and eight or nine lower labials; mental large, triangular; two pairs of chin-shields, median largest and in contact behind the point of the mental. Upper surface of body covered with minute granules intermixed with small trihedral, strongly keeled tubercles arranged in fourteen to eighteen irregular longitudinal series. Abdominal scales small, cycloid imbricate. Males with four to eight præanal pores, and a strong conical tubercle on each side of the base of the tail. Latter cylindrical, slightly depressed, tapering, covered above with minute scales intermixed with large spine-like tubercles forming six longitudinal rows; inferiorly with small irregular imbricate scales. Grey-brown above, uniform or with indistinct darker markings; dirty white beneath.

Total length	111 millim.
Head	17 ,,
Width of head	12 ,,
Body	40 ,,
Fore limb	23 ,,
Hind limb	28 ,,
Tail	54 ,,

West Africa.

a–b. ♂ ♀.	Gaboon.	
c. Many spec.: ♂, ♀, & yg.	Ambriz, Angola.	
d. ♂.	Pungo Adongo.	Dr. Welwitsch [C.].
e–g. ♂ ♀.	Carangigo.	Dr. Welwitsch [C.].

11. Hemidactylus sinaitus.

To be distinguished from *H. turcicus* by the following characters:— Digits very short and with fewer lamellæ, viz. five under the inner digits, seven under the fourth finger, and eight under the fourth toe. Rostral not entering the nostril. No enlarged subcaudals. Nine upper and eight lower labials. Dorsal tubercles in fourteen longitudinal series. Four præanal pores. Grey-brown above, most of the dorsal tubercles white.

Total length	83 millim.
Head	11 ,,
Width of head	7·5 ,,
Body	28 ,,
Fore limb	13 ,,
Hind limb	18 ,,
Tail	44 ,,

Mount Sinai.

a. ♂. Mt. Sinai.

12. Hemidactylus turcicus.

Hemidactylus verruculatus, part., *Gray, Cat.* p. 154.
Lacerta turcica, *Linn. S. N.* i. p. 362.
Gecus cyanodactylus, *Rafin. Caratt. Nuov. Gener. Spec. Sicil.* p. 9.
Gecko verruculatus, *Cuv. R. A.* ii. p. 54.
—— meridionalis, *Risso, Hist. Nat. Eur. Mérid.* iii. p. 87.
Hemidactylus granosus, *Rüppell, Atlas N. Afr., Rept.* p. 17, pl. v. fig. 1.
—— robustus, *Rüpp. l. c.* p. 19.
—— verrucosus, *Gray, Griff. A. K.* ix. *Syn.* p. 50.

Hemidactylus triedrus, *Bonap. Faun. Ital.*
—— verruculatus, *Bibr. in Bory, Expéd. Sc. Morée, Rept.* p. 68, pl. xi. fig. 2; *Dum. & Bibr.* iii. p. 359; *Bonap. Amph. Eur.* p. 29; *De Betta, Faun. Ital., Rett. Anf.* p. 20; *Schreib. Herp. Eur.* p. 487; *Boettger, Ber. Senck. Ges.* 1879–80, p. 192.
—— cyanodactylus, *Strauch, Erp. Alg.* p. 23.
—— turcicus, *Boettger, Ber. Offenb. Ver. Nat.* 1876, p. 57.
—— karachiensis, *Murray, Zool. Sind,* p. 361, pl.—. fig. 2.

Snout rounded, about as long as the distance between the eye and the ear-opening, once and one fourth to once and one third the diameter of the orbit; forehead slightly concave; ear-opening oval, oblique, not quite half the diameter of the eye. Body and limbs moderate. Digits rather variable in the length, the inner always well developed; six to eight lamellæ under the inner digits, eight to ten under the fourth finger, and nine to eleven under the fourth toe. Head anteriorly with large granules, posteriorly with minute granules intermixed with round tubercles. Rostral four-sided, not twice as broad as high, with median cleft above; nostril pierced between the rostral, the first labial, and three nasals; seven to ten upper and six to eight lower labials; mental large, triangular, at least twice as long as the adjacent labials, its point between two large chin-shields, which may be in contact behind it; a smaller chin-shield on each side of the larger pair. Upper surface of body covered with minute granules intermixed with large tubercles; these are generally larger than the interspaces between them, suboval, trihedral, and arranged in fourteen or sixteen pretty regular longitudinal series. Abdominal scales small, smooth, roundish-hexagonal, imbricate. Males with a short angular series of four to ten (exceptionally two) præanal pores. Tail cylindrical, slightly depressed, tapering, covered above with minute scales and transverse series of large keeled tubercles, inferiorly with a series of large transversely dilated plates. Light brown or greyish above, spotted with darker; many of the tubercles white; lower surfaces white.

Total length	90	millim.
Head	14	,,
Width of head	10·5	,,
Body	39	,,
Fore limb	17	,,
Hind limb	23	,,
Tail (reproduced)	37	,,

Borders of the Mediterranean and of the Red Sea; Sind.

a–b. ♀	Mediterranean.	R. Hannah, Esq. [P.].
c, d–g. ♀ & yg.	S. Europe.	P. Webb, Esq. [P.].
h. Yg.	Spain.	Lord Lilford [P.].
i. Yg.	Mt. Santa Cruz, Algeria.	M. F. Lataste [P.].
k. ♀.	Algeria.	M. F. Lataste [P.].
l. ♂.	Marsala, Sicily.	Dr. O. Boettger [P.].
m–o. ♂ & yg.	Sicily.	T. Bell, Esq. [P.].

p. Hgr.	Smyrna.	J. McAndrew, Esq. [P.].
q–t. ♀ & hgr.	Haifa.	Dr. O. Boettger [P.].
u. ♀.	Road to Euphrates.	C. G. Danford, Esq. [P.]
v. ♀.	Egypt.	J. Burton, Esq. [P.].
w–z, a–β. ♂ ♀.	Kurrachee.	J. A. Murray, Esq. [P.].

(As typical of *H. karachiensis.*)

In the latter specimen (*H. karachiensis,* Murray) the body is more slender than usual, but the degree of elongation is a character too variable in *H. turcicus* to be relied upon for distinguishing a species; the tubercles are also rather smaller, and the ground coloration is very light, the markings, especially the streak passing through the eye, being well accentuated. If these characters are constant, they may indicate a distinct race, but, in my opinion, not a species.

13. Hemidactylus brookii.

Hemidactylus brookii, *Gray, Cat.* p. 153.
Hemidactylus verruculatus, part., *Gray, l. c.* p. 154.
Hemidactylus cyanodactylus (*non Rafin.*), *Girard, U. S. Explor. Exped., Herp.* p. 254, pl. xxxv. figs. 17–24.
—— brookii, *Gray, Zool. Erebus and Terror,* pl. xv. fig. 2.
—— affinis, *Steindachn. Sitzb. Ak. Wien,* lxii. i. 1870, p. 328; *Boettg. Abh. Senck. Ges.* xii. 1881, p. 106.
—— guineensis (*non Peters*), *Bocage, Jorn. Sc. Lisb.* iv. 1873, p. 209.

Distinguished from *H. turcicus* in the following points:—Snout more pointed. Infradigital lamellæ less numerous, viz. four to six under the inner digits, six or seven under the fourth finger, and seven or eight under the fourth toe. Tubercles smaller, likewise variable in size, in sixteen to twenty longitudinal series on the back. Ventral scales a little larger. Males with a long, frequently uninterrupted series of femoral pores; these number twenty to forty altogether. Eight to ten upper and seven to nine lower labials. Grey-brown above, with darker spots, which are generally large, quadrangular, and forming longitudinal and transverse series, or a ladder-like pattern on the back; a dark streak on the side of the head, passing through the eye.

This species resembles also very much *H. indicus,* but is distinguished by a smaller head, shorter digits, and a greater number of femoral pores.

Total length	118	millim.
Head	16	,,
Width of head	12	,,
Body	42	,,
Fore limb	19	,,
Hind limb	25	,,
Tail	60	,,

West Africa.

a, b. ♂.	[Borneo.]	Sir E. Belcher [P.].
c. ♂.	[Australia.]	Lord Derby [P.]. } (Types.)
d. ♂.	Cape Verde.	
e–f. ♂ & yg.	Porto Praya.	C. Darwin, Esq. [C.].
g. ♂.	Sierra Leone.	H. C. Hart, Esq. [C.].
h. ♀.	Fernando Po.	Mr. Fraser [C.].
i. ♀.	Old Calabar.	Mr. Rutherford [C.].
k–l. ♂ ♀.	W. Africa.	Mr. Dalton [C.].
m–n, o. ♂ ♀.	W. Africa.	
p. ♂.	——?	T. Bell, Esq. [P.].

14. Hemidactylus gleadovii.

Hemidactylus maculatus, part., *Gray, Cat.* p. 153.
Hemidactylus maculatus, *Kelaart, Prodr. Faun. Zeyl.* p. 158; *Günth. Rept. Brit. Ind.* p. 107; *Stoliczka, Journ. As. Soc. Beng.* xxxix. 1870, p. 164; *Blanford, tom. cit.* p. 361; *Stoliczka, op. cit.* xli. 1872, p. 94; *Anders. Zool. W. Yunnan*, i. p. 800.
? Gecko tytleri, *Tytler, Journ. As. Soc. Beng.* xxxiii. 1864, p. 547.
Hemidactylus gleadowi, *Murray, Zool. Sind*, p. 360, pl. —. fig. 3.

Head oviform; snout longer than the distance between the eye and the ear-opening, once and one fourth or once and one third the diameter of the orbit; forehead concave; ear-opening oval, oblique, about half the diameter of the eye. Body and limbs moderate. Digits moderately dilated, free, inner well developed; infradigital lamellæ with strongly curved anterior border, four to six under the inner digits, six to eight under the median. Head covered anteriorly with convex granular, sometimes keeled, scales, posteriorly with minute granules intermixed with round tubercles; rostral subtetragonal, not twice as broad as high, with median cleft above; nostril pierced between the rostral, the first labial, and three or four nasals; eight to ten upper and seven to nine lower labials; mental large, triangular or pentagonal, twice as long as the adjacent labials; chin-shields generally four, the inner largest, in contact behind the mental. Upper surfaces with minute granular scales intermixed with moderate-sized trihedral strongly-keeled tubercles, the largest not measuring more than one third the diameter of the eye; they form sixteen to twenty pretty regular longitudinal series on the back. Abdominal scales smooth, roundish, imbricate. Males with a series of femoral pores generally interrupted in the middle; altogether sixteen to twenty-four pores. Tail rounded, tapering, depressed, above with small smooth scales and six or eight longitudinal series of large, pointed, strongly keeled tubercles, inferiorly with a median series of transversely dilated plates. Brownish above, spotted with darker; a dark streak on the side of the head, passing through the eye; lower surfaces dirty white.

Total length	117 millim.
Head	16 ,,
Width of head	12 ,,

Body	44 millim.
Fore limb	19 ,,
Hind limb	24 ,,
Tail	57 ,,

East Indian continent; South China; Ceylon.

a. ♂.	India.	Gen. Hardwicke [P.].
b. ♂.	India.	C. Bowring, Esq. [P.].
c–d. ♂ ♀.	India.	
e–f. ♂ ♀.	Sind.	J. A. Murray, Esq. [P.]. (As typical of *H. gleadowi*.)
g. ♀ dry.	Lahore.	Dr. Cantor.
h–k. ♂ ♀.	Bengal.	W. Masters, Esq. [P.].
l. ♀.	Bombay.	Dr. Leith [P.].
m. ♀.	Poona.	Dr. Leith [P.].
n–p. ♂ ♀.	Rajpoor.	W. T. Blanford, Esq. [P.].
q, r–t. ♂ ♀.	Godavery Valley.	W. T. Blanford, Esq. [P.].
u. ♂.	Anamallays.	Col. Beddome [P.].
v. Hgr.	Ceylon.	W. Ferguson, Esq. [P.].
w. ♀.	Birma.	W. Theobald, Esq. [P.].
x. ♂.	China.	J. Reeves, Esq. [P.].
y. ♀.	Ningpo.	

15. Hemidactylus stellatus. (Plate XII. fig. 1.)

Allied to *H. brookii* and *gleadovii*. Snout considerably longer than the distance between the eye and the ear-opening, pointed. Infradigital lamellæ few, four or five under the inner digits, six or seven under the fourth finger, and seven or eight under the fourth toe; only the distal of these lamellæ are well developed, the basal ones being small, tubercle-like. Eight or nine upper and six or seven lower labials. Lepidosis as in *H. brookii*, the tubercles forming eighteen to twenty-two longitudinal series. Males with eight femoral pores on each side, the two series widely separated. Light chocolate-brown above; most of the dorsal tubercles darker, the others pure white; two dark cross bands on scapular region, followed by others less distinct; a dark streak on the side of the head passing through the eye; tail with rings of white tubercles; lower surfaces white.

Total length	105 millim.
Head	14 ,,
Width of head	11 ,,
Body	37 ,,
Fore limb	17 ,,
Hind limb	22 ,,
Tail	54 ,,

West Africa.

a. ♂.	Gambia.	Mr. Dalton [C.].
b–c. ♀.	W. Africa.	Mr. Dalton [C.].
d–g. ♂ ♀.	W. Africa.	

16. Hemidactylus guineensis.

Hemidactylus guineensis, *Peters, Mon. Berl. Ac.* 1868, p. 640.

" Very closely allied to *H. turcicus*: distinguished by the smaller and more numerous tubercles, which form twenty or twenty-two longitudinal rows on the back, and by the small number of infradigital lamellæ, viz. two or three under the inner digits and four or five under the others. Ten or eleven upper and eight lower labials. The young has three broad dark bands across the back."

Guinea.

17. Hemidactylus persicus.

Hemidactylus persicus, *Anderson, Proc. Zool. Soc.* 1872, p. 378, fig. 2; *Blanford, Zool. E. Persia*, p. 342, *and Journ. As. Soc. Beng.* xlv. 1876, p. 18.

Snout rather acuminate, as long as the distance between the eye and the upper border of the ear-opening, once and one third the diameter of the orbit; forehead concave; interorbital space very narrow; upper eyelid strongly fringed; ear-opening large, obliquely crescentic, the concavity being directed forwards and upwards, its diameter equalling three fourths that of the eye. Body and limbs moderate. Digits free, moderately dilated, inner well developed; infradigital lamellæ obliquely curved; ten lamellæ under the thumb, ten under the third finger, nine under the inner toe, and twelve under the third toe. Snout covered with large convex granular scales, largest between the eye and the nostril; hinder part of head with minute granules, and scattered ones of a larger size. Rostral four times as broad as high (having fused with the anterior labial on each side); nostril above the rostral, between the latter and three nasals; ten or eleven upper and nine lower labials; mental large, triangular, twice as long as the adjacent labial, its point between two large chin-shields which are in contact behind it; a small chin-shield on each side of the large pair. Upper surface of body covered with small irregular flat granules and moderately large, trihedral, strongly keeled tubercles arranged in fourteen or sixteen rather irregular longitudinal series; the largest tubercles measure about one fourth the diameter of the eye. Abdominal scales small, smooth, rounded, imbricate. Male with a short angular series of eight præanal pores. Tail cylindrical, tapering, covered above with small, irregular, imbricate smooth scales and scattered pointed tubercles forming four or six longitudinal series, inferiorly with a series of transversely dilated plates. Dr. Anderson describes the colour as "pale yellowish brown, with six faint brownish transverse narrow dorsal bands, the tubercles in these areas being almost black; a dark brown streak from the nostrils through the eye above the ear, with a whitish line above it; lips whitish." The unique specimen in the collection is uniform whitish (bleached?).

Total length	149	millim.
Head	19	,,
Width of head	14	,,

	Body	50 millim.
	Fore limb	22 „
	Hind limb	28 „
	Tail	80 „

Persia; Sind.

♂. Near Rohri, Sind. W. T. Blanford, Esq. [P.].

18. Hemidactylus maculatus.

Hemidactylus maculatus, part., *Gray, Cat.* p. 153.
Hemidactylus maculatus, part., *Dum. & Bibr.* iii. p. 358.
—— sykesii (*Gray*), *Günth. Rept. Brit. Ind.* p. 108, pl. xii. fig. C.

Head large, oviform; snout longer than the distance between the eye and the ear-opening, once and one fourth the diameter of the orbit; forehead concave; canthus rostralis swollen; ear-opening large, oval. Body and limbs moderate. Digits moderately dilated, free, inner well developed; infradigital lamellæ almost perfectly straight, nine to eleven under the inner digits, twelve or thirteen under the median. Head covered anteriorly with convex granular scales, smaller in the frontal concavity, posteriorly with minute granules intermixed with round tubercles; rostral subtetragonal, not twice as broad as high, with median cleft above; nostril pierced between the rostral, the first labial, and three or four nasals; ten to twelve upper and nine or ten lower labials; mental large, triangular or pentagonal, twice as long as the adjacent labials; two pairs of chin-shields, inner largest, elongate, in contact behind the mental. Upper surfaces with minute granular scales intermixed with moderate-sized trihedral, more or less strongly keeled tubercles, the largest not measuring more than one third the diameter of the eye; they are arranged very irregularly on the back, in about twenty longitudinal series. Abdominal scales smooth, roundish, imbricate. Males with a long series of præanal pores, nineteen to twenty-five on each side, interrupted on the præanal region. Tail rounded, tapering, depressed, above with small irregular keeled scales, and six or eight longitudinal series of large trihedral tubercles; inferiorly with a median series of transversely dilated plates. Brown above, with darker spots, generally confluent into transverse undulated bands on the back; two more or less distinct dark streaks on each side of the head, passing through the eye; lower surfaces dirty white.

	Total length	244 millim.
	Head	32 „
	Width of head	24 „
	Body	82 „
	Fore limb	40 „
	Hind limb	51 „
	Tail	130 „

India.

a. ♂.	Deccan.	Col. Sykes [P.]. (Type of *H. sykesii*.)
b–c, d. ♂ ♀.	Matheran.	Dr. Leith [P.].
e–k. ♂ ♀.	Salem.	Col. Beddome [C.].
l–n. ♂ ♀.	Tinnevelly.	Col. Beddome [C.].
o, p. ♂ & yg.	Malabar.	Col. Beddome [C.].

19. Hemidactylus triedrus.

Hemidactylus trihedrus, *Gray, Cat.* p. 152.
Gecko triedrus, *Daud. Rept.* iv. p. 155.
Hemidactylus triedrus, *Lesson, in Bélang. Voy. Ind. Or.* p. 311, pl. v. fig. 1; *Dum. & Bibr.* iii. p. 356, pl. xxviii. fig. 8; *Kelaart, Prodr. Faun. Zeyl.* p. 157; *Günth. Rept. Brit. Ind.* p. 107; *Stoliczka, Journ. As. Soc. Beng.* xli. 1872, p. 93; *Blanford, op. cit.* xlviii. 1879, p. 124.
—— (Tachybates) triedrus, *Fitzing. Syst. Rept.* p. 105.

Head large, oviform; snout longer than the distance between the eye and the ear-opening, once and two fifths the diameter of the orbit; forehead concave; ear-opening large, suboval, oblique, measuring about half the diameter of the eye. Body and limbs moderate. Digits free, moderately dilated, inner well developed; infradigital lamellæ slightly oblique, six or seven under the inner digits, eight to ten under the median digits. Snout covered with convex granules, which may be keeled; hinder part of head with minute granules intermixed with roundish tubercles. Rostral subquadrangular, not twice as broad as high, with median cleft above; nostril pierced between the rostral, the first labial, and three or four nasals; eight to ten upper and seven or eight lower labials; mental large, triangular or pentagonal, at least twice as long as the adjacent labials; four chin-shields, median pair largest and in contact behind the mental. Upper surface of body covered with small flat granular scales, and large trihedral tubercles arranged in sixteen to twenty more or less irregular longitudinal series; these tubercles vary somewhat in size according to specimens, but the largest never exceed two fifths the diameter of the eye. Abdominal scales large, smooth, rounded, imbricate. Males with a series of præanal pores, interrupted medially; six to eight pores on each side. Tail rounded, feebly depressed, tapering, covered above with irregular, small, smooth inbricated scales and rings of large, pointed, keeled tubercles, inferiorly with a median series of transversely dilated plates. Light pinkish brown above, generally with more or less defined transverse darker bands bordered by pure white tubercles surrounded by deep-brown rings; young very regularly barred with dark brown, there being four dark bars between head and hind limbs; a more or less defined dark-brown streak, white-edged above, on the side of the head, passing through the eye; lower surfaces white. Specimen *i* has five white, deep-brown-edged cross bars, one just behind the head, the second on scapula, two on the back, and the last on sacrum.

Total length	192 millim.
Head	21 ,,
Width of head	16 ,,
Body	73 ,,
Fore limb	25 ,,
Hind limb	31 ,,
Tail	98 ,,

India; Ceylon.

a. ♂.	Madras.	Sir W. Elliot [P.].
b. Hgr.	Madras.	Sir J. Boileau [P.].
c–f. ♂, hgr., & yg.	Madras.	T. C. Jerdon, Esq. [P.].
g. Several spec.: hgr. & yg.	Cauvery.	Col. Beddome [C.].
h. Yg.	Sevagherry.	Col. Beddome [C.].
i. ♀.	Ajmeer.	W. T. Blanford, Esq. [P.].
k. ♂.	India.	Sir J. McGregor [P.].
l. Several spec.: ♂,hgr., & yg.	India.	
m. ♂.	India.	Haslar Collection.

20. Hemidactylus subtriedrus.

Hemidactylus subtriedrus, *Jerdon, Journ. As. Soc. Beng.* xxii. 1853, p. 467; *Stoliczka, op. cit.* xli. 1872, p. 93, pl. ii. fig. 1.

Differs from *H. triedrus* in the following points:—Head more depressed. Infradigital lamellæ more numerous, ten under the thumb, eleven under the median finger, nine under the inner toe, and twelve under the median toe. Ventral scales smaller. Ten lower labials. Light brown above, with five undulated dark-brown cross bands on the body, the anterior on the neck, confluent with a dark streak passing through the eye and light-edged above.

Total length	114 millim.
Head	17 ,,
Width of head	12 ,,
Body	40 ,,
Fore limb	19 ,,
Hind limb	26 ,,
Tail	57 ,,

Ellore.

a. ♀.	Kamarmet Sircar, near Bhadrachalam.	W. T. Blanford, Esq. [P.].

21. Hemidactylus depressus.

Hemidactylus depressus, *Gray, Cat.* p. 153.
Nubilia argentii, *Gray, l. c.* p. 273.
Hemidactylus depressus, *Gray, Zool. Misc.* p. 58, *and Zool. Erebus & Terror*, pl. xv. fig. 1.
—— pieresii, *Kelaart, Prodr. Faun. Zeyl.* p. 159; *Stoliczka, Journ. As. Soc. Beng.* xli. 1872, p. 94.

Head oviform; snout longer than the distance between the eye and the ear-opening, once and one third or once and one fourth the diameter of the orbit; forehead concave; ear-opening oblique, nearly half the diameter of the eye. Body and limbs moderate. The skin forms a fold on each side of the belly, from axilla to groin. Digits distinctly webbed at the base, moderately dilated, inner well developed, with curved, scarcely oblique lamellæ; latter, seven (or six) under the inner digits, nine or ten under the median fingers, ten or eleven under the median toes. Snout covered with granular scales, largest in front of the orbits; hinder part of head with minute granules intermixed with round tubercles; rostral subquadrangular, not twice as broad as high, with median cleft above; nostril pierced between the rostral, the first labial, and three or four nasals; ten to twelve upper and eight to ten lower labials; mental large, triangular or pentagonal, twice as long as the adjacent labials; four large chin-shields, median pair largest, forming a long suture behind the point of the mental. Upper surface of body covered with small flat granules intermixed with trihedral tubercles, forming sixteen to twenty very irregular longitudinal series; the largest tubercles measure about one fourth the diameter of the eye. Abdominal scales moderate, smooth, rounded, imbricate. Males with a long series of femoral pores interrupted medially; seventeen or eighteen pores on each side. Tail tapering, much depressed, flat inferiorly, with angular lateral edge; it is covered above with small pointed imbricated scales, which are enlarged and raised on the sides, forming a strong denticulation, and six or eight series of spine-like tubercles forming rings; a median series of regular transversely dilated plates inferiorly. Upper parts light brown, marbled or cross-barred with brown; a dark brown, above white-edged streak on the side of the head, passing through the eye; lower surfaces whitish.

Total length	156 millim.
Head	22 ,,
Width of head	17 ,,
Body	58 ,,
Fore limb	26 ,,
Hind limb	35 ,,
Tail	76 ,,

Ceylon; Malayan peninsula.

a, b. ♂.	—— ?	(Types.)
c, d, e. ♂, ♀, & yg.	Ceylon.	
f. ♀, bad state.	Ceylon.	C. R. Buller, Esq. [P.].
g. ♀.	Singapore.	(Type of *Nubilia argentii*.)

22. Hemidactylus kushmorensis.

Hemidactylus kushmorensis, *Murray, Ann. Mag. N. H.* (5) xiv. 1884, p. 109.

Seven lamellæ under the middle toe. Crown of the head interspersed with numerous round tubercles; nostril between the rostral, the first labial, and three small nasals; ten upper and eight lower labials; two pairs of chin-shields, the first only in contact. Back with rounded tubercles arranged in twenty-two longitudinal series; a few tubercles between the hind legs are subtrihedral; ten to twelve femoral pores on each side. Tail verticillate, each verticil armed laterally with three rows of rather elongate subtrihedral tubercles; inferiorly with a single series of transverse plates. Colours neutral grey or brown, with three rows of squarish dark blotches; forming either longitudinal or obliquely transverse interrupted bands; a few smaller spots on the sides; a dark streak through the eye with a pale line above it; tail with dark cross bands.

Upper Sind.

23. Hemidactylus leschenaultii.

Hemidactylus leschenaultii, *Gray, Cat.* p. 155.
Hemidactylus bellii, *Gray, l. c.*
Hemidactylus leschenaultii, *Dum. & Bibr.* iii. p. 364; *Jerd. Journ. As. Soc. Beng.* xxii. 1853, p. 468; *Günth. Rept. Brit. Ind.* p. 109; *Blanf. Proc. As. Soc. Beng.* 1871, p. 173; *Stoliczka, op. cit.* p. 193, and *Journ. As. Soc. Beng.* xli. 1872, p. 97.
—— (Tachybates) leschenaultii, *Fitzing. Syst. Rept.* p. 105.
—— coctæi (*non D. & B.*), *Kelaart, Prodr. Faun. Zeyl.* p. 160.
—— pustulosus, *Lichtenst. Nomencl. Rept. Mus. Berol.* p. 5.
—— coctæi, part., *Günth. Rept. Brit. Ind.* p. 109.
—— kelaartii, *Theobald, Cat. Rept. As. Soc. Mus.* p. 29.
—— marmoratus (*non Hallow.*), *Blanf. Journ. As. Soc. Beng.* xxxix. 1870, p. 363, pl. xvi. figs. 1–3.

Snout longer than the distance between the eye and the ear-opening, once and one third or once and two fifths the diameter of the orbit; forehead concave, the supraorbital ridges prominent in full-grown specimens; ear-opening rather large, oval, vertical. Body and limbs moderate. A slight fold of the skin on the side of the belly, from axilla to groin. Digits free, strongly dilated, inner well developed; six or seven (rarely five) lamellæ under the inner digits, nine to eleven under the median digits. Head covered with minute granules posteriorly, with larger ones anteriorly; rostral four-sided, not quite twice as broad as high, with median cleft above; nostril pierced between the rostral, three nasals, and generally the first labial; ten to twelve upper and eight or nine lower labials; mental large, triangular or pentagonal; two pairs of chin-shields, inner largest and in contact behind the mental. Upper surface of body covered with small granules, uniform or intermixed with more or less numerous scattered round tubercles. Abdominal scales moderate, cycloid, imbricate. Males with a series of femoral pores interrupted on the præanal region; twelve to sixteen pores on each side. Tail depressed, flat inferiorly, covered above with small smooth scales and six longitudinal series of conical tubercles; inferiorly with a median series of transversely dilated plates. Grey above, with darker markings, forming undulated cross bars, rhomboidal spots on the middle of the back, or regular longitudinal bands; a dark band from the eye to the shoulder; lower surfaces white.

Total length	162 millim.
Head	22 ,,
Width of head	18 ,,
Body	58 ,,
Fore limb	31 ,,
Hind limb	37 ,,
Tail	82 ,,

India; Ceylon; Malay peninsula.

a–b. ♂ ♀.	Ganjam.	F. Day, Esq. [P.].
c. ♂.	Near Ellore.	W. T. Blanford, Esq. [P.].
d. ♂.	Nagpur.	W. T. Blanford, Esq. [P.].
e–h, i, k. ♂ ♀.	Godavery Valley.	W. T. Blanford, Esq. [P.].
l–n. ♂ & hgr.	Madras.	T. C. Jerdon, Esq. [P.].
o. Hgr.	Poona.	Dr. Leith [P.].
p. ♂.	Anamallays.	Col. Beddome [C.].
q. Several spec., ♂ ♀.	Malabar.	Col. Beddome [C.].
r–t. ♂ ♀.	Ceylon.	G. H. K. Thwaites, Esq. [P.].
u. ♂.	Ceylon.	W. Ferguson, Esq. [P.].
v. ♂.	Pinang.	Dr. Cantor.
w. ♂.	——?	T. Bell, Esq. [P.]. (Type of *Hemidactylus bellii*.)

24. Hemidactylus coctæi.

Boltalia sublævis, *Gray, Cat.* p. 158.
Hemidactylus coctæi, *Dum. & Bibr.* iii. p. 365; *Cantor, Cat. Mal. Rept.* p. 23; *Steindachn. Novara, Rept.* p. 13; *Stoliczka, Journ. As. Soc. Beng.* xli. 1872, p. 98; *Günth. Ann. Mag. N. H.* (4) ix. 1872, p. 86, and *Proc. Zool. Soc.* 1875, p. 224; *Blanf. Proc. Zool. Soc.* 1876, p. 636; *Murray, Zool. Sind,* p. 359.
Boltalia sublævis, *Gray, Zool. Misc.* p. 58.
Hoplopodion cocteaui, *Fitzing. Syst. Rept.* p. 104.
Hemidactylus coctæi, part., *Günth. Rept. Brit. Ind.* p. 109.
—— bengalensis, *Anders. Journ. As. Soc. Beng.* xl. 1871, p. 14.

Snout longer than the distance between the eye and the ear-opening, once and one third to once and a half the diameter of the orbit; forehead concave; ear-opening rather large, oval, oblique. Body and limbs moderate. Digits free, strongly dilated, inner well developed, with nearly straight lamellæ inferiorly; latter nine or ten (rarely eight) under the inner digits, and eleven to thirteen under the median digits. Head covered with minute granules posteriorly, with larger ones anteriorly; rostral four-sided, not quite twice as broad as high, with median cleft above; nostril pierced between the rostral, the first labial, and three nasals; twelve to fifteen upper and ten to eleven lower labials; mental large, triangular or pentagonal; two or three pairs of chin-shields, median largest and in contact behind the mental. Upper surfaces covered with small granules, among which some larger ones are sometimes scattered on the sides. Abdominal scales moderate, cycloid, imbricate. Male with a short series of five or six femoral pores under each thigh. Tail depressed, flat inferiorly, covered above with small smooth scales and four or six longitudinal series of conical tubercles; inferiorly with a median series of transversely dilated plates. Grey above, uniform or with indistinct darker marblings; lower surfaces white.

Total length	157	millim.
Head	20	,,
Width of head	16	,,
Body	56	,,
Fore limb	28	,,
Hind limb	34	,,
Tail	81	,,

India; Malay peninsula.

a. ♂.	India.	(Type of *Boltalia sublævis*.)
b, c–d. ♀ & hgr.	India.	
e–f. ♂ ♀.	India.	W. Masters, Esq. [P.].
g–h. ♀.	Patna.	W. Masters, Esq. [P.].
i. ♀.	Calcutta.	Dr. Anderson. (One of the types of *Hemidactylus bengalensis*.)
k. ♀.	Ellore.	W. T. Blanford, Esq. [P.].
l. ♀.	Bombay.	Dr. Leith [P.].
m. ♂.	Pinang.	Dr. Cantor.

25. Hemidactylus giganteus.

Hemidactylus giganteus, *Stoliczka, Journ. As. Soc. Beng.* xli. 1872, p. 99, pl. ii. fig. 2; *Blanf. Proc. Zool. Soc.* 1876, p. 636.
—— coctæi, part., *Günth. Ann. Mag. N. H.* (4) ix. 1872, p. 86.

Snout longer than the distance between the eye and the ear-opening, once and a half the diameter of the orbit; forehead concave; ear-opening rather large, suboval, vertical. Body and limbs stout. Digits free, inner well developed, strongly dilated, with straight transverse lamellæ inferiorly; latter, eleven or twelve under the inner digits, thirteen to fifteen under the median digits. Upper surfaces covered with uniform small granular scales, somewhat larger on the snout, smallest on the hinder part of the head. Rostral quadrangular, not quite twice as broad as high; nostril pierced between the rostral and three nasals; twelve to fifteen upper and eleven to thirteen lower labials; mental large, pentagonal; two pairs of chin-shields, inner largest. Abdominal scales rather small, cycloid, imbricate. Male with a series of femoral pores interrupted medially; nineteen to twenty-two pores on each side. Tail without large tubercles. Olive-grey above, with irregular dark, pale-edged marks in imperfect circles, inclined to form, or forming, four or five transverse undulating bands on the body; below uniform white.

Total length	217	millim.
Head	32	,,
Width of head	26	,,
Body	90	,,
Fore limb	44	,,
Hind limb	52	,,
Tail (reproduced)	95	,,

India.

a. ♂.	Godavery Valley.	W. T. Blanford, Esq. [C.]. (One of the types.)
b, c, d. ♂ ♀.	Godavery Valley.	W. T. Blanford, Esq. [P.].
e. ♂.	Malabar.	Col. Beddome [C.].

26. Hemidactylus bowringii. (PLATE XII. fig. 2.)

Doryura bowringii, *Gray, Cat.* p. 156.
Leiurus berdmorei, *Blyth, Journ. As. Soc. Beng.* xxii. 1853, p. 646.
Doryura berdmorei, *Theob. Journ. Linn. Soc.* x. 1868, p. 29.
Hemidactylus (Doryura) berdmorei, *Stoliczka, Journ. As. Soc. Beng.* xli. 1872, p. 100; *Blanf. Proc. Zool. Soc.* 1876, p. 637.
—— coctæi, part., *Günth. Ann. Mag. N. H.* (4) ix. 1872, p. 86.

Snout longer than the distance between the eye and the ear-opening, once and two fifths the diameter of the orbit; forehead slightly concave; ear-opening small, roundish. Body and limbs moderate; a slight fold of the skin along the flank. Digits free, moderately dilated, inner well developed; infradigital lamellæ obliquely curved, five under the thumb, seven or eight under the fourth finger, five or six under the first toe, and nine or ten under the fourth toe. Upper surfaces covered with uniform small granular scales, largest on the snout, smallest on the occiput. Rostral four-sided, twice as broad as high, with median cleft above; nostril pierced between the rostral, the first labial, and three or four nasals; nine to eleven upper and seven or eight lower labials; mental large, triangular, followed by a pair of chin-shields; an outer pair of much smaller chin-shields. Abdominal scales moderate, cycloid, imbricate. Male with a series of præanal pores, interrupted medially, composed of thirteen pores on each side. Tail depressed, rounded, oval in section, covered above with uniform small scales, inferiorly with a median series of transversely dilated plates. Light brown above, with darker spots, having sometimes a tendency to form four longitudinal bands on the back; frequently small whitish spots on the body and limbs; a dark streak passing through the eye; tail above with small chevron-shaped markings; lower surfaces whitish.

Total length	97	millim.
Head	12	,,
Width of head	8	,,
Body	31	,,

Fore limb	14 millim.
Hind limb	17 ,,
Tail	54 ,,

Birma; Northern and Central India.

a–b. ♂ & yg.	——?	C. Bowring, Esq. [P.]. (Types.)
c. ♂	——?	C. Bowring, Esq. [P.].
d. ♀.	Minhla, Birma.	Marquis G. Doria [P.].
e–f. ♀.	Pegu.	W. Theobald, Esq. [C.].
g. ♀.	Sikkim.	Col. Beddome.
h. ♀.	Godavery Valley.	W. T. Blanford, Esq. [C.].

27. Hemidactylus karenorum.

Doryura karenorum, *Theobald, Journ. Linn. Soc.* x. 1868, p. 30.

Snout longer than the distance between the eye and the ear-opening, once and one third or once and two fifths the diameter of the orbit; forehead slightly concave; ear-opening small, roundish. Body and limbs moderate; a slight fold of the skin along the sides of the belly, and another bordering the thighs posteriorly. Digits free, moderately dilated, inner well developed; infradigital lamellæ obliquely curved, five under the thumb, nine under the fourth finger, five or six under the first toe, and ten to twelve under the fourth toe. Head covered with minute granules posteriorly, with larger ones anteriorly; rostral four-sided, not quite twice as broad as high, with median cleft above; nostril pierced between the rostral, the first labial, and three nasals; eleven or twelve upper and seven to nine lower labials; mental large, triangular; two pairs of chin-shields. Upper surfaces of body covered with minute granules intermixed with numerous small convex round tubercles. Abdominal scales moderate, cycloid, imbricate. Male without præanal or femoral pores (?). Tail depressed, flat inferiorly, with sharp denticulated lateral edge; the scales on the upper surface very small, equal; those on the lower surface larger, imbricate, with a median series of large transversely dilated plates. Light grey-brown above, with rather indistinct darker variegation; lower surfaces whitish.

Total length	106 millim.
Head	14 ,,
Width of head	8·5 ,,
Body	36 ,,
Fore limb	17 ,,
Hind limb	21 ,,
Tail	56 ,,

Birma.

a–b. ♀.	Pegu.	W. Theobald, Esq. [P.].

28. Hemidactylus blanfordii.

Head much depressed; snout rather pointed, longer than the distance between the eye and the ear-opening, once and a half the diameter of the orbit; forehead slightly concave; ear-opening small, suboval. Body and limbs moderate; a very slight fold of the skin along the flanks. Digits free, moderately dilated, inner well developed; infradigital lamellæ obliquely curved, six under the thumb and inner toe, ten under the fourth finger, and eleven under the fourth toe. Upper surfaces covered with uniform minute granules. Rostral four-sided, with median cleft above; nostril pierced between the rostral, the first labial, and three nasals; thirteen upper and eleven lower labials; mental large, triangular, followed by two pairs of chin-shields, one behind the other, the anterior pair forming a suture, the second pair separated. Abdominal scales moderate, cycloid, imbricate. Tail depressed, flat inferiorly, with sharpish denticulated lateral edge, covered above with uniform granules, inferiorly with a median series of transversely dilated plates. Reddish brown above, with small round yellowish spots; lower surfaces whitish.

Total length	72	millim.
Head	11	,,
Width of head	6·5	,,
Body	24	,,
Fore limb	12	,,
Hind limb	15	,,
Tail	37	,,

Himalayas.

a. ♀. Darjeeling. W. T. Blanford, Esq. [P.].

29. Hemidactylus peruvianus.

Hemidactylus peruvianus, *Gray, Cat.* p. 156.
Hemidactylus peruvianus, *Wiegm. Nova Acta Ac. Leop.-Carol.* xvii. i. 1835, p. 240; *Dum. & Bibr.* iii. p. 369.
Hoplopodion (Microdactylus) peruvianum, *Fitz. Syst. Rept.* p. 104.

Body covered above with uniform granules. Digits free, slightly dilated. Two pairs of chin-shields, one behind the other, the anterior elongate-pentagonal, the posterior smaller and roundish. Tail tapering, depressed, the lateral edges obtusely angular, with a denticulation of spine-like scales. Grey above, marbled with blackish, with indistinct whitish round spots; white inferiorly.

Tacna, Peru.

30. Hemidactylus garnotii.

Doryura garnotii, *Gray, Cat.* p. 157.
Hemidactylus garnotii, *Dum. & Bibr.* iii. p. 368; *Bouleng. Proc. Zool Soc.* 1883, p. 118, pl. xxii. fig. 1.
Hoplopodion (Onychopus) garnotii, *Fitz. Syst. Rept.* p. 104.

Doryura vulpecula, *Girard*, Proc. Ac. Philad. 1857, p. 197, *and U.S. Explor. Exped.*, Herp. p. 286, pl. xxiv. figs. 17-24.
Hemidactylus ludekingii, *Bleek.* Nat. Tijds. Ned. Ind. xvi. 1859, p. 27.
Doryura gaudama, *Theob.* Journ. Linn. Soc. x. 1868, p. 30.
Hemidactylus (Doryura) mandellianus, *Stoliczka,* Journ. As. Soc. Beng. xli. 1872, p. 101, pl. iii. figs. 1, 2.

Snout obtusely pointed, much longer than the distance between the eye and the ear-opening, once and two thirds the diameter of the orbit; forehead slightly concave; ear-opening small, rounded. Body and limbs moderate. A slight but distinct fold of the skin along the flanks, and another bordering the hind limb posteriorly. Digits with a very slight rudiment of web, moderately dilated, inner well developed; infradigital lamellæ oblique, six or seven under the inner digits, ten to twelve under the fourth finger, and thirteen or fourteen under the fourth toe. Upper surfaces and throat covered with minute granular scales, a little larger on the snout; abdominal scales moderate, imbricate. Rostral subquadrangular, with median cleft above; nostril pierced between the rostral and three nasals; twelve or thirteen upper and nine or ten lower labials; mental large, triangular, in contact posteriorly with a pair of pentagonal chin-shields, followed by a second smaller pair; the anterior pair of chin-shields in contact with the first infralabial, and with each other medially; the posterior pair separated from each other, and also completely or nearly completely from the labials. Tail depressed, flat inferiorly, with sharp denticulated lateral edge; the scales on the upper surface very small, equal; those on the lower surface larger, imbricate, with a median series of large, transversely dilated plates. Brownish grey above, uniform or with more or less distinct brown and whitish spots; lower surfaces uniform whitish. [Theobald mentions nineteen pores on each thigh in his *Doryura gaudama*, which appears to me, from the short description, identical with, and the male of, the present species.]

Total length	129	millim.
Head	16	,,
Width of head	10	,,
Body	44	,,
Fore limb	20	,,
Hind limb	27	,,
Tail	69	,,

South Pacific islands; Indian archipelago; Philippines; Birma; Sikkim.

a. ♀.	South-Sea Islands.	
b. ♀.	New Caledonia.	E. L. Layard, Esq. [P.].
c. ♀.	Agam.	Dr. Bleeker. (As typical of *H. ludekingii.*)
d. ♀.	Philippines.	
e. ♀.	Sikkim.	(One of the types of *H. mandellianus.*)
f. ♀.	Birma.	F. Day, Esq. [P.].

31. Hemidactylus richardsonii. (Plate XII. fig. 3.)

Velernesia richardsonii, *Gray, Cat.* p. 156.

Snout rounded, a little longer than the distance between the eye and the ear-opening, as long as the diameter of the orbit; ear-opening moderately large, vertical, oval. Body and limbs stout. A fold of the skin from axilla to groin, and another bordering the hind limb posteriorly. Digits one third webbed, strongly dilated, inner well developed; infradigital lamellæ nearly straight, six under the thumb, nine under the fourth finger, seven under the inner toe, and ten under the fourth toe. Upper surfaces covered with small granules, largest on the snout, smallest on the occiput; a few larger granules are scattered among the dorsal ones; abdominal scales very small, cycloid, imbricate, smooth. Rostral four-sided, twice as broad as high, with median cleft; nostril pierced between the rostral, the first labial, and three nasals; eleven upper and nine lower labials; mental broadly triangular; two chin-shields, in contact behind the rostral. Tail depressed, flat inferiorly, with sharp denticulated lateral edge, covered above with small imbricate scales and transverse series of four conical tubercles, inferiorly with a median series of large transversely dilated plates. Grey above, with small brown markings; a dark-brown band on each side of head and body, passing through the eye.

Total length	159 millim.
Head	21 ,,
Width of head	18 ,,
Body	59 ,,
Fore limb	27 ,,
Hind limb	35 ,,
Tail	79 ,,

Hab. ——?

a. ♀. ——? Sir J. Richardson [P.]. (Type.)

32. Hemidactylus platyurus.

Platyurus schneiderianus, *Gray, Cat.* p. 157.
Stellio platyurus, *Schneid. Amph. Phys.* ii. p. 30, *and Denkschr. Ak. Münch.* 1811, p. 62, pl. i. fig. 3.
Lacerta schneideriana, *Shaw, Zool.* iii. p. 278.
Gecko platyurus, *Merr. Syst. Amph.* p. 41.
—— marginatus, *Cuv. R. A.* ii. p. 54.
Hemidactylus platyurus, *Wiegm. Nova Acta Ac. Leop.-Carol.* xvii. i. 1835, p. 288; *Cantor, Catal. Mal. Rept.* p. 24.
—— marginatus, *Gray, Griff. A. K.* ix. *Syn.* p. 51; *Dum. & Bibr.* iii. p. 370, pl. xxx. fig. 2.
Hoplopodion (Cosymbotus) platyurum, *Fitz. Syst. Rept.* p. 104.
Crossurus platyurus, *Girard, U.S. Explor. Exped., Herp.* p. 281.
Nycteridium schneideri, *Günth. Rept. Brit. Ind.* p. 111.
—— himalayanum, *Anders. Journ. As. Soc. Beng.* xl. 1871, p. 15.
—— platyurus, *Stoliczka, Journ. As. Soc. Beng.* xli. 1872, p. 103.

Snout longer than the distance between the eye and the ear-opening, once and a half the diameter of the orbit; forehead concave; ear-opening small, oval, oblique. Body and limbs moderate, much depressed; a cutaneous expansion from axilla to groin, and another bordering the hind limb posteriorly. Digits strongly dilated, about half-webbed, inner well developed; five or six lamellæ under the inner, and seven to nine under the median digits. Upper surfaces covered with uniform small granules, largest on the snout. Rostral four-sided, not twice as broad as high, with median cleft above; nostril pierced between the rostral, the first labial, and three nasals; nine to eleven upper and seven or eight lower labials; mental large, triangular or pentagonal, in contact with two large chin-shields, followed by two smaller ones. Abdominal scales moderate, cycloid, imbricate. Male with an uninterrupted series of femoral pores, seventeen or eighteen on each side. Tail much depressed, flat inferiorly, with sharp denticulated lateral edge, covered above with uniform small granules, inferiorly with a median series of transversely dilated plates. Grey above, marbled with darker; generally a dark streak from eye to shoulder; white beneath.

Total length	113	millim.
Head	15	,,
Width of head	11	,,
Body	41	,,
Fore limb	20	,,
Hind limb	24	,,
Tail	57	,,

Ceylon; North India; South China; Indo-China; Malay peninsula; East-Indian archipelago.

a–b. ♀.	Ceylon.	Dr. Kelaart.
c. ♂.	Sikkim.	Col. Beddome.
d. Yg.	Hong Kong.	C. Bowring, Esq. [P.].
e–f. ♂.	Gamboja.	M. Mouhot [C.].
g. ♀.	Siam.	M. Mouhot [C.].
h. ♀.	Siam.	C. Bowring, Esq. [P.].
i. ♀.	Siam.	W. H. Newman, Esq. [P.].
k–m. ♀ & hgr.	Pinang.	Dr. Cantor.
n–o. ♂ ♀.	Philippines.	
p. ♀.	Dinagat Island.	A. Everett, Esq. [C.].
q. ♀.	Borneo.	Sir E. Belcher [P.].
r–s. ♀.	Java.	Leyden Museum.
t. ♀.	Java.	Dr. Bleeker.
u. Yg.	Java.	Mrs. Lyon [P.].
v. ♀.	Manado.	Dr. A. B. Meyer [C.].

24. **TERATOLEPIS**.

Teratolepis, *Günther, Proc. Zool. Soc.* 1869, p. 504.

Digits dilated, inferiorly with a double series of lamellæ, with elongate compressed distal joint rising from within the extremity

of the dilated part; all clawed. Ear-opening concealed. Body covered with large strongly imbricated scales. Pupil —— ?
India.

1. Teratolepis fasciata.

Homonota fasciata, *Blyth, Journ. As. Soc. Beng.* xxii. 1853, p. 468.
Teratolepis fasciata, *Günth. l. c.* p. 505.

Body somewhat depressed; limbs rather long and slender. Head covered with polygonal flat scales. Seven lower labials; mental large, triangular; two larger anterior chin-shields, in contact behind the mental, followed by others passing gradually into the small gular granules. Dorsal scales large, lozenge-shaped, slightly keeled; ventral scales much smaller, smooth. Tail depressed, swollen, tapering at the end, covered with imbricate irregular scales, some of those of the upper surface being extremely large. Greyish above, with five brown longitudinal bands, which at regular intervals are interrupted by white spots forming cross bands; seven of these cross bands on the neck and trunk. Length of head 13 millim., body 25, tail 25.

Central and North-western India.

a. Very bad state. Sind. Dr. Leith [P.].

25. PHYLLOPEZUS.

Phyllopezus, *Peters, Mon. Berl. Ac.* 1877, p. 415.

"A single series of lamellæ under the base of the fingers and toes, the two distal phalanges of all five fingers and toes narrowed and furnished with a claw."

"Distinguished from *Gehyra* by the first finger and toe being provided with a narrow distal joint and a claw."
Brazil.

1. Phyllopezus goyazensis.

Phyllopezus goyazensis (*Behn*), *Peters, l. c.* pl. —. fig. 1.

"Head rather elongate, with rather flat snout; latter covered with small oval scales to between the eyes, where they gradually pass into the minute granulation of the hinder part of the head. Nine upper labials, the two last small; also nine lower labials, the three last smallest. Mental very large; behind it two larger hexagonal chin-shields, followed by a second transverse row of three other chin-shields, the median one being almost as large as those of the anterior row. Hinder part of head, neck, and base of tail finely granulate, with scattered roundish tubercles which are not larger than the scales on the snout. The ear-cleft, which is obliquely directed downwards and forwards, is furnished anteriorly and posteriorly with short spine-like scales. The reproduced tail, with smooth scales scarcely smaller than the ventrals; on the lower surface of the tail a row of broad band-like plates. On each side near the vent an oblique row of three spine-like prominent scales. Neither præanal nor femoral pores."

" Grey-brown, with brown transverse band-like spots."
Goyaz, Brazil.

26. ARISTELLIGER.

Aristelliger, *Cope, Proc. Ac. Philad.* 1861, p. 496.
Idiodactylus, *Bocourt, Miss. Sc. Mex., Rept.* p. 41.

Digits slender at the base, free, strongly dilated at the extremity, furnished with undivided transverse lamellæ inferiorly; distal phalanges, except of thumbs, free, elongate, compressed, raised, clawed; inner digits short, scarcely dilated at the base, the apex dilated into a small disk with a circular plate inferiorly, and provided with a sheathed retractile claw, the sheath opening laterally and inwards. Upper surfaces covered with granular scales; belly with cycloid imbricated scales. Pupil vertical. Eyelid distinct all round the eye. No femoral nor præanal pores.

West Indies; Central America.

1. Aristelliger præsignis.

Hemidactylus præsignis, *Hallow. Proc. Ac. Philad.* 1856, p. 222.
Aristelliger præsignis, *Cope, l. c.*
Idiodactylus georgeensis, *Bocourt, l. c.* pl. x. fig. 1.

Head moderate, elongate; snout obtusely conical, considerably longer than the distance between the eye and the ear-opening, once and two thirds or once and three fourths the diameter of the orbit; ear-opening moderate, oval, oblique, its greatest diameter about half that of the orbit. Body and limbs moderate; digits unequal; about thirteen lamellæ under the fourth toe. Head and upper surfaces covered with small granular scales; abdominal scales moderate. Upper eyelid with a small spine-like scale. Rostral broad, with median cleft above, entering the nostril; eight upper and as many lower labials; mental large, pentagonal, extending beyond the labials, bordered on each side by a small chin-shield. Tail slender, cylindrical, tapering, compressed distally, longer than head and body, covered above with small, flat, slightly imbricated scales, inferiorly with a central series of transversely dilated plates. Brown above, generally with darker cross bars on the back and tail and a dark spot above axilla; sometimes with whitish dots; a dark streak on each side of the head, passing through the eye; lips white-spotted; flanks reticulated with dark brown; lower surfaces whitish, throat with brown marblings or oblique streaks.

Total length	226	millim.
Head	28	,,
Width of head	21	,,
Body	68	,,

Fore limb .... 33 millim.
Hind limb .... 42 „
Tail .... 130 „

Jamaica; St. George Island, Belize.

a–c, d–g. Adult & hgr.	Jamaica.	P. H. Gosse, Esq. [P.].
h. Adult.	Jamaica.	F. Beckford, Esq. [P.].
i. Yg.	Jamaica.	
k. Adult.	——?	

2. Aristelliger lar.

Aristelliger lar, *Cope, l. c.* p. 497.

Differs from the preceding in the following points:—Anterior mandibulary teeth longer than the median. Subdigital lamellæ more numerous, about twenty-one beneath the fourth toe. Size larger, about 130 millim. from snout to vent. Colour above grey, with numerous brown spots, especially upon the scapular and ischiatic regions; crown and front dark; labial region varied; an indistinct brown band extends posteriorly from the orbit, bordered below by a pale one, not more visible.

S. Domingo.

27. GEHYRA*.

Peropus (non *Beechey*), *Wiegm. N. Acta Ac. Leop.-Carol.* xvii. i. 1835, p. 238; *Fitzing. Syst. Rept.* p. 103; *Gray, Cat. Liz.* p. 159.
Hemidactylus, part., *Wiegm. Herp. Mex.* p. 20; *Dum. & Bibr.* iii. p. 344.
Gehyra, *Gray, Zool. Misc.* p. 57, and *Cat. Liz.* p. 162.
Perodactylus, *Fitzing. l. c.*
Dactyloperus, *Fitzing. l. c.*
Peripia, *Gray, Cat. Liz.* p. 158.
Peripia, part., *Günth. Rept. Brit. Ind.* p. 110.

Digits strongly dilated, free or webbed at the base, inferiorly with undivided or medially divided transverse lamellæ; distal phalanges free, elongate, compressed, clawed, raised from within the extremity of the dilatation; inner digits without free distal phalange, clawless, or with a very indistinct retractile claw. Upper surfaces covered with granular scales; belly with cycloid imbricated scales. Pupil vertical. Males with femoral or præanal pores.

East Indies; Australia; islands of the Indian and South-Pacific Oceans; west coast of Mexico.

---

* *Peripia papuensis*, Macleay, Proc. Linn. Soc. N. S. W. ii. p. 97.—Katow, N. Guinea.
P. *ornata*, id. ibid. p. 98.—Port Moresby.
P. *longicauda*, id. ibid.—Endeavour River.
P. *dubia*, id. ibid.—Cape Grenville.
P. *marmorata*, id. ibid.—Katow.
P. *brevicaudis*, id. ibid.—Darnley Island.

Synopsis of the Species.

I. Digital lamellæ divided by a median groove.

A. A fold of the skin bordering the hind limb posteriorly.

Inner pair of chin-shields very large; rostral quadrangular; 8 or 9 upper and 6 or 7 lower labials; tail normally with a sharpish lateral edge .... 1. *mutilata*, p. 148.
Chin-shields shorter; rostral horseshoe-shaped .... 2. *baliola*, p. 150.
Chin-shields smaller; 10 upper and 9 lower labials; tail without lateral keel ...... 3. *brevipalmata*, [p. 150.
Chin-shields small and polygonal; tail with a sharp serrated lateral edge .... 4. *neglecta*, p. 150.

B. Hind limb without cutaneous fold; digits free or with a very slight rudiment of web.

Scales a good deal smaller on the median line of the back than on the sides; male with about 40 femoral pores .... 5. *insulensis*, p. 150.
Male with 10 to 16 femoral pores .... 6. *variegata*, p. 151.

II. Digital lamellæ undivided.

7 to 9 upper labials; toes free or nearly so . 7. *australis*, p. 152.
11 to 13 upper labials; toes webbed at the base; 25 to 40 femoral pores .... 8. *oceanica*, p. 152.
13 or 14 upper labials; digits webbed at the base; a strong fold of the skin bordering the fore limb anteriorly; 50 to 60 femoral pores .... 9. *vorax*, p. 153.

1. Gehyra mutilata.

Peripia peronii, *Gray, Cat.* p. 159.
Peropus mutilatus, *Gray, l. c.*
Hemidactylus (Peropus) mutilatus, *Wiegm. Nova Acta Ac. Leop.-Carol.* xvii. p. 238.
—— mutilatus, *Dum. & Bibr.* iii. p. 354.
—— peronii, *Dum. & Bibr.* iii. p. 352, pl. xxx. fig. 1; *Jacquin. Voy. Pôle Sud, Saur.* pl. i. fig. 2; *Cantor, Cat. Mal. Rept.* p. 22; *Kelaart, Prodr. Faun. Zeyl.* p. 187.
Peropus (Dactyloperus) peronii, *Fitz. Syst. Rept.* p. 103.
—— mutilatus, *Fitz. l. c.*; *Girard, U.S. Explor. Exped., Herp.* p. 277.
Peripia peronii, *Günth. Rept. Brit. Ind.* p. 110; *Stoliczka, Journ. As. Soc. Beng.* xxxix. 1870, p. 163, & xli. 1872, p. 103.
Gecko pardus, *Tytler, Journ. As. Soc. Beng.* xxxiii. 1864, p. 547.
? Gecko harrieti, *Tytler, l. c.* p. 548.

Hemidactylus (Peripia) mutilatus, *Peters, Mon. Berl. Ac.* 1867, p. 14; *Anders. Zool. W. Yunnan,* i. p. 799.
Peropus packardii, *Cope, Proc. Ac. Philad.* 1868, p. 319.
Peripia mutilata, *Günth. Proc. Zool. Soc.* 1873, p. 168; *Peters & Doria, Ann. Mus. Genov.* xiii. 1878, p. 370.

Head longer than broad; snout longer than the distance between the eye and the ear-opening, about once and one third the diameter of the orbit; forehead with a median groove; ear-opening moderately large, suboval. Body and limbs moderately elongate, depressed; a fold of the skin bordering the hind limb posteriorly. Digits short, more or less webbed at the base; the inferior lamellæ angular, divided by a median groove. Upper surfaces and throat covered with small granular scales, largest and flat on the back. Abdominal scales moderate. Rostral quadrangular, broader than high, with a median cleft superiorly; nostril pierced between the rostral, the first labial, and three nasals, the upper much the largest and generally in contact with its fellow; eight or nine upper and six or seven lower labials; mental moderately large, pentagonal; chin-shields three pairs, inner very large, elongate, outer small, frequently broken up into small scales; the distance from the border of the lip to the extremity of the median pair of chin-shields equals the diameter of the orbit. Femoral pores in a doubly curved line, angular medially, fourteen to nineteen on each side. Tail depressed, normally with a sharpish, minutely serrated lateral edge; its upper surface covered with very small flat scales, its lower surface generally with a median series of large transversely dilated scales. Greyish or reddish-brown above, uniform or dotted or variegated with darker; lower surfaces uniform whitish.

Total length	114	millim.
Head	16	,,
Width of head	12	,,
Body	41	,,
Fore limb	16	,,
Hind limb	21	,,
Tail	57	,,

Mascarene Islands; Seychelles; Ceylon; Birma; Malay peninsula and archipelago; New Guinea; Western Mexico.

a. Hgr.	Mauritius.	
b. Many spec.: ♂, ♀, hgr., & yg.	Rodriguez.	Transit of Venus Exp.
c. ♂.	Rodriguez?	A. Newton, Esq. [P.].
d. ♂.	Seychelles.	Dr. E. P. Wright [P.].
e, f. ♀.	Ceylon.	Dr. Kelaart.
g. ♀.	Ceylon.	W. Ferguson, Esq. [P.].
h. Hgr.	Ceylon.	G. H. K. Thwaites, Esq. [P.].
i. ♀.	S. Ceylon.	
k–m. ♂ ♀.	Pegu.	W. Theobald, Esq. [C.].
n–o. ♂ ♀.	Penang.	Dr. Cantor.
p. ♂.	Indian archipelago.	Dr. Bleeker.

q. ♀.	Negros, Philippines.	Dr. A. B. Meyer [C.].
r–s. ♀.	Manado, Celebes.	Dr. A. B. Meyer [C.].
t. ♀.	Timor Laut.	H. O. Forbes, Esq. [C.].
u. ♂.	San Blas, W. Mexico.	Hr. A. Forrer [C.].
v. ♀.	Presidio, W. Mexico.	Hr. A. Forrer [C.].

2. Gehyra baliola.

Hemidactylus baliolus, *A. Dum. Cat. Méth. Rept.* p. 38, *and Arch. Mus.* viii. p. 461, pl. xvii. fig. 2.

Differs from *G. mutilata* in the following points:—Snout more conical; cheeks not swollen; dorsal scales larger, those on the vertebral line smaller; rostral horseshoe-shaped; chin-shields shorter. Brown above, with lighter or reddish spots.

New Guinea.

3. Gehyra brevipalmata.

Hemidactylus (Peripia) brevipalmata, *Peters, Mon. Berl. Ac.* 1874, p. 159.

Differs from *G. mutilata* in having the scales, especially the ventrals, smaller; a greater number of labials, viz. ten superiorly and nine inferiorly, the decidedly smaller chin-shields, the greater development of the interdigital web, the greater number of lamellæ under the digits (fourteen or fifteen under the fifth toe), and the absence of a keel on the side of the tail. Uniform grey.

Pelew Islands.

4. Gehyra? neglecta.

Peropus neglectus, *Girard, Proc. Ac. Philad.* 1857, p. 197, *and U.S. Explor. Exped., Herp.* p. 278.

Distinguished from *G. mutilata* in the following characters:— Eye much larger; chin-shields small and polygonal; tail more conspicuously serrated laterally; granulation of the upper regions and scales of the inferior regions larger. Upper parts light brown, with a few black dots, spots, or streaks, irregularly disposed; a conspicuous streak may be traced from the nostril to the eye, and from behind the eye, across the ear, to the shoulder, or else to the insertion of the fore limb; lower surfaces uniform whitish.

Habitat unknown.

Perhaps a *Lepidodactylus*.

5. Gehyra insulensis.

Dactyloperus insulensis, *Girard, Proc. Ac. Philad.* 1857, p. 197, *and U.S. Explor. Exped., Herp.* p. 280.

Snout longer than the distance between the eye and the ear-opening, once and a half the diameter of the orbit; ear-opening moderate. Digits free; the inferior lamellæ angular, divided by a median groove. Upper surfaces covered with small granular scales, a good deal smaller along the middle line of the back than on the sides. Rostral pentagonal; four chin-shields, the median pair large and elongated, narrowest posteriorly. A long series of femoral pores, about twenty on each side. Tail depressed, rounded, with a median row of large scutellæ inferiorly. Greyish brown above, minutely speckled with black; whitish beneath. From snout to vent 50 millim.; tail 40 millim.

Sandwich Islands.

6. Gehyra variegata.

Peripia variegata, *Gray, Cat.* p. 159.
Hemidactylus variegatus, *Dum. & Bibr.* iii. p. 353.
Peropus (Dactyloperus) variegatus, *Fitzing. Syst. Rept.* p. 103.
Peripia torresiana, *Günth. Ann. Mag. N. H.* (4) xix. 1877, p. 415.

Head longer than broad; snout longer than the distance between the eye and the ear-opening, about once and a half the diameter of the orbit; forehead with a median groove; ear-opening moderately large, suboval. Body and limbs moderately elongate, depressed, without cutaneous folds. Digits short, free or with a very slight rudiment of web; the inferior lamellæ angular, divided by a median groove. Upper surfaces and throat covered with very small granular scales; abdominal scales moderate. Rostral quadrangular, broader than high, with a median cleft superiorly; nostril pierced between the rostral, the first labial, and three nasals; seven to nine upper and six to eight lower labials; mental moderately large, pentagonal; chin-shields three pairs, inner largest, elongate, outer small, frequently broken up into small scales; these shields considerably shorter than in *G. mutilata.* A short angular series of femoral pores, ten to sixteen altogether. Tail depressed, tapering, the sides rounded; its upper surface covered with very small flat scales, its lower surface with a median series of large transversely dilated scales. Brown above, generally variegated with darker, sometimes dark with lighter blotches, or with dark bands across the back; frequently two dark streaks on the side of the head and neck, the lower passing through the eye; lower surfaces uniform whitish.

Total length	147	millim.
Head	18	,,
Width of head	14	,,
Body	53	,,
Fore limb	20	,,
Hind limb	26	,,
Tail	76	,,

Australia; Polynesia.

a. Hgr.   Houtman's Abrolhos.
b-c. ♂ ♀.   Peak Downs.   Godeffroy Museum

d. ♂.	Champion Bay.	Mr. Duboulay [C.].
e–g. ♂ ♀.	N. Australian Expedition.	J. B. Elsey, Esq. [C.].
h. ♂.	Australia.	Mr. Duboulay [C.].
i–l. ♂ ♀.	Australia.	G. Krefft, Esq. [P.].
m–o. ♂.	Islands of Torres Straits.	Rev. S. Macfarlane [C.]. (Types of *Peripia torresiana*.)
p. Several spec.: ♂, ♀, & hgr.	Murray Island.	Rev. S. Macfarlane [C.].
q–u. ♂ ♀.	Sunday Island.	J. B. Jukes, Esq. [P.].
v. ♂.	——?	Haslar Collection.

7. Gehyra australis.

Gehyra australis, *Gray, Cat.* p. 163.
Gecko grayi, *Steindachn. Novara, Rept.* p. 11.

This species agrees in every respect with *G. variegata* except in lacking the median division under the dilated part of the digits.
Australia.

a–c. ♂.	Swan River.	Earl of Derby [P.].	(Types.)
d. ♂.	Port Essington.	Earl of Derby [P.].
e. ♀.	Port Darwin.	R. G. S. Buckland, Esq. [C.].

8. Gehyra oceanica.

Gehyra oceanica, *Gray, Cat.* p. 163.
Gecko oceanicus, *Lesson, Coquille,* ii. i. p. 42, pl. ii. fig. 3.
Hemidactylus oualensis, *Dum. & Bibr.* iii. p. 350, pl. xxviii. fig. 7.
Gehyra oceanica, *Gray, Zool. Misc.* p. 58; *Girard, U.S. Explor. Exped., Herp.* p. 273; *Peters & Doria, Ann. Mus. Genov.* xiii. 1878, p. 369.
Peropus (Perodactylus) oualensis, *Fitzing. Syst. Rept.* p. 103.
Gehyra papuana, *Meyer, Mon. Berl. Ac.* 1874, p. 129.

Head longer than broad; snout longer than the distance between the eye and the ear-opening, about once and a half the diameter of the orbit; forehead and the space between the nostrils with a well-marked median groove; ear-opening moderately large, suboval, horizontal. Body and limbs moderately elongate, rather depressed; a more or less indistinct fold of the skin along the side of the throat and belly, and another bordering the hind limb posteriorly. Digits moderate, united at the base by a short web; the inferior lamellæ not divided by a median groove, distinctly curved, subangular. Upper surfaces and throat covered with very small granular scales. Abdominal scales moderate. Rostral large, quadrangular, its superior border emarginate; nostril pierced between the rostral, the first labial, and four or five small nasals; eleven to thirteen upper and nine to eleven lower labials; mental small, variable in shape; chin-shields three on each side, inner elongate. Tail rounded, tapering, slightly depressed, covered above with very small juxtaposed scales, inferiorly with larger scales. Femoral pores in an angular series, thirteen to twenty on each side. Brown above, uniform or with darker or lighter markings; lower surfaces uniform whitish.

Total length	162	millim.
Head	24	,,
Width of head	18	,,
Body	65	,,
Fore limb	24	,,
Hind limb	30	,,
Tail	73	,,

Moluccas; New Guinea; Polynesia.

a–c. ♂ ♀.	Wild Island, Admiralty Islands.	'Challenger' Expedition.
d–e. ♂ ♀.	Faro Island, Solomon Group.	H. B. Guppy, Esq. [P.].
f–i.	Shortland Islands.	H. B. Guppy, Esq. [P.].
k. ♂.	Lord Howe's Island.	J. Macgillivray, Esq. [C.].
l, m–n. ♂, ♀, & yg.	Fiji Islands.	J. Macgillivray, Esq. [C.].
o–p. ♀.	Fiji Islands.	E. A. Liardet, Esq. [P.].
q. ♀.	Matuka, Fiji Islands.	'Challenger' Expedition.
r–u. ♂, ♀, & yg.	Tongataboo.	J. Brenchley, Esq. [P.].
v. ♀.	Viti Levu.	Museum Godeffroy.
w, x. ♂ ♀.	Samoa Islands.	Rev. J. Powell [P.].
y. ♀.	Savage Island.	Dr. A. Günther [P.].
z. ♀.	Rarotonga Island.	Sir J. Lubbock [P.].
α. ♂.	Port Essington (?).	

9. Gehyra vorax.

Gehyra vorax, *Girard, Proc. Ac. Philad.* 1857, p. 179, *and U.S. Explor. Exped., Herp.* p. 274, pl. xvi. figs. 1–8; *Bouleng. Proc. Zool. Soc.* 1883, p. 119, pl. xxii. fig. 2.

Closely allied to *G. oceanica.* Differs in the following characters:— Size much greater. Folds on the side of the body very distinct; a fold bordering the fore limb anteriorly. Ear-opening narrower. Digits broader and shorter; the inferior lamellæ very gently curved. Abdominal scales smaller. Twenty-five to thirty femoral pores on each side. Thirteen or fourteen upper and eleven or twelve lower labials. Brown or greyish above, uniform or with darker and lighter markings; lower surfaces uniform whitish.

Total length	237	millim.
Head	38	,,
Width of head	29	,,
Body	114	,,
Fore limb	42	,,
Hind limb	58	,,
Tail (injured)	85	,,

Polynesia.

a. ♀.	Loyalty Islands.	Sir G. Grey [P.].
b. ♀.	Erromango.	
c–d. ♂ ♀.	Fiji Islands.	J. Macgillivray, Esq. [C.].
e–f. ♂ & hgr.	Norfolk Island.	F. N. Rayner, Esq. [C.].

28. PEROCHIRUS.

Digits strongly dilated, slightly webbed, inferiorly with transverse lamellæ, the anterior of which are divided by a median groove, with free, slender, compressed, clawed phalanges raising from within the extremity of the dilated portion; inner finger rudimentary, clawless; inner toe with distinct attached clawed phalange. Body covered above with uniform granular scales, inferiorly with imbricated or juxtaposed scales. Pupil vertical. Males with or without præanal or femoral pores.

Philippines, Carolines, and New Hebrides.

Synopsis of the Species.

I. Rostral higher than broad .... 1. *ateles*, p. 154.
II. Rostral broader than high.
   A. Abdominal scales subhexagonal, slightly imbricate; tail as broad as the body.

Males with 10 præanal pores; no femoral pores .... 2. *guentheri*, p. 155.
Male with 4 præanal pores; no femoral pores; head and body very much depressed .... 3. *depressus*, p. 155.
Male with a series of 50 femoral pores; size very large .... 4. *scutellatus*, p. 156.
   B. Abdominal scales granular, juxtaposed; tail much narrower than the body .... 5. *articulatus*, p. 156.

1. Perochirus ateles.

Hemidactylus ateles, *A. Dum. Arch. Mus.* viii. p. 426, pl. viii. fig. 9.

Head conical, not distinct from neck; eyes moderate. Twelve upper labials on each side of the rostral, which is higher than broad and extends on the upper part of the snout, the extremity of which is rather acute. Nostrils large. Mental narrow, triangular. Enlarged scutes on the chin, some of which are irregularly arranged in rows along the infralabials. Tail depressed, the borders finely denticulated, covered with granulations same above and beneath and arranged in regular rings. No femoral pores (?). Brownish grey, darker on the tail, the borders of which, above and beneath, are of a rather vivid brown. Total length 150 millim., tail 70.

Zamboanga, Philippines.

2. Perochirus guentheri. (Plate XII. fig. 4.)

Head moderately depressed; snout obtusely conical, longer than the distance between the eye and the ear-opening, once and two thirds the diameter of the orbit; forehead concave; ear-opening small, oval, oblique. Body moderately elongate, depressed. Limbs rather short, stout, depressed; digits short, with a rudiment of web, all but the rudimentary thumbs strongly dilated; the free phalanges of the fourth toe measuring the diameter of the eye. Head covered with small granules, larger on the snout; rostral subquadrangular, with median cleft above, nearly twice as broad as high; nostril pierced between the rostral, the first labial, and three small nasals; ten upper and eight or nine lower labials; mental small, narrow, transverse; three or four rows of small hexagonal chin-shields, passing gradually into the minute granules of the throat. Upper surface of body and limbs covered with very small granules; abdominal scales very small. Male with ten femoral pores forming a short angular series. Tail very much depressed, as broad as the body, flat beneath, with sharpish lateral edges; latter denticulated when the organ is intact; the upper surface covered with small juxtaposed scales, considerably larger than the granules of the back; the lower surface covered with larger, flat, slightly imbricated scales; the tail, when intact, is divided in rather distinct segments, each line of separation being indicated on the lateral edge by a large, pointed scale. Brown above, whitish beneath.

Total length	120	millim.
Head	12	,,
Width of head	9	,,
Body	57	,,
Fore limb	22	,,
Hind limb	26	,,
Tail	51	,,

New Hebrides.

a–b. ♂ ♀ . Erromango.

This species is evidently very closely allied to *P. ateles*, from which it appears to differ in having the interdigital membrane shorter, the rostral broader than high, the mental trapezoid, and probably also in other characters when the types can be compared.

3. Perochirus depressus.

Hemidactylus ateles, var. depressus, *Fischer, Arch. f. Nat.* 1882, p. 300, pl. xvii. figs. 31–36 *.

Differs from *P. guentheri* by the more depressed head and body, the still smaller abdominal scales, and the presence of four præanal pores only. From snout to vent 80 millim.

Ruk Island, Carolines.

---

* Type (a ♂) examined.

4. Perochirus scutellatus.

Hemidactylus ateles, var. scutellatus, *Fischer, Arch. f. Nat.* 1882, p. 299, pl. xvii. figs. 26–30 *.

Distinguished from *P. guentheri* in the following points:—Size larger; the free phalanges longer, that of the fourth toe measuring more than the diameter of the eye; anterior row of chin-shields larger, much longer than broad; a long series of fifty femoral pores, forming a right angle medially.

Total length	216	millim.
Head	31	,,
Width of head	24	,,
Body	92	,,
Fore limb	40	,,
Hind limb	52	,,
Tail	93	,,

Greenwich Islands, Carolines.

5. Perochirus articulatus.

Hemidactylus ateles, var. articulatus, *Fischer, l. c.* p. 297, pl. xvii. figs. 20–25 *.

Distinguished from the three preceding in the following points:— Snout a little shorter and more convex; dorsal granules larger and coarser; belly covered with large juxtaposed granules; tail much narrower than the body; no femoral or præanal pores (?). Digits as in *P. guentheri* and *depressus*. Reddish brown above, uniform or marbled with dark brown.

Total length	110	millim.
Head	16	,,
Width of head	12	,,
Body	46	,,
Fore limb	19	,,
Hind limb	23	,,
Tail	48	,,

Ponape, Carolines.

29. SPATHOSCALABOTES.

Spathodactylus (*non Pictet*), *Günth. Proc. Zool. Soc.* 1872, p. 594.

Digits slender, and subcylindrical at the base, the penultimate joint bearing a strong discoid terminal dilatation with two series of regular oblique lamellæ below, separated by a median groove; the thumb rudimentary and clawless; the other digits with a free,

---

* Types examined.

clawed terminal phalange projecting considerably beyond the discoid dilatation. Upper surfaces covered with juxtaposed granular scales; lower surfaces with imbricate scales. Pupil vertical.

East-Indian archipelago.

1. Spathoscalabotes mutilatus. (Plate XIII. fig. 1.)

Spathodactylus mutilatus, *Günth. l. c.*

Head regularly oviform, much longer than broad; snout as long as the distance between the eye and the ear-opening, once and a half the diameter of the eye; ear-opening very small, oval, oblique. Rostral broad, subpentagonal; nostril pierced between the rostral, the first labial, and three small nasals; eleven upper and as many lower labials; mental small, subtriangular; no chin-shields. Body elongate. Scales of upper surfaces small, granular, not larger on the snout; abdominal scales small, cycloid, imbricate, smooth. Digits, especially fingers, very unequal, free; inner rudimentary, tubercle-like; four pairs of lamellæ under the other digits. Tail subcylindrical, slender, covered with small equal scales. Brown, finely marbled with darker; a dark streak from the tip of the snout to the shoulder, passing through the eye; a series of round whitish spots commences behind the eye, and is continued along each side of the back to the tail; lower parts whitish, finely speckled with brown.

Total length	82 millim.
Head	10 ,,
Width of head	6·5 ,,
Body	34 ,,
Fore limb	11 ,,
Hind limb	15 ,,
Tail	38 ,,

East-Indian archipelago.

a. ♀. Dr. Bleeker. (Type.)

30. MICROSCALABOTES.

Microscalabotes, *Bouleng. Ann. Mag. N. H.* (5) xi. 1883, p. 174.

Digits of very unequal size, free, slender at the base, strongly dilated at the end, with free distal clawed phalange; thumbs rudimentary, not dilated, with strong, very distinct claw; the digital dilatations bearing inferiorly two series of regular oblique lamellæ, separated by a median groove. Upper surfaces covered with juxtaposed granular scales; lower surfaces with imbricate scales. Pupil round. Eyelid distinct all round the eye. Males with præanal pores.

Madagascar.

1. Microscalabotes cowani. (Plate XIII. fig. 2.)

Microscalabotes cowanii, *Bouleng. l. c.*

Head small, much longer than broad, not distinct from neck; snout obtuse, as long as the distance between the eye and the ear-opening, not quite once and a half the diameter of the orbit; ear-opening very small, roundish. Body rather elongate, moderately depressed, not distinct from the tail, which is almost as thick at the base; the latter tapers into a point, and is slightly depressed, suboval in section. Limbs rather feeble; digits gradually increasing in size from first to fourth; fifth half the length of latter; seven transverse lamellæ under the dilated part of the third and fourth digits, the two basal undivided; the slender part of the digits with narrow transverse lamellæ. Upper surfaces covered with uniform small granular scales, considerably larger on the snout and interorbital space. Rostral broad, subpentagonal; nostril pierced between the rostral, the first upper labial, and two nasals, the upper being the largest and separated from its fellow on the other side by two granules; six or seven upper and as many lower labials; mental rather larger, subtriangular; irregular small chin-shields gradually passing into the gular scales; latter relatively large, though considerably smaller than the ventrals, which are large, smooth, and distinctly imbricate. Six or eight præanal pores, forming a short angular series. Caudal scales uniform, cycloid, imbricate, largest on the lower surface. Reddish golden brown above, with darker, greyish-brown vertebral and lateral bands; tail with more or less distinct darker angular markings; labials brown-dotted; lower surfaces whitish, the throat with a few scattered brown dots.

Total length	73	millim.
Head	9	„
Width of head	6	„
Body	22	„
Fore limb	10	„
Hind limb	13	„
Tail	42	„

Madagascar.

a–b. ♂. East Betsileo. Rev. W. Deans Cowan [C.]. (Types.)

31. LYGODACTYLUS *.

Lygodactylus, *Gray, Proc. Zool. Soc.* 1864, p. 59.
Scalabotes, *Peters. Mon. Berl. Ac.* 1880, p. 795.

---

* The following species, from Central Madagascar, have been too shortly described by Peters, Sitzb. Naturf. Fr. 1883, p. 28 :—

SCALABOTES PICTUS.

Stouter, with broader shorter snout and rather less elongate limbs than in

31. LYGODACTYLUS.

Digits of very unequal length, slender and subcylindrical at the base, free, with strong discoid terminal dilatation bearing two series of regular oblique lamellæ below, separated by a median groove; thumbs rudimentary, not dilated, with a very small, frequently indistinct, sheathed retractile claw; the other digits with a recurved terminal clawed phalange, the claw being retractile between the anterior pair of lamellæ. Upper surfaces covered with juxtaposed granular scales; lower surfaces with imbricate scales. Pupil round. Eyelid distinct all round the eye. Males with præanal pores.

Africa; Madagascar.

Synopsis of the Species.

I. Tail without a median series of large transversely dilated scales inferiorly.

Nostril posterior to the suture of the rostral and first labial .... 1. *capensis*, p. 160.
Nostril anterior to the suture of the rostral and first labial .... 2. *madagascariensis*, [p. 160.

II. Tail with a median series of large transversely dilated scales inferiorly.

Nostril anterior to the suture of the rostral and first labial .... 3. *thomensis*, p. 161.
Nostril posterior to the suture of the rostral and first labial; the males' throat with black chevron-shaped markings .... 4. *gutturalis*, p. 161.
Nostril posterior to the suture of the rostral and first labial; the males' throat black .... 5. *picturatus*, p. 161.

S. thomensis. Above brown, with black and yellow markings. On the snout, between the eyes, and on the nape black cross lines, and on the side of the neck two or three longitudinal black lines. Yellow, black-edged spots forming irregular transverse rows on the neck and back; similar spots on the limbs. Yellow, black-edged cross-lines on the sacral region and on the tail. Below dirty yellow, black-dotted on the lower lip and chin.

SCALABOTES BIVITTIS.

Three small scales between the nasals; three scales in a cross row behind the mental; snout twice and a half as long as the eye. Dark brown above, with an ochreous yellow lateral band originating from the supero-posterior part of the eye, and continued to the base of the tail. Ochreous yellow inferiorly, dotted with bluish black.

SCALABOTES HILDEBRANDTI.

A single scale between the nasals; three scales behind the mental, the posterior lateral sides of which are shorter than the median; snout twice as long as the eye. Above marbled with dark and light brown; sides darker, yellow-dotted. Limbs cross-barred. Lower surface yellow, dotted with dark.

1. Lygodactylus capensis.

Hemidactylus capensis, *Smith, Ill. Zool. S. Afr., Rept.* pl. lxxv. fig. 3.
Lygodactylus strigatus, *Gray, Proc. Zool. Soc.* 1864, p. 59, & 1865, p. 642.
Hemidactylus (Peropus) capensis, *Peters, Reise n. Mossamb.* iii. p. 28.

Head oviform, much longer than broad; snout longer than the distance between the eye and the ear-opening, about once and a half the diameter of the orbit; ear-opening very small, roundish. Rostral broad; nostril pierced above and behind the suture of the rostral and first labial, between the latter and three small nasals; the rostral entering scarcely the nostril; upper labials seven or eight; lower labials six or seven; mental large, subtriangular, extending beyond the posterior margin of the adjacent labials, bordered on each side by a row of small irregular chin-shields; behind the latter, large granular scales passing gradually into the smaller ones which cover the throat. Scales of upper surfaces small, granular, larger on the snout; abdominal scales large, hexagonal, imbricate, smooth. Digits very unequal, free; inner rudimentary, tubercle-like; four or five pairs of lamellæ under the other digits. Four or six præanal pores, forming an angular line. Tail tapering, rounded, feebly depressed, covered above with small juxtaposed or subimbricate scales, inferiorly with larger imbricate scales. Upper surfaces greyish-brown or olive, variegated with darker; a blackish lateral streak, passing through the eye, generally broken up on the sides of the body; lower surfaces uniform yellowish.

Total length	75	millim.
Head	9	,,
Width of head	6·5	,,
Body	25	,,
Fore limb	12	,,
Hind limb	15	,,
Tail	41	,,

S.E. Africa.

a. Hgr.   S.E. Africa.   Sir J. Kirk [C.]. (Type of *L. strigatus*.)
b. ♀.   S. Africa.
c, d–e. ♂.   ——?

2. Lygodactylus madagascariensis.

Scalabotes madagascariensis, *Boettger, Zool. Anz.* 1881, p. 360, *and Abh. Senck. Ges.* xii. p. 469, pl. ii. fig. 8.

This species, as well as the following, is extremely closely allied to *L. capensis*, differing only in a few points. Ear-opening rather larger. Nostril pierced above and in front of the suture of the rostral and first labial; the rostral entering considerably the nostril. Upper labials six or seven; lower labials five or six. Seven præanal pores. Upper caudal scales distinctly imbricate. Greyish brown above, variegated with darker and lighter; one or two blackish transverse lines between the eyes; a dark lateral band, passing through the eye, less distinct on the sides of the body; a blackish, light-edged horseshoe-shaped marking on the base of the tail; latter with angular transverse markings; lower surfaces whitish; throat generally with scattered blackish dots; lower labials blackish-spotted, the spots generally confluent into a band bordering the lower jaw. Total length 72 millim.

Madagascar.

a–e. ♂ ♀.　　　　　East Betsileo.　　　　Rev. W. Deans Cowan [C.].

3. Lygodactylus thomensis.

Scalabotes thomensis, *Peters, Mon. Berl. Ac.* 1880, p. 795, pl. —. fig. 1.

Nostril as in *L. madagascariensis*. Tail with a central series of transversely dilated plates inferiorly. Olive-brown above, with small dark brown spots; a dark transverse streak between the anterior angles of the eyes; a dark streak along the side of the head and neck, passing through the eye; tail with alternating brown and greenish transverse bands; lower surfaces greenish yellow, variegated with brown; gular region dark brown with two irregular **V**-shaped yellow markings. Total length 69 millim.

S. Thomé, W. Africa.

4. Lygodactylus gutturalis.

Hemidactylus gutturalis, *Bocage, Jorn. Sc. Lisb.* iv. 1873, p. 211.

The rostral does not enter at all the nostril, being separated from it by a nasal scale. Other differences between this species and *L. capensis* are the broader head, which is more distinct from neck, the shorter mental, the posterior borders of which form a very open angle, and the presence in the male of two large concentric chevron-shaped black markings on the gular region. From snout to vent 38 millim.; tail (reproduced) 28 millim.

West Africa.

a. ♂.　　　　　Bissao.　　　　Prof. Barboza du Bocage [P.]. (One of the types.)

5. Lygodactylus picturatus.

Hemidactylus capensis (*non Smith*), *Peters, Mon. Berl. Ac.* 1865, p. 455.
—— variegatus (*non Dum. & Bibr.*), *Peters, Mon. Berl. Ac.* 1868, p. 449, *and in Decken's Reise O. Afr.* iii. p. 13, pl. ii.
Liurus capensis, *Cope, Proc. Ac. Philad.* 1868, p. 320.
Hemidactylus picturatus, *Peters, Mon. Berl. Ac.* 1870, p. 115.

Snout more pointed. Scales of upper and lower surfaces smaller. Nostril as in *L. capensis*. Tail with a median series of transversely dilated plates inferiorly. Præanal pores eight or nine. Head and anterior part of body bright yellow, with dark brown or black lines and spots; posterior part of body, limbs, and tail grey, generally with black spots arranged in longitudinal lines on the back; these spots sometimes confluent; lower surfaces yellowish; throat of male deep black, that of the female marbled with grey. Total length 67 millim.

Zanzibar.

a. ♂. Magiba, Pangani. Sir J. Kirk [C.].

32. LEPIDODACTYLUS*.

Platydactylus, part., *Dum. & Bibr. Erp. Gén.* iii. p. 290.
Lepidodactylus, *Fitzing. Syst. Rept.* p. 98.
Amydosaurus, *Gray, Cat. Liz.* p. 162.
Peripia, part., *Günth. Rept. Brit. Ind.* p. 110.

Digits more or less dilated, free or with a rudiment of web, inferiorly with transverse lamellæ divided by a median groove, with very short compressed distal clawed joint raising from the extremity of the digit; inner digit clawless. Body covered above with granular scales, inferiorly with juxtaposed or subimbricate scales. Pupil vertical. Males with præanal or femoral pores.

East Indies; Polynesia; South-west Australia.

Synopsis of the Species.

I. Thumb rudimentary.

Four or five divided lamellæ under the median toes; digits very strongly denticulated on the borders ...... 1. *crepuscularis*, p. 163.
Four or five divided lamellæ under the median toes; digits less strongly denticulated .... 2. *ceylonensis*, p. 164.
Two divided lamellæ under the median toes .... 3. *aurantiacus*, p. 164.

II. Thumb well developed.

A. Tail flat inferiorly, with sharpish lateral edge.

Four transverse rows of small chin-shields; male with femoral pores .. 4. *lugubris*, p. 165.

B. Tail cylindrical.

1. Three or four transverse rows of small chin-shields.

14 upper and 15 lower labials; male with a series of præanal pores .... 5. *labialis*, p. 166.
10 or 11 upper and 9 or 10 lower labials; 18 to 20 lamellæ under the median toes; male with a series of præanal pores .... 6. *pulcher*, p. 166.

---
* *Peropus roseus*, Cope, Proc. Ac. Philad. 1863, p. 321. *Hab.* —— ?

11 or 12 upper and as many lower labials; 11 lamellæ under the median toes .... 7. *guppyi*, p. 166.

2. Two transverse rows of small chin-shields.

Tail swollen...................... 8. *pusillus*, p. 167.

3. A median chin-shield behind the mental.

Digits strongly dilated; male with a double series of præanal pores .... 9. *cyclurus*, p. 167.
Digits feebly dilated; male with a single series of præanal pores .... 10. *sauvagii*, p. 168.

1. Lepidodactylus crepuscularis.

Platydactylus crepuscularis, *Bavay, Cat. Rept. N. Caléd.* p. 8.
Lepidodactylus crepuscularis, *Bouleng. Proc. Zool. Soc.* 1883, p. 122, pl. xxii. fig. 6.

Head oviform, much longer than broad; snout as long as the distance between the eye and the ear-opening, about once and a half the diameter of the orbit; ear-opening small, vertically oval. Body much elongate; limbs short, the fore limb not measuring half the distance between axilla and groin. Digits short, free, inner very small, rudimentary; inferior lamellæ four or five; the basal part of the digit broad, with small rounded scales; the borders of the digital expansions strongly denticulated. Upper surfaces and throat covered with very small granular scales, scarcely larger on the snout; abdominal scales larger, subimbricate. Rostral quadrangular, twice as broad as high; nostril pierced between the rostral, the first upper labial and three small nasals, the upper separated from its fellow by five small scales; twelve upper and eleven lower labials; mental small, trapezoid; no chin-shields. Male with ten præanal pores forming a short angular series. Tail cylindrical, covered with small, rhomboidal, imbricated scales; these are distinctly raised on the sides, forming a slight denticulation. Greyish brown above, variegated with darker, and with small round white spots; a dark streak from the tip of the snout to the shoulder, passing through the eye; above this streak and behind the eye, two round white spots; a round white spot on each of the digits; lower surfaces dirty white, the belly dotted with brown.

Total length	65	millim.
Head	8	,,
Width of head	5	,,
Body	28	,,
Fore limb	8	,,
Hind limb	11	,,
Tail	29	,,

New Caledonia.

a-b. ♀.   New Caledonia.   J. Brenchley, Esq. [P.].

2. Lepidodactylus ceylonensis. (PLATE XIII. fig. 3.)

This species resembles exactly the preceding in proportions, pholidosis, and coloration. The only difference I can detect is that the digits are more slender, less strongly dilated at the base, and less strongly denticulated on the sides. Head and body 40 millim., tail 34.

Ceylon.

a. ♀.   Gampola.   Col. Beddome [C.].

3. Lepidodactylus aurantiacus. (PLATE XIII. fig. 4.)

Hemidactylus aurantiacus, *Beddome, Madras Journ. Med. Sc.* 1870; *Stoliczka, Journ. As. Soc. Beng.* xli. 1872, p. 99.

Head oviform, longer than broad; snout rounded, very convex, slightly shorter than the distance between the eye and the ear-opening, once and one third the diameter of the orbit; ear-opening very small, round. Body elongate, more so in females than in males; limbs short, the fore limb measuring half the distance between axilla and groin, or rather less. Digits short, free, inner very small, rudimentary; only two large chevron-shaped divided lamellæ under the distal part of the digits, followed by transverse undivided lamellæ, decreasing in width. Head covered with very minute granules; rostral and mental very small, former four-sided, latter pentagonal or triangular; nostril pierced between the rostral, the first labial, and several granules; labials very small, nine or ten upper and as many lower; no chin-shields. Back covered with very small granular scales; abdominal scales a little larger, flat, imbricate. Male with an angular series of seven to nine præanal pores. Tail cylindrical, tapering, covered with small imbricated smooth scales, larger inferiorly. Grey-brown above, with dark-brown undulated lines along the head and back, which may be broken up into spots; a dark brown streak from the tip of the snout to the fore limb, passing through the eye; whitish dots scattered on the head and back; tail with darker spots or annuli and two large whitish black-edged spots at the base, frequently confluent medially. Lower surfaces whitish, more or less speckled with brownish.

Total length	70	millim.
Head	10	,,
Width of head	7	,,
Body	27	,,
Fore limb	10	,,
Hind limb	12	,,
Tail	33	,,

Southern India.

a. Many spec.: ♂, ♀, hgr., & yg.   Shevaroys.   Col. Beddome [C.].   (Types.)

b–e. ♂, ♀, & hgr.   S. India.   Col. Beddome [C.].

4. Lepidodactylus lugubris.

Amydosaurus lugubris, *Gray, Cat.* p. 162.
Platydactylus lugubris, *Dum. & Bibr.* iii. p. 304; *Jacquin. Voy. Pôle Sud, Saur.* pl. i. fig. 1; *Cantor, Cat. Mal. Rept.* p. 16.
Lepidodactylus lugubris, *Fitzing. Syst. Rept.* p. 98; *Bouleng. Proc. Zool. Soc.* 1883, p. 120, pl. xxii. fig. 3.
Hemidactylus meyeri, *Bleek. Nat. Tijds. Nederl. Ind.* xvi. 1859, p. 47.
Peripia cantoris, *Günth. Rept. Brit. Ind.* p. 110; *Stoliczka, Journ. As. Soc. Beng.* xli. 1872, p. 103.
Gecko mœstus, *Peters, Mon. Berl. Ac.* 1867, p. 13.
Gymnodactylus candeloti, *Bavay, Cat. Rept. N. Caléd.* p. 13.
Peripia meyeri, *Günth. Proc. Zool. Soc.* 1872, p. 594.
—— mysorensis, *Meyer, Mon. Berl. Ac.* 1874, p. 129.
—— lugubris, *Peters & Doria, An. Mus. Genov.* xiii. 1878, p. 371.
Platydactylus (Lepidodactylus) crepuscularis (*non Bav.*), *Sauvage, Bull. Soc. Philom.* (7) iii. 1878, p. 69.

Head much longer than broad; snout subacuminate, longer than the distance between the eye and the ear-opening, about once and a half the diameter of the orbit; forehead with a median groove; ear-opening small, round. Body and limbs moderate. Digits moderate, inner well developed, with a rudiment of web; inferior lamellæ numerous, seven or eight under the fingers, eight or nine under the toes. Upper surfaces and throat covered with very minute granules, a little larger on the snout; scales on the belly much larger, flat, juxtaposed, or slightly imbricate. Rostral quadrangular, broad; nostril pierced between the rostral, the first upper labial, and two or three nasal shields, the upper separated from its fellow by one or three small shields; eleven to thirteen upper and ten or eleven lower labials; mental small, smaller than the adjacent labials; four transverse rows of small chin-shields. Femoral pores in a long series, angular medially, twenty-five altogether. Tail flat inferiorly, with sharpish, sometimes feebly serrated, lateral edge; caudal scales small, flat, equal. Upper surfaces light pinkish-grey or brownish, generally with a series of small blackish or purplish-brown spots on each side of the vertebral line; a purplish-brown streak from the end of the snout to the ear, passing through the eye; labials generally finely dotted with brown; lower surfaces white.

Total length	81	millim.
Head	11	,,
Width of head	8	,,
Body	33	,,
Fore limb	12	,,
Hind limb	17	,,
Tail	37	,,

Malay peninsula and archipelago; New Guinea; Polynesia.

a. ♀. Penang. Dr. Cantor. (Type of *Peripia cantoris.*)
b. ♀. Bintang. Dr. Bleeker. (Type of *Hemidactylus meyeri.*)

c. ♀.	Amboyna.	H.M.S. 'Challenger.'
d. ♂.	Pelew Islands.	Godeffroy Museum.
e. ♀.	Murray Island.	Rev. S. Macfarlane [C.].
f. ♀.	Tahiti.	H.M.S. 'Alert.'
g. ♀.	Isld. of Havannah, New Hebrides.	Dr. Corrie [P.].
h. ♀.	Mallicollo, N. Hebrides.	W. Wykeham Perry, Esq. [P.]
i–k. ♀.	Island of Onio, Fiji.	F. M. Rayner, Esq. [C.].
l–n. ♀.	Fiji Islands.	E. A. Liardet, Esq. [P.].
o. ♀.	Island of Vati, S. Pacific.	Dr. Corrie [P.].

5. Lepidodactylus labialis.

Geckō labialis, *Peters, Mon. Berl. Ac.* 1867, p. 14.

Very closely allied to *L. lugubris*, from which it differs in the following points: fourteen upper and fifteen lower labials; no femoral pores, but on each side nine præanal pores forming a doubly arched series, angular medially. Tail cylindrical.

Mindanao.

6. Lepidodactylus pulcher. (Plate XIII. fig. 5.)

Differs from *L. lugubris* in the following characters:—Snout rather shorter. Seventeen præanal pores, forming a doubly arched series, angular medially. Tail cylindrical. Pinkish brown above, head with small black spots; limbs and sides of the back with minute blackish specks; tail with dark transverse blotches; throat and breast minutely speckled with brown.

Total length	105	millim.
Head	12	,,
Width of head	9	,,
Body	39	,,
Fore limb	15	,,
Hind limb	19	,,
Tail	54	,,

Admiralty Islands.

a–c. ♂ ♀.   Wild Island.   H.M.S. 'Challenger.'

7. Lepidodactylus guppyi.

Lepidodactylus guppyi, *Bouleng. Proc. Zool. Soc.* 1884, p. 211.

Head small, oviform, very convex; snout once and one third the diameter of the orbit, which equals the distance between the latter and the ear-opening; forehead slightly concave; ear-opening very small, round. Body elongate, limbs moderate. Digits moderate, inner well developed, webbed at the base; eleven lamellæ under the median digits, the two or three proximal divided. Scales uniformly granular, the granules larger on the snout, largest and flat on the belly. Rostral quadrangular, more than twice as broad as high;

nostril pierced between the rostral, the first upper labial, and three nasals; eleven or twelve upper and as many lower labials; three or four rows of very small chin-shields. Tail cylindrical, tapering, covered with small equal flat scales. Pinkish brown above, sides with darker spots; a dark streak on the side of the head, passing through the eye; tail with dark annuli; lower surfaces whitish, throat speckled with reddish brown.

Total length	93 millim.
Head	12 ,,
Width of head	7 ,,
Body	36 ,,
Fore limb	11·5 ,,
Hind limb	17·5 ,,
Tail	45 ,,

Solomon Islands.

a. ♀. Faro Island. H. B. Guppy, Esq. [P.]. (Type.)

8. Lepidodactylus pusillus.

Peropus pusillus, *Cope, Proc. Acad. Philad.* 1868, p. 319.

Two cross rows of ovate chin-shields, those behind graduating through several rows to the gulars; two scales between the upper nasals; nine superior labials to below the pupil. Tail cylindric, swollen. Colour light brown, with a much paler dorsal shade: a brown band through orbit to axilla, and band across muzzle; tail with a series of pale rounded spots on the median line above.

S.W. Australia.

9. Lepidodactylus cyclurus. (Plate XIII. fig. 6.)

Peripia cyclura, *Günth. Ann. & Mag. N. H.* (4) x. 1872, p. 422, *and in Brenchley's 'Curaçoa,' Rept.* p. 407.
Lepidodactylus neocaledonicus, *Bocage, Jorn. Sc. Lisb.* iv. 1873, p. 206.
Hemidactylus (Peripia) bavayi, *Sauvage, Bull. Soc. Philom.* (7) iii. 1878, p. 71.
Lepidodactylus cyclurus, *Bouleng. Proc. Zool. Soc.* 1883, p. 121, pl. xxii. fig. 4.

Head oviform, longer than broad; snout a little longer than the distance between the eye and the ear-opening, about once and one third the diameter of the orbit; ear-opening moderate, roundish. Body and limbs moderate. Digits moderate, inner well developed, with a slight rudiment of web; inferior lamellæ numerous, all divided by a median groove, ten or eleven. Upper surfaces and throat covered with very small granular scales, larger on the snout; abdominal scales subimbricate. Rostral quadrangular, twice as broad as high; nostril pierced between the rostral, the first upper labial, and four or five small nasals, the upper separated from its fellow by three or five small shields; nine to eleven upper and nine or ten lower labials; mental small, subtriangular, shorter than the adjacent labials, followed by a median chin-shield; a few other irregular chin-shields gradually passing into the granules of the throat. Præanal pores in two angular series; these series in contact and containing each eleven to sixteen pores. Tail cylindrical, covered with small, equal, flat scales arranged in verticils. Upper surfaces brownish grey, with more or less irregular dark brown bands across the back; sides with more or less distinct small round whitish spots; a more or less indistinct dark streak on the side of the head, passing through the eye; lower surfaces uniform whitish.

Total length	117	millim.
Head	17	,,
Width of head	13	,,
Body	49	,,
Fore limb	19	,,
Hind limb	26	,,
Tail	51	,,

New Caledonia.

a–b, c–e. ♂ & hgr.   New Caledonia.   J. Brenchley, Esq. [P.]. (Types.)

10. Lepidodactylus sauvagii.

Hemidactylus (Peripia) cyclura (*non Günth.*), *Sauvage, Bull. Soc. Philom.* (7) iii. 1878, p. 72.
Lepidodactylus sauvagii, *Bouleng. Proc. Zool. Soc.* 1883, p. 122 pl. xxii. fig. 6.

This species resembles exactly the preceding in the proportions, scutellation, size, and colour, but differs in the much narrower digits, and the presence of a single series of præanal pores; this is composed of twenty-three pores.

New Caledonia.

33. NAULTINUS.

Naultinus, part., *Gray, Zool. Misc.* p. 72, and *Cat. Liz.* p. 169.
Heteropholis, *Fischer, Abh. Naturw. Ver. Brem.* vii. 1882, p. 236.

Digits free, feebly dilated, gradually narrowing distally, clawed, with a series of transverse lamellæ under their entire length. Dorsal scales uniformly granular or intermixed with enlarged tubercles. Pupil vertical. Males with præanal and femoral pores.
New Zealand.

1. **Naultinus elegans.** (Plate XIV. fig. 3 [*]).

Naultinus elegans, *Gray, Cat.* p. 169.
Naultinus grayii, *Gray, l. c.* p. 170.
Naultinus punctatus, *Gray, l. c.* p. 170.

---

[*] Lower surface of foot.

Naultinus elegans, *Gray, Zool. Miscell.* p. 72, *and in Dieffenb. N. Zeal.* ii. p. 203; *Steindachn. Novara, Rept.* p. 19; *Buller, Trans. N. Zeal. Instit.* iii. 1871, p. 7, pl. ii. fig. 1; *Hutton, op. cit.* iv. 1872, p. 170.
—— punctatus, *Gray, in Dieffenb. N. Zeal.* ii. p. 204; *Girard, U.S. Explor. Exped., Herp.* p. 309, pl. xvi. figs. 17–26; *Steindachn. l. c.* p. 20; *Buller, l. c.* p. 8; *Hutton, l. c.* p. 171.
—— grayi, *Bell, Zool. 'Beagle,' Rept.* p. 27, pl. xiv. fig. 2; *Buller, l. c.* p. 7.
Gymnodactylus elegans, *A. Dum. Cat. Méth. Rept.* p. 43, *and Arch. Mus.* viii. p. 477, pl. xviii. fig. 14.
? Naultinus lineatus, *Gray, Ann. & Mag. N. H.* (4) iii. 1869, p. 243.
Naultinus sulphureus, *Buller, l. c.* p. 8; *Hutton, l. c.* p. 172.
—— silvestris (*Buller*), *Field, N. Zeal. Journ. Sc.* i. 1882, p. 177.

Head small, short: snout obtusely subtriangular, short, measuring slightly more than the distance between the eye and the ear-opening, and twice the diameter of the orbit, with vertical loreal region and obtuse canthus rostralis; no trace of a concavity on the upper surface of the head; eye very small, with distinct circular lid; ear-opening very small, oval, horizontal. Body and limbs moderate, slightly depressed. Head covered with granular scales posteriorly, with polygonal, flat, or more or less convex, considerably larger scales on the snout; rostral about twice as broad as high, with distinct median cleft superiorly; nostril pierced between the first upper labial and three or four small nasals; eleven or twelve upper and ten or eleven lower labials; mental twice as broad as high anteriorly, narrowed posteriorly; the mental and labials followed by very small chin-shields gradually passing into the minute granules of the throat. Upper surfaces covered with small granules; lower surfaces with very small, slightly imbricated scales. Male with a large median patch of præanal pores and two series of femoral pores. Tail long, cylindrical, covered with very small subequal juxtaposed scales. The coloration varies considerably: the specimens in the collection may be referred to the following varieties of coloration:—

A. (*N. elegans*, Gray.) Brownish, purplish, or dark olive above, yellow beneath; a streak on each side of the crown, another on the lower lip, and generally also one from axilla to groin, and large paired spots on the back and tail, yellow, black-margined. (Specs. *a, m, n, o, q, r, s, x, z, β, γ, δ.*)

B. (*N. punctatus*, Gray.) Green, upper parts minutely dotted with black; hands and feet yellow inferiorly. (Specs. *c, a.*)

C. (*N. grayi*, Bell.) Uniform green, lighter beneath; hands and feet yellow inferiorly. (Specs. *d, f, g, h, i, k, l.*)

D. (*N. sulphureus*, Buller.) Uniform lemon-coloured. (Specs. *p, v, x.*)

E. Pink above, yellow beneath, uniform. (Spec. *ε.*)

F. Green above, yellow beneath, uniform. (Specs. *t, w.*)

G. Green, lighter beneath, with a few distant, paired yellow spots on the back, and a rather indistinct yellow streak from axilla to groin; hands and feet yellow. (Specs. *e, u.*)

Total length	188 millim.
Head	20 ,,
Width of head	17·5 ,,
Body	61 ,,
Fore limb	32 ,,
Hind limb	38 ,,
Tail	107 ,,

New Zealand.

a. ♀.	Auckland.	Sir R. Owen [P.]. (Type.)
b. Yg.	Bay of Islands.	Antarctic Expedition.
c. ♂.	N. Zealand.	Sir J. Richardson [P.].
		(Type of *N. punctatus*.)
d. ♀.	N. Zealand.	C. Darwin, Esq. [C.].
		(Type of *N. grayi*.)
e, f, g. ♀.	N. Zealand.	Mr. Egerley.
h. ♀.	N. Zealand.	Antarctic Expedition.
i–l, m–p, q. ♂, ♀, hgr., & yg.	N. Zealand.	Sir G. Grey [P.].
r, s–t, u, v, w. ♀ & hgr.	N. Zealand.	Dr. Sinclair [P.].
x, y. ♂.	N. Zealand.	Capt. Drury [P.].
z. ♀.	Interior of N. Zealand.	G. F. Angas, Esq. [P.].
α. ♂.	N. Zealand.	
β–δ. ♂ ♀.	N. Zealand.	E. S. Ellisdon, Esq. [P.].
ε. ♀.	——?	Haslar Collection.

2. Naultinus rudis.

Heteropholis rudis, *Fischer, Abh. Naturw. Ver. Brem.* vii. 1882, p. 236, pl. xvi.

Head elongate oviform, very distinct from neck; eye moderate, the eyelid not distinct inferiorly; ear-opening small, linear, oblique. Body and limbs moderate. Two lateral folds of the skin, enclosing a groove. Head covered with granular scales intermixed with larger flat scales, largest and most numerous on the snout. Rostral four times as broad as high; nostril pierced between the first upper labial and three nasals; ten upper and as many lower labials; mental three times as broad as high anteriorly, narrowing posteriorly; no chin-shields. Back covered with small granular scales, intermixed, especially on the sides, with large roundish, flat or keeled tubercles; lower surfaces covered with imbricated scales, the throat granular. A large patch of præanal pores and a single series of femoral pores. Tail cylindrical, covered with irregular scales. Greenish grey above, with irregular longitudinal and transverse purplish bands on the back; uniform light grey beneath.

Total length	145 millim.
From tip of snout to ear-opening	17 ,,
Fore limb	20 ,,
Hind limb	26 ,,
Tail	82 ,,

New Zealand.

34. HOPLODACTYLUS.

Platydactylus, part., *Dum. & Bibr.* iii. p. 290.
Hoplodactylus, *Fitzing. Syst. Rept.* p. 100.
Pentadactylus, *Gray, Cat. Liz.* p. 160.
Naultinus, part., *Gray, l. c.* p. 169.
Dactylocnemis (*Fitz.*), *Steindachn. Novara, Rept.* p. 11.

Digits free or shortly webbed at the base, more or less dilated; the distal phalanges slender, elongate, clawed, forming an angle with the basal portion; a series of transverse lamellæ under the latter. Scales small, granular, equal. Pupil vertical. Males with præanal, or præanal and femoral pores.

South-Pacific islands; Southern India.

Synopsis of the Species.

I. The length of the slender distal phalanges equals one fourth the length of the digit, or the width of the dilated part.

Rostral not entering the nostril .... 1. *maculatus*, p. 171.

II. The length of the slender distal phalanges equals one third the length of the digit.

Digits much dilated, with a distinct web at the base; 16 lamellæ under the fourth toe 2. *duvaucelii*, p. 172.
Digits not much dilated, with a distinct web at the base; 10 to 12 lamellæ under the fourth toe, the anterior slightly angular . 3. *pacificus*, p. 173.
Digits very feebly dilated, quite free; 10 to 12 lamellæ under the fourth toe, all straight .... 4. *granulatus*, p. 174.
Digits feebly dilated, free; 7 or 8 lamellæ under the fourth toe .... 5. *anamallensis*, p. 175.

1. Hoplodactylus maculatus. (PLATE XIV. fig. 1.)

Naultinus pacificus, part. (N. maculatus), *Gray, Cat.* p. 273.

Head short, oviform; snout as long as, or slightly longer than, the distance between the eye and the ear-opening, once and two thirds the diameter of the orbit; ear-opening large, oval, oblique, two thirds or three fourths the diameter of the eye. Body and limbs moderate. Digits relatively much dilated; the length of the slender distal part equals the width of the dilated portion, which is contained about three times in the length of the same; the anterior inferior lamellæ chevron-shaped; ten to twelve lamellæ under the fourth toe; a very slight web at the base of the digits, absent between the two outer toes. Head covered with granular scales, minute on the posterior half, considerably larger on the snout. Rostral broad, subquadrangular or subpentagonal, with trace of median cleft above; nostril pierced between the first upper labial and four or five small nasals; eleven or twelve upper and nine or ten lower labials; mental small, triangular or trapezoid; small irregular chin-shields, passing gradually into the minute granules of the throat. Dorsal scales minutely granular; abdominal scales very small, juxtaposed or subimbricate. Three or four angular series of præ-anal pores, the two upper extending sometimes on the thighs; the number of pores very variable. Tail cylindrical, tapering, covered with small equal subhexagonal juxtaposed scales arranged in verticils; the base of the tail strongly swollen in the males, the swollen part covered with larger scales. Brown above, with small blackish spots, and more or less distinct irregular transverse dark brown bands on the back and tail; a more or less distinct dark streak on each side of the head, passing through the eye; lower surfaces dirty white, sometimes with a few brown specks.

Total length	156	millim.
Head	22	,,
Width of head	18	,,
Body	54	,,
Fore limb	28	,,
Hind limb	35	,,
Tail	80	,,

New Zealand.

a–b. ♂ & yg.	N. Zealand.	(Types.)
c. ♀.	N. Zealand.	
d–f. ♀.	N. Zealand.	Capt. Stokes [P.].
g. ♀.	N. Zealand.	Capt. Drury [P.].
h–k. ♂, ♀, & hgr.	N. Zealand.	E. S. Ellisdon, Esq. [P.].
l. ♂.	—— ?	

Var. A broad light band on each side of the back.

m. ♀.	N. Zealand.	E. S. Ellisdon, Esq. [P.].
n–o. ♂ ♀.	N. Zealand.	Capt. Stokes [P.].

2. Hoplodactylus duvaucelii.*

Pentadactylus duvaucelii, *Gray, Cat.* p. 160.
Platydactylus duvaucelii, *Dum. & Bibr.* iii. p. 312; *A. Dum. Arch. Mus.* viii. p. 456, pl. xviii. fig. 1.
Hoplodactylus duvaucelii, *Fitzing. Syst. Rept.* p. 100.
Pentadactylus duvaucelii, *Günth. Rept. Brit. Ind.* p. 118.

Head oviform; snout slightly longer than the distance between the eye and the ear-opening; forehead slightly concave; ear-opening rather large, oval, oblique. Body and limbs moderate. Digits relatively much dilated, the width of the dilated part contained about twice and a half in the length of same; the length of the slender distal part equals one third the length of the digit; sixteen lamellæ under the fourth toe; a distinct web at the base of the digits, absent between the two outer toes. Head covered with granular scales, minute on the posterior half, considerably larger on the snout.

---

* Described from one of the types (♂).

Rostral pentagonal, with trace of median cleft above; nostril pierced between the rostral, the first upper labial, and four small nasals; fourteen upper and twelve lower labials; mental trapezoid, a little shorter than the adjacent labials, followed by a small median chin-shield; behind the labials small irregular scales passing gradually into the minute granules of the throat. Dorsal scales minutely granular; abdominal scales very small, subimbricate. Five angular series of præanal pores, the two upper extending as long femoral series; fifty pores in the upper series, from one end to the other. Tail rounded, tapering, covered with very small subquadrangular juxtaposed scales arranged in verticils. Greyish above, indistinctly marbled with darker; lower surfaces dirty white, immaculate.

Total length	205	millim.
Head	32	,,
Width of head	25	,,
Body	93	,,
Fore limb	44	,,
Hind limb	53	,,
Tail	80	,,

Bengal.

a. Skeleton.

3. Hoplodactylus pacificus.

Naultinus pacificus, *Gray, Cat.* p. 169.
Naultinus pacificus, *Gray, in Dieffenb. N. Zeal.* ii. p. 203; *Buller, Trans. N. Zeal. Instit.* iii. 1871, p. 7; *Hutton, op. cit.* iv. 1872, p. 172.
Platydactylus pacificus, *A. Dum. Cat. Méth. Rept.* p. 35, *and Arch. Mus.* viii. p. 455.
Hoplodactylus pomarii, *Girard, Proc. Ac. Philad.* 1857, p. 197, *and U.S. Explor. Exped., Herp.* p. 294, pl. xviii. figs. 10–16.
Dactylocnemis pacificus (*Fitz.*), *Steindachn. Novara Rept.* p. 11, pl. i. fig. 1.

Head oviform; snout distinctly longer than the distance between the eye and the ear-opening, once and two thirds the diameter of the orbit; forehead slightly concave; ear-opening rather large, oval, oblique, half or three fifths the diameter of the eye. Body and limbs rather slender. Digits not much dilated, the width of the dilated part being one third the length of same; the length of the slender distal part equals one third the length of the digit, and is consequently much greater than the width of the dilated part; the anterior inferior lamellæ slightly angular; ten to twelve lamellæ under the fourth toe; a distinct web at the base of the digits, absent between the two outer toes. Head covered with granular scales, minute on the posterior half, considerably larger on the snout. Rostral broad, subquadrangular or subpentagonal, with trace of median cleft above; nostril pierced between the rostral, the first upper labial, and four or five small nasals; eleven or twelve upper and nine or ten lower labials; mental small, trapezoid or triangular, generally followed by a small median chin-shield; small irregular chin-shields passing gradually into the minute granules of the throat. Dorsal scales minutely granular; abdominal scales very small, juxtaposed or subimbricate. Three or four short angular series of præanal pores, forming together a small subtriangular patch; twenty to thirty-five pores altogether. Tail cylindrical, tapering, covered with small equal subquadrangular juxtaposed scales arranged in verticils; the base of the tail strongly swollen in the males, the swollen part covered with larger scales. Brown above, with irregular transverse darker bands on the back and tail, and frequently a dark band on each side, commencing from the tip of the snout, and passing through the eye and above the ear; a short, dark, oblique streak directed posteriorly from the inferior border of the eye; lower surfaces whitish, immaculate.

Total length	163 millim.
Head	20 ,,
Width of head	15 ,,
Body	57 ,,
Fore limb	28 ,,
Hind limb	37 ,,
Tail	86 ,,

New Zealand; (Tasmania?).

a. ♀, bleached.		(Type.)
b-c, d-e. ♂ & yg.	N. Zealand.	Admiralty [P.].
f, g. ♂ ♀.	N. Zealand.	Sir G. Grey [P.].
h-i, k. ♂ ♀.	N. Zealand.	Dr. Sinclair [P.].
l-m. ♂ ♀.	N. Zealand.	Capt. Drury [P.].
n. ♂, skeleton.	N. Zealand.	
o. ♀.	Tasmania.	Zoological Society.

4. Hoplodactylus granulatus. (PLATE XV. fig. 1.)

Naultinus granulatus, *Gray, Cat.* p. 273.

General characters and proportions as in the preceding; distinguished in the following important characters:—Ear-opening smaller, not more than half the diameter of the eye. Digital dilatation very narrow; all the lamellæ straight; no trace of web. Scales of back and belly larger. Four to seven angular series of præanal pores, the three or four upper extending as femoral series; thirty to forty pores in the upper series, from one end to the other. Greyish or brown above, with dark brown or reddish-brown vermiculations and irregular cross bands, light-edged in front; two dark streaks from the eye, as in *H. pacificus*, separated by a yellowish interspace; lower surfaces whitish, generally with dark dots or variegations, especially on the gular region.

Total length	192 millim.
Head	22 ,,
Width of head	17 ,,

Body	68	millim.
Fore limb	31	,,
Hind limb	40	,,
Tail	102	,,

New Zealand.

a, b. ♀.	New Zealand.	Sir R. Owen [P.]. (Types.)
c–e. ♀ & yg.	New Zealand.	Sir G. Grey [P.].
f. ♀.	New Zealand.	Dr. Sinclair [P.].
g. ♀.	New Zealand.	Capt. Drury [P.].
h–k, l. ♂ ♀.	New Zealand.	

5. Hoplodactylus anamallensis. (Plate XIV. fig. 2.)

Gecko anamallensis, *Günth. Proc. Zool. Soc.* 1875, p. 226.

Head shortly oviform; snout a little longer than the distance between the eye and the ear-opening, once and one third the diameter of the orbit; forehead very slightly concave; ear-opening small, oval, oblique. Body and limbs moderate. Digits free, not very strongly dilated, inner well developed: infradigital lamellæ openly angular, five under the inner digits, six or seven under the median fingers, and seven or eight under the median toes. Upper surfaces uniformly granular, the granules scarcely larger on the snout than on the back. Rostral four-sided, more than twice as broad as high; nostril pierced beween the rostral, the first labial, and three nasals; nine or ten upper and seven or eight lower labials; mental broadly triangular or pentagonal; two pairs of larger chin-shields anteriorly, followed by smaller ones passing gradually into the granules of the throat; abdominal scales cycloid, imbricated, smooth. Male with an uninterrupted series of femoral pores, twenty or twenty-one on each side. Tail cylindrical, tapering, slightly depressed at the base, above with small imbricated scales, inferiorly with a median series of transversely dilated plates. Grey-brown above, spotted or marbled with darker; a few scattered lighter dots; lower surfaces brownish white.

Total length	82	millim.
Head	12	,,
Width of head	8·5	,,
Body	33	,,
Fore limb	14	,,
Hind limb	18	,,
Tail	37	,,

Southern India.

a. ♀.	Anamallays.	Col. Beddome [C.]. (Type.)
b. Several specs.: ♂, ♀, hgr., & yg.	Tinnevelly.	Col. Beddome [C.].

35. RHACODACTYLUS.

Platydactylus, part., *Wiegm. Herp. Mex.* p. 19; *Dum. & Bibr. Erp. Gén.* iii. p. 290.
Rhacodactylus, *Fitzing. Syst. Rept.* p. 100; *Bouleng. Proc. Zool. Soc.* 1883, p. 123.
Platydactylus, *Gray, Cat.* p. 160.
Correlophus, *Guichen. Mém. Soc. Sc. Nat. Cherb.* xii. 1866, p. 249.
Ceratolophus, *Bocage, Jorn. Sc. Lisb.* iv. 1873, p. 205.
Chameleonurus, *Bouleng. Bull. Soc. Zool. France*, 1878, p. 68.

Digits strongly dilated, more or less webbed, with undivided lamellæ below, all furnished with a retractile claw supported by the compressed distal joint. Upper surfaces covered with juxtaposed granular scales; lower surfaces with juxtaposed granules or slightly imbricated scales. Tail prehensile. Pupil vertical. Males with præanal pores.

New Caledonia.

Synopsis of the Species.

I. Dorsal scales uniform.

  A. Digits half-webbed.

Rostral separated from the nostril; scales on upper surface of head considerably larger than the granules of the back . . 1. *leachianus*, p. 176.
Rostral separated from the nostril; upper surface of head covered with minute granules, not larger than those of the back .... 2. *aubryanus*, p. 177.
Rostral entering the nostril; upper surface of head covered with small granular scales, almost as small as those of the back .... 3. *chahoua*, p. 177.

  B. Digits not more than one-third webbed.

Upper surface of head covered with very large, unequal, subconical tubercles.. 4. *trachyrhynchus*, p. 178.
Hinder part of head with knob-like prominences .... 5. *auriculatus*, p. 179.

II. A ciliated crest on each side, from the upper eyelid to the back.. 6. *ciliatus*, p. 180.

1. Rhacodactylus leachianus [*].

Platydactylus leachianus, *Gray, Cat.* p. 160.
Platydactylus leachianus, *Cuv. R. A.* 2nd edit. ii. p. 54; *Dum. & Bibr.* iii. p. 315, pl. xxviii. fig. 6; *Bavay, Cat. Rept. N. Caléd.* p. 3.
Hoplodactylus (Rhacodactylus) leachianus, *Fitzing. Syst. Rept.* p. 100.
Rhacodactylus leachianus, *Bocage, Jorn. Sc. Lisb.* iv. 1873, p. 201; *Bouleng. Proc. Zool. Soc.* 1883, p. 124.

---

[*] This description is taken from the young specimen in the Collection, and from two adult female specimens, one in the Lisbon Museum, the other in the Brussels Museum.

Head large, oviform, elongate; forehead concave; snout longer than the distance between the orbit and the ear-opening, a little more than twice the diameter of the orbit; ear-opening narrow, horizontal. Body and limbs moderately elongate, rather depressed; digits moderately elongate, broad, not very unequal, half-webbed; limbs bordered on each side by cutaneous folds, most developed on the anterior side of the fore limbs and on the posterior side of the hind limbs; other cutaneous folds below the rami mandibulæ, on the sides of the neck, and from axilla to groin; throat with a few cross folds. Upper surface of head covered with small irregular polygonal convex scales, larger than the granules of the back, elongated and more or less distinctly keeled on the forehead and snout; rostral twice as broad as high, with more or less distinct median cleft above; nostril pierced between seven or eight small nasals and the first upper labial; upper labials seventeen or eighteen; lower labials fourteen to seventeen, anterior very high; mental narrow, elongate, wedge-shaped, about the size of the proximal labials; no regular chin-shields, but a few larger scales, passing gradually into the minute granules of the throat. Scales small, granular, smallest on the lower surfaces. Tail cylindrical, tapering, covered with uniform small flat juxtaposed scales arranged in verticils. Grey or pinkish grey above, dotted and indistinctly marbled with brown, sometimes with large whitish spots; lower surfaces white, uniform.

Total length *	326	millim.
Head	59	„
Width of head	46	„
Body	152	„
Fore limb	70	„
Hind limb	84	„
Tail	115	„

New Caledonia.

a. Yg.   Isle of Pines.   J. Macgillivray, Esq. [C.].

2. Rhacodactylus aubryanus.

Rhacodactylus aubryanus, *Bocage, Jorn. Sc. Lisb.* iv. 1873, p. 202, & 1881, p. 127; *Bouleng. Proc. Zool. Soc.* 1883, p. 125.

Agrees in every respect with the preceding, except in having the upper surface of the head covered with minute granules, not larger than those covering the back. From snout to vent 195 millim.

New Caledonia.

3. Rhacodactylus chahoua.

Platydactylus chahoua, *Bavay, Cat. Rept. N. Caléd.* p. 3.
Rhacodactylus chahoua, *Bouleng. Proc. Zool. Soc.* 1883, p. 125, pl. xxi. fig. 1.

---

* Of the specimen in the Brussels Museum.

General proportions of *R. trachyrhynchus*. Head oviform; snout longer than the distance between the orbit and the ear-opening, once and a half the diameter of the orbit; forehead concave; ear-opening narrow, horizontal. Body and limbs moderately elongate, not much depressed; digits broadly dilated, not very unequal, their border very distinctly denticulated, half-webbed; the web very short, though distinct, between the two outer toes; limbs bordered by cutaneous folds, most developed on the anterior side of the fore limbs and on the posterior side of the hind limbs; other cutaneous folds below the rami mandibulæ, on the sides of the neck, and from axilla to groin; the latter feeble; throat with irregular cross folds. Upper surfaces covered with very small equal granular scales, scarcely larger on the snout; gular scales minute, granular; abdominal scales very small, flat, juxtaposed. Rostral twice as broad as high; nostril pierced between the rostral, the first upper labial, and six or seven small nasals; thirteen or fourteen upper and twelve or thirteen lower labials; mental elongate wedge-shaped, as long as the adjacent labials, which it separates; no regular chin-shields, but a few larger scales passing gradually into the minute granules of the throat. Male with three short series of femoral pores on each side; these do not extend beyond the inner half of the thighs, and are separated from the corresponding series on the other side; altogether seventy-nine pores. Grey above, whitish beneath, marbled with grey, forming transverse bars on the belly.

Head	33 millim.
Width of head	22 ,,
Body	91 ,,
Fore limb	38 ,,
Hind limb	51 ,,

New Caledonia.

4. Rhacodactylus trachyrhynchus.

Platydactylus duvaucelii, (*non D. & B.*) *Bavay, Cat. Rept. N. Caléd.* p. 6.

Rhacodactylus trachyrhynchus, *Bocage, Jorn. Sc. Lisb.* iv. 1873, p. 203; *Bouleng. Proc. Zool. Soc.* 1883, p. 126, pl. xxi. fig. 2.

Chameleonurus trachycephalus, *Bouleng. Bull. Soc. Zool. France*, 1878, p. 68, pl. ii.

Platydactylus (Rhacodactylus) chahoua, (*non Bavay*) *Sauvage, Bull. Soc. Philom.* (7) iii. 1878, p. 66.

Chameleonurus chahoua, (*non Bavay*) *Bouleng. Bull. Soc. Zool. France*, 1879, p. 142; *Bocage, Jorn. Sc. Lisb.* xxx. 1881, p. 126.

Head oviform; snout longer than the distance between the orbit and the ear-opening, once and two thirds the diameter of the orbit; ear-opening narrow, horizontal. Body and limbs moderately elongate, rather depressed; digits broadly dilated, not very unequal, with a distinct rudiment of web, absent between the two outer toes. Sides of neck with irregular folds. Snout and interorbital space with very large, unequal, rough tubercles confluent with the cranial ossification; hinder part of head, body and limbs with very small, equal granular scales; granules smallest on the throat. Rostral broad, quadrangular, with distinct median cleft above; nostril pierced between the first labial and five or six small nasals, the anterior being much the largest; nine to eleven upper labials; mental small, triangular; nine to eleven infralabials, gradually decreasing in size, inner pair nearly three times as high as broad, in contact behind the mental; a row of large scales behind the labials. Body covered with uniform small granular scales. Tail cylindrical, with uniform small flat scales arranged in verticils. Male with an irregular patch of very numerous præanal pores. Head brown; the rest of the upper surfaces grey, dotted with darker; lower surfaces dirty white, with scattered grey dots.

Total length	308	millim.
Head	38	,,
Width of head	24	,,
Body	115	,,
Fore limb	50	,,
Hind limb	66	,,
Tail	155	,,

New Caledonia.

a–b. ♂. New Caledonia. G. A. Boulenger, Esq. [P.].

5. Rhacodactylus auriculatus.

Platydactylus auriculatus, *Bavay, Cat. Rept. N. Caléd.* p. 6.
Ceratolophus hexaceros, *Bocage, Jorn. Sc. Lisb.* iv. 1873, p. 205.
Platydactylus (Ceratolophus) auriculatus, *Sauvage, Bull. Soc. Philom.* (7) iii. 1878, p. 67.
Ceratolophus auriculatus, *Bocage, eod. loc.* xxx. 1881, p. 130.
Rhacodactylus auriculatus, *Bouleng. Proc. Zool. Soc.* 1883, p. 127.

Head subpyramidal; snout longer than the distance between the orbit and the ear-opening, once and two thirds the diameter of the orbit, slightly swollen at the end; interorbital space and forehead deeply concave; hinder part of head with knob-like prominences, viz. one above the ear-opening, formed by the free end of the quadrate, and five others formed by the extremities of the parietal bones; the borders and sutures of the latter form prominent ridges; the ends of the postfrontal bones and mandible also prominent; ear-opening large, oval, oblique. Body and limbs moderately elongate, rather depressed; digits not very broadly dilated, not very unequal, with a very slight rudiment of web, which is altogether absent between the two outer toes. Throat and sides of neck with a few irregular folds; a slight fold bordering the hind limb posteriorly. Head and body covered with subequal small granular scales, flattened on the belly, smallest on the throat. Rostral quadrangular, not quite twice as broad as high, with a small notch in the middle of its upper border; nostril pierced between the rostral, the first upper labial, and seven small nasals, the anterior being much the largest; sixteen to eighteen upper and fourteen or fifteen lower labials; mental small, subtriangular, separating the inner labials; no regular chin-shields, but larger scales passing gradually into the granules of the throat. Tail cylindrical, covered with uniform small flat scales arranged in verticils. Male with an irregular patch of very numerous præanal pores, and a sort of pouch in the integument of the thigh posteriorly, near the tibia. Yellowish grey, with darker and lighter longitudinal bands on the back; flanks and limbs marbled with darker; lower surfaces dirty white, with scattered blackish dots.

Total length	205	millim.
Head	34	,,
Width of head	27	,,
Body	91	,,
Fore limb	49	,,
Hind limb	60	,,
Tail	80	,,

New Caledonia.

6. Rhacodactylus ciliatus.

Correlophus ciliatus, *Guichen.* ˙*Mém. Soc. Sc. Nat. Cherb.* xii. 1866, p. 249, pl. viii.; *Bavay, Cat. Rept. N. Caléd.* p. 12.
Rhacodactylus ciliatus, *Bouleng. Proc. Zool. Soc.* 1883, p. 128.

Head very large, oviform, very distinct from neck; forehead deeply concave; snout much longer than the distance between the orbit and the ear-opening, once and two thirds the diameter of the orbit; ear-opening moderately large, suboval, horizontal, slightly oblique. Body and limbs moderately elongate, depressed; digits moderate, not very unequal, strongly denticulated on the sides, half-webbed; the web, however, almost entirely absent between the two outer toes. A strong fold of the skin bordering the hind limb posteriorly; a fold of the skin extends from the eye to the anterior part of the back, where it is separated from its fellow by a rather narrow space; a slight fold from axilla to groin. Head covered with granular scales, largest on the snout; rostral quadrangular, twice as broad as high, with median cleft above; nostril pierced between the rostral, the first upper labial, and six small nasals; upper labials thirteen, lower twelve or thirteen; mental small, wedge-shaped; no chin-shields. Dorsal scales granular, intermixed with larger conical ones on the sides of the vertebral line; the upper eyelid and the fold on the side of the head and anterior part of the back with long ciliated scales. Throat covered with small granular scales; belly with small, slightly imbricated smooth roundish scales.

Tail long, slender, cylindrical, covered with small juxtaposed scales, its distal extremity with a rounded flap of skin, placed horizontally. Yellowish or reddish-brown, lighter beneath.

Total length (tail reproduced, rudimentary)	109	millim.
Head	33	,,
Width of head	24	,,
Body	68	,,
Fore limb	36	,,
Hind limb	46	,,

New Caledonia.

36. LUPEROSAURUS.

Luperosaurus, *Gray, Cat.* p. 163.

Digits strongly dilated, half-webbed, with undivided, angularly curved lamellæ below; all but the thumb and inner toe with a very short compressed distal phalanx with retractile claw. Limbs bordered with cutaneous lobes. Upper and lower surfaces covered with juxtaposed granular scales. Pupil vertical. Males with præ-anal pores.

Philippine Islands.

1. Luperosaurus cumingii. (PLATE XV. fig. 2.)

Luperosaurus cumingii, *Gray, l. c.*

Head regularly oviform; snout longer than the distance from the eye to the ear-opening, about once and a half the diameter of the orbit; forehead concave; ear-opening very small, subcircular. Body and limbs moderate; digits short, not very unequal, half-webbed; the limbs bordered on each side by a cutaneous fringe, that extending from the vent to the tarsus being much developed; an angular lobe of skin closes the vent in the female. Tail shorter than the body, convex superiorly, flat inferiorly, probably prehensile. Head, back, and limbs covered with extremely minute granules; rostral large, square; nostril pierced between the latter, the first labial, and three or four small nasals; sixteen upper and fourteen or fifteen lower labials; rostral trapezoid, not larger than the nearest labials; no chin-shields; throat minutely granulate; belly covered with small flat juxtaposed granules. Tail indistinctly annulate, covered with small flat granules, rather larger inferiorly, the lateral keel with a slight fringe formed by projecting elongate scales. Purplish-brown above, marbled with darker; lower surfaces whitish, belly with narrow transverse brown lines.

Total length	129	millim.
Head	20	,,
Width of head	17	,,

Body	63	millim.
Fore limb	23	„
Hind limb	30	„
Tail (reproduced)	46	„

Philippine Islands.

a–b. ♀ & hgr.   Philippines.   Mr. Cuming [C.]. (Types.)

37. GECKO.

Gekko, part., *Laur. Syn. Rept.* p. 44.
Platydactylus, part., *Cuv. R. A.* ii. p. 45.
Gecko, *Gray, Ann. Phil.* (2) x. 1825, p. 199, *and Cat. Liz.* p. 160; *Günth. Rept. Brit. Ind.* p. 101.
Platydactylus, part., *Fitzing. N. Classif. Rept.* p. 13; *Wagl. Syst. Amph.* p. 142; *Wiegm. Herp. Mex.* p. 19; *Dum. & Bibr. Erp. Gén.* iii. p. 290.
Scelotretus, *Fitzing. Syst. Rept.* p. 101.
Platydactylus, *Fitz. l. c.*

Digits strongly dilated, free or webbed at the base, with undivided lamellæ below, all but the thumb and inner toe with a very short compressed distal phalanx with retractile claw. Head and back covered with juxtaposed granular scales or tubercles, belly with small flat imbricated scales. Pupil vertical. Males with præanal or femoral pores.

Japan, China, East Indies, New Guinea and neighbouring islands.

Synopsis of the Species.

I. Granules of the back intermixed with larger tubercles.

A. Rostral not entering the nostril.

The width of the head equals twice the distance from the end of the snout to the orbit; throat covered with small flat granules .... 1. *verticillatus*, p. 183.
The width of the head equals less than twice the distance from the end of the snout to the orbit; throat minutely granulate.... 2. *stentor*, p. 184.

B. Rostral entering the nostril.

Granules on the throat intermixed with larger ones; greatest diameter of ear-opening half that of orbit; toes with a rudiment of web .... 3. *vittatus*, p. 185.
Granules on the throat equal; greatest diameter of ear-opening one third that of orbit; toes with a rudiment of web .... 4. *monarchus*, p. 187.
Granules on the throat equal; greatest diameter of ear-opening not quite one third that of orbit; toes with a rudiment of web 5. *japonicus*, p. 188.

Granules on the throat equal; greatest diameter of ear-opening not quite one third that of orbit; toes without a rudiment of web; dorsal tubercles few .... 6. *swinhonis*, p. 189.

II. Back uniformly granulate.

A rudiment of web between the toes .... 7. *subpalmatus*, p. 189.

1. Gecko verticillatus.

Gecko verus, *Gray, Cat.* p. 160.
Gecko reevesii, *Gray, l. c.* p. 161.
Lacerta gecko, *Linn. S. N.* i. p. 365.
Gekko verticillatus, *Laur. Syst. Rept.* p. 44.
—— teres, *Laur. l. c.*
Stellio gecko, *Schneid. Hist. Amph.* ii. p. 12.
Gecko guttatus, *Daud.* iv. p. 122, pl. xlix.; *Günth. Rept. Brit. Ind.* p. 102.
—— verus, *Merrem, Tent.* p. 42.
—— annulatus, *Kuhl, Beitr. Zool. Vergl. Anat.* p. 132.
—— reevesii, *Gray, Griff. A. K.* ix. *Syn.* p. 48; *Richards. Zool. Herald*, p. 151, pl. xxvii.
Platydactylus guttatus, *Guérin, Icon. R. A., Rept.* pl. xiii.; *Dum. & Bibr.* iii. p. 321, pl. xxviii. fig. 4.
—— reevesii, *Fitzing. Syst. Rept.* p. 101.
Gekko indicus, *Girard, U.S. Exped., Herp.* p. 290, pl. xvi. figs. 9–16.

Head large; its width equals twice the distance from the end of the snout to the orbit, or from the orbit to the ear-opening; snout subtriangular, obtuse, once and three fifths or once and two thirds the diameter of the orbit; forehead concave; ear-opening narrow, oblique, its vertical diameter at least half the diameter of the orbit. Body and limbs moderate; digits free. Head covered with small convex polygonal scales; upper labials twelve to fourteen, first entering the nostril; lower labials ten or eleven, gradually decreasing in size; mental very variable in shape; chin-shields four or five on each side, smaller than the labials, the inner pair generally not longer than the mental. Back covered with small juxtaposed flat granules and about twelve longitudinal series of mamilliform tubercles; throat with flat granules; abdominal scales moderately large. Pores in a short angular series on the præanal region, thirteen to twenty-four altogether. Tail (when intact) slightly depressed, tapering, distinctly annulate, covered with subquadrangular smooth scales, much larger beneath, arranged in transverse series; each annulus is composed of five or six transverse rows of scales above, three beneath: also, on the upper surface, large conical tubercles, wide apart and symmetrically arranged. Slaty grey above, with red spots or vermiculations; tail annulate with darker and lighter; lower surfaces whitish, frequently indistinctly variegated with grey.

Total length	304	millim.
Head	43	,,
Width of head	36	,,

```
Body ................  117 millim.
Fore limb ...........   50    ,,
Hind limb ...........   67    ,,
Tail ................  144    ,,
```

Southern China; North-eastern India; Birma; Anam; Siam; Malay peninsula; Indian archipelago.

a. ♂.	China.	J. Reeves, Esq. [P.]. (Type of *Gecko reevesii.*)
b, c. ♂ ♀.	China.	J. Reeves, Esq. [P.].
d–i, k–l. ♂ ♀.	Canton.	Haslar Collection.
m. ♂.	India.	Dr. Packman [C.].
n. ♂.	India.	Capt. Stafford [P.].
o. ♀.	Patna.	W. Masters, Esq. [P.].
p. ♂.	Calcutta.	Messrs. v. Schlagintweit [C.].
q. ♂.	Calcutta.	
r. ♂.	Birma.	F. Day, Esq. [P.].
s. ♂.	Pegu.	E. Hamilton, Esq. [P.].
t. ♀.	Tenasserim.	Haslar Collection.
u. Yg.	Tenasserim.	W. T. Blanford, Esq. [P.].
v–x. ♂ ♀.	Siam.	C. Bowring, Esq. [P.].
y. ♂.	Siam.	M. Mouhot [C.].
z–β. ♂.	Malay peninsula.	Dr. Cantor.
γ. Yg.	Java.	
δ, ε–η. ♂ & yg.	Java.	G. Lyon, Esq. [P.].
θ. Yg.	Philippines.	Mr. Cuming [C.].
ι–κ. ♂.	Gumiara, Ilo ilo, Philippines.	H.M.S. 'Challenger.'
λ–μ. ♀.	Sooloo Island.	Capt. Belcher [P.].
ν. ♂.	Negros.	Dr. A. B. Meyer [C.].
ξ, ο. ♂.	S. Negros.	Dr. A. B. Meyer [C.].
π. ♂.	Manado.	Dr. A. B. Meyer [C.].
ρ. ♂.	Timor Laut.	H. O. Forbes, Esq. [C.].
σ. ♂.	Port Essington (imported).	
τ. Skeleton.	—— ?	East-India Company.
υ. Skeleton.	—— ?	

2. Gecko stentor.

Gecko smithii, *Gray, Cat.* p. 162.
Gecko smithii, *Gray, Zool. Misc.* p. 54 (no proper description); *Stoliczka, Journ. As. Soc. Beng.* xxxix. 1870, p. 161; *Anders. Proc. Zool. Soc.* 1871, p. 159; *Stoliczka, eod. loc.* xli. 1872, p. 92.
Platydactylus stentor, *Cantor, Cat. Mal. Rept.* p. 18.
Gecko stentor, *Günth. Rept. Brit. Ind.* p. 102, pl. xi. fig. A; *Stoliczka Journ. As. Soc. Beng.* xxxix. 1870, p. 160.
—— verreauxi, *Tytler, Journ. As. Soc. Beng.* xxxiii. 1864, p. 546.
—— albofasciolatus, *Günth. Ann. & Mag. N. H.* (3) xx. 1867, p. 50.

Head large, its width less than twice the distance from the end of the snout to the orbit; snout longer than in *G. verticillatus*, measuring once and two thirds or once and three fourths the diameter of the orbit and considerably more than the distance from the orbit to the ear-opening; forehead concave; ear-opening oval, oblique, its vertical diameter generally less than half the diameter of the orbit. Body and limbs rather more elongate than in *G. verticillatus*; digits free. The scales on the head and back similar to those of *G. verticillatus*, but smaller; upper labials twelve to fourteen, first entering the nostril; lower labials ten to twelve, gradually decreasing in size; chin-shields three to five on each side, smaller than the labials, the inner pair generally longer than the mental. Back with ten or twelve longitudinal series of mamilliform tubercles. Throat covered with exceedingly small granules; ventral scales larger than in *G. verticillatus*. Præanal pores eleven to sixteen altogether. Tail like that of *G. verticillatus*, but the scales on the upper surface much smaller, the annuli being composed of ten or eleven transverse rows of scales; the large scales on the lower surface of the tail are much dilated transversely, forming generally two symmetrical series. Brown or brownish grey above, marbled with darker, with or without transverse rows of whitish spots,—a character of the young in this and the preceding species, but which may here persist throughout life; sometimes a linear, λ-shaped dark brown mark on the head; tail annulate with darker and lighter; lower surfaces whitish, generally dotted or variegated with greyish.

Total length	367	millim.
Head	48	,,
Width of head	36	,,
Body	141	,,
Fore limb	57	,,
Hind limb	76	,,
Tail	178	,,

Birma; Malay peninsula; Andaman Islands; Sumatra; Java; Borneo.

a. ♂.	Pinang.	Dr. Cantor. (Type.)
b-c. ♀.	Andamans.	W. Theobald, Esq. [C.].
d. Hgr.	Padang.	Dr. Bleeker.
e. Hgr.	Banjermassing.	Dr. Bleeker.
f. Yg.	Labuan.	Dr. Collingwood [P.].
g. ♂.	——?	Dr. Bleeker.
h. ♂.	——?	(Type of *Gecko albofasciolatus*.)

3. Gecko vittatus.

Gecko vittatus, *Gray, Cat.* p. 162.
Gekko vittatus, *Houttuyn, Verh. Genotsch. Vlissing.* ix. 1782, p. 325, pl. —. fig. 2; *Brongn. Bull. Soc. Philom.* ii. 1800, pl. vi. fig. 3; *Daud. Rept.* iv. p. 136, pl. 1.; *Peters & Doria, Ann. Mus. Genov.* xiii. 1878, p. 368.
Lacerta unistriata, *Shaw, Natur. Miscell.* iii. pl. lxxxix.
Platydactylus vittatus, *Dum. & Bibr.* iii. p. 331; *Duvernoy, R. A. Rept.* pl. xx. fig. 1.
—— (Scelotretus) vittatus, *Fitzing. Syst. Rept.* p. 101.

Head regularly oviform; snout longer than the distance between the eye and the ear-opening, about once and two thirds the diameter of the orbit; forehead concave; ear-opening oval, vertical or slightly oblique, its greatest diameter half that of the orbit. Body and limbs elongate; digits with a rudiment of web. Head covered with granular scales, larger than the granules of the back; nostril pierced between the rostral and the first labial; twelve to fourteen upper labials; ten to twelve lower labials; mental small, subtriangular; no chin-shields, the chin covered with small irregular polygonal scales. Back covered with very small juxtaposed flat granules intermixed with larger lentiform smooth tubercles; these are very numerous and irregularly arranged. Throat covered with very small granules intermixed with larger ones; abdominal scales moderately large. Femoral pores in a long series, angular on the præanal region; the number varies from twenty-five to twenty-nine on each side. Tail slender, subcylindrical, very distinctly annulate, covered above with very small flat granular scales intermixed with round flat tubercles, inferiorly with tessellated scales; each annulus is composed of twelve to fourteen transverse rows of scales above, four or five beneath. Upper surfaces fulvous; a whitish, dark-edged vertebral line, forked on the neck, extending to each eye; tail with distant whitish annuli; lower surfaces whitish.

Total length	261	millim.
Head	32	,,
Width of head	24	,,
Body	96	,,
Fore limb	41	,,
Hind limb	57	,,
Tail	133	,,

Moluccas; New Guinea; Solomon Islands.

a. ♂.	Amboyna.	
b, c. ♂ ♀.	N. Ceram.	
d. ♀.	Mysol.	
e–f. ♂.	Faro Island, Solomon Group.	H. B. Guppy, Esq. [P.].
g. Yg.	—— ?	College of Surgeons.

Var. bivittatus.

Gecko bivittatus, *Gray, Cat.* p. 162.
Platydactylus bivittatus, *Dum. & Bibr.* iii. p. 334.
—— (Scelotretus) bivittatus, *Fitzing. Syst. Rept.* p. 101.
Gecko trachylæmus, *Peters, Mon. Berl. Ac.* 1872, p. 774.

Messrs. Peters and Doria, who have examined a very great number of this and the preceding form, have come to the conclusion that the differences between them are not sufficiently important nor sufficiently constant to warrant specific distinction. These differences are:—Tubercles larger, less numerous, mamilliform, or feebly keeled; femoral pores generally fewer (ten to twenty-one on each side); each caudal annulus composed of ten to twelve transverse rows of scales above, three or four inferiorly. Greyish or reddish brown above, variegated with darker; a more or less distinct, lighter vertebral band, frequently quite indistinct on the front of the back, sometimes bordered by a blackish band.

New Guinea; Pelew and Solomon Islands.

a–b, c. ♂	Pelew Islands.	Godeffroy Museum.
d–f. ♂ ♀.	Duke of York Island.	Rev. G. Brown [C.].
g–i. ♂ ♀.	San Christoval.	Voyage of the 'Herald.'
k. ♂.	San Christoval.	Museum of Economic Geology.
l. ♀.	Santa Anna Island.	H. B. Guppy, Esq. [P.].

4. Gecko monarchus.

Gecko monarchus, *Gray, Cat.* p. 161.
Platydactylus monarchus (*Schleg.*), *Dum. & Bibr.* iii. p. 335; *Cantor, Cat. Mal. Rept.* p. 19.
—— (Scelotretus) monarchus, *Fitzing. Syst. Rept.* p. 101.
Gecko monarchus, *Girard, U.S. Explor. Exped., Herp.* p. 292; *Günth. Rept. Brit. Ind.* p. 103.

Head moderately large, oviform; snout longer than the distance between the eye and the ear-opening, about once and two thirds the diameter of the orbit; forehead concave; ear-opening oval, oblique, its greatest diameter one third that of the orbit. Body and limbs moderately elongate; digits feebly dilated for the genus, with a rudiment of web. Snout covered with rather large granular scales, the rest of the head with very small granules intermixed with scattered larger ones; nostril pierced between the rostral and first labial; upper labials ten or eleven, lower labials nine or ten; mental subtriangular; a large median pair of chin-shields, about three times as long as broad, bordered by a few irregular smaller shields. Back and limbs covered with very small granules intermixed with numerous, irregularly arranged, rounded conical tubercles. Throat with very small granules; abdominal scales moderate. Femoral pores in a long series, forming a very open angle in the middle, sixteen to twenty on each side. Tail slightly depressed, distinctly annulate, covered above with very small flat granular scales and transverse series of posteriorly directed conical tubercles, inferiorly with large, rather irregular dilated scales: each annulus is composed of twelve to fourteen transverse rows of scales superiorly, three or four inferiorly. Brown or grey, spotted with blackish, a median double series of spots along the middle of the back being constant; tail with more or less marked darker and lighter rings; lower surfaces whitish.

Total length	190	millim.
Head	23	,,
Width of head	18	,,
Body	60	,,
Fore limb	29	,,
Hind limb	37	,,
Tail	107	,,

East Indies, from Ceylon and the Malay peninsula to Amboyna.

a. ♂.	Ceylon.	Dr. Kelaart.
b–c. Dried spec.	Pinang.	Dr. Cantor.

d. ♂.	Singapore.	Voy. of H.M.S 'Alert.'
e. ♂.	Nias.	Hr. H. Sandemann [C.].
f. ♂.	Padang, Sumatra.	
g–l, m. ♂, ♀, & hgr.	Borneo.	Sir E. Belcher [P.].
n. ♂.	Borneo.	L. L. Dillwyn, Esq. [P.].
o. ♂.	Matang.	
p. ♂.	Philippines.	
q–r. ♂.	Placer, N.E. Mindanao.	Dr. A. B. Meyer [C.].
s–u. ♂♀.	Dinagat Island.	
v–z. ♂♀.	Manado, Celebes.	Dr. A. B. Meyer [C.].
α–β. ♂♀.	Mysol.	
γ. ♀.	Amboyna.	Leyden Museum.
δ. ♂.	Amboyna.	

5. Gecko japonicus.

Gecko chinensis, *Gray, Cat.* p. 161.
Platydactylus japonicus, *Dum. & Bibr.* iii. p. 337.
—— jamori, *Schleg. Faun. Japon., Rept.* p. 103, pl. ii. figs. 1–4.
Gecko chinensis, *Gray, Zool. Miscell.* 1842, p. 57.
Hemidactylus nanus, *Cantor, Ann. & Mag. N. H.* ix. 1842, p. 482.
Gecko japonicus, *Günth. Rept. Brit. Ind.* p. 103.

Head moderately large, oviform; snout longer than the distance between the eye and the ear-opening, about once and two thirds the diameter of the orbit; forehead concave; ear-opening suboval, oblique, its vertical diameter not quite one third that of the orbit. Body and limbs moderately elongate; digits rather feebly dilated for the genus; a slight but distinct rudiment of web between the three median toes. Snout covered with rather large granular scales, the rest of the head with very small granules; nostril pierced between the rostral and first labial; ten to twelve upper and nine to eleven lower labials; mental pentagonal or subtriangular; generally a median pair of chin-shields, longer than broad, bordered by other, smaller and less regular shields. Back and limbs covered with small granular scales intermixed with numerous small, subconical, irregularly arranged tubercles. Throat minutely granulate; abdominal scales moderate. A short, slightly angular præanal series of pores, six to sixteen on each side. Tail slender, slightly depressed, not distinctly annulate, covered above with small subequal scales, inferiorly with a central series of transversely dilated scales. Greyish brown, variegated with darker; a series of dark spots along the middle of the back; lower surfaces dirty white.

Total length	109	millim.
Head	14	,,
Width of head	10	,,
Body	39	,,
Fore limb	18	,,
Hind limb	23	,,
Tail	56	,,

Southern Japan; China.

a. ♀.	Japan.	Leyden Museum.
b. ♂.	China.	J. Reeves, Esq. [P.]. (Type of *Gecko chinensis*.)
c–d. ♂ ♀.	China.	Dr. Cantor. (Types of *Hemidactylus nanus*.)
e. ♂.	China.	East-India Company.
f. ♂.	China.	C. Bowring, Esq. [P.].
g. ♀.	Kiukiang Mountains.	Mr. C. Maries [C.].
h. Hgr.	Chefoo.	R. Swinhoe, Esq. [C.].
i. ♂.	Shanghai.	R. Swinhoe, Esq. [C.].
k. ♂.	Sze Chuen.	R. Swinhoe, Esq. [C.].
l. ♀.	Ningpo.	A. Michie, Esq. [P.].
m–n, o. ♂ ♀.	Formosa.	R. Swinhoe, Esq. [C.].
p. Hgr.	Hongkong.	C. Bowring, Esq. [P.].

6. Gecko swinhonis.

Gecko swinhonis, *Günth. Rept. Brit. Ind.* p. 104, pl. xii. fig. A.

Very closely allied to the preceding. Distinguished by the following characters:—The dorsal tubercles are very few and the inner pair of chin-shields is constantly small, widely separated, and there is no trace of web between the toes.

Northern China.

a. ♂.	Pekin.	R. Swinhoe [C.]. (Type.)
b–c. ♂ ♀.	Pekin.	J. Brenchley, Esq. [P.].
d–e. ♂.	Pekin.	S. W. Bushell, Esq. [P.].
f–g. ♀.	Western hills of Pekin.	S. W. Bushell, Esq. [P.].

7. Gecko subpalmatus.

Gecko subpalmatus, *Günth. Rept. Brit. Ind.* p. 104, pl. xii. fig. B.

Very closely allied to the two preceding. Distinguished in having the back uniformly granulate, without any tubercles. Chin-shields small. A distinct rudiment of web between the three inner toes.
China.

a. ♀.	Chikiang.	Mr. Fortune [C.]. (Type.)

38. PTYCHOZOON.

Ptychozoon. *Kuhl, Isis*, 1822, p. 475; *Wagler, Syst. Amph.* p. 141; *Wiegm. Herp. Mex.* p. 20; *Fitzing. Syst. Rept.* p. 100; *Gray, Cat.* p. 164; *Günth. Rept. Brit. Ind.* p. 105.
Pteropleura, *Gray, Phil. Mag.* (2) iii. 1827, p. 54; *Fitzing. l. c.* p. 101.
Platydactylus, part., *Dum. & Bibr. Erp. Gen.* iii. p. 290.

Digits strongly dilated, entirely webbed, with undivided lamellæ below; all but the thumb and inner toe with a compressed curved distal phalanx with retractile claw, originating a little before the extremity of the digital expansion. Limbs and sides of head, body and tail with much developed membranous expansions acting as parachutes. Upper surfaces covered with juxtaposed granular scales and tubercles : lower surfaces with small, slightly imbricated scales ; the parachute-membrane covered above with imbricated square scales arranged like the bricks of a wall, scaleless inferiorly. Pupil vertical. Males with præanal pores.

East Indies.

1. Ptychozoon homalocephalum.

Ptychozoon homalocephala, *Gray, Cat.* p. 164.
Lacerta homalocephala, *Creveldt, Mag. Naturf. Fr. Berl.* iii. 1809, p. 266, pl. viii.
Gecko homalocephalus, *Tilesius, Mém. Acad. St. Pétersb.* vii. 1820, pl. x.
Ptychozoon homalocephalum, *Kuhl, Isis,* 1822, p. 475 ; *Cantor, Cat. Mal. Rept.* p. 20 ; *Günth. Rept. Brit. Ind.* p. 105 ; *Stoliczka, Journ. As. Soc. Beng.* xxxix. 1870, p. 159.
Pteropleura horsfieldii, *Gray, Phil. Mag.* (2) iii. p. 54.
Platydactylus homalocephalus, *Dum. & Bibr.* iii. p. 339, pl. xxviii. fig. 6, & pl. xxix. figs. 1, 2.

Head oviform ; snout longer than the distance from the eye to the ear-opening, once and a half the diameter of the orbit ; forehead concave ; ear-opening rather large, subcircular. Body and limbs moderate, depressed ; digits short, not very unequal, webbed to the tips ; the fore limbs bordered on each side by a broad dermal expansion ; a similar expansion bordering the tibia on each side, the femur posteriorly only. The parachutes on the sides of the body nearly as broad as the latter. Another dermal lobe on the side of the head, below the ear-opening, from the angle of the mouth to the neck. Tail elongate, depressed, the sides fringed with a series of rounded lobes, confluent into a broad rounded flap at the extremity. Head and back covered with small granular scales. Rostral large, quadrangular ; nostral pierced between the rostral, the first labial, and three small nasals ; ten to twelve upper and as many lower labials ; rostral small, subtriangular ; three or four small chin-shields on each side, inner elongate. Back with scattered convex tubercles (absent in specimen *k*). Throat covered with small granules, belly with moderate scales. An angular series of about twenty-five præanal pores. Tail distinctly annulate, covered above with small flat granular scales and transverse series of conical tubercles, inferiorly with imbricated scales. Greyish or reddish brown above, with transverse undulated dark brown bands ; a dark brown streak from the eye to the first dorsal band.

Total length	188	millim.
Head	27	,,
Width of head	21	,,
Body	69	,,
Fore limb	34	,,
Hind limb	43	,,
Tail	92	,,

Java ; Borneo ; Sumatra ; Malay peninsula.

a-b. ♂.	Java.	Leyden Museum.
c, d. ♀.	Java.	
e-f. Dried specimens.	Java.	Dr. Horsfield [P.].
g. ♀.	Borneo.	L. L. Dillwyn, Esq. [P.].
h-i. Dried specimens.	Singapore.	Dr. Cantor.
k. ♀.	Singapore.	Gen. Hardwicke [P.]. (Scales uniformly granular, without tubercles.)
l. ♂.	——?	

39. HOMOPHOLIS.

Digits strongly dilated, with slight rudiment of web, with transverse undivided lamellæ below, all but the thumb and inner toe armed with a retractile claw. Body covered with small flat imbricated scales. Pupil vertical. Males without præanal nor femoral pores.

South Africa.

1. Homopholis wahlbergii.

Gecko wahlbergii, *Smith, Ill. S. Afr., Rept.* pl. lxxv. fig. 1.

Head thick, oviform; snout short, very convex, measuring the distance between the eye and the ear-opening, once and one fourth the diameter of the orbit; ear-opening very small, round. Body and limbs moderate; digits not very unequal, with a slight rudiment of web, absent between the two outer toes. Head above and beneath minutely granulate. Rostral small, subquadrangular, twice as broad as high; nostril pierced between the first upper labial and five small nasals; eleven upper and eleven or twelve lower labials; mental small, trapezoid; latter and front labials bordered posteriorly by a row of very small chin-shields. Body and limbs covered above and below with equal, small, subhexagonal imbricated scales, similar to those covering the abdominal region of most Geckos. Tail cylindrical, tapering, covered with small smooth irregular scales. Grey above, with dark brown variegations; a dark brown band on each side, from behind the eye to a little beyond the scapular region; lower surfaces whitish, with a few scattered brown dots.

Total length	177	millim.
Head	25	,,
Width of head	22	,,
Body	77	,,
Fore limb	33	,,
Hind limb	40	,,
Tail	75	,,

Kaffraria.

a. ♂.	Natal.	Sir A. Smith [P.]. (Type.)

40. GECKOLEPIS.

Geckolepis, *Grandid. Rev. et Mag. Zool.* (2) xix. 1867, p. 233.

Digits strongly dilated, free, with transverse undivided lamellæ below, all furnished with a retractile claw. Body covered with large flat imbricated cycloid scales. Pupil vertical.

Madagascar.

1. Geckolepis maculata.

Geckolepis maculata, *Peters, Mon. Berl. Ac.* 1880, p. 509, pl. —. fig. 3;
Boettger, Abh. Senck. Ges. xii. 1881, p. 457.

Habit clumsy. Head short. Limbs short, thick; digits short, subequal. Seven or eight upper and as many lower labials; four chin-shields, inner largest, pentagonal, posteriorly truncate. Scales in twenty-five longitudinal series; thirty-six transverse series from chin-shields to vent. Tail rounded, conical, covered with scales similar to, but rather smaller than, those of the body. Greyish-brown above, with irregular black, and fewer white spots; lips spotted; lower surfaces dirty white. Total length about 130 millim.

N.W. Madagascar; Nossi Bé.

2. Geckolepis typica.

Geckolepis typicus, *Grandid. l. c.*

"Fire-red above, greyish beneath. Tail depressed, not denticulated on the sides. Body with rounded, tail with oval scales, all with distinct minute black specks. Mental scute pointed, with two unequal scutes on each side; the other scales similar to the dorsals. From tip of snout to base of tail 40 millim.; length of tail 40 millim."

Sancta Maria.

41. EURYDACTYLUS.

Eurydactylus, *Sauvage, Bull. Soc. Philom.* (7) iii. 1878, p. 70.

Digits strongly dilated, free, with undivided lamellæ below, without compressed distal phalanx, all furnished with a retractile claw. Upper surfaces covered with large flat scales, largest and subsymmetrical on the head; lower surfaces with juxtaposed flat granules. Tail prehensile. Pupil vertical. Males with præanal pores.

New Caledonia.

1. Eurydactylus vieillardi.

Platydactylus vieillardi, *Bavay, Cat. Rept. N. Caléd.* p. 10.
Eurydactylus vieillardi, *Sauvage, l. c.; Bouleng. Proc. Zool. Soc.* 1883, p. 129, pl. xxii. fig. 7.

Head not depressed, much longer than broad; snout long, obtuse, with slightly distinct canthus rostralis, and scarcely oblique loreal region, measuring nearly twice the distance between the eye and the ear-opening or once and three fourths the diameter of the orbit; ear-opening minute, horizontal. Body slightly compressed, the back keeled. Limbs short. Scales of upper surfaces large, separated by intervals forming reticulated lines, much like crocodile-skin; the shields on the head larger, subsymmetrically arranged. Rostral twice as broad as high, with distinct cleft above; nostril pierced between the rostral, the first labial, and two nasals; ten upper and nine lower labials; mental small, triangular, the adjacent labials meeting behind; a few irregular chin-shields. Throat covered with small granules; two strong transverse and two longitudinal folds. Belly covered with rather large flat granules. [Male with a triangular patch of about fifty præanal pores.] Tail cylindrical, covered with squarish juxtaposed scales arranged in verticils. Upper surfaces reddish yellow; the interspaces between the scales forming blackish reticulated lines; lower surfaces uniform whitish.

Total length	91	millim.
Head	15	,,
Width of head	9	,,
Body	39	,,
Fore limb	14	,,
Hind limb	18	,,
Tail	37	,,

New Caledonia.

42. ÆLURONYX.

Platydactylus, part., *Dum. & Bibr. Erp. Gén.* iii. p. 290.
Ailuronyx, *Fitzing. Syst. Rept.* p. 98.
Theconyx, *Gray, Cat.* p. 159.

Digits strongly dilated, free, with undivided lamellæ below, all clawed; claws retractile, ensheathed, the sheath opening inferiorly and laterally, the three inner claws being turned inwards, the two outer outwards. Upper surfaces covered with granular scales; belly covered with cycloid imbricated scales or juxtaposed granules. Pupil vertical. Males with præanal pores.

Seychelles; Madagascar.

1. Æluronyx seychellensis.

Theconyx seychellensis, *Gray, Cat.* p. 159.
Platydactylus seychellensis, *Dum. & Bibr.* iii. p. 310, pl. xxviii. fig. 1; *A. Dum. Arch. Mus.* viii. pl. xviii. fig. 5.
Thecodactylus (Ailuronyx) seychelliensis, *Fitzing. Syst. Rept.* p. 98.

Head longer than broad, with swollen cheeks, not distinct from neck; snout conical, considerably longer than the distance from the eye to the ear-opening, once and three fourths the diameter of the orbit; ear-opening suboval, oblique, its greatest diameter about half that of the orbit. Body and limbs short, thick; a groove along the vertebral line; digits short, not very unequal. Head covered with small granular scales of irregular size; nostril pierced between eight or nine small scales; rostral broad; thirteen to sixteen upper labials; thirteen to fifteen infralabials; mental small, narrow, not longer than the neighbouring labials; six or seven small chin-shields on each side, inner pair elongate. Back and sides with large, juxtaposed, mamilliform or conical tubercles, generally very irregular in size; the vertebral groove with small granules. Limbs and throat granulate. Abdominal scales small, cycloid, imbricate, smooth. Præanal pores in a ◡◡-shaped line, twenty-nine altogether. Tail round, with rather indistinct annuli, covered with small subquadrangular smooth scales. Yellowish or reddish brown above, uniform or variegated with blackish; lower surfaces yellowish white, immaculate.

Total length	202 millim.
Head	29 ,,
Width of head	23 ,,
Body	76 ,,
Fore limb	32 ,,
Hind limb	41 ,,
Tail	97 ,,

Seychelles.

a–c. ♂ ♀.	Seychelles.	G. Nevill, Esq. [P.].
d, e. ♂ ♀.	Seychelles.	Dr. E. P. Wright [P.].
f. Skeleton.	Seychelles.	Dr. E. P. Wright [P.].

2. Æluronyx trachygaster.

Platydactylus trachygaster, *A. Dum. Cat. Méth. Rept.* p. 35, *and Arch. Mus.* viii. p. 452, pl. xvii. fig. 1.

Head broad; snout very obtuse, rounded. No vertebral groove; a slight fold along the side of the body. Three rows of chin-shields. Dorsal granules more or less prominent, not conical. Belly covered with suboval granules, arranged regularly, each granule being surrounded with smaller ones. Fulvous brown, lighter beneath, darkest on the vertebral line. Length of head and body 160 millim.

Madagascar.

43. TARENTOLA*.

Platydactyles, part., *Cuv. Règne Anim.* ii. p. 45.
Tarentola, *Gray, Ann. Phil.* (2) x. 1825, p. 199, *and Cat. Liz.* p. 164.
Platydactylus, part., *Wagler, Syst. Amph.* p. 142; *Dum. & Bibr. Erp. Gén.* iii. p. 290.
Ascalabotes, *Bonap. Amph. Eur.* p. 13; *Fitzing. Syst. Rept.* p. 102.

Digits strongly dilated, free, with undivided lamellæ below and a flat nail-like scute on their upper surface near the tip; third and fourth clawed, others clawless. Pupil vertical. No femoral or præanal pores.

W. Africa and borders of Mediterranean; one species in the West Indies (and another stated to be from the United States).

All the species agree in the following characters:—The snout is obtuse and convex; the circular lid is well developed, most so superiorly and anteriorly; the ear-opening is narrow and vertical. The limbs are stout, the digits short and subequal. The body is depressed. The tail when not reproduced is depressed and formed of distinct annuli; its length equals about the distance between the eye and the vent. Behind the latter are two small transverse slits, but there are neither præanal nor femoral pores. The head is covered with polygonal convex scales, larger than the granules on the body; the nostril is pierced between the first labial and two small shields; a trace of median cleft is seen in the upper part of the rostral; the mental is large, wedge-shaped, and separates entirely the chin-shields. The upper surfaces of the body, limbs,

---

* The following species are not sufficiently well established to enter the system:—

1. TARENTOLA AMERICANA.

Tarentola americana, *Gray, Cat.* p. 165.
Platydactylus (Tarentola) americanus, *Gray, in Griff. A. K.* ix. *Synops.* p. 48.
—— milbertii, *Dum. & Bibr.* iii. p. 325.

Anterior border of ear denticulated. Back with twelve longitudinal series of large, smooth, convex tubercles close together.
New York (?).

2. TARENTOLA CUBANA.

Platydactylus (Tarentola) americanus, var. cubanus, *Gundl. & Peters, Mon. Berl. Ac.* 1864, p. 384.

Anterior border of ear denticulated. Back with twenty longitudinal series of strongly keeled large tubercles. Back and tail with broad black transverse bands; a black streak from the eye to the shoulder.
Cuba.

3. TARENTOLA CLYPEATA. Gray, Zool. Misc. p. 57.

"Head flattened and circumscribed by a cross rib behind, covered with larger convex scales; back granular with close cross ridges of tubercles; greyish in spirits: limbs and tail slender."
Hab. ——?

and tail are covered with granules of subequal size, intermixed with larger scales arranged symmetrically on the back and tail; the lower surfaces are covered with flat, hexagonal, slightly imbricate scales; these are largest, and transversely dilated on the lower surface of the tail. I am not able to distinguish the sexes externally.

Synopsis of the Species.

I. A supraorbital bone.

Dorsal tubercles strongly keeled; anterior border of ear not denticulated .... 1. *mauritanica*, p. 196.
Dorsal tubercles not keeled; anterior border of ear denticulated; all the chin-shields in contact with the infralabials........ 2. *annularis*, p. 197.
Dorsal tubercles not keeled; anterior border of ear not denticulated; only the innermost chin-shield in contact with the infralabials...................... 3. *ephippiata*, p. 198.

II. No supraorbital bone.

Snout a little longer than the distance between the eye and the ear-opening; mental not thrice as long as it is broad in the middle .... 4. *delalandii*, p. 199.
Snout as long as the distance between the eye and the ear-opening; mental thrice as long as it is broad in the middle .... 5. *gigas*, p. 200.

1. Tarentola mauritanica.

Tarentola mauritanica, *Gray, Cat.* p. 164.
Lacerta mauritanica, *Linn. S. N.* i. p. 202.
Gekko muricatus, *Laur. Syn. Rept.* p. 44.
Stellio mauritanicus, *Meyer, Syn. Rept.* p. 31.
Gecko fascicularis, *Daud. Rept.* iv. p. 144.
Gekko stellio, *Merr. Tent.* p. 43.
Tarentola stellio, *Gray, Ann. Phil.* (2) x. 1825, p. 199.
Gecko mauritanicus, *Risso, Hist. Nat. Eur. Mér.* p. 87.
Platydactylus fascicularis, *Wagl. Syst. Amph.* p. 142.
—— (Tarentola) fascicularis, *Gray, in Griff. A. K.* ix. *Syn.* p. 48.
—— muralis, *Dum. & Bibr.* iii. p. 319.
Ascalabotes mauritanicus, *Bonap. Amph. Eur.* p. 28, *and Faun. Ital.*
Platydactylus facetanus (*Aldrov.*), *Strauch, Erp. Alg.* p. 22; *Schreib. Herp. Eur.* p. 490.
—— mauritanicus, *Boettger, Abh. Senck. Ges.* ix. 1874, p. 16.

A supraorbital bone. Snout longer than the distance between the eye and the ear-opening. Rostral twice as broad as high; ten upper labials, the last minute; mental not twice as long as it is broad in the middle, its posterior border not quite half as broad as its anterior; chin-shields two or three on each side, in contact with the lower labials; latter eight or nine. Anterior border of ear-opening not denticulated. Sides of neck and body and upper surface of limbs with conical tubercles. Back with seven or nine longitudinal series of very prominent, strongly keeled, large tubercles. Anterior part of tail with posteriorly directed spine-like tubercles. Greyish brown above, more or less distinctly marbled with darker and lighter; a more or less distinct dark streak on each side of the head, through the eye.

Total length	155 millim.
Head	23 ,,
Width of head	17 ,,
Body	54 ,,
Fore limb	25 ,,
Hind limb	36 ,,
Tail	78 ,,

Mediterranean District.

a–c. Ad.	Valencia, Spain.	Lord Lilford [P.].
d–h. Ad.	Spain.	Lord Lilford [P.].
i–m. Ad. & hgr.	Near Lisbon.	H. O. Forbes, Esq. [P.].
n–o. Yg.	Sicily.	T. Bell, Esq. [P.].
p. Skeleton.	Sardinia.	
q–r. Ad. & hgr.	Mogador.	J. McAndrew, Esq. [P.].
s. Ad.	Tunis.	P. L. Sclater, Esq. [P.].
t. Ad.	Tripoli.	J. Ritchie, Esq. [P.].
u–w. Ad. & hgr.	Susa, Tripoli.	Mr. Fraser [C.].
x–β. Ad. & hgr.	Egypt.	T. Burton, Esq. [P.].
γ–ε. Ad. & hgr.	N. Africa.	
ζ–η. Ad.	Mediterranean.	J. Miller, Esq. [P.].
θ–ι. Ad.	Mediterranean.	
κ. Ad.	——?	

2. Tarentola annularis.

Tarentola ægyptiaca, *Gray, Cat.* p. 165.
Gecko annularis, *Geoffr. Descr. Egypte, Rept.* p. 130, pl. v. figs. 6 & 7.
—— savignyi, *Aud. eod. loc. Suppl.* p. 164, pl. i. fig. 1.
Platydactylus ægyptiacus, *Cuv. R. A.* 2nd edit. ii. p. 53; *Dum. & Bibr.* iii. p. 322, pl. xxviii. fig. 3.

A supraorbital bone. Snout as long as the distance between the eye and the ear-opening. Rostral twice as broad as high; ten or eleven upper labials, the last minute; mental about twice as long as it is broad in the middle, its posterior border half as broad as its anterior; chin-shields two or three on each side, in contact with the lower labials; latter nine or ten. Anterior border of ear with a denticulation formed by small conical tubercles. Sides of neck and body and upper surface of limbs with convex or conical tubercles. Back with eight or ten longitudinal series of slightly convex large tubercles. Anterior part of tail with posteriorly directed conical tubercles. Light greyish brown above, with five more or less distinct transverse brown bands on the back; the anterior unites with a dark streak on each side of the head, passing through the eye; four equidistant white spots on scapular region, each situated behind and at the end of the two anterior dark dorsal cross bands.

Total length	208	millim.
Head	33	,,
Width of head	28	,,
Body	97	,,
Fore limb	41	,,
Hind limb	54	,,
Tail	78	,,

Egypt; Arabia; Abyssinia.

a–c. Ad.	Egypt.	A. Christy, Esq. [P.].
d–h. Ad. & yg.	Egypt.	Sir J. G. Wilkinson [P.].
i–k. Ad. & hgr.	Medinet Abu, Upper Egypt.	Dr. Anderson [P.].
l–m. Ad. & yg.	Mount Sinai.	

3. Tarentola ephippiata. (PLATE XVI. fig. 1.)

Tarentola ephippiata, *O'Shaughn. Ann. Mag. N. H.* (4) xvi. 1875, p. 264.

A supraorbital bone. Snout as long as the distance between the eye and the ear-opening. Scales on upper surface of snout and supraorbitals larger than in the other species. Rostral not twice as broad as high; nine or ten upper labials, the last minute; mental twice as long as it is broad in the middle, its posterior border not half as broad as its anterior; chin-shields three on each side, the innermost in contact with the labial, the others separated from the labials by a row of small shields; lower labials eight. Anterior border of ear-opening not denticulated. Back with twelve longitudinal series of large oval flat tubercles; similar tubercles on the sides and limbs. Anterior part of tail with rows of large, not very prominent tubercles. Pale reddish brown above; a reddish-brown streak on each side of the head, passing through the eye, gradually broadening and finally joining its fellow on the anterior part of the back; hinder part of back and tail with transverse dark patches.

Total length	122	millim.
Head	19	,,
Width of head	15	,,
Body	44	,,
Fore limb	21	,,
Hind limb	28	,,
Tail	59	,,

West Africa.

a, b. Adult.   W. Africa.   (Types.)

4. Tarentola delalandii.

Tarentola delalandii, *Gray, Cat.* p. 165.
Tarentola borneensis, *Gray, l. c.*
Platydactylus delalandii, *Dum. & Bibr.* iii. p. 324; *Boettger, Abh. Senck. Ges.* ix. 1874, p. 59, pl. —. fig. 2.
Ascalabotes delalandii, *Girard, U. S. Explor. Exped., Herp.* p. 289.

No supraorbital bone. Snout a little longer than the distance between the eye and the ear-opening. Rostral twice as broad as high; ten upper labials, the last minute; mental at least twice as long as it is broad in the middle, its posterior border not quite half as broad as its anterior; chin-shields two or three on each side, the innermost twice as long as it is broad in the middle, in contact with the lower labials; latter eight or nine. Anterior border of ear-opening not denticulated. Sides of neck and body and upper surface of limbs with convex subconical tubercles. Back with twelve longitudinal series of smooth or very feebly keeled large tubercles. Anterior part of tail with large prominent tubercles, obtusely pointed, the point directed posteriorly. Reddish grey above, with more or less distinct transverse dark reddish-brown bands on the back and tail; these dark dorsal bands generally bordered posteriorly by a row of white tubercles; a more or less distinct dark streak on each side of the head, through the eye.

Total length	137	millim.
Head	25	,,
Width of head	19	,,
Body	56	,,
Fore limb	26	,,
Hind limb	34	,,
Tail	56	,,

West Africa, from Madeira to Guinea.

a. Several spec.: ad. & hgr.	Madeira.	P. B. Webb, Esq. [P.]
b. Ad.	Madeira.	Sir A. Smith [P.].
c–d. Ad.	Teneriffe.	'Challenger' Exped.
e–f. Ad. & hgr.	Canaries.	Rev. R. T. Lowe [P.].
g. Hgr.	St. Vincent.	Rev. R. T. Lowe [P.].
h–i. Ad.	St. Vincent.	J. Macgillivray, Esq. [C.].
k–l. Ad. & hgr.	St. Vincent.	Dr. Cunningham [C.].
m. Ad.	Senegambia.	Baron v. Maltzan [C.].
n–o. Ad.	Porto Praya, Cape Verde.	C. Darwin, Esq. [C.].
p. Ad.	Porto Praya, Cape Verde.	'Challenger' Exped.
q. Many spec.: ad., hgr., & yg.	Niger.	
r–s. Ad.	——?	Sir E. Belcher [P.].

(Types of *Tarentola borneensis.*)

5. Tarentola gigas.

Ascalabotes gigas, *Bocage, Jorn. Sc. Lisb.* v. 1875, p. 108.

No supraorbital bone. Snout as long as the distance between the eye and the ear-opening. Rostral twice as broad as high; ten upper labials, the last minute; mental nearly three times as long as it is broad in the middle, its posterior border not half as broad as its anterior; chin-shields three on each side, in contact with the labials, the innermost twice as long as broad; infralabials eight. Anterior border of ear-opening not denticulated. Body with sixteen longitudinal series of flat or very feebly keeled large tubercles; similar tubercles on the limbs and on the anterior part of the tail. Brownish grey above, with more or less defined dark transverse bands on the back and tail; sometimes a lighter vertebral band; a dark streak, bordered above with lighter, on each side of the head, passing through the eye.

Total length	191	millim.
Head	30	,,
Width of head	24	,,
Body	72	,,
Fore limb	36	,,
Hind limb	50	,,
Tail	89	,,

Cape Verde Islands.

a. Adult.   Ilheo Raso.   Prof. Barboza du Bocage [P.].

44. PACHYDACTYLUS*.

Platydactyles, part., *Cuv. Règne Anim.* ii. p. 45.
Platydactylus, *Gray, Ann. Phil.* (2) x. 1825, p. 199.
Anoplopus, part., *Wagl. Syst. Amph.* p. 142.
Pachydactylus, *Wiegm. Herp. Mex.* p. 19; *Fitzing. Syst. Rept.* p. 99; *Gray, Cat. Liz.* p. 167.
Platydactylus, part., *Dum. & Bibr. Erp. Gén.* iii. p. 290.
Homodactylus, *Gray, Proc. Zool. Soc.* 1864, p. 59.

Digits more or less dilated, free, clawless, with undivided angular lamellæ under their terminal part. Pupil vertical. No præanal or femoral pores.

Africa.

The eye is large and the pupil vertical; the circular lid distinct above, with scales forming a more or less distinct denticulation. The digits are subequal; their borders are more or less distinctly toothed; a flat scale, like a human nail, on the upper part of the extremity; their lower surface covered anteriorly with transverse lamellæ, the remainder with small scales arranged in longitudinal series. Two small transverse slits behind the vent. The nostril is pierced between three scales, and the rostral shows no trace of cleft. There are no chin-shields.

---

* *Pachydactylus tristis*, Hallow. Proc. Ac. Philad. 1854, p. 98.—Liberia.

Synopsis of the Species.

I. Dorsal scales intermixed with larger tubercles.

A. Dorsal tubercles keeled, not conical.

Dorsal tubercles strongly keeled; digits with eight to ten transverse lamellæ inferiorly...... 1. *bibronii*, p. 201.

Dorsal tubercles feebly keeled; digits with five or six transverse lamellæ inferiorly...... 2. *capensis*, p. 202.

Dorsal tubercles strongly keeled; digits with four or five transverse lamellæ inferiorly...... 3. *formosus*, p. 203.

Dorsal tubercles feebly keeled; digits with four transverse lamellæ inferiorly ...... 9. *mentomarginatus*, [p. 207.

B. Dorsal tubercles conical.

Head-scales conical; abdominal scales granular, subconical ...... 4. *rugosus*, p. 204.

Head-scales simply convex; abdominal scales imbricated...... 5. *oshaughnessyi*, p. 204.

C. Dorsal tubercles distinguished from the granules only by their larger size.

Conical tubercles on the tibia...... 8. *maculatus*, p. 206.

II. Dorsal scales equal.

Digits with four or five transverse lamellæ inferiorly; scales small.... 6. *ocellatus*, p. 205.

Scales on the snout three or four times as large as those on the back of the head ...... 7. *punctatus*, p. 206.

Digits with three transverse lamellæ inferiorly; snout very short, hardly as long as the diameter of the orbit. 10. *mariquensis*, p. 207.

1. Pachydactylus bibronii.

Tarentola bibronii, *Smith, Ill. S. Afr., Rept.* pl. 1. fig. 1.
Pachydactylus bibronii, *Smith, l. c. Errata*; *Peters, Reise n. Mossamb.* iii. p. 25.
Homodactylus turneri, *Gray, Proc. Zool. Soc.* 1864, p. 59, pl. ix. fig. 2.
—— bibronii, *Gray, Proc. Zool. Soc.* 1865, p. 612.

In general habit similar to *Tarentola mauritanica*. Snout obtuse, convex, longer than the diameter of the orbit; cheeks swollen. Ear-opening narrow, vertical. Body depressed. Limbs stout, short; digits short, considerably broader at the extremity than at the base, the dilated part with eight to ten lamellæ inferiorly. Tail rather depressed, tapering, distinctly annulate. Head covered with small convex scales, largest on the occiput; naso-rostrals generally in contact; rostral broader than high; upper labials eight or nine, the anterior as large as the mental; latter twice as long as broad, narrowing posteriorly; cheeks with large keeled tubercles. Upper parts covered with small keeled scales intermixed with large strongly keeled ones arranged in longitudinal series; flanks with conical tubercles. Abdominal scales imbricate, largest on the sides of the belly. Upper surface and sides of tail with imbricated keeled scales and transverse rows of large, posteriorly directed, strongly keeled, spinose scales; lower surface of tail with rather large, flat, hexagonal, smooth scales. Brown or brownish grey above; a dark brown line on each side of the head, passing through the eye, and another from the nostril to the upper border of the orbit; back and tail with more or less distinct transverse, narrow, dark brown bands; sometimes a few scattered white spots on the back; lower surfaces whitish or brownish.

Total length	164	millim.
Head	25	,,
Width of head	23	,,
Body	70	,,
Fore limb	27	,,
Hind limb	37	,,
Tail	69	,,

Africa, from Benguela and Mozambique to the Cape of Good Hope.

a. Ad.	Benguela.	Prof. Barboza du Bocage [P.].
b–c. Ad.	Benguela.	J. J. Monteiro, Esq. [P.].
d–g. Ad.	Free State, Orange River.	F. P. M. Weale, Esq. [P.].
h. Several spec.: ad., hgr., & yg.	Cape of Good Hope.	Sir A. Smith [P.]. (Types.)
i. Ad.	S. Africa.	
k–o,p–q. Ad. & hgr.	Tette.	Sir J. Kirk [C.]. (Types of *Homodactylus turneri*.)

2. Pachydactylus capensis.

? Pachydactylus elegans, *Gray, Cat.* p. 168.
Tarentola capensis, *Smith, Ill. S. Afr., Rept.* pl. 1. fig. 2.
Pachydactylus capensis, *Smith, l. c., Errata.*

Head regularly oviform, not much broader than the neck; snout a little longer than the diameter of the orbit. Ear-opening oval, oblique. Body depressed. Limbs moderate; digits short, rather slender, distinctly broader at the end than at the base, the dilated part with five or six lamellæ inferiorly. Tail rather depressed, tapering, distinctly annulate. Snout covered with convex scales which are much larger than the granules of the back; hinder part of head covered with granules intermixed with large round tubercles;

naso-rostrals in contact; rostral broader than high; seven or eight upper labials; six or seven infralabials, the anterior as large as the mental; latter about twice as long as broad, slightly narrowing posteriorly. Upper parts covered with very small granules intermixed with large, feebly keeled, roundish or suboval tubercles, arranged symmetrically on the back; abdominal scales moderate, increasing in size from throat to groin. Upper surface and sides of tail with slightly imbricate smooth scales and transverse rows of large feebly keeled tubercles; lower surface of tail with imbricate smooth scales. Light brown above, variegated with dark brown and whitish; a dark brown streak on each side of the head, passing through the eye; lower surfaces immaculate.

Total length	112	millim.
Head	16	,,
Width of head	12	,,
Body	47	,,
Fore limb	18	,,
Hind limb	24	,,
Tail	49	,,

South Africa.

a. Many spec.: ad., hgr., & yg.	S. Africa.	Sir A. Smith [P.]. (Types.)
b. Ad.	S. Africa.	Prof. Peters [P.].
c. Ad.	Karroo.	The Trustees of the South-African Museum [P.].
d. Ad.	Free State, Orange River.	F. P. M. Weale, Esq. [P.].

The following specimen, type of Gray's *P. elegans*, is in too bad a state to be described; it appears to differ little, if at all, from *P. capensis*; however, the dorsal tubercles seem to be rather larger.

Yg., very bad state.	S. Africa.	Dr. J. Lee [P.].

3. Pachydactylus formosus. (PLATE XVI. fig. 2.)

Pachydactylus formosus, *Smith, Ill. S. Afr., Rept. App.* p. 4.

Agrees in general with *P. capensis*. Differs in having the digits rather shorter and furnished with four or five lamellæ inferiorly, the large dorsal tubercles strongly keeled and the small scales between them more irregular, and the ventral scales smaller. Scales on the upper part of the tail keeled. Yellowish, with brown variegations; a broad, brown, crescent-shaped band bordering the head posteriorly; four broad, brown, transverse bars on the back, broader than the interspaces between them.

Interior of South Africa.

a. Adult.	S. Africa.	Sir A. Smith [P.]. (Type.)
b. Adult.	S. Africa.	Sir A. Smith [P.].

4. Pachydactylus rugosus.

Pachydactylus rugosus, *Smith, Ill. S. Afr., Rept.* pl. lxxv. fig. 2.

Head oviform, very distinct from neck; snout moderate, obtuse, a little longer than the diameter of the orbit. Ear-opening oval, oblique. Body not much depressed. Limbs rather long; digits short, of subequal width throughout their length, with four or five lamellæ inferiorly. Tail (reproduced?) short, thick. Head covered with conical tubercles intermixed with small granules; naso-rostrals separated by a granule; rostral a little broader than high; ten upper labials; nine lower labials, gradually decreasing in size; mental subtriangular. Upper parts with small granules of irregular size, intermixed with large, conical, spinose tubercles arranged irregularly. Scales on lower surfaces granular, subconical. Head brownish white, with a large, crescent-shaped, greyish-brown mark extending from eye to eye; back greyish brown, with four transverse, dentated, whitish bars, narrower than the intervals; a whitish band from the mouth to the anterior cross bars; lower surfaces whitish, speckled with brown.

Total length	84 millim.
Head	17 ,,
Width of head	12 ,,
Body	39 ,,
Fore limb	21 ,,
Hind limb	28 ,,
Tail	28 ,,

Interior of South Africa.

a. Adult.   S. Africa.   Sir A. Smith [P.].   (Type.)

5. Pachydactylus oshaughnessyi. (PLATE XVI. fig. 3.)

Head regularly oviform, not much broader than the neck; snout obtuse, subtriangular, a little longer than the diameter of the orbit. Ear-opening very small, oval, oblique. Limbs moderate; digits short, of subequal width throughout their length, with four or five lamellæ inferiorly. Body not much depressed. Tail rounded, very thick in its anterior half, ending in a thin point. Snout covered with convex scales which are much larger than the granules of the back; the remainder of the head covered with very small granules intermixed with larger ones; naso-rostrals in contact; rostral broader than high; eight upper labials; seven lower labials, first as long as, and broader than, the mental; latter a little narrowed posteriorly, twice as long as broad. Back covered with small granules intermixed with large, keeled, conical tubercles arranged rather irregularly. Outer side of femur and tibia with conical tubercles. Throat granulate, belly covered with moderate-sized imbricated scales. Tail covered with rather large, equal, strongly imbricate, cycloid, smooth scales. Head yellowish, bordered by a blackish streak extending from one nostril to the other, passing through the eye and conturning the occiput; this streak bordered posteriorly by a broader cream-coloured band; back and upper surface of tail light brown, the former with two, the latter with seven, large, transverse, cream-coloured, blackish-edged spots; limbs and lower surfaces uniform cream-coloured.

Total length	79 millim.
Head	13 ,,
Width of head	8·5 ,,
Body	28 ,,
Fore limb	13 ,,
Hind limb	17 ,,
Tail	38 ,,

Lake Nyassa.

a. Adult.	Lake Nyassa.	Mr. B. Thelwall [C.].
b. Young.	Lake Nyassa.	Mr. A. A. Simons [C.].

6. Pachydactylus ocellatus.

Pachydactylus ocellatus, *Gray, Cat.* p. 167.
Gecko ocellatus (*Oppel*), *Cuv. R. A.* ii. p. 46.
? Gecko inunguis, *Cuv. l. c.*
Platydactylus (Phelsuma) ocellatus, *Gray, in Griff. A. K.* ix. *Syn.* p. 47.
Pachydactylus bergii, *Wiegm. Herp. Mex.* p. 19.
Platydactylus ocellatus, *Dum. & Bibr.* iii. p. 298.
Pachydactylus ocellatus, *Steindachn. Novara, Rept.* p. 10.

Snout very short, scarcely longer than the diameter of the orbit, obtuse, convex. Ear-opening subcircular. Body not much depressed. Limbs moderate; digits short, subequal, the end not much dilated, with four or five lamellæ inferiorly. Tail thick, rounded, tapering. Upper surface of head, body, and limbs covered with granules of nearly equal size. Naso-rostrals generally separated; rostral scarcely broader than high; upper labials eight or nine; lower labials seven or eight, the anterior as long as, and broader than, the mental; latter quadrangular, not, or but little, narrowed posteriorly. Abdominal scales imbricate, equal. Caudal scales smooth, slightly imbricate, subequal, about twice as large as the granules of the back. Male with a series of three or four conical scales on each side of the base of the tail. Grey or brown above, with small, white, dark-edged ocelli; upper lip white; a dark streak on the side of the head and neck, passing through the eye; whitish beneath, throat dotted with brown.

Total length	73 millim.
Head	10 ,,
Width of head	8·5 ,,
Body	33 ,,
Fore limb	14 ,,

Hind limb . . . . 16 millim.
Tail . . . . 30 „

South Africa.

a–c. ♂ ♀.	Island of Ascension.	H.M.S. 'Chanticleer.'
d–f. ♂ ♀.	Cape of Good Hope.	Dr. J. Lee [P.].
g. ♂.	Cape of Good Hope.	
h. ♂.	Benguela.	Prof. Barboza du Bocage [P.].
i. ♀.	——?	

7. Pachydactylus punctatus.

Pachydactylus punctatus, *Peters, Mon. Berl. Ac.* 1854, p. 615, *and Reise n. Mossamb.* iii. p. 26, pl. v. fig. 2.

Differs from *P. ocellatus* in having the snout somewhat longer and more pointed, the granules of the back and the scales of the tail larger, and the scales on the snout three or four times as large as those on the back of the head. Seven upper and six lower labials. Brown above, spotted with blackish brown; behind the eye a yellow band, blackish-edged above; lower surfaces pure white.

Mozambique.

8. **Pachydactylus maculatus.** (Plate XVI. fig. 4.)

Pachydactylus maculatus, *Gray, Cat.* p. 167.
Pachydactylus maculatus, *Smith, Ill. S. Afr., Rept. App.* p. 4.

Agrees in size, proportions, and scutellation with *P. ocellatus*, except that the granules of the back are intermixed with scattered larger tubercles, twice or three times the size of the granules, that there are a few conical tubercles on the outer side of the tibia, and that there is one lamella less under the disks of the digits. Greyish or brownish above; a broad dark brown streak on the side of the head, passing through the eye, converging towards its fellow on the occiput; back with four longitudinal series of large dark brown spots, those of the outer series generally, those of the inner series constantly confluent into a band; tail spotted and variegated with dark brown; lower surfaces dotted with brown.

South Africa.

a. ♀.	S. Africa.	Saffron Walden Nat. Hist. Soc. [P.]. (Type.)
b. Many spec.: ♂, ♀, & yg.	S. Africa.	F. P. M. Weale, Esq. [P.].
c, d–e. ♂.	S. Africa.	
f. Young.	Cape Colony.	F. P. M. Weale, Esq. [P.].
g. ♂.	Port Elizabeth.	
h. ♂ (anomalously large specimen, 56 millim. from snout to vent).	Karroo.	The Trustees of the South-African Museum [P.].
i. ♂.	——?	Prof. St. George Mivart [P.].
k. ♂.	——?	

9. Pachydactylus mentomarginatus. (PLATE XVI. fig. 5.)

Pachydactylus mentomarginatus, *Smith, Ill. S. Afr., Rept. App.* p. 5.

Snout short, as long as the diameter of the orbit, convex, rounded. Ear-opening oval. Body not much depressed. Limbs rather long; digits short, subequal, of subequal width, with four lamellæ inferiorly; the outer toe inserted a good deal below the fourth. Tail ——? Head covered with granules intermixed with larger ones posteriorly; naso-rostrals separated; rostral broader than high; eight or nine upper labials; eight or nine infralabials, gradually decreasing in size, the anterior as long as, and broader than, the mental; latter square, not narrowed posteriorly. Back granular, intermixed with large, feebly-keeled, roundish tubercles, arranged irregularly. Outer part of femur and tibia with scattered oval, strongly-keeled, large tubercles. Lower surfaces covered with very small scales. Reddish brown above; head variegated with whitish. Back with five transverse, dentated, whitish, dark-brown-edged bands, much narrower than the interspaces; lower surfaces dirty white (wine-yellow according to Smith), the throat freckled with brownish, and the lower edge of the infralabials margined with brown.

Total length (tail missing)	29 millim.
Head	9 ,,
Width of head	6·5 ,,
Body	20 ,,
Fore limb	11 ,,
Hind limb	13 ,,

Interior of South Africa.

a–b. Hgr.?   S. Africa.   Sir A. Smith [P.].   (Types.)

10. Pachydactylus mariquensis. (PLATE XVI. fig. 6.)

Pachydactylus mariquensis, *Smith, Ill. S. Afr., Rept. App.* p. 3.

Head very convex; snout very short, hardly as long as the diameter of the orbit, round. Ear-opening oval or subcircular. Body not much depressed. Limbs long; digits rather short, the end slightly dilated, with three lamellæ inferiorly; the outer toe inserted a good deal below the fourth. Tail cylindrical, tapering, without annuli. Upper surface of head, body, and limbs covered with granules of subequal size; naso-rostrals generally in contact; rostral broader than high; upper labials seven or eight; lower labials six or seven, gradually decreasing in size, the anterior as long as, and broader than, the mental, which is very slightly narrowed posteriorly. Abdominal scales equal. Caudal scales equal, smooth, slightly imbricated, much larger than the granules of the back. Male with a series of four conical scales on each side of the base of the tail. Grey above, with reddish-brown blackish-margined markings; these markings are a spot on the nose, another on the forehead, a semi-circular broad bar surrounding the back of the head, stretching from eye to eye, and posteriorly directed angular broad bars on the back and tail; lower surfaces whitish, immaculate.

Total length	87 millim.
Head	11 ,,
Width of head	10 ,,
Body	37 ,,
Fore limb	18 ,,
Hind limb	21 ,,
Tail	39 ,,

Interior of South Africa.

a–d. ♂ ♀. S. Africa. Sir A. Smith [P.]. (Types.)

45. COLOPUS.

Colopus, *Peters, Mon. Berl. Ac.* 1869, p. 57.

Tips of fingers slightly dilated, of toes rather narrowed; no claws; two enlarged scales under the extremity of the digits and a nail-like scale above, the rest being covered with small granules. Dorsal scales uniform, granular; abdominal scales imbricate. Pupil vertical. No præanal or femoral pores.

South Africa.

1. Colopus wahlbergii.

Colopus wahlbergii, *Peters, l. c.* pl. —. fig. 1.

Head short, convex; ear-opening small, oblique. Convex scales on the snout hardly larger than those on the back; the latter equal those of the submental region, whilst those on the gular region are the smallest. Rostral broader than mental; eight or nine upper and seven lower labials; nostril between three nasals, the anterior largest. Limbs moderate; digits short. Tail cylindrical, tapering, covered with uniform, smooth, imbricate scales, which are about twice as large as the ventrals. Olive-green above, with large yellow, darker-edged spots, which are confluent into irregular cross bands on the back; a yellow median streak from snout to occiput; lower surfaces light yellow.

Total length	135 millim.
Head to ear-opening	14 ,,
From snout to vent	52 ,,
Fore limb	25 ,,
Hind limb	30 ,,

Damaraland.

46. DACTYCHILIKION.

Dactychilikion, *Thominot, Bull. Soc. Philom.* (7) ii. 1878, p. 254.

Digits dilated at the apex only, spatulate, inferiorly with transverse undivided lamellæ furnished on their hinder edge with fine fringes giving them a felt-like appearance. No claws (?). Back covered with small hexagonal scales. Abdominal scales large, hexagonal.

South Africa.

1. Dactychilikion braconnieri.

Dactychilikion braconnieri, *Thom. l. c.*

Limbs long and slender. Five lamellæ under the digits; seven plates of different shapes on the upper side. Seven upper and eight lower labials; three chin-shields. Tail very slender and flat at its extremity, as long as head and body. Olive-green above, marbled with blackish; three small transverse black spots near the insertion of the thighs; limbs with reddish-brown chevron-shaped bands; tail annulate with black; abdomen yellowish white. Total length 112 millim.

Near Lake N'Gami.

47. PHELSUMA.

Platydactylus, part., *Cuv. Règne Anim.* ii. p. 45.
Phelsuma, *Gray, Ann. Phil.* (2) x. 1825, p. 199, *and Cat. Liz.* p. 166; *Günth. Rept. Brit. Ind.* p. 112.
Anoplopus, part., *Wagl. Syst. Amph.* p. 142.
Platydactylus, part., *Dum. & Bibr.* iii. p. 290.
Anoplopus, *Fitzing. Syst. Rept.* p. 99.

Digits strongly dilated, free, clawless, inner rudimentary, with undivided lamellæ below. Pupil round; eyelid distinct all round the eye. Males with præanal or femoral pores.

Madagascar, Comoro, Seychelles, Mauritius, Bourbon, Rodriguez, and Andaman Islands.

All the species agree in the following characters:—The head is rather elongate and the snout obtusely conical; the eye is small for the family, with circular pupil and a well-developed lid all round; the ear is round or vertically oval. The limbs are stout; the digits very unequal in size; the inner appears as a small tubercle; the longest or fourth is narrow in its basal half. The body is depressed and covered with small, subequal, granular scales. The males have two series of pores uniting in an angular line in front of the vent; in both sexes two small transverse slits behind the anal cleft. The tail is more or less depressed, tapering, and formed of more or less marked segments. The head is covered with granular scales, largest on the snout. The mental is subtriangular and does not separate the inner chin-shields. The scales on the belly are rather large, flat, imbricate.

Synopsis of the Species.

I. The suture between the rostral and first upper labial falls below the centre of the nostril.

A. Ventral scales smooth.

Snout twice as long as the distance between the orbit and the ear-opening; latter oval, its diameter half or three fifths that of the orbit; femoro-præanal pores 14 to 22 altogether .... 1. *cepedianum*, p. 211.

Snout twice as long as the distance between the orbit and the ear-opening; latter round, one fourth or one third the diameter of the orbit; 24 to 30 pores altogether .... 3. *andamanense*, p. 212.

Snout not twice as long as the distance between the orbit and the ear-opening; latter oval, its diameter not half that of the orbit; gular granules as large as the ventral scales; 10 to 12 pores altogether .... 4. *newtonii*, p. 212.

Snout nearly twice as long as the distance between the orbit and the ear-opening; latter oval, measuring one third the diameter of the orbit; 45 or 46 pores altogether .... 5. *guentheri*, p. 213.

B. Ventral scales keeled.

Snout twice as long as the distance between the orbit and the ear-opening; latter oval, its diameter half that of the orbit; 17 pores .... 2. *trilineatum*, p. 212.

II. Nostril pierced above the first upper labial only.

A. Ventral scales smooth.

Snout twice as long as the distance between the orbit and the ear-opening; femoro-præanal pores 33 to 50 altogether........................ 6. *madagascariense*, [p. 214.

Snout not twice as long as the distance between the orbit and the ear-opening; tail very much depressed, nearly as broad as the body; 26 to 28 pores altogether .... 7. *laticauda*, p. 215.

B. Ventral scales keeled.

Snout not twice as long as the distance between the orbit and the ear-opening; 24 to 30 pores altogether .... 8. *lineatum*, p. 216.

1. Phelsuma cepedianum.

Phelsuma cepedianus, *Gray, Cat.* p. 166.
Gecko cépédien, *Cuv. R. A.* ii. p. 46.
Gecko cepedianus, *Merr. Tent.* p. 43.
Phelsuma cepedianum, *Gray, Ann. Phil.* (2) x. 1825, p. 199.
Platydactylus cepedianus, *Cuv. R. A.* 2nd edit. ii. p. 52; *Geoffr. Mag. Zool.* iii. 1833, pl. iii.
Phelsuma ornatum, *Gray, in King's Voy. Austral.* ii. p. 428.
Platydactylus cepedianus, part., *Dum. & Bibr.* iii. p. 301.
Anoplopus cepedeanus, *Fitzing. Syst. Rept.* p. 99.

Snout twice as long as the distance between the orbit and the ear-opening or the diameter of the orbit. Upper part of rostral with a median cleft. Nostril pierced above the rostral and first labial, the suture between those two plates entering its lower border; eight to ten upper labials; six to eight infralabials; chin-shields three on each side, gradually decreasing in size, inner pair considerably larger than outer; three scales between the naso-rostrals. Ear-opening large, its vertical diameter half or three fifths that of the orbit. Dorsal scales all granular, perfectly smooth. Ventral scales smooth. Femoral pores seven to eleven on each side, nine being the usual number. Tail not much depressed, narrower than the body; segments of tail rather indistinctly marked, composed each of six or seven transverse rows of scales on the side, eight or nine on the upper surface; lower surface of tail (when intact) with a double tessellated series of larger scales (the reproduced tail shows no trace of segments, and the lower surface bears generally a series of strongly transversely dilated scales); all the caudal scales perfectly smooth. Upper parts bluish or purplish in spirits, with reddish markings; of these a ∩-shaped band on the snout, extending from eye to eye, and a band from the eye to the shoulder, sometimes extending along the side of the body, passing through the ear, are constant; there are frequently two or three longitudinal bands on the nape; a variously shaped interocular marking; dorsal and caudal spots irregular, numerous; lower surfaces whitish, the throat sometimes greyish, but never spotted.

Total length	118	millim.
Head	16	„
Width of head	12	„
Body	41	„
Fore limb	20	„
Hind limb	26	„
Tail	21	„

Mauritius and Bourbon.

a. ♂.	Mauritius.	Capt. P. P. King [P.]. (Type of *P. ornatum.*)
b–l. ♂, ♀, & yg.	Mauritius.	Sir H. Barkly [P.].
e. Hgr.	Mauritius.	Sir W. Hooker [P.].
f. ♂.	Mauritius.	E. Newton, Esq. [P.].
g. ♂.	Bourbon.	

2. Phelsuma trilineatum.

Phelsuma lineatum, part., *Gray, Cat.* p. 166.
Phelsuma trilineatum, *Gray, Zool. Misc.* p. 57.

Agrees in every respect with *P. cepedianum*, but the ventral scales are keeled, though less strongly than in *P. lineatum*, and the segments of the tail are composed of seven transverse rows of scales above and five or six on the side.

Madagascar.

a. ♂.	Madagascar.	Dr. J. E. Gray [P.].	(Type.)
b. ♂.	—— ?		

3. Phelsuma andamanense.

Phelsuma andamanense, *Blyth, Journ. As. Soc. Beng.* xxix. 1860, p. 108; *Günth. Rept. Brit. Ind.* p. 112; *Stoliczka, Journ. As. Soc. Beng.* xxxix. 1870, p. 162; *Anderson, Proc. Zool. Soc.* 1871, p. 160; *Stoliczka, Journ. As. Soc. Beng.* xlii. 1873, p. 163.
Gecko chameleon, *Tytler, Journ. As. Soc. Beng.* xxxiii. 1864, p. 548.

Snout twice as long as the distance between the orbit and the ear-opening, or the diameter of the orbit. Upper part of rostral with a median cleft. Nostril pierced above the rostral and first labial, the suture between the two plates entering its lower border; nine or ten upper labials; eight or nine infralabials; chin-shields irregular, scarcely distinct from the surrounding scales; two or three scales between the naso-rostrals. Ear-opening small, round, one fourth or one third the diameter of the orbit. Dorsal scales all granular, perfectly smooth. Ventral scales smooth. Femoral pores twelve to fifteen on each side. Tail not much depressed, narrower than the body; segments of tail rather indistinctly marked, composed of seven transverse series on the side, of nine above; lower surface of tail with a median series of transversely dilated scales, two narrower ones alternating with a broader one; all the caudal scales perfectly smooth. Greenish above, with more or less numerous orange spots, which, however, may be absent; an orange streak from the ear to the nostril, passing through the eye, and another, $\wedge$-shaped, from eye to eye, are generally present; lower surfaces whitish, throat not spotted. Size and proportions of *P. cepedianum*.

Andaman Islands.

a. Several spec.: ♂ ♀. Andaman Islands. Col. R. C. Tytler [P.].

4. Phelsuma newtonii. (PLATE XVII.)

Phelsuma newtonii, *Bouleng. Proc. Zool. Soc.* 1884, p. 2.

Snout once and two thirds as long as the distance between the orbit and the ear-opening. Upper part of rostral with a median cleft; nostril pierced above and between the rostral and the first labial; eight upper labials; seven infralabials; chin-shields, only the inner pair well distinguished from the gular scales, which are larger than in the other species, equalling the ventral scales in size; three scales between the naso-rostrals. Ear-opening small, its vertical diameter not half that of the orbit. All the scales on the back and sides very small, granular, without trace of a keel; ventrals smooth. Digits shorter than in the other species. Tail much depressed, as broad as the pelvic region, with extremely strongly marked segments; these composed of seven or eight transverse rows of scales above, of five or six beneath; all the scales of the tail perfectly smooth, those on the lower surface larger, subequal, without median transversely dilated scales. Præanal pores forming a short angular series in front of the vent, six on each side. Upper parts blackish olive, lower bluish grey; lips and throat whitish in specimen *a*.

Total length	223	millim.
Head	25	,,
Width of head	22	,,
Body	80	,,
Fore limb	32	,,
Hind limb	41	,,
Tail	118	,,

Rodriguez.

a. ♂.	Rodriguez.	E. Newton, Esq. [P.].
b. ♂.	Rodriguez.	J. C. O'Halloran, Esq. [P.].	(Types.)

5. Phelsuma guentheri.

A large species with the physiognomy of *P. madagascariense*. Snout nearly twice as long as the distance between the orbit and the ear-opening. Rostral with median cleft above; nostril pierced above the rostral and the first labial, the suture between those two plates entering its lower border; eleven to fourteen upper and nine or ten lower labials; chin-shields rather small, three or four on each side, decreasing in size from inner to outer, well distinguished from the gular granules, which are minute; a single scale between the anterior nasals. Ear-opening oval, about one third the diameter of the orbit. All the scales of the back and sides very small, granular, without trace of a keel; ventrals smooth. Tail depressed, much narrower than the body, divided into distinct segments, each of which contains six transverse rows of scales on the side, eight or nine on the upper surface; lower surface with enlarged scales, not dilated transversely; all the caudal scales perfectly smooth. Femoral pores forming a long series, as in *P. madagascariense*, composed of forty-five or forty-six pores altogether. Upper surfaces uniform grey (epidermis lost); lower surfaces whitish.

Total length	223	millim.
Head	32	,,
Width of head	24	,,
Body	93	,,
Fore limb	40	,,
Hind limb	49	,,
Tail (reproduced)	98	,,

Round Island, near Mauritius.

a–c. ♂ ♀.   Round Island.   Dr. A. Günther [P.].

6. Phelsuma madagascariense.

Phelsuma madagascariensis, *Gray, Cat.* p. 166.
Platydactylus cepedianus, part., *Dum. & Bibr.* iii. p. 301.
—— cepedianus, var., *A. Dum. Cat. Méth. Coll. Rept.* p. 34.
Phelsuma grandis, *Gray, Ann. Mag. N. H.* (4) vi. 1870, p. 191.
Pachydactylus cepedianus, var. madagascariensis, *Boettg. Abh. Senck. Ges.* xii. 1881, p. 458.

Snout twice as long as the distance between the orbit and the ear-opening or the diameter of the orbit. Upper part of rostral with a median cleft. Nostril pierced above and bordered beneath by the first labial; upper labials seven to nine; lower labials six to eight; chin-shields three on each side, gradually decreasing in size, inner pair about twice as large as outer. Generally a single scale between the naso-rostrals. Ear-opening small, its vertical diameter not half that of the orbit. Scales on hinder half of back distinctly, though very feebly, keeled in adult specimens. Ventral scales smooth. Femoral pores seventeen to twenty-five on each side in the adult, a few less in the young. Tail not very much depressed, narrower than the body; segments of tail rather indistinct, composed each of five or six transverse rows of scales on the side, seven to nine on the upper surface; lower surface of tail (when intact) with a median series of transversely dilated scales, two narrower ones alternating with a broader one; the reproduced tail shows no trace of annuli, and the lower surface bears a series of very strongly dilated scales, some of which may be dilated along the middle; the scales on the upper surface of the tail, at least in its anterior part, keeled. Upper parts greenish, bluish, or purple in spirits, uniform or with bright (red) markings; these form generally a band from eye to nostril, and a ∧-shaped mark on the forehead; no band on temple or side of neck; on the back the markings may form irregular roundish spots, transverse bands, or a large vermiculation; throat with more or less distinct blackish markings.

Total length	203	millim.
Head	26	,,
Width of head	21	,,
Body	80	,,
Fore limb	32	,,
Hind limb	45	,,
Tail (reproduced)	97	,,

Madagascar; Seychelles (and probably Comoro Islands).

a. ♂.	Madagascar.	Dr. J. E. Gray [P.]. (Type.)
b. ♂.	Madagascar.	(Type of *Phelsuma grandis.*)
c–e, f–i. ♂, ♀, & yg.	Seychelles.	Dr. E. P. Wright [C.].
k. Yg., dried.	Zungomero.	Capt. Speke [P.].
l–m. Yg.	Quellimane. (Probably imported.)	Sir J. Kirk [C.].

Pachydactylus dubius, Boettger, Zool. Anz. 1881, p. 46, and Abh. Ver. Brem. vii. p. 179, and Abh. Senck. Ges. xii. p. 464, from Nossi Bé, is perhaps a young *P. madagascariense*; the characters upon which that form now rests seem hardly sufficient to admit it as a species. The following is an abstract of the description:—

In habit intermediate between *P. madagascariense* and *P. laticauda*. Rostral with a median cleft above ; nostril pierced above the first upper labial ; naso-rostrals separated by three scales ; chin-shields four on each side, of equal size, scarcely larger than the surrounding gular scales. Dorsal scales rather large, smooth ; abdominal scales smooth. Caudal scales smooth ; the caudal segments composed of five or six transverse rows of scales above. Femoral pores twelve or thirteen on each side. Reddish brown above, spotted and variegated with yellowish anteriorly, with bluish posteriorly ; two blackish lateral streaks ; lower surfaces whitish, immaculate.

7. Phelsuma laticauda.

Pachydactylus cepedianus, var., *Boetty. Abh. Senck. Ges.* xi. 1879, p. 480.

——— laticauda, *Boetty. Zool. Anz.* 1880, p. 280, and *Abh. Senck. Ges.* xii. 1881, p. 461.

Snout once and two thirds as long as the distance between the eye and the ear-opening, twice the diameter of the orbit. Rostral without median cleft; nostril pierced above, and bordered beneath by the first upper labial ; eight to ten upper labials ; seven to nine lower labials ; chin-shields eight, subequal, the inner pair hardly larger than the outer; generally a single scale between the naso-rostrals. Vertical diameter of the ear-opening about half the diameter of the orbit. Dorsal scales smooth or more or less distinctly keeled. Ventral scales smooth. Femoral pores thirteen or fourteen on each side. Tail much depressed, nearly as broad as the back; segments of tail not very distinct, composed each of six or seven transverse rows of scales on the side, eight to ten on the upper surface ; lower surface generally very irregularly plated ; the scales on the upper surface generally keeled. Green above ; head with two or three rather indistinct reddish cross bars, the anterior V-shaped, extending from one eye to the other; sometimes a few large reddish spots on the hinder part of the back ; a blackish band from the eye along the side of the neck and body, passing through the ear and over the shoulder, darkest as it approaches the groin ; limbs and tail dotted or vermiculated with blackish ; lower surfaces whitish, throat generally greenish, immaculate.

Total length	139	millim.
Head	17	,,
Width of head	12	,,
Body	47	,,
Fore limb	19	,,

Hind limb	26 millim.
Tail	75 ,,

Madagascar; Comoro Islands.

a–b. ♂.   Nossi Bé.
c. Several spec.: ♂, ♀, & hgr.   Johanna.   C. E. Bewsher, Esq. [C.].

Pachydactylus quadriocellatus, Peters, Sitzb. Ges. Naturf. Fr. 1883, p. 28, from Central Madagascar, is no doubt very near to, and probably only a variety of, *P. laticauda*. The only difference I can find, from the very short description, is the presence of large round black spots edged with light blue at axilla and groin.

8. Phelsuma lineatum. (Plate XVIII. fig. 1.)

Phelsuma lineatum, part., *Gray, Cat.* p. 166.
Phelsuma lineatum, *Gray, Zool. Misc.* p. 57.

Snout once and a half or once and two thirds as long as the distance between the eye and the ear-opening, twice the diameter of the orbit. Rostral generally without median cleft; nostril pierced above, and bordered beneath by the first upper labial; eight or nine upper labials; seven or eight lower labials; chin-shields eight, scarcely distinct from the surrounding scales, which are quite or almost as large; generally a single scale between the nasorostrals. Vertical diameter of the ear-opening about half the diameter of the orbit. Dorsal scales smooth, or very feebly keeled on the hinder part of the back. Ventral scales strongly keeled. Femoral pores twelve to fifteen, the number being generally thirteen. Tail depressed, considerably narrower than the body; segments distinct, limited by a row of larger scales, composed each of seven or eight transverse rows on the side, eight or nine on the upper surface; scales on the upper surface of the tail feebly keeled, those on the lower surface strongly keeled, larger, subequal, not transversely dilated. Colour much as in *A. laticauda*; the lateral blackish band is constantly bordered inferiorly by another pure white band, and it extends along each side of the tail.

Total length	115 millim.
Head	14 ,,
Width of head	10 ,,
Body	37 ,,
Fore limb	15 ,,
Hind limb	20 ,,
Tail	64 ,,

Madagascar.

a. ♂.   Madagascar.   Dr. J. E. Gray [P.].   (Type.)
b–e. ♂ ♀.   Madagascar.   Prof. A. Newton [P.].
f–i. ♂ ♀.   Madagascar.
k, l. ♂ ♀.   Madagascar.

48. RHOPTROPUS.

Rhoptropus, *Peters, Mon. Berl. Ac.* 1869, p. 58.

Digits slender, free, dilated at the end, with undivided transverse lamellæ below, furnished with extremely small, very indistinct, retractile claw; upper surface with a flat, nail-like distal scale. Upper surfaces covered with granular scales; belly covered with cycloid imbricated scales. Pupil vertical. Eyelid distinct all round the eye. No præanal nor femoral pores.

South-western Africa.

1. Rhoptropus afer.

Rhoptropus afer, *Peters, l. c.* p. 59, pl. —. fig. 2; *Bocage, Jorn. Sc. Lisb.* iv. 1873, p. 212.

Habit of a *Ptyodactylus*. Snout depressed, broad, and rounded; ear-opening rather large, horizontal, concealed by a narrow dermal fold. Nostril pierced in the centre of a swelling formed by three or four nasal scutes; ten or eleven upper and eight or nine lower labials; mental and the contiguous labials much elongate; no chin-shields. Upper surfaces covered with minute granular scales, larger on the snout. Limbs elongate. Tail rounded, tapering, somewhat depressed, covered with small flat smooth scales. Yellowish or greenish-olive above, uniform or with small scattered dark spots; lower surfaces greenish white.

Total length	135	millim.
From snout to vent	52	,,
Head to ear-opening	14	,,
Fore limb	25	,,
Hind limb	30	,,

Damaraland; Mossamedes.

49. SPHÆRODACTYLUS.

Sphærodactylus, part., *Wagler, Syst. Amph.* p. 143.
Sphæriodactylus, *Gray, Griff. An. Kingd., Syn.* p. 52; *Wiegm. Herp. Mex.* p. 20; *Dum. & Bibr.* iii. p. 401; *Bocourt, Miss. Sc. Mex. Rept.* p. 44.
Sphærodactylus, *Fitzing. Syst. Rept.* p. 93; *Gray, Cat. Liz.* p. 168; *Cope, Proc. Ac. Philad.* 1861, p. 497.

Digits narrow, slender, free, with transverse lamellæ inferiorly, the apex dilated into a disk, with a circular undivided plate inferiorly; all digits with a sheathed retractile claw, the sheath opening laterally and inwards. Scales granular or imbricate. Pupil round or subelliptical. Eyelid nearly circular. No præanal or femoral pores.

West Indies; Central America; Colombia.

Synopsis of the Species.

I. Dorsal scales very small, granular, smooth.

  A. Rostral large (see Plate XVIII. fig. 2 *a*).

   1. Scales subequal.

Five brown transverse bands between the end of the snout and the hind limbs .... 1. *sputator*, p. 219.
Ten brown transverse bands between the end of the snout and the hind limbs .... 2. *elegans*, p. 220.
Body not barred .... 3. *punctatissimus*, p. 220.

   2. Scales much larger on the flanks than on the middle of the back .... 4. *nigropunctatus*, p. 220.

  B. Rostral moderately large (see Plate XVIII. fig. 3).

Tail with two yellow black-edged spots near the tip .... 5. *glaucus*, p. 221.

II. Scales very small, juxtaposed, keeled.

Ear-opening smaller than digital pallet .... 6. *lineolatus*, p. 221.
Ear-opening larger than digital pallet . 7. *casicolus*, p. 222.
Snout very acute: tail much longer than body .... 8. *alopex*, p. 222.

III. Scales of medium size, slightly imbricate, keeled.

Snout much longer than the distance between the eye and the ear-opening . 9. *oxyrrhinus*, p. 222.
Snout as long as the distance between the eye and the ear-opening ...... 10. *argus*, p. 223.

IV. Scales very large, at least as large as those on the belly, strongly keeled and imbricate.

  A. Vertebral line with granular scales.

   1. Ventral scales quite smooth.

Dorsal scales not larger than ventrals . 11. *fantasticus*, p. 223.

Dorsal scales much larger than ventrals .... 14. *anthracinus*, p. 225.

2. Ventral scales feebly keeled.

Dorsal scales not larger than ventrals. 12. *microlepis*, p. 224.
Dorsal scales rather larger than ventrals, very strongly keeled ...... 13. *copii*, p. 225.

B. No granular scales on the vertebral line.

1. Scales on the breast keeled .. 15. *macrolepis*, p. 226.

2. Scales on the breast smooth.

Snout rather short; four upper and four lower labials .... 16. *notatus*, p. 226.
Snout short; three upper and three lower labials .... 17. *gilvitorques*, p. 227.
Snout longer than the distance between the eye and the ear-opening .... 18. *richardsonii*, p. 227.

1. Sphærodactylus sputator.

Sphærodactylus sputator, *Gray, Cat.* p. 168.
? Lacerta sputator, *Sparrman, Vetensk. Acad. Handl.* v. 1784, p. 164, pl. iv. fig. 1.
Anolis sputator, var. 1, *Daud. Rept.* iv. p. 99.
Gecko sputator, *Merr. Tent.* p. 43.
Sphærodactylus sputator, *Gray, in Griff. A. K., Syn.* p. 52; *Dum. & Bibr.* iii. p. 402; *Cope, Proc. Ac. Philad.* 1861, p. 498.

Snout pointed, slightly longer than the distance between the eye and the ear-opening, about once and two thirds the diameter of the orbit; ear-opening small, oval, vertical. Rostral large, covering the end of the snout, with longitudinal cleft above; nostril pierced between the rostral, the first labial, and one or two small nasals; five upper and as many lower labials, the anterior very long; mental large, truncate posteriorly; no regular chin-shields, but polygonal scales passing gradually into the granules of the throat. Upper eyelid with an exceedingly small spine-like scale above the middle of the eye. Upper surfaces covered with equal, small, flat, granular scales, largest on the snout; abdominal scales larger, imbricate. Tail cylindrical, tapering, covered with small scales arranged in verticils, and a median series of transversely dilated scales inferiorly. Yellowish brown, upper surfaces with transverse dark-brown bands, as wide as the interspaces between them; five brown bands between the end of the snout and the hind limbs, the three anterior extending across the throat and neck; the bars on the tail complete, forming rings; snout dark brown, with two longitudinal yellowish streaks.

Total length (tail injured). 45 millim.
Head .... 7 ,,

                Width of head..........   4 millim.
                Body 21   ,,
                Fore limb   8   ,,
                Hind limb 11   ,,
Cuba; San Domingo.

a–e. Ad. & hgr.   —— ?

2. Sphærodactylus elegans.

Anolis sputator, var. 2, *Daud. Rept.* iv. p. 99.
Sphærodactylus sputator, *Cocteau, in R. de la Sagra, Hist. Cuba,*
   p. 160, pl. xvii.
—— elegans, (*McLeay*) *Reinh. & Lütk. Vidensk. Meddel.* 1862, p. 275.

Very closely allied to the preceding. The scales are smaller, and there are as many as ten transverse dark bands from the eyes to the hind limbs; these bars much narrower than the interspaces between them. From snout to vent 19 millim.; tail 18 millim.

Cuba; San Domingo.

a. Ad.   San Domingo.

3. Sphærodactylus punctatissimus.

Sphærodactylus punctatissimus, *Gray, Cat.* p. 168.
Sphærodactylus punctatissimus, *Dum. & Bibr.* iii. p. 405; *Reinh. &
   Lütk. Vidensk. Meddel.* 1862, p. 276.
—— cinereus, *Cocteau, in R. de la Sagra, Hist. Cuba, Rept.* p. 166,
   pl. xviii.; *Cope, Proc. Ac. Philad.* 1861, p. 498.

Also closely allied to *S. sputator.* Snout still more pointed, longer, measuring twice the diameter of the orbit and distinctly more than the distance between the eye and the ear-opening. Grey above, uniform or dotted with white; in specimen *b* the snout is white. From snout to vent 35 millim.; tail 30 millim.

Cuba; San Domingo; Martinique; Caraccas.

a. Ad.   Martinique.   Paris Museum [P.].
b. Ad.   Caraccas.
c. Ad.   —— ?

4. Sphærodactylus nigropunctatus. (Plate XVIII. fig. 2.)

Sphærodactylus nigropunctatus, *Gray, Cat.* p. 168.

Agrees with *S. sputator* in general characters, but the granules on the flanks are considerably larger than those on the middle of the back. Brown (the very numerous minute black specks mentioned by Gray being no longer visible), tail and lower surfaces lighter. From snout to vent 35 millim.; tail 32 millim.
       —— ?

a. Ad.   —— ?   (Type.)

5. Sphærodactylus glaucus. (Plate XVIII. fig. 3*.)

Sphærodactylus glaucus, *Cope, Proc. Ac. Philad.* 1865, p. 192.
? Sphærodactylus inornatus, *Peters, Mon. Berl. Ac.* 1873, p. 738.

General characters as in *S. sputator*, but the rostral is considerably smaller. Light brown above, with darker vermiculations; a dark streak on the side of the head, passing through the eye; a dark brown transverse spot on middle of upper side of neck; tail with two yellow black-edged spots near the tip; digits annulated with yellow; lower surfaces whitish, the sides and throat with brownish vermiculations. From snout to vent 30 millim.; tail 30 millim.

Central America.

a. Ad.	Vera Paz, low forest.	F. D. Godman, and O. Salvin, Esqrs. [P.].
b. Hgr.	Vera Cruz.	

6. Sphærodactylus lineolatus.

Sphærodactylus fantasticus, part., *Gray, Cat.* p. 168.
Sphærodactylus lineolatus, *Lichtenst. Nomencl. Rept. Mus. Berol.* p. 6; *Bocourt, Miss. Sc. Mex., Rept.* p. 46.
—— millepunctatus, *Hallow. Proc. Ac. Philad.* 1860, p. 480; *Cope, Proc. Ac. Philad.* 1861, p. 499.

Snout pointed, as long as the distance between the eye and the ear-opening, once and a half the diameter of the orbit; ear-opening small, oval, vertical. Rostral moderately large, with longitudinal cleft above; nostril pierced between the rostral, the first labial, and three scales; five upper and as many lower labials, the anterior very long; mental large, truncate posteriorly, followed by polygonal scales passing gradually into the granular gular scales. A small spine-like scale on the border of the upper eyelid, above the middle of the eye. Upper surfaces covered with small, equal, slightly keeled, juxtaposed granular scales; on the head the granules are smooth, largest on the snout; abdominal scales larger, imbricate. Tail cylindrical, tapering, covered with imbricate smooth scales; a median series of large scales inferiorly. Two varieties of colour.

Total length	56	millim.
Head	8	,,
Width of head	4·5	,,
Body	20	,,
Fore limb	8	,,
Hind limb	10	,,
Tail	28	,,

Central America.

Var. A.—Brownish above; head with five dark brown streaks, the central one ending on occiput, the others extending to the neck; the outer passing through the eye; lower surfaces whitish.

a. Ad.	—— ?	E. Laforest, Esq. [P.].

---

* End of snout, upper view, × 2.

VAR. B.—Brown above, dotted with darker; a dark line on the side of the head, passing through the eye; a dark band on each side of neck, and from axilla to groin; yellowish dark-edged ocelli on the back, limbs, and tail; digits annulate with yellowish; end of tail with two blackish annuli, separated by a yellowish interspace; lower surfaces whitish, with minute brown dots.

b. Ad. Santa Cruz.
c. Ad. Chontales, Nicaragua.

7. Sphærodactylus casicolus.

Sphærodactylus casicolus, *Cope, Proc. Ac. Philad.* 1861, p. 499.

Dorsal scales very minute, keeled; occipital granular, frontal keeled. Inferior labials (anterior to posterior border of orbit) five. Supra-nasal plates as long as broad. Muzzle elongate. Auricular aperture larger than digital pallet. Dark-brown rufous, with three distant, transverse, dorsal blotches, bordered with lighter; the anterior or interscapular indistinct. A dark spot upon the nape, bounded by two light dots. Numerous short longitudinal white lines upon the dorsal and lateral regions; none upon the head. A loreal and three postocular dark lines. Beneath whitish, chin and sides of neck punctulated with rufous.

Region of the Truando, Colombia.

8. Sphærodactylus alopex.

Sphærodactylus alopex, *Cope, l. c.*

Dorsal scales keeled, smaller than in *S. oxyrrhinus*. Muzzle very acute, profile sloping regularly from the frontal region. Inferior labials (anterior to posterior border of orbit) six. Supra-nasal plates separated. Pupil apparently elliptic. Tail much longer than body. Above rufous grey, closely vermiculated with longitudinal rufous lines; tail and extremities spotted with the same. Beneath pale brownish, faintly vermiculated with rufous brown on the gular region and the sides of the neck; many of the abdominal and femoral plates margined with the same.

San Domingo.

9. Sphærodactylus oxyrrhinus. (PLATE XVIII. fig. 4.)

Sphærodactylus oxyrhinus, *Gosse, Ann. Mag. N. H.* (2) vi. 1850, p. 347; *Cope, l. c.*

Very closely allied to *S. argus*, from which it differs in the much longer and sharper snout and the larger rostral plate; the length of the snout equals nearly twice the diameter of the orbit, and is much more than the distance between the eye and the ear-opening. Pale brown above, covered with more or less confluent darker specks; lower surfaces whitish, the sides and throat speckled with brown. From snout to vent 32 millim.

Jamaica.

a. Ad.	St. Elizabeth's, Jamaica.	P. H. Gosse, Esq. [P.].	(Type.)

10. Sphærodactylus argus. (PLATE XVIII. fig. 5.)

Sphærodactylus argus, *Gosse, Ann. Mag. N. H.* (2) vi. 1850, p. 347; *Cope, Proc. Ac. Philad.* 1861, p. 498.

Snout pointed, as long as the distance between the eye and the ear-opening, once and a half the diameter of the orbit; ear-opening small, oval, vertical. Rostral moderately large, with longitudinal cleft above; nostril pierced between the rostral, the first labial, and three scales; five upper and as many lower labials, the anterior very long; mental large, truncate posteriorly, followed by granules larger than those on the throat. A small spine-like scale on the border of the upper eyelid, above the middle of the eye. Head and throat covered with very small granular scales, largest on the snout; back covered with medium-sized, flat, imbricate, keeled scales; these dorsal scales about two thirds the size of the ventrals, which are smooth and imbricate. Tail cylindrical, tapering, covered with imbricate smooth scales, the lower median series of which are enlarged. Pale brown above, with small, yellowish, dark-edged ocelli; on the head these ocelli are generally lengthened, and frequently more or less confluent into six longitudinal lines, which sometimes extend along the back. Lower surfaces whitish, the throat variegated with brown.

Total length	58	millim.
Head	9	,,
Width of head	5	,,
Body	22	,,
Fore limb	8	,,
Hind limb	10	,,
Tail	27	,,

Jamaica.

a–c, d–e. Ad.	Jamaica.	P. H. Gosse, Esq. [P.].	(Types.)
f–i. Yg.	Jamaica.		
k. Ad.	—— ?	Sir W. Hooker [P.].	
l–m, n, o, p. Ad.	—— ?		

11. Sphærodactylus fantasticus.

Sphærodactylus fantasticus, part., *Gray, Cat.* p. 168.
Sphærodactylus fantasticus (*Cuv.*), *Dum. & Bibr.* iii. p. 406, pl. xxxii. fig. 2; *A. Dum. Arch. Mus.* viii. p. 469, pl. xvii. fig. 3; *Cope, Proc.*

Ac. Philad. 1861, p. 500; *Reinh. & Lütk. Vidensk. Meddel.* 1862, p. 277; *Bocourt, Miss. Sc. Mex., Rept.* p. 44, pl. x. fig. 4.

Snout pointed, as long as the distance between the eye and the ear-opening, once and a half the diameter of the orbit; ear-opening roundish, about as large as digital expansion. Rostral large, with longitudinal cleft above; nostril pierced between the rostral, the first labial, and three scales; three upper labials, median smallest; three lower labials, first very large, third very small; mental large, truncate posteriorly. A small spine-like scale on the border of the upper eyelid, above the centre of the eye. Upper surface of head covered with small, granular, slightly keeled scales, largest on the snout. Back covered with subrhomboidal imbricate scales about the size of the ventral scales; the vertebral line covered with minute granular scales. Scales on the limbs small, slightly keeled. Gular scales granular, abdominal scales large, imbricate, smooth. Tail cylindrical, tapering, covered with smooth imbricate scales, with a series of large transversely dilated plates inferiorly. Light brown above, the head generally darker, vermiculated with white lines; whitish beneath.

Total length	54	millim.
Head	7	,,
Width of head	5	,,
Body	20	,,
Fore limb	8	,,
Hind limb	10·5	,,
Tail	27	,,

West Indies; Mexico; Venezuela.

a. Ad. Caraccas.
b. Ad. Antigua.

12. Sphærodactylus microlepis.

Sphærodactylus fantasticus, var., *A. Dum. Arch. Mus.* viii. p. 469, pl. xvii. fig. 4.
—— microlepis, *Reinh. & Lütk. Vidensk. Meddel.* 1862, p. 278.
—— melanospilos (*A. Dum.*), *Bocourt, Miss. Sc. Mex., Rept.* p. 44 (footnote).

Distinguished from *S. fantasticus* in having the snout shorter and more obtuse, the dorsal scales smaller, and the granules forming a less distinct zone on the middle of the back larger. Bocourt adds that the ventral scales are keeled and that there are no enlarged inferior caudal scales. Back with dark spots and large transverse markings; throat lineolate with brown.

West Indies (Sa. Lucia, St. Croix).

13. Sphærodactylus copii.

Sphærodactylus copei, *Steindachn. Novara, Rept.* p. 18, pl. i. fig. 5.

Closely allied to *S. fantasticus*. The dorsal scales are rather larger and much more strongly keeled; all the head-scales keeled; ventral scales slightly keeled.

a. Ad. ———? E. Laforest, Esq. [P.].

Probably the species described under this name by Steindachner, though no mention is made of the slight keel on the ventral scales.

14. Sphærodactylus anthracinus.

Sphærodactylus anthracinus, *Cope, Proc. Ac. Philad.* 1861, p. 500.

Snout acuminate, elongate, longer than the distance between the eye and the ear-opening, once and two thirds the diameter of the eye. Ear-opening oval, oblique, nearly as large as digital pallet. Rostral large, with median cleft above; nostril pierced between the rostral, the first labial, and three scales; four upper and as many lower labials, anterior very long; mental large, truncate posteriorly. Upper eyelid with a small spine-like scale. Head covered with granular scales, larger and slightly keeled on the snout. Dorsal scales very large, much larger than ventrals, rhomboidal, imbricate, strongly keeled, in ten rows on each side, separated on the median dorsal line by a narrow zone of minute granular scales. Gular scales granular, minute; ventral scales moderate, imbricate, smooth. Caudal scales strongly imbricate, rather pointed, those on the anterior half of the tail keeled; inferiorly a median series of transversely enlarged scales. Cope describes the colour as black, the large dorsal scales tinged with blue. The present specimen, which agrees well in other points with Cope's diagnosis, is light brownish above, with dark brown spots on the back, limbs, and tail; head with whitish spots and three whitish streaks, one longitudinally between the orbits and the others from the nostril to the orbit; inferiorly brownish white.

Total length	79	millim.
Head	10	,,
Width of head	6	,,
Body	27	,,
Fore limb	11	,,
Hind limb	15	,,
Tail	42	,,

Mexico; West Indies.

a. Ad. San Domingo.

15. Sphærodactylus macrolepis.

Sphærodactylus macrolepis, *Günth. Ann. Mag. N. H.* (3) iv. 1859, p. 215, pl. iv. fig. B; *Cope, Proc. Ac. Philad.* 1861, p. 500; *Reinh. & Lütk. Vidensk. Meddel.* 1862, p. 279.
—— imbricatus, *Fischer, Abh. Nat. Ver. Bremen,* vii. 1882, p. 234, pl. xv. figs. 4–10.

Snout obtusely pointed, as long as the distance between the eye and the ear-opening, hardly once and a half the diameter of the eye. Ear-opening roundish, as large as digital expansion. Rostral moderately large, with median cleft above; nostril pierced between the rostral, the first labial, and two or three scales; three or four upper and as many lower labials, anterior longest; mental large, truncate posteriorly. Upper eyelid with a small spine-like scale. Head covered with small, juxtaposed, slightly keeled scales, largest on the snout. Dorsal scales larger than ventrals, rhomboidal, pointed behind, strongly keeled and imbricate. Scales on the breast keeled, on the belly smooth. Caudal scales similar to the dorsals; a series of larger scales on the lower side of the tail. Light brown above; head with dark brown lines; sometimes a large interscapular black spot with two white dots anteriorly; lower surfaces dirty white, throat generally striolated with brown.

Total length	54 millim.
Head	7 ,,
Width of head	4 ,,
Body	19 ,,
Fore limb	7 ,,
Hind limb	9 ,,
Tail	28 ,,

West Indies (St. Croix, St. Thomas).

a–b. Ad.	St. Croix.	A. Newton, Esq. [P.].	(Types.)
c–d. Ad.	St. Thomas.	Hr. Riise [C.].	

16. Sphærodactylus notatus.

Sphærodactylus notatus, *Baird, Proc. Ac. Philad.* 1858, p. 254, *and U. S. Mex. Bound. Surv., Rept.* p. 24, pl. xxiv. figs. 29–37; *Cope, Proc. Ac. Philad.* 1861, p. 500.

Snout rather short. Four upper and as many lower labials. Upper eyelid with a spine-like scale. Frontal scales keeled. Dorsal scales large, keeled. Gular scales smooth. Tail longer than body. Colour reddish brown, with faint darker markings on the back, indicating longitudinal streaks. A median longitudinal line upon the head, which expands posteriorly; a supraciliary, three postocular, and a loreal line; the first-mentioned extends from a greater or less distance posteriorly; in some specimens these lines are broken into small spots.

Florida; West Indies (Bahamas, Cuba).

17. Sphærodactylus gilvitorques.

Sphærodactylus gilvitorques, *Cope, Proc. Ac. Philad.* 1861, p. 500.

Snout short. Three upper and three lower labials. Mental narrowed behind. Frontal scales keeled. Dorsal scales large, keeled, intermaxillary in about sixteen rows. Colour dark brown above, with a yellow collar just anterior to the interscapular region. Head darker, marked with narrow lines as follow:—one median, one supraciliary, one loreal, two postauricular. Beneath a little paler.

Jamaica.

18. Sphærodactylus richardsonii. (PLATE XVIII. fig. 6.)

Sphærodactylus richardsonii, *Gray, Cat.* p. 168.
Sphærodactylus richardsonii, *Cope, Proc. Ac. Philad.* 1861, p. 499.

Snout pointed, longer than the distance between the eye and the ear-opening, once and two thirds the diameter of the eye. Ear-opening roundish, nearly as large as digital expansion. Rostral large, with median cleft above; nostril pierced between the rostral, the first labial, and two rather large nasals; four upper and three or four lower labials, anterior longest; mental large, truncate posteriorly. Upper eyelid with a very small spine-like scale. Head covered with granular scales, which are not larger, and slightly keeled on the snout. Dorsal scales very large, much larger than ventrals, rhomboidal, strongly keeled, and imbricate; they form about eighteen longitudinal series. Gular scales minutely granular; abdominal scales large, imbricate, smooth. Tail cylindrical, tapering, covered with large strongly imbricate scales; those on the upper surface keeled; a series of feebly enlarged scales inferiorly. Light pinkish-brown above, with six lighter, dark-edged, transverse bands on the body; head with two white streaks from the tip of the snout to the neck, and another on each side from eye to ear; tail completely encircled by eight brown annuli, becoming gradually darker towards the end, as broad as the interspaces between them; limbs also annulate. Lower surfaces lighter, the throat with brown chevron-shaped bands.

Total length	76	millim.
Head	11	,,
Width of head	7	,,
Body	26	,,
Fore limb	11	,,
Hind limb	13	,,
Tail	39	,,

Jamaica.

a. Ad. —— ? Sir J. Richardson [P.]. (Type.)
b. Ad. Montego Bay.

The following genus, characterized by Gray in 1842, has never been rediscovered; the type is not to be found in the Collection. If the description is correct, this Gecko constitutes a distinct genus.

PHYRIA, Gray, Zool. Misc. p. 53.

"Toes moderate, base scaly above and below, dilated, ends expanded into a rounded disk, with two series of diverging plates beneath, last joint free, compressed, clawed; thumb of fore and hind feet smaller, clawless; back covered with small granular, belly with rather larger scales; labial and chin-shields distinct; tail round, tapering, with a series of large hexagonal plates beneath; præanal pores distinct, in an angular line. Like *Ptyodactylus*, but the toes are shorter, thicker, dilated at the base, and it has præanal pores."

Phyria punctulata, Gray, *l. c.*

"Olive-grey, in spirit; scales minutely black-speckled; lips, sides of throat, chest, belly, and underside of tail yellow."

Port Essington.

The following bones of an apparently extinct Lacertilian, from Rodriguez, have been referred to *Gecko* (*Gecko newtonii*, Günther, Journ. Linn. Soc., Zool. xiii. 1878, p. 324), which generic identification seems to me extremely doubtful, the parietal bone being single, whereas it is double in all known Geckoids.

a. Two parietals, posterior half of right ramus of lower jaw, right humerus, right half of pelvis, five left and two right femora.   Rodriguez.   Transit-of-Venus Exped.

Fam. 2. EUBLEPHARIDÆ.

Geckonidæ, part., *omn. auct.*
Eublepharidæ, *Boulenger, Ann. Mag. N. H.* (5) xii. 1883, p. 308, and *eod. loc.* (5) xiv. 1884, p. 119.

Except in the procœlian vertebræ and the single parietal bone, the skeleton is similar to that of the preceding family, to which the *Eublepharidæ* are closely affined. The teguments are also very similar, and of a soft kind; the upper surfaces are covered with small scales or granules, which are usually intermixed with enlarged tubercles, and the lower surface of the body with small, cycloid, imbricated scales. The skin of the head is free from the skull. The eyes are moderately large, with elliptico-vertical pupil, and are protected by thick, movable, connivent lids. The nostril is rather large, directed slightly upwards, though lateral, and separated from the rostral and labial plates. The tympanum is exposed. The limbs are weak and the digits short and cylindrical; they are all provided with a small, sharp, retractile claw which, in *Coleonyx*, is entirely concealed in a much-developed compressed sheath; this sheath, which differs only in size according to the genera, is composed of two lateral plates, the superior suture of which is covered by a third narrower one, a structure which we have already met with in the Geckoid genus *Ælurosaurus*. As in the Geckos, the tail is extremely fragile. Males have a few præanal pores, forming an angular series.

Three species are natives of Central America, one of the southern parts of the United States, two of Southern Asia, and one of West Africa; the genus *Eublepharis* occurs in America as well as in Asia. This extraordinary distribution seems to indicate that the few representatives of this small family are the remnants of some ancient, more generally dispersed, group; it nevertheless remains a matter of wonder how forms, now so widely separated, have retained so great a resemblance, not only in structure, but also in the pattern of coloration.

Synopsis of the Genera.

Claws partly exposed; lower surface of digits granular.
1. **Psilodactylus,** p. 229.

Claws partly exposed; lower surface of digits lamellar.
2. **Eublepharis,** p. 230.

Claws entirely concealed into a large compressed sheath; lower surface of digits lamellar . . . . 3. **Coleonyx,** p. 234.

1. PSILODACTYLUS.

Psilodactylus, *Gray, Proc. Zool. Soc.* 1864, p. 61.

Digits granular inferiorly; claw-sheath small, the claws only partially retractile.
West Africa.

1. Psilodactylus caudicinctus.

Stenodactylus caudicinctus, *A. Dum. Rev. Mag. Zool.* 1851, p. 478, pl. xiii. *and Cat. Méth. Rept.* p. 48, *and Arch. Mus.* viii. p. 489, pl. xviii. fig. 15.
Psilodactylus caudicinctus, *Gray, l. c.*

Body stout; limbs short, weak; digits very short. Snout as long as the distance between the orbit and the ear-opening; latter large, oval, slightly oblique, its longitudinal diameter half the diameter of the orbit. Head covered with flat, irregular, polygonal tubercles, largest on the temples; rostral quadrangular, twice as broad as high, with median cleft above; nostril between several scales, the anterior of which is largest and widely separated from its fellow on the other side; eight to ten upper, and about ten lower labials; mental broadly pentagonal, in contact with two enlarged chin-shields, surrounded by irregular smaller ones passing gradually into the flat granules of the gular region. Body covered above with small, irregular, flat scales intermixed with numerous suboval, obtusely keeled tubercles, which on the sides unite three to three. Male with thirteen præanal pores. Tail swollen, rounded, tapering at the end, circularly plaited, above with small flat scales and rows of enlarged obtusely keeled tubercles, inferiorly with larger flat scales arranged regularly. Cream-coloured above, with reddish-brown bands, viz. a horseshoe-shaped band from eye to eye across the nape, two broad transverse bands across the back, and three round the tail; these bands are as broad as the interspaces between them; slight reddish marblings on the lips and sides of neck; lower surfaces white.

Total length	210	millim.
Head	31	,,
Width of head	25	,,
Body	99	,,
Fore limb	33	,,
Hind limb	45	,,
Tail	80	,,

West Africa.

a, b. ♂ ♀ .   W. Africa.

2. EUBLEPHARIS.

Eublepharis, *Gray, Phil. Mag.* (2) iii. 1827, p. 56, *and Zool. Journ.* iii. 1828, p. 223, *and Cat.* p. 170; *Wagl. Syst. Amph.* p. 143; *Fitzing. Syst. Rept.* p. 90; *Günth. Rept. Brit. Ind.* p. 119.
Gymnodactylus, part., *Wiegm. Herp. Mex.* p. 19.

Digits with a row of lamellæ inferiorly; claw-sheaths small, the claws only partially retractile.

Southern Asia; Central America; southern North America.

Synopsis of the Species.

I. Back with enlarged tubercles.

Tubercles larger than the interspaces between them .... 1. *hardwickii*, p. 231.
Tubercles not larger than the interspaces between them; mental pentagonal, much broader than long, followed by a pair of chin-shields .... 2. *macularius*, p. 232.
Tubercles not larger than the interspaces between them; mental subquadrangular, as long as broad; no chin-shields .... 3. *dovii*, p. 233.

II. Back uniformly granular.

Snout as long as the distance between the orbit and the ear-opening; eight labials to below the centre of the eye .... 4. *variegatus*, p. 233.
Snout longer than the distance between the orbit and the ear-opening; six labials to below the centre of the eye .... 5. *fasciatus*, p. 234.

1. Eublepharis hardwickii.

Eublepharis hardwickii, *Gray, Cat.* p. 170.
Eublepharis hardwickii, *Gray, ll. cc. and Ill. Ind. Zool.*; *Günth. l. c.* p. 119, pl. xi. fig. B.
Gymnodactylus lunatus, *Blyth, in Cantor, Cat. Mal. Rept.* p. 27, *and Journ. As. Soc. Beng.* xxiii. 1854, p. 210.

Body stout; limbs rather short; digits short. Snout as long as the distance between the orbit and the ear-opening; latter large, suboval, vertical. Head covered with irregular polygonal scales, intermixed with enlarged tubercles on the temples and occiput; rostral subpentagonal, twice as broad as high, with median cleft above; three or four internasals; about ten upper and as many lower labials; mental broadly pentagonal, in contact with two enlarged chin-shields, surrounded by irregular smaller ones passing gradually into the flat granules of the gular region. Body covered above with small, irregular, flat scales, intermixed with numerous roundish, subconical tubercles; these tubercles larger than the interspaces between them. Male with fourteen to eighteen præanal pores. Tail swollen, rounded, tapering at the end, circularly plaited, above with small flat scales and rows of enlarged subconical tubercles, inferiorly with larger flat scales arranged regularly. Above reddish brown and cream-coloured; the former colour occupies the head and forms two broad bands across the back, the anterior broadest, and three round the tail; the latter borders the upper lip and extends as a horseshoe-shaped band to the other side, passing across the neck, and occupies the interspace between the dorsal and caudal brown bands, which are by far the widest; lower surfaces white.

Total length	199 millim.
Head	35 ,,
Width of head	23 ,,
Body	77 ,,
Fore limb	39 ,,
Hind limb	47 ,,
Tail	87 ,,

Eastern and Southern India.

a. ♂.	Penang, Chittagong.	Gen. Hardwicke [P.].	(Type.)
b–c. ♂.	Russelconda.	Dr. Traill [P.].	
d. ♀.	Anamallays.	Col. Beddome [C.].	

2. Eublepharis macularius.

Cyrtodactylus macularius, *Blyth, Journ. As. Soc. Beng.* xxiii. 1854, p. 738.
Eublepharis fasciolatus, *Günth. Ann. Mag. N. H.* (3) xiv. 1864, p. 429.
—— macularius, *Theob. Cat. Rept. As. Soc. Mus.* p. 32; *Anders. Proc. Zool. Soc.* 1871, p. 163; *Theob. Cat. Rep. Brit. Ind.* p. 94; *Murray, Zool. Sind,* p. 366.
—— hardwickii, (*non Gray*) *Murray, l. c.*

Differs from the preceding by the smaller and less numerous tubercles, the granular interspaces being as wide as the tubercles themselves; these are subconical or slightly keeled. The body and the digits are rather more elongate. Nine to fourteen præanal pores in the male. Young with five chestnut-brown transverse bands from head to sacrum, slightly broader than their interspaces, which are whitish, and similar rings round the tail; the first band is horse-shoe-shaped, and encircles the occiput, each branch advancing to the eye. In the adult these bands become more indistinct, and the head and body become spotted or largely vermiculated with chestnut-brown and whitish; in some only the brown edges of the dorsal bands remain. Specimen *i* has the two broad dorsal bands of *E. hardwickii* with the variegations of *E. macularius*. Lower surfaces white.

Total length	199 millim.
Head	32 ,,
Width of head	24 ,,
Body	87 ,,
Fore limb	40 ,,
Hind limb	50 ,,
Tail	80 ,,

North-western India, probably ranging through Baluchistan and Persia to the Euphrates*.

a. ♀.	Between Cashmere and Murree.	A. Kinloch, Esq. [P.].
b–d. ♂ & yg.	Rajampore, Punjab.	C. Tufnell, Esq. [P.].

---

* I have examined a specimen, belonging to the Paris Museum, collected at Nineveh by M. de Saulcy.

e-f. ♂ & yg.	Hydrabad, Sind.	R. T. Riddell, Esq. [P.]. (Types of *E. fasciolatus*.)
g. ♂.	Sind.	Dr. Leith [P.].
h. ♂.	Kurrachee.	Dr. Leith [P.].
i. ♂.	Kandesh.	Dr. Leith [P.].

3. Eublepharis dovii.

Habit rather slender, as in *Coleonyx elegans*. Snout as long as the distance between the orbit and the ear-opening; latter rather large, oval, vertical. Head covered with granules intermixed with round subconical tubercles; rostral pentagonal, not quite twice as broad as high; a pair of large internasals forming a suture behind the rostral; seven upper and as many lower labials; mental large, squarish, as broad as high; no chin-shields. Body covered above with small granules intermixed with round, mostly trihedral, tubercles, which are not larger than the interspaces between them. Abdominal scales strongly imbricate. Seven præanal pores. Tail cylindrical, tapering gradually, covered with flat, squarish scales arranged regularly, those on the lower surface larger, above with rows of enlarged tubercles. Brownish white above, marbled with chocolate-brown, which colour forms irregular transverse bands on the tail; a rather indistinct horseshoe-shaped mark from eye to eye, over the nape; lower surfaces whitish.

Total length	148	millim.
Head	20	,,
Width of head	13	,,
Body	50	,,
Fore limb	24	,,
Hind limb	30	,,
Tail	78	,,

Panama.

a. ♂.	Panama.	Capt. T. C. Dow [P.].

4. Eublepharis variegatus*.

Stenodactylus variegatus, *Baird, Proc. Ac. Philad.* 1858, p. 254, *and U. S. Mex. Bound. Surv., Rept.* pl. xxiii. figs. 9-27, & xxiv. figs. 11-19.

Coleonyx variegatus, *Cope, Proc. Ac. Philad.* 1866, p. 125.

Habit of *Coleonyx elegans*. Snout as long as the distance between the orbit and the ear-opening; latter small, oval, very oblique, almost horizontal. Head and body covered with uniform granules, which are largest on the snout; rostral pentagonal, broader than high; anterior nasal largest, separated from its fellow by a granule

---

* Described from a female specimen, from Texas, in M. F. Lataste's collection.

behind the rostral; eight upper and as many lower labials to below the centre of the eye; mental large, subtrapezoid, slightly broader than long, longer than the adjacent labials. Abdominal scales strongly imbricate. Tail cylindrical, tapering, covered with uniform, small, imbricate scales, which are slightly enlarged inferiorly. Pale brown above, with round darker spots and six light wavy cross bands on the neck and back; the two anterior bands on the nape horseshoe-shaped, reaching to below the eye and to below the ear respectively; tail with alternate dark and light annuli of equal width; lips with brown and white spots; lower surfaces white, throat spotted with brown.

Total length	107	millim.
Head	12	,,
Width of head	8	,,
Body	39	,,
Fore limb	17	,,
Hind limb	22	,,
Tail	56	,,

Texas and California.

5. Eublepharis fasciatus.

Closely allied to the preceding, from which it differs in the larger and more elongate head, the length of the snout exceeding the distance between the orbit and the ear-opening. Six labials to below the centre of the eye. Brown above, with five or six black-edged, whitish, transverse bands, the anterior on the nape horseshoe-shaped, extending to the eye; tail with alternate, broader dark, and narrower light annuli; lower surfaces whitish.

Total length	113	millim.
Head	16	,,
Width of head	9	,,
Body	40	,,
Fore limb	20	,,
Hind limb	25	,,
Tail	57	,,

Mexico.

a. ♀. Ventanas. Hr. A. Forrer [C.].

3. COLEONYX.

Coleonyx, *Gray, Ann. & Mag. N. H.* xvi. 1845, p. 162; *Bocourt, Miss. Sc. Mex., Rept.* p. 49.
Brachydactylus, *Peters, Mon. Berl. Ac.* 1863, p. 41.

Digits with a row of lamellæ inferiorly; claws concealed into very large compressed sheaths.

Central America.

1. Coleonyx elegans.

Coleonyx elegans, *Gray, l. c.* p. 163; *Cope, Proc. Ac. Philad.* 1866, p. 125; *Bocourt, l. c.* p. 49, pl. x. fig. 7.
Gymnodactylus scapularis, *A. Dum. Cat. Méth. Rept.* p. 45.
—— coleonyx, *A. Dum. Arch. Mus.* viii. p. 483, pl. xvi. fig. 6.
Brachydactylus mitratus, *Peters, l. c.* p. 42.

Body and limbs rather elongate, latter slender; digits slender, short. Snout longer than the distance between the orbit and the ear-opening; latter large, oval, slightly oblique. Snout covered with large, slightly keeled granules: the rest of the head with small granules intermixed with small round tubercles; rostral large, pentagonal, once and a half as broad as high; anterior nasal very large, forming a suture with its fellow; six or seven upper and seven or eight lower labials; mental large, trapezoid, a little broader than long; no chin-shields. Body covered above with small granules intermixed with numerous round, smooth or very slightly keeled tubercles. Abdominal scales strongly imbricate. Male with eight or nine præanal pores. Tail cylindrical, tapering, covered with small, flat scales, slightly enlarged inferiorly, above with rows of enlarged tubercles. Brownish white above, variegated with chestnut-brown; a U-shaped band on the nape, and four cross bands on the body, light, bordered with brown; head largely vermiculated with brown; tail with brown rings; lower surfaces whitish.

Total length	124	millim.
Head	17	,,
Width of head	11	,,
Body	50	,,
Fore limb	23	,,
Hind limb	28	,,
Tail	57	,,

Specimen *b*, in which the tail is lost, measures 87 millim. from snout to vent.

Central America.

a. ♂.	Belize.	Mr. Dyson [C.].	(Type.)
b. ♂.	Yucatan.		

Fam. 3. UROPLATIDÆ.

Geckonidæ, part., *omn. auct.*
Uroplatidæ, *Boulenger, Ann. & Mag. N. H.* (5) xiv. 1884.

Tongue moderately elongate, simply papillose, slightly nicked anteriorly. Dentition strictly pleurodont; teeth numerous, small, obtusely conical, with long cylindrical shafts. No pterygoid teeth. Skull thin, much depressed; a rather strong ligamentous postorbital arch; no frontosquamosal arch; premaxillary single; nasal single; frontal single; parietals two. Vertebræ amphicœlian; abdominal ribs. Limbs well developed. Clavicles slender, not dilated proximally; interclavicle minute. Skin of head free from the skull; teguments soft, granular and tubercular.

This family contains a single genus, the aberrant *Uroplates* of Madagascar, which combines with a Geckoid structure a peculiar sternal apparatus and the union of the nasal bones.

1. UROPLATES*.

Uroplates, *Gray, Ann. Phil.* (2) x. 1825, p. 198, *and Cat. Liz.* p. 151.
Uroplatus, *Fitzing. N. Classif. Rept.* p. 13.
Sarruba, *Fitzing. l. c.*
Thecadactylus, part., *Gray, Phil. Mag.* (2) iii. 1827, p. 54.
Rhacoëssa, *Wagl. Syst. Amph.* p. 142; *Fitzing. Syst. Rept.* p. 97.
Chiroperus, *Wiegm. Herp. Mex.* p. 20; *Fitzing. l. c.* p. 96.
Ptyodactylus, part., *Dum. & Bibr.* iii. p. 375.
Oiacurus (*non Leuck.*), *Fitzing. l. c.* p. 97.
Lonchurus, *Fitzing. l. c.*
Caudiverbera, *Gray, Cat.* p. 152.

Digits depressed, more or less webbed, with very small equal scales inferiorly, the extremity strongly dilated, with two diverging series of lamellæ inferiorly; all the digits clawed, the claw retractile in the anterior notch of the distal expansion. Body covered with small juxtaposed scales. Pupil vertical. No eyelids. Ear-opening distinct. No præanal nor femoral pores.

Madagascar.

---

* The Gecko described by Feuillée in 1714 (Journ. des Observ. phys.-math. &c. i. p. 319) is probably mythical. Its synonymy runs as follows:—

Lacerta caudiverbera, *Linn. S. N.* i. p. 359.
Caudiverbera peruviana, *Laur. Syst. Amph.* p. 43; *Gray, Cat.* p. 152.
Gecko cristatus, *Daud. Rept.* iv. p. 167.
— — caudiverbera, *Merr. Syst. Amph.* p. 40.
Ptyodactylus feuillæi, *Dum. & Bibr.* iii. p. 386
Oiacurus feuillæi, *Fitz. Syst. Rept.* p. 97.

Hab. Chili.

Synopsis of the Species.

A strong denticulated fold bordering the body
and limbs .... 1. *fimbriatus*, p. 237.
Limbs without folds or spines; upper eyelid
with a spine-like scale .... 2. *lineatus*, p. 238.
Head and limbs with spine-like scales; tail
not longer than the head .... 3. *ebenaui*, p. 238.

1. Uroplates fimbriatus*.

Uroplates fimbriatus, *Gray, Cat.* p. 151.
Flacourt, *Hist. Madag.* p. 155.
Stellio fimbriatus, *Schneid. Amph. Phys.* ii. p. 32.
Gecko fimbriatus, *Latr. Rept.* ii. p. 54; *Daud. Rept.* iv. p. 160, pl. lii.
Ptyodactylus fimbriatus, *Dum. & Bibr.* iii. p. 381.
—— (Rhacoëssa) fimbriatus, *Fitzing. Syst. Rept.* p. 97.
Rhacoëssa hypoxantha, *Wagl. Icon. Descr. Amph.* pl. xxxvi.

Head large, depressed; snout much longer than the distance between the eye and the ear-opening, once and two thirds the diameter of the orbit; a distinct canthus rostralis; interorbital space deeply concave; ear-opening very small, oval, vertical. Body depressed. Limbs slender, very long; digits half-webbed. A denticulated dermal fold bordering the lower jaw, the neck, the body, and the limbs. Upper surfaces covered with small, irregular-sized, juxtaposed, flat scales; throat with minute granules; abdominal scales very small, flat, subhexagonal, juxtaposed. Rostral small, subquadrangular; nostril pierced between numerous small scales, separated from the rostral by three rows of scales; labials minute, about forty upper and thirty-five lower; mental minute; no chin-shields. Tail short, depressed, covered with small juxtaposed scales, surrounded by a broad membrane with rounded outline. Reddish brown above, variegated with darker; lower surfaces dirty white, uniform.

Total length	234	millim.
Head	49	,,
Width of head	34	,,
Body	120	,,
Fore limb	74	,,
Hind limb	92	,,
Tail	65	,,

Madagascar.

a, b, c, d. Ad. & hgr.   Madagascar.
e. Skeleton.   Madagascar.

---

* The following synonyms must also probably be referred to this species:—
Stellio tetradactylus, *Schneid. l. c.* p. 33.
Gecko sarroubea, *Daud. l. c.* p. 176.
—— tetradactylus, *Merr. Tent.* p. 41.
Ptyodactylus sarrube, *Fitz. Syst. Rept.* p. 96.

2. Uroplates lineatus.

Uroplates lineatus, *Gray, Cat.* p. 152.
Ptyodactylus lineatus, *Dum. & Bibr.* iii. p. 384, pl. xxxi. figs. 1-3 *;
A. Dum. Cat. Méth. Rept. p. 40.
Oiacurus (Lonchurus) lineatus, *Fitzing. Syst. Rept.* p. 97.

Differs from *U. fimbriatus* in the following characters:—Head narrower; interorbital space not concave; upper eyelid with two projecting spine-like scales. Limbs rather shorter, without cutaneous fringe. A slight, not denticulated, dermal fold from axilla to groin. About twenty-five upper and as many lower labials. Tail rounded, bordered by a narrower membrane, narrowing posteriorly. Head and back with five or six parallel blackish longitudinal lines.

Madagascar.

3. Uroplates ebenaui.

Uroplates ebenaui, *Boettger, Abh. Senck. Ges.* xi. 1878, p. 273, pl. i. fig. 1.
—— boettgeri, *Fischer, Jahresb. Naturh. Mus. Hamb. f.* 1883, p. 33, pl. iii. fig. 9.

Head large, very convex posteriorly; snout not much longer than the distance between the eye and the ear-opening; latter very small. Body not depressed. Limbs shorter, and interdigital web less developed than in *U. fimbriatus.* Upper surfaces with small granular scales and soft spine-like scales or lobes, the largest of which are at the posterior corner of the upper eyelid, a pair on each side on the posterior border of the head, and one at the knee. Tail not longer than the head, with a dermal lobe or a series of spine-like scales on each side. Light brownish grey above, with darker markings; the spines and oblique lines on the back and limbs white.

Total length	6 2·5	millim.
Head	17	,,
Width of head	13	,,
Body	30	,,
Tail	15·5	,,

Nossi Bé.

---

* The figures represent the digits nearly free, whilst the description says "Pattes demi-palmées."

Fam. 4. **PYGOPODIDÆ**.

Scincoïdiens, part., *Cuvier, Règne Anim.* ii. 1817.
Gymnophthalmoidea, part., *Fitzinger, Neue Classif. Rept.* 1826.
Autarchoglossæ, part., *Wagler, Syst. Amph.* 1830.
Gymnophthalmi, part., *Wiegmann, Herp. Mex.* 1834.
Scincoïdiens, part., *Duméril & Bibron, Erp. Gén.* v. 1839.
Pygopidæ, *Gray, Cat. Liz.* 1845.
Aprasiadæ, *Gray, l. c.*
Lialisidæ, *Gray, l. c.*
Pygopodidæ, *Boulenger, Ann. & Mag. N. H.* (5) xiv. 1884.

The premaxillary is single, narrowed, and much produced posteriorly between the nasals, in the long-snouted *Lialis* quite as much as in the *Varanidæ*; the nasals are distinct; the frontal is single; the præ- and postfrontals are in contact, separating the frontal from the orbit; the parietals remain distinct, except in *Lialis*; the jugal is rudimentary, there being no postorbital arch; a postfronto-squamosal arch is also absent; the pterygoids are widely separated and toothless. The mandible contains only four bones, the angular, supra-angular, and articular having coalesced. The dentition is pleurodont. The teeth are small, numerous, and closely set; in *Lialis* they are recurved posteriorly, very acute, and swollen at the base, thus resembling those of the *Varanidæ*, whilst in the other genera they do not diverge from the normal pleurodont type, being obtusely pointed and with long cylindrical shafts. The skin of the head is quite free from the cranial ossification, and there are no supraorbital bones. The serpentiform body is destitute of functional limbs; the fore limb is entirely absent, while the hind pair is visible externally as a scaly flap, most developed in *Pygopus*, in which the bones of the limbs may be felt more or less distinctly; when the skin is removed, in *Pygopus*, the foot, with five ossified toes, is seen most plainly, especially in the males; the ischium appears externally as a small spur on each side behind the anal cleft. The sternal apparatus exists in a rudimentary state. The tail is long and fragile. The eyes are rather small, with elliptico-vertical or subelliptical pupil, and not protected by movable lids. The ear is either exposed or concealed under the scales. The tongue is fleshy, papillose, elongate, more or less feebly incised anteriorly, and extensible. The body is covered with roundish imbricate scales, and the head is more or less regularly plated. Præanal pores are frequently present.

To place this family in the system is a matter of some difficulty. Though formerly associated with, or placed near, the Skinks, the Pygopods have nothing in common with them except superficial appearance. The structure of the skull is most similar to that of Geckos, but differs in two points:—(1) the separation of the frontal from the orbit by the union of the præ- and postfrontal, a character which is shared by *Heloderma*; (2) the reduction of the number of bones in the mandible, in which respect they resemble the Snakes. The vertical pupil, though of secondary importance, deserves special notice, for it occurs in none of the other Skink-like Lizards.

Synopsis of the Genera.

I. Parietal bones distinct; head with large symmetrical shields.

A. Præanal pores.

Scales keeled .... 1. **Pygopus**, p. 240.

Scales smooth .... 2. **Cryptodelma**, p. 242.

B. No præanal pores.

Scales smooth; parietal plates large; two rows of enlarged ventral plates .... 3. **Delma**, p. 243.

Scales bicarinate .... 4. **Pletholax**, p. 245.

Scales smooth, subequal; no parietal plates; ear concealed.
5. **Aprasia**, p. 245.

II. Parietal bone single; head covered with small scales.
6. **Lialis**, p. 246.

1. PYGOPUS.

Sheltopusik (*non Latr.*), *Oppel, Ordn.* p. 40.
Pygopus, *Merr. Tent. Syst. Amph.* p. 77; *Fitzing. N. Class. Rept.* p. 26; *Wagl. Syst. Amph.* p. 160; *Wiegm. Herp. Mex.* p. 11; *Gray, Cat.* p. 67.
Hysteropus, *Dum. & Bibr.* v. p. 826.

Parietal bones distinct. Tongue slightly nicked at the tip, with rows of large round papillæ inferiorly. Ear exposed. Rudiments of hind limbs externally. Head with large symmetrical plates. Scales cycloid-hexagonal, imbricate, those on the back keeled, the two median series on the belly, and the median series under the tail transversely enlarged, hexagonal. Præanal pores.

Australia.

1. Pygopus lepidopus.

Pygopus lepidopodus, *Gray, Cat.* p. 67.
Pygopus squamiceps, *Gray, l. c.* p. 68.
Bipes lepidopodus, *Lacép. Ann. Mus.* iv. 1804, p. 209, pl. lv. fig. 1; *Guérin, Icon. R. A., Rept.* pl. lxi. fig. 1; *Duvern. R. A., Rept.* pl. xxii. bis. fig. 2.
Sheltopusik novæ-hollandiæ, *Oppel, l. c.*
Pygopus lepidopus, *Merr. Tent.* p. 77; *Günth. Ann. Mag. N. H.* (3) xx. 1867, p. 45.
Hysteropus novæ-hollandiæ, *Dum. & Bibr.* v. p. 828, pl. lv.
Pygopus squamiceps, *Gray, Zool. Erebus and Terror, Rept.* pl. viii. fig. 3.

Snout scarcely prominent, rounded, as long as the distance between the orbit and the ear-opening; canthus rostralis obtuse; eye small, with rudimentary circular scaly lid; ear-opening oval, oblique. Tail, when intact, at least twice as long as the body. Rudimentary hind limbs measuring about the distance between eye and end of snout in females, more than the distance between the posterior border of the eye and the end of the snout in males. Ten to fourteen præanal pores. Rostral low, from twice and a half to thrice and a half as broad as high; nostril between the first labial and three nasals, the two anterior of which are band-like and extend across the upper surface of the snout, where they form a suture with their fellows, or are separated by one or two small azygos plates; a large polygonal præfrontal, separated from the nasals by two (or one) pairs of small transverse plates, its transversely truncate posterior border forming a suture with the frontal, which is pentagonal and about once and two thirds as long as broad; the posterior angle of the latter plate wedged in between the pair of parietals, which are nearly as large as the frontal, and subhexagonal; sometimes a narrow band-like plate on the outer side of the parietals; two large supraorbitals; loreal region with numerous small polygonal plates, from four to seven in a row, between the orbit and the nasal; five to seven upper labials, separated from the orbit by a row of scales; mental large, broadly trapezoid; four to six lower labials, the first or the two first much dilated vertically. Keels of the dorsal scales forming regular lines on the body, alternate on the tail. Twenty-two or twenty-three (in one specimen twenty-one) longitudinal series of scales round the middle of the body, ten smooth and twelve or thirteen (or eleven) keeled. The enlarged ventral scales twice as broad as long, in 70 to 85 longitudinal series. Two enlarged anal scales separated from the perforated præanal scales by one or two rows of scales. Coppery grey above, uniform or with three or five longitudinal series of blackish dots or elongate quadrangular spots; lower surfaces more or less marbled or pulverated with grey.

	♂	♀
Head	16	16 millim.
Width of head	10	10 ,,
Body	165	155 ,,
Tail	400	345 ,,
Hind limb	11	6 ,,

Australia and Tasmania.

a, b–d. ♂ & hgr.	Champion Bay, N.W. coast of Australia.	Mr. Duboulay [C.].
e–f. ♂ ♀.	Swan River.	Sir J. Richardson [P.].
g. ♀.	Garden Island, Sydney.	J. B. Jukes, Esq. [P.].
h. ♀.	Sydney.	R. Schütte [C.].
i, k, l, m, n. ♂, ♀, & yg.	N. S. Wales.	G. Krefft, Esq. [P.].
o. ♀.	Australia.	Dr. Mair [P.]. (Type of *P. squamiceps*.)

p–q, r–s, t–u. ♂, ♀, & yg.	Australia.	Dr. Fletcher [P.].
v. ♀.	Australia.	Rev. N. Wilton [P.].
w, x, y, z, a, β–γ, δ–ϵ. ♂, ♀, hgr. & yg.	Australia.	
ζ. Hgr.	Tasmania.	R. Gunn, Esq. [P.].

2. CRYPTODELMA.

Cryptodelma, *Fischer, Arch. f. Nat.* xlviii. 1882, p. 289.

Parietal bones distinct. Tongue slightly nicked at the tip, with rows of large round papillæ inferiorly. Ear exposed. Rudiments of hind limbs externally. Head with large symmetrical plates. Scales smooth, cycloid-hexagonal, imbricate, the two median series on the belly and the median series under the tail transversely enlarged, hexagonal. Præanal pores.

Australia.

1. Cryptodelma nigriceps.

Cryptodelma nigriceps, *Fischer, l. c.* p. 290, pl. xvi. figs. 5–9.

Tail as long as the head and body. Eleven præanal pores. Rostral broad, low; nostril pierced between the first labial and two nasals, the anterior of which is largest and forms a suture with its fellow on the snout; præfrontal broken up in the unique specimen, separated from the nasals by two pairs of transverse plates; parietals narrow, elongate; supraorbitals two, rather small; loreal region with numerous small plates, about six in a row between the eye and the nasal; seven upper labials, separated from the orbit by a row of scales; mental very large; eight lower labials, first vertically enlarged. Twenty-six or twenty-eight longitudinal series of scales round the body. Two large anal scales; four rows of small scales between the latter and the perforated præanal scales. Flesh-coloured above, the head black.

The single young specimen measures, from snout to vent, 64 millim., tail 60 millim.

Nicol Bay, West Australia.

2. Cryptodelma orientalis. (Plate XIX.)

Delma orientalis, *Günth. Journ. Mus. Godeffr.* xii. 1876, p. 45.

Snout not prominent, rounded, as long as the distance between the orbit and the ear-opening; canthus rostralis rounded; eye small, without distinct circular lid; ear-opening small, suboval, oblique. Tail as long as, or not very much longer than, the body. Rudimentary hind limbs slightly developed, measuring hardly two thirds the length of the snout, four præanal pores. Rostral pentagonal, about twice as broad as high; nostril between the first labial and two nasals, the anterior of which is largest and forms a suture with its fellow on the snout; a pair of fronto-nasals; præfrontal large, seven-sided, the antero-lateral sides very short, in contact with the upper loreal; frontal much smaller than the præfrontal or parietals, broader than long, seven-sided; parietals subpentagonal; two large supraorbitals; four small plates in a row between the eye and the nasal; six upper labials, separated from the orbit by a row of scales; mental trapezoid; four or five lower labials, the two first much dilated vertically. Eighteen longitudinal series of scales round the middle of the body. The enlarged ventral scales twice as broad as long, in 97 to 109 pairs. Two enlarged anal scales and a row of four scales between the latter and the perforated præanal scales. Brown above, with a darker line along each series of scales; occipital region yellowish, edged behind with a blackish band; lower surfaces yellowish.

Head	13 millim.
Width of head	9·5 ,,
Body	185 ,,
Tail	175 ,,
Hind limb	5 ,,

New South Wales.

a–b. Ad. & hgr.	Peak Downs.	Godeffroy Museum.	(Types.)
c. Ad.	Gayndah.	Godeffroy Museum.	

3. DELMA.

Delma, *Gray, Zool. Misc.* p. 14, *and Cat.* p. 68.
Nisara, *Gray, Liz. Austr.* p. 3.
Pseudodelma, *Fischer, Arch. f. Nat.* xlviii. 1882, p. 286.

Parietal bones distinct. Tongue slightly nicked at the tip, with rows of large round papillæ inferiorly. Ear exposed. Rudiments of hind limbs externally. Head with large symmetrical plates. Scales smooth, cycloid hexagonal, imbricate, the two median series on the belly and the median series under the tail transversely enlarged, hexagonal. No præanal pores.

Australia.

1. Delma fraseri.

Delma fraseri, *Gray, Cat.* p. 68.
Delma fraseri, *Gray, Zool. Misc.* p. 14, *and in Grey's Trav. Austral.* ii. p. 427, pl. iv. fig. 3; *Günth. Ann. & Mag. N. H.* (4) xii. 1873, p. 145.
—— grayii, *Smith, Ill. S. Afr., Rept.* pl. lxxvi. fig. 2.
—— mölleri, *Lütken, Vidensk. Meddel.* 1862, p. 296, pl. i. fig. 2.
Nisara grayii, *Gray, Liz. Austr.* p. 3.

Snout not prominent, as long as the distance between the orbit and the ear-opening; canthus rostralis obtuse; eye with distinct circular scaly lid; ear-opening elliptical, oblique, its diameter equal to that of the eye. Tail, when intact, three or four times as long as the body. The rudimentary hind limbs measure about the length of the snout in males, considerably less in females. Rostral triangular or pentagonal, nearly twice as broad as high; nostril pierced between the first nasal and three nasals (two in the specimen described as *D. mölleri*, in which the naso-rostral and upper nasal have fused), the two anterior of which form a suture with their fellows on the snout; exceptionally, however, the upper nasal is separated from the nostril; a pair of fronto-nasals; præfrontal large, a little broader than long, seven-sided, the antero-lateral sides very short, in contact with a large loreal; frontal as broad as or a little narrower than the præfrontal, longer than broad, seven-sided, its posterior angle wedged in between the pair of parietals, which are considerably larger than the frontal; a pair of enlarged scales on the outer side of the parietals; two large supraorbitals; a large loreal, and four or five small plates between the orbit and the nasal; five or six upper labials, fourth much elongate and situated under the orbit, from which it is separated by a row of small scales; mental large, triangular, broader than long; four lower labials, the two anterior much dilated vertically, the first forming a suture with its fellow behind the mental. Sixteen longitudinal rows of scales round the middle of the body. The enlarged ventral scales vary considerably in width, being sometimes not quite twice as broad as long, whilst in most specimens they are more than twice as broad as long; they form 45 to 60 pairs. Two large and a smaller median anal scales. Olive above; head generally with four more or less confluent black cross bands, which may be separated by whitish bands; in specimens *c* and *k* these bands are indistinct, and the sides of the head and body are vertically barred with darker and whitish; specimen *i* is uniform olive, without any markings. Lower surfaces yellowish.

	♂.
Head	13 millim.
Width of head	7 ,,
Body	85 ,,
Tail	355 ,,
Hind limb	4·5 ,,

Western Australia.

a–b. Hgr.	W. Australia.	J. Hunter, Esq. [P.].	(Types.)
c–f. ♂, ♀, & hgr.	Perth.	Mr. Duboulay [C.].	
g, h. ♂ ♀.	Champion Bay.	Mr. Duboulay [C.].	
i. ♂.	Nicol Bay.	Mr. Duboulay [C.].	
k. ♂.	—— ?	Sir A. Smith [P.].	(Type of *Delma grayi*.)

2. Delma impar.

Pseudodelma impar, *Fischer, Arch. f. Nat.* xlviii. 1882, p. 287, pl. xvi. figs. 1–4.

Tail twice as long as head and body. Rudimentary limbs small.

Rostral pentagonal; nostril pierced in the lower portion of the nasal, which forms a suture with its fellow on the snout; a pair of large plates between the nasals and the præfrontal; latter seven-sided, a little larger than the frontal, which is also seven-sided and smaller than the parietals; a band-like plate on the outer side of latter; two supraorbitals; a large loreal and four small plates between the orbit and the nasal; seven upper labials, fourth elongate and situated below the orbit, from which it is separated by a row of small scales; mental large, triangular; six lower labials, the first forming a suture with its fellow behind the mental. Fifteen longitudinal rows of scales round the middle of the body. Two enlarged præanal scales. Olive-green, lighter beneath; on each side of the back two light, dark-edged longitudinal lines. From snout to vent 80 millim.; tail 167 millim.

Melbourne.

4. PLETHOLAX.

Pletholax, *Cope, Proc. Ac. Philad.* 1864, p. 229.

Rudiments of hind limbs externally. Head with large symmetrical plates. All the scales imbricated, with two keels and a groove between; no larger abdominal series. No præanal pores.

Australia.

1. Pletholax gracilis.

Pletholax gracilis, *Cope, l. c.*
Pygopus gracilis, *Günth. Zool. Erebus & Terror, Rept.* p. 10.

Rostral oval, prominent; nostril between the first labial and a nasal, which is followed by a second; one transverse præfrontal; parietals broad, acuminate, as long as frontal and præfrontal; three supraorbitals, posterior largest; temporal scales large, keeled; four lower labials, the two anterior large, following the very large mental; gular scales keeled. Sixteen rows of scales. Pale brown; a paler median dorsal band, two scales wide, bordered with dark brown.

South-west Australia.

5. APRASIA.

Aprasia, *Gray, Ann. & Mag. N. H.* ii. 1839, p. 331, *and in Grey's Trav. Austral.* ii. p. 438, *and Cat.* p. 68.

Parietal bones distinct. Tongue rounded and slightly nicked at the end. Ear concealed. Slight rudiments of hind limbs externally. Head with large symmetrical plates; no parietals. Scales smooth, cycloid, imbricate, those on the belly scarcely enlarged. No præanal pores.

Australia.

1. Aprasia pulchella.

Aprasia pulchella, *Gray, Cat.* p. 68.
Aprasia pulchella, *Gray, Ann. & Mag. N. H.* ii. 1839, p. 332, *and in Grey's Trav. Austral.* ii. pl. iv. fig. 2; *Lütken, Vidensk. Meddel.* 1862, p. 300, pl. i. fig. 3; *Günth. Ann. Mag. N. H.* (4) xii. 1873, p. 145.
—— octolineata, *Peters, Mon. Berl. Ac.* 1863, p. 233.

Head very small, with very prominent rounded snout; eyes well developed, with circular scaly rudimentary lid. Body calamiform. Tail shorter than the body, of subequal diameter throughout, its end obtuse, rounded. Rudiments of hind limbs extremely small, hardly distinct. Rostral very high, narrow, the portion seen from above the snout triangular; nostril pierced between the first labial and a very large nasal, which forms a suture with its fellow on the snout: a pair of large præfrontals, forming a suture with the second labial; a large hexagonal frontal, the posterior angle of which is rounded off; four or five enlarged occipital scales, but no parietals; a supraorbital; a narrow præorbital; no loreal; five or six upper labials, third and fourth entering the orbit; mental large, broadly trapezoid; two or three lower labials, anterior very large. Twelve series of scales round the body. Three slightly enlarged anal scales. Yellowish or pinkish, with eight dark-brown lines above following the longitudinal series of scales, or with series of brown dots arranged in four widely separated longitudinal series on the back and very crowded on the sides.

Head	6 millim.
Body	112 ,,
Tail	64 ,,
Diameter of body	3·5 ,,

Western Australia.

a–b.	W. Australia.	(Types.)
c.	W. Australia.	Mr. Duboulay [C.].
d.	Swan River.	Sir A. Smith [P.].
e–g, h.	Australia.	

6. LIALIS.

Lialis, *Gray, Proc. Zool. Soc.* 1834, p. 134; *Dum. & Bibr.* v. p. 830; *Gray, Cat.* p. 69.

Parietal bones coalesced. Teeth sharply pointed, directed backwards. Tongue elongate, narrowing towards the end, bifid. Ear exposed. Slight rudiments of hind limbs externally. Head covered with small plates. Scales soft, smooth, cycloid, imbricate, the two median series on the belly and the median series under the tail transversely enlarged, hexagonal. Præanal pores.

Australia and New Guinea.

1. Lialis burtonii.

Lialis burtonii, *Gray, Cat.* p. 69.
Lialis bicatenata, *Gray, l. c.*
Lialis punctulata, *Gray, l. c.*

Lialis burtonii, *Gray, Proc. Zool. Soc.* 1834, p. 134; *Dum. & Bibr.* v. p. 831; *Gray, in Grey's Trav. Austral.* ii. p. 437, pl. iii. fig. 1, *and Zool. Misc.* p. 52, *and Zool. Ereb. & Terr., Rept.* p. 5, pl. viii. fig. 2; *A. Dum. Cat. Méth. Rept.* p. 194; *Günth. Ann. & Mag. N. H.* (3) xx. 1867, p. 46.

—— bicatenata, *Gray, Zool. Misc.* p. 52, *and Zool. Ereb. & Terr.* p. 5; *Peters, Mon. Berl. Ac.* 1873, p. 606.

—— punctulata, *Gray, Zool. Misc.* p. 52, *and Zool. Ereb. & Terr.* p. 5, pl. viii. fig. 1; *Günth. l. c.*

—— leptorhyncha, *Peters, l. c.* p. 605.

Snout narrow, depressed, long, acuminate, truncate at the tip, with angular canthus rostralis; eye small, with circular scaly rudimentary lid; ear-opening elliptical, oblique. Tail, when intact, nearly as long as head and body, gradually tapering to a fine point. Rudiments of hind limbs extremely small, scarcely distinct, especially in females. Four præanal pores, frequently indistinct in females. Snout covered with small plates, variable in number and arrangement; three supraorbitals, median large; loreal region covered with numerous small scales; the rest of the head with equal scales; rostral very low; nostril pierced in the posterior portion of a nasal; thirteen to seventeen upper labials, all very small, separated from the orbit by two or three rows of scales; mental rather large, trapezoid or pentagonal; twelve to sixteen lower labials; a series of dilated gular scales on each side, separated from the lower labials by one or two rows of scales. 19 or 21 (occasionally 20, according to Peters) longitudinal rows of scales round the middle of the body; the dilated ventral scales in 70 to 100 pairs. Three or five anal scales. Ground-colour brown, grey, reddish, or yellowish, variously marked or uniform.

Head	27	millim.
Body	220	,,
Tail	270	,,

Australia; New Guinea.

The rich series before me shows that this Lizard varies extremely in the degree of elongation of the snout, in the scutellation of the head, in the number of rows of scales, and in colour; but I am satisfied that the several forms hitherto described should be united into one species, which I divide into numerous varieties of coloration.

* A white labial band, extending along the side of the body.

† The band very narrow and bordering the lower lip only.

VAR. A. The sides of the head between the labial white line and a more or less distinct narrow line from the tip of the snout along the canthus rostralis and above the eye to the nape, dark brown; the belly darker towards the narrow lateral white band, which is edged inferiorly by a fine brown line; the rest of the body uniform, with minute dark dots. (Gray, Zool. Ereb. & Terror, pl. viii. fig. 1.) All have 21 series of scales.

a. ♀.	Port Essington.	Mr. Gilbert [C.]. (Type o *L. punctulata*.)
b. ♀.	W. Australia.	
c. ♀.	Australia.	Sir J. Richardson [P.].
d. Hgr.	Australia.	Capt. Stokes [P.].
e. ♂.	New South Wales.	G. Krefft, Esq.

(Contains a young *Amphibolurus* in its stomach.)

VAR. B. As in the preceding, but the white lateral line edged above with a darker band of the same tint as the belly, which band is again edged with a rather indistinct lighter band; belly scarcely darker than the back, with white dots or shafts. 21 series of scales.

f. ♀.	Islands of Torres Straits.	Rev. S. Macfarlane [C.].
g. ♂.	Thursday Island, Torres Straits.	H.M.S. 'Alert.'

†† The white labial band is broad and occupies both the upper and the lower lip, entering the eye.

VAR. C. In all other respects like var. A. (Doubtless the form originally described by Gray under the name of *L. burtonii*, P. Z. S. 1833.) Scales in 21 rows.

h. ♀.	——?	Haslar Collection.

VAR. D. Light grey, the lateral band passing through the eye and the lower surface of the head dark brown; the white lateral band soon splits up into several, separated by darker bands; belly with darker longitudinal bands and whitish dots. 21 rows of scales.

i–m. ♂ ♀.	Cornwallis Island.	Rev. S. Macfarlane [C.].
n–o. ♂ ♀.	Islands of Torres Straits.	Rev. S. Macfarlane [C.].

VAR. E. Five or seven regular longitudinal dark bands on the body, the central one dividing into two on the nape, uniting again on the tip of the snout. (Gray, in Grey's Austr. pl. iii. fig. 1.)

a. 21 series of scales.

p. ♀.	Houtman's Abrolhos.	

b. 19 series of scales.

q, r. ♂.	W. Australia.	
s. ♂.	Swan River.	
t. ♂.	——?	Haslar Collection.

** No white labial band.

VAR. F. Pale grey, abdomen a little darker; throat dark brown, whitish-spotted on the sides; lips with small dark vertical bars; back with several rather indistinct darker longitudinal bands; belly white-spotted. 21 series of scales.

u. ♂.	Cornwallis Island.	Rev. S, Macfarlane [C.].
v. ♂.	Port Darwin.	R. G. S. Buckland, Esq. [C.].

VAR. G. Like var. E, but with spotted lips as in var. F. (*L. burtonii*, Gray, Zool. Ereb. & Terr. pl. viii. fig. 2.) 19 series of scales.

w. ♀.	W. Australia.
x–z. ♂ & hgr.	Swan River.

VAR. H. No longitudinal bands, but generally a series of small distant black spots along each side of the back; lips spotted as in F and G; lower surfaces white-spotted. (*L. bicatenata*, Gray.)

a. 21 series of scales.

a. ♀.	Port Essington.	Mr. Gilbert [C.]. (Type of *L. bicatenata*.)
β–γ. ♂.	W. Australia.	Sir J. Macgregor [P.].
δ–ζ. ♂.	Islands of Torres Straits.	Rev. S. Macfarlane [C.].
η. Hgr.	Australia.	Sir J. Richardson [P.].
θ. ♀.	Australia.	Capt. Stokes [P.].
ι. ♂.	——?	Haslar Collection.

b. 19 series of scales.

κ. ♂.	Australia.

VAR. I. Uniformly light-coloured, with a few small dark dots above and beneath. (*L. punctulata*, var. *concolor*, Peters, *l. c.*)

a. 21 series of scales.

λ. ♂.	Cape York.	
μ–ν. ♂.	Australia.	G. Krefft, Esq.

b. 19 series of scales.

ξ. ♂.	Australia.	Capt. Stokes [P.].
ο–π. ♂ ♀.	Australia.	
ρ. ♂.	Champion Bay, N.W. coast of Australia.	Mr. Duboulay [C.].

Fam. 5. AGAMIDÆ.

Iguaniens, part., *Cuvier, Règne Anim.* ii. 1817.
Ascalabotæ, part., *Merrem, Tent. Syst. Amph.* 1820.
Stellionidæ, part., *Gray, Ann. Phil.* (2) x. 1825.
Pneustoidea, Draconoidea, Agamoidea, part., *Fitzinger, Neue Classif. Rept.* 1826.
Agamidæ, *Gray, Phil. Mag.* (2) ii. 1827.
Pachyglossæ, part., *Wagler, Syst. Amph.* 1830.
Dendrobatæ emphyodontes *and* Humivagæ emphyodontes, *Wiegmann, Herp. Mex.* 1834.
Iguaniens acrodontes, *Duméril & Bibron, Erp. Gén.* iv. 1837.
Gonyocephali, Calotæ, Semiophori, Otocryptæ, Lophuræ, Dracones, Trapeli, Stelliones, Leiolepides, Phrynocephali, *Fitzinger, Syst. Rept.* 1843.
Agamidæ, *Gray, Cat. Liz.* 1845.
Agamidæ, *Cope, Proc. Ac. Philad.* 1864.
Agamidæ *and* Uromasticidæ. *Theobald, Journ. Linn. Soc.* x. 1868.
Agamidæ, *Boulenger, Ann. & Mag. N. H.* (5) xiv. 1884.

The chief character by which the Lizards of this family are at once distinguished from their allies is the acrodont dentition. The teeth may usually be divided into three kinds, viz. incisors, canines, and molars. The latter are more or less compressed, frequently tricuspid; regular canine teeth are present, one or two on each side, in most of the genera; in *Uromastix* and *Aporoscelis*, however, these enlarged teeth are absent, and the anterior lateral teeth wearing out with age, a toothless cutting-edge is left between the molars and the incisors, which, in those two genera, unite in the adult to form a large single or divided cutting-tooth.

The skull is less depressed and more strongly ossified than in the preceding families, and postorbital and postfronto-squamosal bone-arches are well developed; in *Lyriocephalus*, as in several Iguanoids, another arch is formed by processes of the præ- and postfrontals which unite surrounding a large supraorbital fossa. The premaxillary is single, the nasals are double, and the frontal and parietal single; the pterygoids are usually widely separated and constantly toothless; the os transversum is strongly developed; a columella cranii is present. Dermal ossifications on the skull are constantly absent. The fore limbs are well developed, and, except in *Sitana*, which lacks the outer toe, pentadactyle. The clavicle is not dilated, and the interclavicle is T-shaped or anchor-shaped, frequently small; the sternum usually presents two fontanelles, which, however, are missing in *Lyriocephalus* and *Moloch*, the latter genus being especially remarkable for having the sternum divided longitudinally.

The tympanum is either exposed or concealed under the skin. The eye is small and the pupil round; eyelids well developed. The tongue is thick, entirely attached or slightly free in front, not or but slightly nicked anteriorly; it is more free, protractile, and more distinctly incised in the herbivorous genera *Lophura*, *Liolepis*, and *Uromastix*.

Femoral and præanal pores are absent in the majority of the genera; it is a remarkable fact that they exist, at least in the males, in all Australian genera but one (*Chelosania*, known as yet from a unique specimen), whereas they are missing in all others except *Uromastix* and *Liolepis*. There are no large symmetrical plates on the head or on the belly; and ornamental appendages, such as crests, gular pouches, &c., are frequently present, either in the males only or in both sexes. The tail is usually long and not fragile; it is prehensile only in the genus *Cophotis*, and some *Phrynocephali* have the curious faculty of curling upwards the extremity of that organ. The digits are usually keeled inferiorly or denticulated laterally. The shape of the body as well as the scaling vary considerably according to the genera and in adaptation to the modes of life. Generally speaking, ground Agamoids have the body depressed and arboreal compressed; but a division of the genera into terrestrial and arboreal, which has hitherto been almost generally accepted, must be given up as impracticable and unnatural. Most Agamoids are exclusively insectivorous; *Lophura*, *Liolepis*, and *Uromastix* are herbi- or frugivorous, while some species of *Agama* have a mixed diet; again, a systematic division into insectivorous and herbivorous, as has been proposed by Theobald, would be as unsatisfactory as that into terrestrial and arboreal.

Leaving out the strongly specialized genera *Draco* and *Moloch*, all the forms pass very gradually one into another in different directions, rendering a sharp generic division, and still more a serial arrangement, a matter of great difficulty.

The *Agamidæ* inhabit Africa, Asia, Australia, and Polynesia: they are most numerous in species as well as in genera in the Indian Region; in Africa they are represented by only three genera, viz. *Agama*, *Aporoscelis*, and, in the northern parts, *Uromastix*. Four species extend slightly beyond the limits of Asia and Africa into South-eastern Europe. They are absent from Madagascar and New Zealand.

Synopsis of the Genera.

I. Mouth large; teeth erect in both jaws.
 A. Incisors small, conical.
  1. No true § præanal or femoral pores.
   a. Ribs much prolonged, supporting a wing-like dermal expansion .... 1. **Draco**, p. 253.
   b. No wing-like lateral expansion.
    α. Body not depressed.
     * Four toes only .... 2. **Sitana**, p. 270.

---

§ "True" by opposition to the callous pore-like swellings of the præanal scales of the males in the genera *Agama* and *Aporoscelis*.

** Five toes.

† Tympanum hidden.

Fifth toe short, not longer than first; no dorsal crest.
.... 3. **Otocryptis**, p. 271.

Three parallel longitudinal folds on each side of the middle of the throat, curved and converging backwards, forming a U-shaped figure .... 4. **Ptyctolæmus**, p. 273.

A dorsal crest §; scales small; no fold across the throat nor in front of the shoulder.......... 5. **Aphaniotis**, p. 274.

A dorsal crest: scales very large, subequal, irregular; tail prehensile .... 7. **Cophotis**, p. 275.

No dorsal crest; a large rostral appendage, at least in the male.
.... 8. **Ceratophora**, p. 277.

No dorsal crest; dorsal scales small, intermixed with very large conical tubercles .... 10. **Phoxophrys**, p. 280.

A dorsal crest; a V-shaped gular fold; a bony supraorbital arch.
.... 11. **Lyriocephalus**, p. 281.

A dorsal crest; an oblique fold in front of the shoulder.
.... 14. **Japalura**, p. 307.

†† Tympanum exposed.

Digits not keeled inferiorly ...... 6. **Lophocalotes**, p. 274.

Snout ending in a long compressed appendage.
.... 9. **Harpesaurus**, p. 279.

A strong fold across the throat.... 12. **Gonyocephalus**, p. 282.

No fold across the throat; dorsal scales unequal; no gular pouch.
.... 13. **Acanthosaura**, p. 299.

No fold across the throat; dorsal scales large, unequal; males with a gular pouch................ 15. **Salea**, p. 312.

No fold, or a very feeble one, across the throat; dorsal scales equal.
.... 16. **Calotes**, p. 314.

No fold across the throat; scales minute, equal; a gular pouch.
.... 17. **Chelosania**, p. 331.

β. Body more or less depressed.

Tympanum exposed; males without calose præanal scales.
.... 18. **Charasia**, p. 332.

---

§ The presence in the male of a dorsal crest is indicated in the female by a regular vertebral series of strongly keeled scales.

Tympanum exposed; males with calose præanal scales.
19. **Agama**, p. 334.

Tympanum concealed .... 20. **Phrynocephalus**, p. 369.

2. True præanal or femoral pores, at least in the males.
Body depressed; tympanum distinct; femoral and præanal pores.
21. **Amphibolurus**, p. 380.

Body depressed; tympanum hidden. 22. **Tympanocryptis**, p. 392.

Body slightly depressed; no femoral pores.
23. **Diporophora**, p. 393.

Body compressed; toes denticulated laterally.
24. **Physignathus**, p. 395.

Body slightly compressed; neck with a large frill-like expansion.
25. **Chlamydosaurus**, p. 401.

Body compressed; toes lobate .... 26. **Lophura**, p. 402.

Body depressed; no præanal pores. 27. **Liolepis**, p. 403.

B. Incisors united into one or two large cutting-teeth; tail short, with whorls of spines.

Femoral and præanal pores ...... 28. **Uromastix**, p. 405.

No true pores .... 29. **Aporoscelis**, p. 410.

II. Mouth very small; teeth in the upper jaw horizontal, directed inwards; body covered with large spines.
30. **Moloch**, p. 411.

1. DRACO.

Draco, *Linn. S. N.* i. p. 358, *et omn. auct.*
Dracunculus, *Wiegm. Herp. Mex.* p. 14; *Gray, Cat.* p. 235.
Rhacodracon, *Fitzing. Syst. Rept.* p. 50.
Pterosaurus, *Fitz. l. c.* p. 51.
Pleuropterus, *Fitz. l. c.*
Dracontoidis, *Fitz. l. c.*
Dracocella, *Gray, Cat.* p. 234.

Body depressed, with a large lateral wing-like membrane, supported by the much-produced five or six posterior ribs, folding like a fan. A gular appendage, and a lateral smaller one on each side. Tympanum distinct or covered with scales. Tail long. No femoral or præanal pores.

East Indies.

Synopsis of the Species.

I. Nostril lateral, directed outwards.
　A. The adpressed hind limb reaches at the utmost slightly beyond the elbow of the adpressed fore limb; tympanum naked.

1. No orbital spines.

Tympanum smaller than the eye-opening; wing-membranes black-spotted below.	1. *volans*, p. 256.
Tympanum smaller than the eye-opening; wing-membranes unspotted below ..	2. *reticulatus*, p. 257.
Tympanum as large as the eye-opening; snout a little longer than the diameter of the orbit......................	3. *guentheri*, p. 257.

2. A large spine-like scale on the supraciliary border.

The largest median dorsal scales at least twice as large as the ventrals; tympanum as large as the eye-opening ..	4. *everetti*, p. 258.
Dorsal scales hardly larger than the ventrals; tympanum smaller than the eye-opening	5. *cornutus*, p. 258.

B. The adpressed hind limb reaches at least to halfway between the elbow of the adpressed fore limb and the axil.

1. The largest median dorsal scales at least twice as large as the ventrals.

    a. Tympanum, if distinct, smaller than the eye-opening.

Wing-membranes above marbled or reticulated with black, enclosing round light spots	6. *ornatus*, p. 259.
Wing-membranes above with small round black spots; snout not longer than the diameter of the orbit	7. *spilopterus*, p. 260.
Wing-membranes above with small round black spots; snout considerably longer than the diameter of the orbit	8. *rostratus*, p. 261.

   b. Tympanum naked, as large as the eye-opening.

A series of strongly enlarged keeled scales along each side of the vertebral line, separated by a row of small smooth scales	9. *timorensis*, p. 261.

2. Dorsal scales little larger than, or as large as, the ventrals.

    a. On each side of the back a series of enlarged, keeled, distant scales.

Tympanum naked; the male's gular appendage much longer than the head.	10. *maculatus*, p. 262.
Tympanum scaly; the male's gular appendage shorter than the head	11. *bimaculatus*, p. 263.

1. DRACO. 255

b. No enlarged lateral scales.

Wing-membranes above dark brown or with dark brown and light cross bands, with white longitudinal lines ...... 12. *lineatus*, p. 264.
Wing-membranes above yellowish, with a black longitudinal band near their outer margin .... 13. *beccarii*, p. 264.
Wing-membranes above yellowish, with small black spots on their basal half .. 14. *spilonotus*, p. 265.

3. Dorsal scales all smaller than the ventrals.

A feeble caudal crest. .... 15. *fimbriatus*, p. 265.
A caudal crest composed of long distant scales; the male's gular appendage with a brown anterior margin ...... 16. *cristatellus*, p. 266.

II. Nostril pierced vertically, directed upwards.

A. Tympanum naked.

1. The adpressed hind limb does not reach beyond the axil.

Tympanum smaller than the eye-opening; wing-membranes black above with round light spots, inferiorly immaculate; male without nuchal fold, with a large black spot on each side of the throat........................ 17. *hæmatopogon*, p. 267.
Tympanum smaller than the eye-opening; wing-membranes above marbled with dark brown, with lighter spots and lines, inferiorly immaculate; male with a nuchal fold .... 18. *blanfordii*, p. 267.
Tympanum as large as the eye-opening; wing-membranes black above with round light spots, inferiorly with a series of large black spots near the margin .... 19. *dussumieri*, p. 268.

2. The adpressed hind limb reaches beyond the axil.

Wing-membranes above with five transverse black bands, inferiorly without markings .... 20. *tæniopterus*, p. 269.

B. Tympanum scaly.

Wing-membranes above and below with [p. 269.
five regular transverse black bands .. 21. *quinquefasciatus*,

1. Draco volans.

Draco volans, *Gray, Cat.* p. 233.
Draco volans, *Linn. S. N.* i. p. 358; *Latr. Rept.* ii. p. 3; *Cantor, Cat. Mal. Rept.* p. 37; *Günth. Rept. Brit. Ind.* p. 124; *Stoliczka, Journ. As. Soc. Beng.* xxxix. 1870, p. 182.
—— præpos, *Linn. l. c.*
—— major, *Laur. Syn. Rept.* p. 50.
—— minor, *Laur. l. c.* p. 51.
—— viridis, *Daud. Rept.* iii. p. 301, pl. xli.; *Kuhl, Beitr. Zool. u. Vergl. Anat.* p. 102; *Schleg. Abbild.* p. 89, pl. xxiv. fig. 1.
—— fuscus, *Daud. l. c.* p. 307; *Kuhl, l. c.*
—— daudinii, *Dum. & Bibr.* iv. p. 451.

Head small; snout short, as long as, or scarcely longer than, the diameter of the orbit; nostril lateral, directed outwards; tympanum naked, smaller than the eye-opening. Upper head-scales very unequal, strongly keeled; a more or less distinct ⋀-shaped series of enlarged scales on the forehead; a small subconical tubercle at the posterior corner of the orbit; seven to ten upper labials, the last twice or thrice as large as the preceding. The male's gular appendage longer than the head. Male with a small nuchal crest. Dorsal scales irregular, keeled, the largest at least twice as large as the ventrals; on each side of the back a series of enlarged keeled distant scales. The fore limb stretched forwards extends beyond the tip of the snout; the adpressed hind limb reaches the elbow of the adpressed fore limb or not quite so far. Upper surfaces metallic, with small dark spots and undulated cross bands; a black spot between the orbits and another on the nape; wing-membranes orange, with black marblings or irregular cross bands; throat speckled with black; gular appendage orange in the male, bluish in the female.

Total length	210	millim.
Head	15	,,
Width of head	11	,,
Body	70	,,
Fore limb	30	,,
Hind limb	36	,,
Tail	125	,,

Malay peninsula; Sumatra; Java; Borneo.

a–c, d. ♂, ♀, & yg.	Penang.	Dr. Cantor.
e. ♂.	Singapore.	Gen. Hardwicke [P.].
f–i. ♂ ♀.	Nias.	Hr. H. Sandemann [C.].
k, l. ♀.	Sumatra.	H. O. Forbes, Esq. [C.].
m. ♀.	Java.	J. B. Jukes, Esq. [P.].
n–q. ♂, ♀, & yg.	Java.	C. Bowring, Esq. [P.].
r. ♀.	Java.	Dr. Ploem [C.].
s. ♂.	Java.	Leyden Museum.
t. Several spec.: ♂ ♀.	Java.	
u. ♀.	Borneo.	Sir E. Belcher [P.].
v–x. ♂, ♀, & hgr.	Borneo.	Rev. R. T. Lowe [P.].
y–z, a–β. ♂ ♀.	Borneo.	
γ, δ–ε. ♂.	East Indies.	East-India Company.
ζ. Skeleton.	East Indies.	

2. Draco reticulatus. (Plate XX. fig. 1, the head.)

Draco reticulatus, *Günth. Rept. Brit. Ind.* p. 125; *Peters & Doria, Ann. Mus. Genov.* xiii. 1878, p. 374.

Head small; snout as long as the diameter of the orbit; nostril directed outwards and slightly upwards; tympanum naked, smaller than the eye-opening. Upper head-scales very unequal, keeled; a ⋏-shaped series of enlarged scales on the forehead; a compressed prominent scale on the posterior part of the supraciliary region; eight upper labials, the last twice as large as the preceding. The male's gular appendage as long as the head. Male with a distinct nuchal crest and with a similar crest along each side of the neck. Dorsal scales irregular, very feebly keeled, the largest at least twice as large as the ventrals; on each side of the posterior half of the trunk a continuous series of keeled scales, some of which are larger than the rest. The fore limb stretched forwards extends beyond the tip of the snout; the adpressed hind limb reaches slightly beyond the elbow of the adpressed fore limb. Brownish, with metallic lustre above, with darker reticulation; wing-membranes above with a dark-brown meshwork enclosing round greenish-white spots, below immaculate; throat and wattles with purplish-brown reticulation.

Total length	225	millim.
Head	16	„
Width of head	11	„
Body	69	„
Fore limb	35	„
Hind limb	41	„
Tail	140	„

Philippines; Sanghir Islands.

a. ♂.   Philippines.   (Type.)

3. Draco guentheri. (Plate XX. fig. 2, the head.)

Head larger than in *D. volans*; snout a little longer than the orbit; nostril lateral, directed outwards; tympanum naked, as large as the eye-opening. Upper head-scales very unequal, keeled; a ⋏-shaped series of enlarged scales on the forehead; a compressed prominent scale on the posterior part of the supraciliary region; twelve upper labials, subequal. Dorsal scales irregular, keeled, the largest at least twice as large as the ventrals; on each side of the back a series of enlarged keeled distant scales. The fore limb stretched forwards extends to the tip of the snout; the adpressed hind limb reaches the elbow of the adpressed fore limb. Greenish grey with metallic lustre above, with very indistinct darker markings; a black spot between the orbits; wing-membranes largely black-marbled above, immaculate inferiorly except near the outer border, where there are one or two small black spots; lower surface of neck whitish, blackish-dotted.

Total length	192	millim.
Head	16	,,
Width of head	11	,,
Body	55	,,
Fore limb	28	,,
Hind limb	34	,,
Tail	121	,,

Philippine Islands.

a. ♀.   Philippines.   A. Everett, Esq. [C.].

4. Draco everetti. (Plate XX. fig. 3, the head.)

Head larger than in *D. volans*; snout a little longer than the diameter of the orbit; nostril lateral, directed outwards; tympanum naked, as large as the eye-opening. Upper head-scales very unequal, strongly keeled; a ⋏-shaped series of enlarged scales on the forehead; a large compressed spine-like scale on the posterior part of the supraciliary region; nine or ten upper labials, the last thrice as large as the preceding. Dorsal scales irregular, keeled, the largest at least twice as large as the ventrals; on each side of the back a series of enlarged keeled distant scales. The fore limb stretched forwards extends beyond the tip of the snout; the adpressed hind limb nearly reaches the elbow of the adpressed fore limb. Upper surfaces grey with metallic lustre, with small dark markings forming five short cross lines on the back; three blackish cross bars on the anterior part of the head, the posterior of which includes the interorbital black spot; a dark band from the eye to the ear; a small nuchal black spot; wing-membranes azure-blue, black-marbled above, immaculate inferiorly; throat and lower surface of neck white, speckled and reticulated with blackish.

Total length	208	millim.
Head	17	,,
Width of head	12	,,
Body	66	,,
Fore limb	30	,,
Hind limb	36	,,
Tail	125	,,

Philippine Islands.

a. ♀.   Placer, N.E. Mindanao.   A. Everett, Esq. [C].
b. ♀.   Dinagat Island.   A. Everett, Esq. [C].

5. Draco cornutus. (Plate XX. fig. 4, the head.)

Draco cornutus, *Günth. Rept. Brit. Ind.* p. 125.

Head small; snout as long as the diameter of the orbit; nostril lateral, directed outwards; tympanum naked, smaller than the eye-opening. Upper head-scales very unequal, keeled; a ⋏-shaped series of enlarged scales on the forehead; a large compressed spine-like scale on the posterior part of the supraciliary region; eight or nine upper labials, the last twice or thrice as large as the preceding. The male's gular appendage as long as the head. A slight indication of a nuchal crest in both sexes. Dorsal scales subequal, keeled, small, hardly larger than the ventrals; a more or less distinct lateral series of enlarged keeled distant scales. The fore limb stretched forwards extends to the tip of the snout or slightly beyond; the adpressed hind limb does not reach the elbow of the adpressed fore limb. Upper surfaces variegated olive and reddish-brown; a black spot between the orbits, another on the nape, and another on each side of the neck before the shoulders; wing-membranes above spotted or marbled with black and with the margin black, inferiorly with a few black spots or cross bands; lower surface of limbs and a broad zone along the belly blue in both sexes; throat with blue-green variegations; the male's gular appendage red.

Total length	222	millim.
Head	16	,,
Width of head	11	,,
Body	76	,,
Fore limb	30	,,
Hind limb	36	,,
Tail	130	,,

Borneo.

a. ♂.	Borneo.	Sir E. Belcher [P.].	(Types.)
b. ♀.	Borneo.		
c. ♀.	N. Borneo.		
d. ♀.	Sarawak.		

6. Draco ornatus.

Dracunculus ornatus, *Gray, Cat.* p. 235.
Draco ornatus, *Günth. Proc. Zool. Soc.* 1873, p. 167.

Head larger than in *D. volans*; snout a little longer than the diameter of the orbit; nostril lateral, directed outwards; tympanum scaly. Upper head-scales very unequal, strongly keeled; a more or less distinct ⋏-shaped series of enlarged scales on the forehead; eight to eleven upper labials, the last usually twice or thrice as large as the preceding. The male's gular appendage as long as the head. Male with a small nuchal crest. Dorsal scales unequal, keeled, the largest twice as large as the ventrals; on each side of the back a series of enlarged keeled distant scales. The fore limb stretched forwards extends beyond the tip of the snout; the adpressed hind limb reaches beyond the elbow of the adpressed fore limb, sometimes nearly to the axil. Greenish-grey above with metallic gloss, dark transverse markings and lighter spots; generally a dark cross band between the eyes, another on the snout, and black lines radiating from the eye; wing-membranes above marbled or reticulated with black, enclosing round light spots, inferiorly with a few large black spots near the outer border; throat with blue-grey reticulation and blackish dots; belly sometimes with blackish dots.

Total length	260	millim.
Head	20	,,
Width of head	14	,,
Body	70	,,
Fore limb	40	,,
Hind limb	50	,,
Tail	170	,,

Philippine Islands.

a-b. ♀.	Philippines.	Mr. Cuming [C.]. (Types.)
c-d. ♀.	Philippines.	
e-g. ♀.	Philippines.	H. J. Veitch, Esq. [P.].
h. ♀.	Manilla.	H.M.S. 'Challenger.'
i-n. ♀ & yg.	Luzon.	A. Everett, Esq. [C.].
o-u. ♂ ♀.	Dinagat Island.	A. Everett, Esq. [C.].
v. ♀.	S. Negros.	A. Everett, Esq. [C.].

7. Draco spilopterus.

Dracunculus spilopterus, *Gray, l. c.* p. 236.
Dracunculus spilopterus, *Wiegm. Nova Acta Ac. Cæs.-Leop.* xvii. i. 1835, p. 216, pl. xv.
Draco spilopterus, *Dum. & Bibr.* iv. p. 461; *Gerv. in Eyd. Voy. Favorite, Zool.* pl. xxvii.; *Schleg. Abbild.* p. 92; *Günth. Rept. Brit. Ind.* p. 124.
Dracontoidis personatus, *Fitz. Syst. Rept.* p. 51.

Head larger than in *D. volans*; snout as long as, or slightly longer than, the diameter of the orbit; nostril lateral, directed outwards; tympanum usually scaly. Upper head-scales very unequal, very strongly keeled; a longitudinal series of enlarged scales on the snout; eight to ten upper labials, the last twice or thrice as large as the preceding. The male's gular appendage longer than the head. Male with a very distinct nuchal crest. Dorsal scales rather regular, strongly keeled, the largest twice as large as the ventrals; on each side of the posterior half of the trunk a more or less distinct continuous series of strongly keeled scales, some of which are larger than the rest. The fore limb stretched forwards extends beyond the tip of the snout; the adpressed hind limb reaches half way between the elbow of the adpressed fore limb and the axil, or nearly to the latter. Upper parts metallic, more or less distinctly spotted with blackish; a black interorbital spot; wing-membranes above with small round black spots, inferiorly immaculate except a few large black spots near the outer margin; throat and base of gular appendage dotted or reticulated with blackish.

Total length	225 millim.
Head	17 ,,
Width of head	11 ,,
Body	68 ,,
Fore limb	35 ,,
Hind limb	42 ,,
Tail	140 ,,

Philippine Islands.

a. ♂.	Philippines.	Mr. Cuming [C.].
b. ♂.	Philippines.	
c–g. ♂.	Luzon.	Dr. A. B. Meyer [C.].
h–i. ♂.	Manilla.	H.M.S. 'Challenger.'
k. ♂.	S. Negros.	A. Everett, Esq. [C.].
l. ♂.	Java (?).	C. Bowring, Esq. [P.].

8. Draco rostratus. (Plate XX. fig. 5, the head.)

Draco rostratus, *Günth. Rept. Brit. Ind.* p. 127.

Head rather small; snout longer than the diameter of the orbit, sloping very gently; nostril lateral, directed outwards; tympanum scaly. Upper head-scales unequal, strongly keeled; a Λ-shaped series of enlarged scales on the forehead; ten upper labials, the last twice as large as the preceding. The male's gular appendage a little longer than the head. Male with a very distinct nuchal crest. Dorsal scales irregular, keeled, those on the vertebral region considerably larger than the ventrals. The fore limb stretched forwards extends considerably beyond the tip of the snout; the adpressed hind limb reaches the axil. Greyish above with metallic reflections; wing-membranes above with narrow longitudinal lines of white scales and small round black spots; throat and base of gular appendage brown-dotted.

Total length	219 millim.
Head	16 ,,
Width of head	10 ,,
Body	73 ,,
Fore limb	35 ,,
Hind limb	44 ,,
Tail	130 ,,

Borneo?

| a. ♂. | Borneo? | Sir E. Belcher [P.]. (Type.) |

9. Draco timorensis.

Draco timorensis, part., *Gray, Cat.* p. 233.
Draco timorensis (*Péron*), *Kuhl, Beitr. Zool. u. Vergl. Anat.* p. 103; *Dum. & Bibr.* iv. p. 454; *Günth. Rept. Brit. Ind.* p. 123.
—— viridis, var. timoriensis, *Schleg. Abbild.* p. 91.

Head larger than in *D. volans*; snout longer than the diameter of the orbit; nostril lateral, directed outwards; tympanum naked, as large as the eye-opening. Upper head-scales very unequal, rather large, keeled; a small conical tubercle at the posterior corner of the orbit; nine upper labials, the last twice as large as the preceding. A slight nuchal crest. Dorsal scales larger than ventrals, smooth, except a series of strongly enlarged scales running along each side of the vertebral line; these large scales, the keels of which form two continuous lines, are separated by a row of small smooth scales; on each side of the back a series of enlarged keeled distant scales. The fore limb stretched forwards extends beyond the tip of the snout; the adpressed hind limb reaches midway between the elbow of the adpressed fore limb and the axil. Grey-brown above, with metallic gloss and irregular darker markings; wing-membranes rather indistinctly blackish-marbled above, with a few black spots inferiorly.

Total length	240 millim.
Head	18 ,,
Width of head	12 ,,
Body	72 ,,
Fore limb	37 ,,
Hind limb	44 ,,
Tail	150 ,,

Timor.

a. ♀.   Timor.   Leyden Museum.

10. Draco maculatus.

Dracunculus maculatus, *Gray, Cat.* p. 236.
Draco maculatus, *Cantor, Cat. Mal. Rept.* p. 39; *Günth, Rept. Brit. Ind.* p. 125, pl. xiii. fig. C; *Anders. Zool. W. Yunnan*, p. 802.

Head small; snout a little longer than the diameter of the orbit; nostril lateral, directed outwards; tympanum scaly. Upper head-scales unequal, strongly keeled; a compressed prominent scale on the posterior part of the superciliary region; seven to eleven upper labials. The male's gular appendage very large, always much longer than, and frequently twice as long as, the head; female also with a well-developed but smaller gular sac. Male with a very small nuchal crest. Dorsal scales little larger than the ventrals, irregular, smooth or very feebly keeled; on each side of the back a series of large trihedral keeled distant scales. The fore limb stretched forwards reaches beyond the tip of the snout; the adpressed hind limb reaches a little beyond the elbow of the adpressed fore limb, or the axilla. Greyish above, with more or less distinct darker markings; a more or less distinct darker interorbital spot; wing-membranes above with numerous small round black spots, which are seldom confluent, inferiorly immaculate or with a few black spots; a blue spot on each side the base of the gular appendage.

Total length	197 millim.
Head	16 ,,
Width of head	12 ,,
Body	66 ,,
Fore limb	33 ,,
Hind limb	41 ,,
Tail	115 ,,

From Yunnan to Singapore.

a–b. ♂.	——?	Sir J. McGregor [P.]. } (Types.)
c. ♂.	Penang.	
d–g. ♂ ♀.	Penang.	Dr. Cantor.
h–k. ♂ ♀.	Tenasserim.	Dr. Packman.
l. ♂.	Tavoy.	W. T. Blanford, Esq. [P.].
m, n, o. ♂ ♀.	Pachebone.	M. Mouhot [C.].
p–q. ♂ ♀.	Camboja.	M. Mouhot [C.].
r–s, t. ♂ ♀.	Pegu.	W. Theobald, Esq. [C.].
u. ♂.	Assam.	Indian Museum.

11. Draco bimaculatus. (Plate XX. fig. 6, the head.)

Draco bimaculatus, *Günth, Rept. Brit. Ind.* p. 127.

Head small; snout very short, hardly as long as the diameter of the orbit; nostril lateral, directed outwards; tympanum naked, much smaller than the eye-opening. Upper head-scales unequal, strongly keeled; eight to ten upper labials. The male's gular appendage shorter than the head. Male with a very slight nuchal crest. Dorsal scales rather regular, feebly keeled, a little larger than the ventrals; on each side a more or less distinct series of enlarged keeled distant scales. The fore limb stretched forwards extends considerably beyond the tip of the snout; the adpressed hind limb reaches the axil or a little beyond. Upper surfaces greenish grey with metallic gloss, and more or less defined broad blackish cross bars; a black interorbital spot; a large round black spot, with a white scale in the centre, behind the angle of the mouth; wing-membranes above blue-green, reticulated with black, with four or five more or less indistinct broad blackish transverse bands, and numerous pale longitudinal lines; wing-membranes inferiorly with irregular large black spots; throat and base of gular appendage with dark reticulation; female with two black cross bars under the neck; belly sometimes with a few blackish dots.

Total length	197 millim.
Head	12 ,,
Width of head	9 ,,
Body	55 ,,
Fore limb	34 ,,
Hind limb	41 ,,
Tail	130 ,,

Philippine Islands.

a. ♂.   Philippines.   Mr. Cuming [C.]. (Type.)
b–g. ♂ ♀.   Dinagat Island.   A. Everett, Esq. [C.].

12. Draco lineatus.

Dracunculus lineatus, *Gray, Cat.* p. 235.
Draco lineatus, *Daud. Rept.* iii. p. 298; *Kuhl, Beitr. Zool. u. Vergl. Anat.* p. 102; *Dum. & Bibr.* iv. p. 459; *Schleg. Abbild.* p. 93, pl. xxiv. fig. 5; *Günth. Rept. Brit. Ind.* p. 124.
—— amboinensis, *Lesson, Ill. Zool.* pl. xxxviii.
Dracunculus lineatus, *Wiegm. Herp. Mex.* p. 14, *and Nova Acta Ac. Leop. Carol.* xvii. i. 1835, p. 217.
Dracontoidis lineatus, *Fitz. Syst. Rept.* p. 51.

Head rather small; snout as long as, or slightly longer than, the diameter of the orbit; nostril lateral, directed outwards; tympanum usually scaly. Upper head-scales very unequal, keeled; five to eight upper labials. The male's gular appendage shorter than the head. Male with a slight nuchal crest. Dorsal scales subequal, feebly keeled, not larger than ventrals; no lateral series of enlarged scales. The fore limb stretched forwards extends beyond the tip of the snout; the adpressed hind limb reaches the axil or a little beyond. Brown or olive above, with lighter spots and undulated cross bands: wing-membranes above dark brown, or with dark brown and light cross bands, constantly with white longitudinal lines; throat and sides of neck with wide-meshed dark network enclosing round light spots.

Total length	240	millim.
Head	17	,,
Width of head	11	,,
Body	65	,,
Fore limb	37	,,
Hind limb	47	,,
Tail	158	,,

Moluccas; Celebes; Java.

a. ♂.   Amboyna.   Leyden Museum.
b–c. ♀.   Mysol.
d–g, h–k. ♂ ♀.   N. Ceram.
l–o. ♂ ♀.   Java.   Sir E. Belcher [P.].
p. ♂.   Java.

13. Draco beccarii.

Draco beccarii, *Peters & Doria, Ann. Mus. Genov.* xiii. 1878, p. 373.

Nostril lateral; tympanum scaly. The male's gular appendage nearly as long as the head. Nuchal crest rudimentary. Dorsal scales small, feebly keeled; no lateral series of enlarged scales. The adpressed hind limb reaches the axil. Olive-green above, irregularly marbled with darker; a quadrangular black nuchal spot; wing-membranes yellowish white, with white longitudinal lines, and a few black spots near the outer margin confluent into a longitudinal band occupying a little more than a third of the length of the wing; throat and sides of neck reticulated with blackish.

Total length	185	millim.
Head	15	,,
Body	47	,,
Fore limb	30	,,
Hind limb	35	,,
Tail	123	,,

Kandari, Celebes.

14. Draco spilonotus.

Draco spilonotus, *Günth. Proc. Zool. Soc.* 1872, p. 592, pl. xxxv. fig. B, & 1873, p. 167.

Head small; snout as long as the diameter of the orbit; nostril lateral, directed outwards; tympanum naked (probably not constantly). Upper head-scales very unequal, keeled; six or seven upper labials. The male's gular appendage much shorter than the head. A very slight nuchal crest. Dorsal scales subequal, smooth, slightly larger than ventrals; no lateral series of enlarged scales. The fore limb stretched forwards extends much beyond the tip of the snout; the adpressed hind limb reaches a little beyond the axil. Greyish above, marbled with brown; a black nuchal spot; sides of the neck reticulated; wing-membranes yellowish white, with small black spots on their basal half.

Total length	182	millim.
Head	13	,,
Width of head	9	,,
Body	47	,,
Fore limb	28	,,
Hind limb	34	,,
Tail	122	,,

Celebes.

a-b. ♂. Celebes. Dr. A. B. Meyer [C.]. (Types.)

15. Draco fimbriatus.

Draco fimbriatus, *Gray, Cat.* p. 234.
Draco fimbriatus, *Kuhl, Beitr. Zool. u. Vergl. Anat.* p. 101; *Guér. Icon. R. A., Rept.* pl. x. fig. 1; *Dum. & Bibr.* iv. p. 448; *Duvern. R. A., Rept.* pl. xvi. fig. 1; *Schleg. Abbild.* p. 92, pl. xxiv. figs. 2–4; *Günth. Rept. Brit. Ind.* p. 123; *Stoliczka, Journ. As. Soc. Beng.* xlii. 1873, p. 119.
—— abbreviatus, *Gray, Zool. Journ.* iii. 1828, p. 219; *id. in Hardw. Ill. Ind. Zool.*
—— (Rhacodracon) fimbriatus, *Fitz. Syst. Rept.* p. 50.

Head large; snout slightly longer than the diameter of the orbit; nostril lateral directed outwards and slightly upwards; tympanum naked, smaller than the eye-opening. Upper head-scales very small, keeled; a more or less distinct ⋏-shaped series of enlarged scales on the forehead; a conical tubercle at the posterior corner of the orbit; ten to thirteen upper labials, the last largest. The male's gular appendage longer than the head; female also with a well-developed, but shorter, appendage. Male with a small nuchal crest. Dorsal scales smaller than ventrals, smooth or very feebly keeled; on each side of the back a series of small subconical distant scales. The fore limb stretched forwards extends beyond the tip of the snout; the adpressed hind limb reaches the axil or nearly so far. Greyish above, marbled with reddish brown, with a few equidistant angular transverse dark markings, the anterior of which encloses the black interorbital spot; wing-membranes above dark brown with lighter lines, inferiorly immaculate or with a few scattered black spots; throat bluish, with a more or less marked dark network enclosing white spots; gular appendage red.

Total length	273	millim.
Head	25	,,
Width of head	19	,,
Body	83	,,
Fore limb	46	,,
Hind limb	57	,,
Tail	165	,,

Malay peninsula; Sumatra; Java; Borneo.

a. ♀.	Singapore.	
b–d. ♂ ♀.	Sumatra.	Leyden Museum.
e. ♂.	Java.	H. O. Forbes, Esq. [C.].
f–g. ♀ & hgr.	Matang.	
h. ♂.	E. Indies.	East-India Company.

16. Draco cristatellus.

Draco cristatellus, *Günth. Proc. Zool. Soc.* 1872, p. 592, pl. xxxv. fig. A.

Head rather large; snout as long as the diameter of the orbit; nostril lateral, directed outwards and slightly upwards; tympanum partly scaly. Upper head-scales very small, keeled; a distinct ⋏-shaped series of enlarged scales on the forehead; a subconical tubercle at the posterior corner of the orbit; eleven upper labials. The male's gular appendage much longer than the head. A well-marked nuchal fold. Dorsal scales smaller than ventrals, smooth or very feebly keeled; no lateral series of enlarged scales. The fore limb stretched forwards extends beyond the tip of the snout; the adpressed hind limb reaches the axil. Tail with a crest composed of long distant triangular scales. Reddish brown above, black-spotted; a black interorbital spot; wing-membranes above dark brown, with irregular light longitudinal lines, inferiorly whitish, with a few blackish spots on their margins; gular appendage yellow, with a brown anterior edge.

Total length	190	millim.
Head	16	,,
Width of head	11	,,
Body	62	,,
Fore limb	36	,,
Hind limb	44	,,
Tail	112	,,

Borneo.

a. ♂.   Sarawak.   (Type.)

17. Draco hæmatopogon.

Dracocella hæmatopogon, *Gray, Cat.* p. 234.
Draco hæmatopogon (*Boie*), *Gray, Griff. A. K.* ix., *Syn.* p. 59; *Dum. & Bibr.* iv. p. 458; *Schleg. Abbild.* p. 95, pl. xxiv. figs. 6-9; *Günth. Rept. Brit. Ind.* p. 124.
Draco (Pleuropterus) hæmatopogon, *Fitz. Syst. Rept.* p. 51.

Head small; snout as long as the diameter of the orbit; nostril directed upwards, perfectly vertical; tympanum naked, smaller than the eye-opening: upper head-scales unequal, strongly keeled; nine to twelve upper labials. The male's gular appendage considerably longer than the head. No nuchal fold or crest. Dorsal scales smooth, equal, not larger than ventrals; a lateral series of widely distant enlarged keeled scales. The fore limb stretched forwards extends considerably beyond the tip of the snout; the adpressed hind limb reaches the axil. Greenish above, with metallic gloss and with more or less irregular darker and lighter cross bands; wing-membranes black above, enclosing numerous roundish yellow spots, inferiorly immaculate; the male's gular appendage yellow; throat red, with a large black spot on each side of the gular appendage.

Total length	264	millim.
Head	17	,,
Width of head	12	,,
Body	72	,,
Fore limb	43	,,
Hind limb	52	,,
Tail	175	,,

Java; Borneo; Sumatra.

a-b. ♂ ♀.   Sumatra.   Leyden Museum.

18. Draco blanfordii. (PLATE XX. fig. 7, the head.)

Draco major (*non Laur.*), *Blanf. Journ. As. Soc. Beng.* xlvii. 1878, p. 125, & xlviii. 1879, p. 128.

Head small; snout constricted, slightly longer than the diameter of the orbit; nostril directed upwards, perfectly vertical; tympanum naked, smaller than the eye-opening. Upper head-scales unequal, keeled; a prominent tubercle at the posterior corner of the orbit; nine upper labials. The male's gular appendage longer than the head, very thin, covered with large scales. Male with a slight nuchal fold. Dorsal scales equal, smooth or very feebly keeled, not larger than ventrals; a series of widely separated enlarged keeled scales along the side of back. The fore limb stretched forwards extends considerably beyond the tip of the snout; the adpressed hind limb reaches nearly the axil. Grey-brown above, with small dark spots; wing-membranes above marbled with dark brown, with lighter spots and lines, inferiorly immaculate; throat unspotted, greenish, pale scarlet beneath the lateral wattles.

Total length	356 millim.
Head	21 ,,
Width of head	14 ,,
Body	100 ,,
Fore limb	51 ,,
Hind limb	59 ,,
Tail	235 ,,

Tenasserim.

a, b. ♂.   Tavoy.   W. T. Blanford, Esq. [C.]. (Two of the types.)

19. Draco dussumieri.

Dracocella dussumieri, *Gray, Cat.* p. 234.
Draco dussumieri, *Dum. & Bibr.* iv. p. 456; *Jerdon, Journ. As. Soc. Beng.* xxii. 1853, p. 474; *Günth. Rept. Brit. Ind.* p. 125, pl. xiii. fig. D; *Anders. Proc. Zool. Soc.* 1871, p. 164.
Draco (Pterosaurus) dussumieri, *Fitz. Syst. Rept.* p. 51.

Head small; snout constricted, as long as the diameter of the orbit; nostril directed upwards, perfectly vertical; tympanum naked, as large as the eye-opening. Upper head-scales unequal, keeled; a spinose conical scale at the posterior corner of the orbit; nine to twelve upper labials, the last twice or thrice as large as the preceding. The male's gular appendage much longer than the head. Male with a slight nuchal fold. Dorsal scales scarcely larger than ventrals, unequal, smooth, or very slightly keeled; on each side of the back a series of small tubercular prominences, each being composed of several small scales. The fore limb stretched forwards extends beyond the tip of the snout; the adpressed hind limb reaches the axil or not quite so far. Grey-brown above; a series of more or less distinct dark circles on the back; wing-membranes above purplish black, enclosing round light spots, inferiorly with a series of large black spots near the margin; throat with irregular dark spots.

Total length	200 millim.
Head	16 ,,
Width of head	11 ,,

Body	62	millim.
Fore limb	33	,,
Hind limb	42	,,
Tail	122	,,

Malabar; Cochin; Travancore.

a–b.	♂ ♀.	India.	
c–g.	♂ ♀, bad state.	Madras.	J. E. Boileau, Esq. [P.].
h–i.	♂ ♀.	Madras.	T. C. Jerdon, Esq. [P.].

20. Draco tæniopterus.

Draco tæniopterus, *Günth. Proc. Zool. Soc.* 1861, p. 187, *and Rept. Brit. Ind.* p. 126, pl. xiii. fig. E.

Head small; snout as long as the diameter of the orbit; nostril directed upwards, perfectly vertical; tympanum naked, smaller than the eye-opening. Upper head-scales unequal, strongly keeled: six to nine upper labials. The male's gular appendage slightly longer than the head, covered with very large scales. Male with a slight nuchal fold. Dorsal scales equal, very feebly keeled, not larger than ventrals; a more or less distinct lateral series of enlarged keeled distant scales. The fore limb stretched forwards extends considerably beyond the tip of the snout; the adpressed hind limb reaches beyond the axil. Greyish or brownish above, with metallic gloss, without distinct spots; wing-membranes above with five arched transverse black bands, about as wide as the interspaces between them; some of these bands may be forked at the base or enclose large light spots; wing-membranes inferiorly uniform.

Total length	206	millim.
Head	14	,,
Width of head	10	,,
Body	57	,,
Fore limb	38	,,
Hind limb	45	,,
Tail	135	,,

Siam and Tenasserim.

a.	♂.	Chartaboum, Siam.	M. Mouhot [C.]. (Type.)
b.	Yg.	Tenasserim.	Dr. Packman.
c.	♀.	Tavoy.	W. T. Blanford, Esq. [P.].

21. Draco quinquefasciatus. (Plate XX. fig. 8, the head.)

Dracunculus quinquefasciatus, *Gray, Cat.* p. 235.
Draco quinquefasciatus, *Gray, Zool. Journ.* iii. 1827, p. 219, *and in Hardw. Ill. Ind. Zool.*; *Dum. & Bibr.* iv. p. 455; *Günth. Rept. Brit. Ind.* p. 126; *Stoliczka, Journ. As. Soc. Beng.* xlii. 1873, p. 118.

Head small; snout longer than the diameter of the orbit; nostril directed upwards, perfectly vertical : tympanum scaly. Upper headscales very small, subequal, keeled; fifteen upper labials. The male's gular appendage nearly twice as long as the head. A slight nuchal fold. Dorsal scales smooth, equal, not larger than ventrals; lateral series of enlarged scales very indistinct. The fore limb stretched forwards extends considerably beyond the tip of the snout; the adpressed hind limb reaches a little beyond the elbow of the adpressed fore limb. Reddish-brown above, brown-dotted all over; wing-membranes beautifully marked above with five regular curved black bands as broad as the light interspaces between them, inferiorly with three much narrower black bands occupying the same position as the three median dorsal bands; a black band across the neck, behind the gular appendage; throat grey, brown-dotted.

Total length (tail broken)	205	millim.
Head	20	,,
Width of head	13	,,
Body	85	,,
Fore limb	46	,,
Hind limb	52	,,

Penang.

a. ♂. Penang. Gen. Hardwicke [P.]. (Type.)

2. SITANA.

Sitana, *Cuv. Règne Anim.* 2nd ed. ii. p. 43 ; *Dum. & Bibr.* iv. p. 435 ; *Gray, Cat.* p. 236 ; *Günth. Rept. Brit. Ind.* p. 134.
Semiophorus, *Wagl. Syst. Amph.* p. 152 ; *Wiegm. Herp. Mex.* p. 14 ; *Fitzing. Syst. Rept.* p. 47.

Body slightly compressed, limbs long. Fifth toe absent. Scales all keeled, regular, smallest on the flanks. No dorsal crest. Male with a slight nuchal fold and a large folding gular appendage extending backwards to the belly and covered with large scales. No gular fold. Ear exposed. No præanal or femoral pores.

India; Ceylon.

1. Sitana ponticeriana.

Sitana ponticeriana, *Gray, Cat.* p. 236.
Sitana ponticeriana, *Cuv. l. c.*; *Guér. Icon. R. A., Rept.* pl. x. fig. 2; *Dum. & Bibr.* p. 437; *Duvern. R. A. Rept.*, pl. xvi. fig. 2; *Jerdon, Journ. As. Soc. Beng.* xxii. 1853, p. 473; *Kelaart, Prodr. Faun. Zeyl.* p. 164; *Günth. l. c.* p. 135; *Jerdon, Proc. As. Soc. Beng.* 1870, p. 76; *Blanf. Journ. As. Soc. Beng.* xxxix. 1870, p. 365; *Stoliczka, Journ. As. Soc. Beng.* xli. 1872, p. 108; *Theob. Cat. Rept. Brit. Ind.* p. 102.
Semiophorus ponticerianus, *Wagl. l. c.*
Sitana minor, *Günth. l. c.* pl. xiv. fig. A; *Steind. Novara, Rept.* p. 26; *Anders. Proc. Zool. Soc.* 1871, p. 166.
—— deccanensis, *Jerdon, Proc. As. Soc. Beng.* 1870, p. 76; *Blanf. l. c.* p. 367.

Upper head-scales small, sharply keeled: canthus rostralis and supraciliary edge sharp, with strongly enlarged scales. Dorsal scales larger than ventrals, with sharp keels forming straight longitudinal lines; lateral scales smallest, uniform or intermixed with scattered enlarged ones. Limbs above with uniform strongly keeled scales. The length of the limbs varies very much; in some specimens the hind limb stretched forwards does not extend beyond the orbit, in others (*S. minor*, Gthr.) it reaches the tip of the snout or even considerably beyond. Tail round, slender, once and a half to twice as long as the head and body, covered with equal keeled scales. Olive-brown above, with a series of rhomboidal spots along the middle of the back: a more or less distinct light band along each side of the back; gular appendage tricoloured—blue, black, and red.

Total length	198	millim.
Head	21	,,
Width of head	15	,,
Body	60	,,
Fore limb	40	,,
Hind limb	68	,,
Tail	117	,,

India; Ceylon.

a, b. ♂.	India.		
c-d. ♂ ♀.	India.	T. C. Jerdon, Esq. [P.].	
e-f, g-h. ♂ ♀.	India.	Dr. J. E. Gray [P.].	(Types of
i. Yg.	India.	Mr. Mathers.	*S. minor*.)
k-m. ♂ ♀.	Madras.	T. C. Jerdon, Esq. [P.].	
n-o. Hgr.	S.E. Berar.	Indian Museum, Calcutta.	
p-t. Hgr.	Godavery Valley.	W. T. Blanford, Esq. [P.].	
u-y. ♂ & hgr.	W. of Raitsur.	W. T. Blanford, Esq. [P.].	
z-a. Yg.	Sevagherry.	Col. Beddome [C.].	
β. Yg.	Malabar.	Col. Beddome [C.].	
γ-δ. Hgr.	Ceylon.	W. Holdsworth, Esq. [C.].	

3. OTOCRYPTIS.

Otocryptis, *Wiegm. Isis,* 1831, p. 293, *and Herp. Mex.* p. 14; *Wagl. Syst. Amph.* p. 150; *Dum. & Bibr.* iv. p. 430; *Fitzing. Syst. Rept.* p. 48; *Gray, Cat.* p. 237; *Günth. Rept. Brit. Ind.* p. 127.

Body compressed, limbs very long. Fifth toe very short, not longer than first. All the scales keeled, the dorsals heterogeneous. No dorsal crest. No gular fold. Male with a low nuchal crest and a large folding gular appendage extending backwards to the belly, and covered with large scales. Ear concealed. No præanal or femoral pores.

Ceylon; Southern India.

1. Otocryptis bivittata.

Otocryptis bivittata, *Gray, Cat.* p. 238.
Otocryptis bivittata, *Wiegm. l. c.*; *Dum. & Bibr.* p. 432; *Günth. l. c.*
—— wiegmanni, *Wagl. l. c.*

Upper head-scales sharply keeled; canthus rostralis and supraciliary edge sharp, with strongly enlarged scales; supraorbital scales large, the inner series forming, with some enlarged scales on the snout, a regular ⋏-shaped figure; interorbital region with four or five longitudinal series of very small scales, nine to eleven upper and as many lower labials. Dorsal scales unequal, the enlarged ones sometimes forming regular longitudinal series; lateral scales small, with scattered enlarged ones; ventral scales larger than dorsals. Limbs covered with large subequal scales; the hind limb stretched forwards reaches far beyond the tip of the snout, the heel reaching the eye or the posterior border of the orbit. Tail round, slender, twice and a half as long as head and body, covered with equal strongly keeled scales. Brownish olive above, sides darker; a dark-brown, light-edged cross band between the eyes, and more or less distinct dark cross bands along the middle of the back; generally a light oblique band from below the eye to the angle of the mouth; males generally with a light band along each side of the back; limbs and tail with more or less distinct dark cross bars; lower surfaces whitish, the throat brownish in the females and young.

Total length	246	millim.
Head	18	,,
Width of head	12	,,
Body	48	,,
Fore limb	38	,,
Hind limb	89	,,
Tail	180	,,

Ceylon.

a, b, c, d, e, f–g. ♂, ♀, hgr., & yg.   Ceylon.

2. Otocryptis beddomii. (Plate XXII. fig. 1.)

Head-scales as in the preceding, but the interorbital scales only a little smaller than the others, in two or three longitudinal series; the ⋏-shaped figure formed by the keels of some of the scales generally indistinct; canthus rostralis less prominent; nine or ten upper and as many lower labials. A small pit on each side of the neck, in front of the shoulder. Dorsal scales unequal, the enlarged ones sometimes forming regular chevrons on the back, with the point directed backwards, or a lateral longitudinal series, the latter always distinct on the sacral region; lateral scales a little smaller, with scattered enlarged ones; ventral scales much larger than dorsals. Limbs above with large equal keeled scales; the hind limb stretched forwards reaches beyond the tip of the snout, the heel reaches the tympanum. Tail round, slender, not twice as long as head and body, covered with equal strongly keeled scales. Light brownish-olive above, uniform, or with small scattered dark-brown spots on the back and limbs; a more or less distinct light, dark-edged oblique band from below the eye to the mouth; lower surfaces whitish, the throat brownish in the young.

	Total length	121 millim.
	Head	14 ,,
	Width of head	9 ,,
	Body	31 ,,
	Fore limb	24 ,,
	Hind limb	50 ,,
	Tail	76 ,,

S. India.

a-e. ♀ & hgr.   Sevagherry Ghat.   Col. Beddome [C.].

4. PTYCTOLÆMUS.

Ptyctolæmus, *Peters, Mon. Berl. Ac.* 1864, p. 386.

Body compressed. Fifth toe much longer than first. All the scales keeled, the dorsals heterogeneous. No dorsal crest. Three parallel longitudinal folds on each side of the middle of the throat, curved and converging backwards, forming a U-shaped figure. Ear concealed. No præanal or femoral pores.
Northern India.

1. Ptyctolæmus gularis.

Otocryptis (Ptyctolæmus) gularis, *Peters, l. c.*

Head rather elongate, the snout longer than the diameter of the orbit; canthus rostralis and supraciliary edge angular; upper head-scales unequal, keeled; eight upper and as many lower labials. A slight indication of a nuchal crest (♀). Dorsal and lateral scales small, feebly keeled, with irregularly scattered enlarged strongly keeled ones; ventrals larger, strongly keeled. Limbs above with subequal scales; the hind limb stretched forwards reaches the posterior border of the orbit; fifth toe as long as third. Tail roundish, slender, a little more than twice as long as head and body, covered with subequal keeled scales. Olive-brown above, with darker transverse spots; two curved dark-brown cross bands, separated by a light one of equal width, between the eyes; an oblique dark-brown band from below the eye to the angle of the mouth; limbs and tail above with dark cross bands; the skin in the gular folds black.

	Total length	227 millim.
	Head	19 ,,
	Width of head	11·5 ,,
	Body	50 ,,
	Fore limb	31 ,,
	Hind limb	54 ,,
	Tail	158 ,,

N. India.

a. ♀.   Sadya.   W. T. Blanford, Esq. [P.].

5. APHANIOTIS.

Aphaniotis, *Peters, Mon. Berl. Ac.* 1864, p. 385.

Body compressed, limbs very long and slender. Fifth toe much longer than first. All the scales keeled, the dorsals heterogeneous. A dorsal crest. Probably a gular pouch in the male. No gular fold. Ear concealed. No præanal or femoral pores.

Malayan district.

1. Aphaniotis fusca.

Otocryptis (Aphaniotis) fusca, *Peters, Mon. Berl. Ac.* 1864, p. 385.

Snout pointed, a little longer than the diameter of the orbit, with sharp canthus rostralis; upper head-scales sharply keeled, largest on the supraorbital region; seven or eight upper and eight lower labials. The enlarged dorsal scales forming longitudinal series; lateral scales small and feebly keeled, intermixed with scattered enlarged ones; ventral scales largest. Limbs above with large equal keeled scales; the hind limb stretched forwards reaches beyond the tip of the snout, the heel reaches the posterior border of the head; fifth toe as long as third. Tail roundish, slender, twice as long as head and body, covered with equal strongly keeled scales. Brown above, with rather indistinct darker transverse bars; throat brown, the rest of the lower surface whitish.

Total length	159	millim.
Head	15	,,
Width of body	9	,,
Body	39	,,
Fore limb	31	,,
Hind limb	52	,,
Tail	105	,,

Malacca; Nias; Borneo.

a. ♀. Nias. Hr. Sandemann [C.].

Peters's description of *A. fusca* being rather vague, I am not quite satisfied as to the correctness of my identification. At any rate, Peters's lizard must be very closely allied.

6. LOPHOCALOTES.

Lophocalotes, *Günth. Proc. Zool. Soc.* 1872, p. 593.

Tympanum distinct. Body compressed, covered with large subequal keeled scales. A dorsal crest. No gular sac; a transverse gular fold. Digits not keeled inferiorly. No femoral or præanal pores.

East-Indian archipelago.

1. Lophocalotes interruptus.

Lophocalotes interruptus, *Günth. l. c.* pl. xxxvii. fig. A.

Head large, swollen below the ears, very distinct from neck; snout longer than the diameter of the orbit; tympanum much larger than the eye-opening. Upper head-scales irregular, faintly keeled; supraorbital scales small, strongly keeled, bordered inwards by a row of enlarged scales; three large scales in a series on the temple; a small slightly prominent tubercle behind the supraciliary edge and another on each side of the nape; seven upper and eight lower labials. Gular scales subrhomboidal, smooth or feebly keeled, those on the median line the smallest, the lateral ones larger than the ventrals. Nuchal crest formed of lanceolate spines not quite as long as the diameter of the tympanum; dorsal crest not continuous with nuchal, formed by about fourteen triangular spines, only every alternate median scale being modified into a spine. Dorsal scales large, rhomboidal, rather feebly keeled, the points of the upper scales directed upwards and backwards; ventral scales smaller than dorsals, strongly keeled. Limbs above with subequal keeled scales, some of which are spinose; fourth finger and toe slightly longer than third; the adpressed hind limb reaches hardly the tympanum. Tail compressed at the base, covered with equal keeled scales; its length equals nearly once and three fourths that of head and body. Green above, with rather irregular yellowish markings on the head and body and cross bars on the limbs.

Total length	244	millim.
Head	29	,,
Width of head	18	,,
Body	63	,,
Fore limb	40	,,
Hind limb	56	,,
Tail	152	,,

East-Indian archipelago.

a. ♂. ——? Dr. Bleeker. (Type.)

7. COPHOTIS.

Cophotis, *Peters, Mon. Berl. Ac.* 1861, p. 1103; *Günth. Rept. Brit. Ind.* p. 131.

Tympanum hidden. Body compressed, covered with large subequal irregular scales. A dorsal crest. A very small gular sac in both sexes; a slight transverse gular fold. Tail prehensile. No femoral or præanal pores.

Ceylon; Sumatra.

1. Cophotis ceylanica.

Cophotis ceylanica, *Peters, l. c.*; *Günth. l. c.* p. 132, pl. xiii. fig. H.

Snout nearly twice as long as the diameter of the orbit; upper head-scales rather large, unequal, tubercular; male with a small tubercle on the tip of the snout; eight to ten upper and as many lower labials; gular scales feebly keeled, smallest on the median line. Nuchal crest composed of three or four lanceolate spines, the longest of which about equals the diameter of the orbit; dorsal crest non-continuous with the nuchal, composed of twelve to fourteen similar lobes separated from one another; in the female the lobes of the crest are shorter and not raised, but bent sideways on the back. Dorsal scales very large, irregular, imbricate, smooth or shortly keeled, the keels pointing downwards and backwards; ventral scales small, lanceolate, strongly keeled, mucronate. Scales on the limbs irregular, keeled, some slightly spinose; fingers and toes not very long, third slightly shorter than fourth; infradigital scales feebly keeled; the adpressed hind limb reaches hardly the axil. Tail feebly compressed, covered with keeled scales which are smaller inferiorly; its length is not once and a half that of head and body. Olive above, with irregular dark-brown cross bands; a light, reddish-brown to cream-coloured band from the end of the snout on the upper lip to above the shoulder; a cream-coloured spot on the nape and a cross band of the same colour on the anterior part of the back; lower lip with a broad dark-brown margin; more or less distinct oblique brown lines on the sides; tail with dark annuli.

Total length	140	millim.
Head	19	,,
Width of head	10	,,
Body	47	,,
Fore limb	28	,,
Hind limb	33	,,
Tail	74	,,

Ceylon.

a. ♂.	Ceylon.	G. H. K. Thwaites, Esq. [P.].
b. ♀.	Ceylon.	
c–e. ♂ & yg.	Ceylon.	W. H. Holdsworth, Esq. [C.].

2. Cophotis sumatrana.

Cophotis sumatrana, *Hubrecht, Notes Leyd. Mus.* i. 1879, p. 243.

Very closely allied to the preceding, from which it is to be distinguished by the following characters:—A rostral appendage consisting of a small pointed horn-like scale; a fringe of elongated scales along the supraciliary edge. Nuchal crest composed of nine larger and several smaller lobes; dorsal crest composed of eighteen lobes; the crest extends on more than three fourths of the length of the tail. Caudal scales as large ventrally as dorsally. Body and tail with broad brown cross bands. Head marmorated with brown. Total length 180 millim.

Sumatra.

8. CERATOPHORA.

Ceratophora, *Gray, Ill. Ind. Zool.*; *Dum. & Bibr.* iv. p. 433; *Fitzing. Syst. Rept.* p. 47; *Gray, Cat.* p. 237; *Günth. Rept. Brit. Ind.* p. 129.

Tympanum hidden. Body more or less compressed, covered with unequal scales. No dorsal crest; a nuchal crest present or absent. No gular sac; no gular fold. A large rostral appendage, at least in the males. No femoral or præanal pores.

Ceylon.

Synopsis of the Species.

Gular scales larger than the ventrals, smooth; lateral scales large, unequal; rostral appendage scaleless .... 1. *stoddartii*, p. 277.

Gular scales larger than the ventrals, feebly keeled; lateral scales large, equal; rostral appendage scaly .... 2. *tennentii*, p. 278.

Gular scales smaller than the ventrals, strongly keeled; lateral scales small; rostral appendage scaly .... 3. *aspera*, p. 278.

1. Ceratophora stoddartii.

Ceratophora stoddartii, *Gray. Cat.* p. 237.
Ceratophora stoddartii, *Gray, Ind. Zool.*; *Dum. & Bibr.* iv. p. 434; *Kelaart, Prodr. Faun. Zeyl.* p. 129; *Günth. l. c.* pl. xiii. fig. F.

Upper head-scales small, irregular, keeled or tubercular; occiput concave, with a short raised ridge on each side; interorbital space concave; rostral appendage scaleless, flexible, pointed; its length varies considerably and it is short or entirely absent in the female; ten to twelve upper and nine or ten lower labials. Gular scales smooth, large, subquadrangular, forming regular longitudinal series, those on the median line smaller. A low denticulated nuchal crest. Dorsal scales irregular and unequal in size, those on the sides large, strongly imbricate, and pointing upwards and backwards, intermixed with smaller ones; some of the dorsal scales feebly keeled, the others smooth; ventral scales small, smooth or feebly keeled. Limbs above with unequal keeled scales; the adpressed hind limb reaches the posterior border of the orbit or not quite so far. Tail slightly compressed, covered with equal keeled scales, intermixed with a few enlarged ones at the base; its length is once and two thirds to twice that of head and body. Olive above, with more or less distinct irregular darker cross bars on the back and limbs; frequently a white streak behind the orbit and a white spot or longitudinal band on the side of the neck; a white line along the hinder side of the thigh; rostral appendage and throat white (in spirit).

Total length .... 252 millim.
Head .... 26 ,,
Width of head .... 16 ,,

	Body	56 millim.
	Fore limb	45 "
	Hind limb	68 "
	Tail	170 "

Ceylon.

a. ♂.	Ceylon.	Col. Stoddart [P.]. (Type.)
b, c, d, e–f, g–i. ♂, ♀, & hgr.	Ceylon.	
k. ♂, dry.	Ceylon.	Dr. Kelaart.
l. Skeleton.	Ceylon.	

2. Ceratophora tennentii.

Ceratophora tennentii, *Günth. in Tenn. Nat. Hist. Ceyl.* p. 281, fig., and *Rept. Brit. Ind.* p. 130.

Upper head-scales small, irregular, keeled; interorbital space and occiput slightly concave; a short raised ridge on each side of the occiput; rostral appendage large in both sexes, fleshy, compressed, suboval, covered with small scales and granules; ten upper and nine or ten lower labials. Gular scales feebly keeled, large, subquadrangular, forming regular longitudinal series, those on the median line smaller. A low denticulated nuchal crest. Upper dorsal scales irregular and unequal in size, the larger ones feebly keeled; lateral scales equal, large, strongly imbricate and pointing upwards and backwards, smooth or very feebly keeled; ventral scales smaller, keeled. Limbs above with subequal keeled scales; the adpressed hind limb reaches the eye or a little beyond. Tail slightly compressed, covered with equal keeled scales; its length is not quite twice that of head and body. Olive above, irregularly marbled with brownish; young with more distinct dark brown markings, an angular cross band between the eyes being constant, sometimes with light longitudinal lines; a more or less distinct white line along the hinder side of the thighs.

	Total length	260 millim.
	Head	25 "
	Width of head	15 "
	Body	65 "
	Fore limb	48 "
	Hind limb	76 "
	Tail	170 "

Ceylon.

a, b–e. ♂ & yg.	Ceylon.	(Types.)
f, g. ♂ & yg.	Ceylon.	
h–k. ♀ & hgr.	Ceylon.	G. H. K. Thwaites, Esq. [P.].
l. Hgr.	Ceylon.	Col. Beddome [C.].

3. Ceratophora aspera.

Ceratophora aspera, *Günth. Rept. Brit. Ind.* p. 131, pl. iii. fig. G.

Upper head-scales small, irregular, tubercular; interorbital space deeply concave; occiput with a pair of low ridges, convergent anteriorly; rostral appendage large in the male, cylindrical, pointed, covered with small imbricate strongly keeled scales, absent or rudimentary in the female; ten to twelve upper and as many lower labials. Gular scales smaller than ventrals, very strongly keeled. No nuchal crest. Dorsal scales small, irregular, unequal, strongly keeled; ventral scales larger, very strongly keeled. Limbs above with strongly keeled unequal scales; digits very strongly keeled; fifth toe shorter than third; the adpressed hind limb reaches between the shoulder and the orbit. Tail not compressed, covered with subequal strongly keeled scales; its length not more than once and a half that of head and body. Brown, with lighter and darker markings or longitudinal lines; generally a rhombic mark on the sacral region; males with a large white spot or cross band on the gular region.

Total length	83	millim.
Head	11	,,
Width of head	6·5	,,
Body	26	,,
Fore limb	17	,,
Hind limb	26	,,
Tail	46	,,

Ceylon.

a-b. ♂ ♀.	Ceylon.	(Types.)
c-f, g-l. ♂, ♀, & yg.	S. Ceylon.	
m-p, q-t. ♂, ♀, & yg.	Ceylon.	
u-z. ♂, ♀, & hgr.	Ceylon.	G. H. K. Thwaites, Esq. [P.].

9. HARPESAURUS.

Arpephorus, (non *Fischer de Waldh.*) *A. Dum. Cat. Méth. Rept.* p. 92.

Tympanum distinct. Snout ending in a long compressed cutaneous appendage. Body and tail compressed, covered above with equal smooth scales of moderate size. A dorsal crest. No gular sac; no transverse gular fold. No femoral or præanal pores.

Java.

1. Harpesaurus tricinctus.

Arpephorus tricinctus, *A. Dum. l. c.* p. 93, and *Rev. Mag. Zool.* (2) iii. 1851, p. 214, pl. vii., and *Arch. Mus.* viii. 1856, p. 371.

Rostral appendage longer than the head, formed of a compressed falciform scale curved upwards, surrounded at the base by a few large scales. Upper head-scales small, slightly tubercular; a large prominent triangular scale on the snout; canthus rostralis forming a serrated ridge. Gular scales tubercular. Dorsal crest a low serrated ridge. Dorsal scales smooth, equal, forming regular transverse series; ventral scales keeled. Scales on the limbs keeled.

Tail compressed, crested above, the crest a little higher than the dorsal; caudal scales keeled, the inferior spinose. Brown, with three broad transverse yellow bands on the body, the anterior narrowest on the scapular region.

Total length (the rostral appendage included)	168 millim.
Head (without the appendage)	19 ,,
Body	45 ,,
Tail	83 ,,

Java.

10. PHOXOPHRYS.

Phoxophrys, *Hubrecht, Notes Leyd. Mus.* iii. 1881, p. 51.

" Tympanum hidden. No femoral pores. Back and sides covered with small smooth scales, intermixed with larger keeled ones, and with very large multicarinate conical tubercles. No dorsal crest. A row of longer crest-scales above the eye. Upper surface of head covered with conical tubercles."

Sumatra.

1. Phoxophrys tuberculata.

Phoxophrys tuberculata, *Hubr. l. c.*

" Head tetrahedral, with a sharp canthus rostralis; above the eye the scales of the canthus become large and erect, giving rise to a sort of horn-like appendage, more than half as high as the diameter of the eye. The head is covered with tubercular scales and carinate tubercles, the interorbital space is deeply concave. At the back of the head a row of scales larger than the surrounding ones forms a transverse bridge between the posterior borders of the eyes, at the same time limiting the interorbital cavity from behind. Ten upper labial shields and as many lower labials. Larger multicarinate tubercles behind the eye and at the angle of the mouth; throat covered with sharply carinated keels; no gular appendage. The small imbricate scales on the sides have their free margins turned upwards. Of the larger multicarinate tubercles on the back, three or four on each side of the median line are more especially prominent. The scales on the belly are larger than those on the sides, and strongly keeled. Strongly keeled scales on the limbs, which on the upper surface are intermixed with larger tubercular multicarinate ones. Tail much longer than the body, all its scales keeled, those on the inferior surface the largest. The hind limb laid forwards extends to the angle of the mouth, the hind limb laid backwards does not reach to the thigh."

From snout to vent 43 millim.

W. Sumatra.

11. LYRIOCEPHALUS.

Lyriocephalus, *Merrem, Tent. Syst. Amph.* p. 49; *Wagler, Syst. Amph.* p. 150; *Wiegm. Herp. Mex.* p. 14; *Dum. & Bibr.* iv. p. 425; *Fitzing. Syst. Rept.* p. 44; *Gray, Cat.* p. 237; *Günth. Rept. Brit. Ind.* p. 128.

Tympanum hidden. Body compressed, covered with small scales intermixed with enlarged ones. A nuchal and a dorsal crest. A gular sac and a V-shaped gular fold. Adult with a globular hump on the nose. No femoral or præanal pores.

Præ- and post-orbital bones forming an arch limiting a supraorbital fossa.

Ceylon.

1. Lyriocephalus scutatus.

Lyriocephalus scutatus, *Gray, Cat.* p. 237.
Lacerta scutata, *Linn. S. N.* i. p. 360; *Shaw, Zool.* iii. p. 221, pl. lxviii.
Iguana clamosa, *Laur. Syn. Rept.* p. 49.
—— scutata, *Latr. Rept.* i. p. 267.
Agama scutata, *Daud. Rept.* iii. p. 345.
Lyriocephalus margaritaceus, *Merr. l. c.*; *Guér. Icon. R. A., Rept.* pl. viii. fig. 2; *Dum. & Bibr.* p. 427; *Duvern. R. A., Rept.* pl. xv. fig. 2.
—— scutatus, *Wagl. l. c.*; *Kelaart, Prodr. Faun. Zeyl.* p. 166; *Günth. Rept. Brit. Ind.* p. 128.
—— macgregorii, *Gray, Ill. Ind. Zool.*

Rostral hump large and globular in the adult, absent in the young; it is covered with subequal smooth scales; canthus rostralis and supraciliary edge sharp, projecting, the latter ending behind in a triangular compressed spine; a pair of small spines on the occiput; upper head-scales irregular, unequal, feebly keeled; temple granular, with enlarged tubercles; fourteen or fifteen upper and as many lower labials. Gular sac large in the male, small in the female; gular scales larger than ventrals, keeled, those on the gular sac separated from one another by granules. Body strongly compressed. A low nuchal crest, formed of a cutaneous fold with small triangular scales forming a denticulation. Dorsal crest composed of small triangular compressed tubercles, separated from one another. Dorsal scales very small, smooth, intermixed with flat, smooth or feebly keeled tubercles which on the nape, and sometimes also on the back, form regular longitudinal series; these tubercles irregularly scattered on the flanks. Ventral scales moderate, strongly keeled. Limbs with keeled scales, with enlarged tubercles on the thighs; the adpressed hind limb reaches the neck or the temple. Tail strongly compressed, with a crest similar to the dorsal; upper caudal scales unequal, feebly keeled, lower equal and strongly keeled; the length of the tail nearly equals that of the head and body. Greenish above, whitish inferiorly.

Total length .... 335 millim.
Head .... 43 ,,
Width of head .... 25 ,,

Body	122 millim.
Fore limb	80 ,,
Hind limb	105 ,,
Tail	170 ,,

Ceylon.

a, b, c–d. ♂ ♀. Ceylon.
e. Skin. ———? Sir A. Smith [P.].
f. Skeleton. Ceylon.
g. Skull. Ceylon.

12. GONYOCEPHALUS.

Gonyocephalus, *Kaup, Isis,* 1825, p. 590 *and* 1827, p. 614; *Wagler, Syst. Amph.* p. 150; *Wiegm. Herp. Mex.* p. 14; *Fitzing. Syst. Rept.* p. 45; *Gray, Cat.* p. 230.
Lophyrus (*Dum.*), *Fitzing. N. Class. Rept.* p. 17.
Lophyrus, part. (*non Latr.*), *Dum. & Bibr.* iv. p. 410.
Lophosaurus, *Fitz. Syst. Rept.* p. 45.
Dilophyrus, *Gray, l. c.* p. 238; *Günth. Rept. Brit. Ind.* p. 136.
Tiaris, *Gray, l. c.* p. 239; *Günth. l. c.* p. 151.
Coryphophylax (*Fitz.*), *Steindachn. Novara, Rept.* p. 30.
Hypsilurus, *Peters, Mon. Berl. Ac.* 1867, p. 707.
Arua, *Doria, Ann. Mus. Genov.* vi. 1874, p. 345.
Lophosteus, *Peters & Doria, Ann. Mus. Genov.* xiii. 1878, p. 377.

Tympanum distinct. Body compressed. Dorsal scales small, uniform or intermixed with enlarged ones. A dorsal crest. A strong transverse gular fold. Males with a gular sac. No præanal or femoral pores.

East Indies, Papuasia and Polynesia, from the Nicobars and Andamans to the Fiji Islands; North-eastern Australia.

Synopsis of the Species.

I. Supraciliary border strongly raised, forming an angular projection posteriorly.

  A. Dorsal crest almost as much developed as the nuchal. .... 1. *doriæ*, p. 284.

  B. Dorsal crest much less developed than the nuchal.

Ventral scales perfectly smooth; nuchal crest not higher than the snout is long, extending forwards to the occiput .... 2. *chamæleontinus*, p. 285.
Ventral scales keeled............. 3. *kuhlii*, p. 286.
Ventral scales perfectly smooth; nuchal crest higher than the snout is long, extending forwards to the interorbital region............... 4. *sumatranus*, p. 286.

II. Supraciliary border normal.

  A. Enlarged scales scattered among the dorsals.

    1. Nuchal and dorsal crests continuous.

      a. Ventral scales perfectly smooth.

        5. *liogaster*, p. 286.

      b. Ventral scales feebly keeled.

Enlarged dorsal scales forming a lateral series; none scattered on the sides .... 6. *miotympanum*, p. 287.

Enlarged dorsal scales irregularly scattered....................... 7. *borneensis*, p. 288.

      c. Ventral scales strongly keeled.

Gular scales smooth .... 8. *bellii*, p. 288.
Gular scales strongly keeled .... 9. *sophiæ*, p. 288.

    2. Nuchal crest strongly separated from the dorsal.

      a. Ventral scales keeled.

Enlarged dorsal scales forming a regular lateral series parallel to the dorsal crest .... 10. *semperi*, p. 289.

Enlarged scales forming irregular vertical series on the flanks; tympanum smaller than the eye-opening .. 11. *interruptus*, p. 290.

Enlarged scales scattered or forming vertical series; tympanum quite as large as the eye-opening; a high caudal crest .... 12. *dilophus*, p. 290.

      b. Ventral scales smooth ...... 13. *tuberculatus*, p. 291.

    3. Nuchal and dorsal crests subcontinuous.

Limbs above with unequal-sized scales, the largest being spinose .... 14. *spinipes*, p. 292.

Limbs above with subequal scales; the longest spines of the nuchal crest shorter than the greatest diameter of the tympanum .... 15. *subcristatus*, p. 292.

Limbs above with unequal scales; nuchal crest much higher than dorsal .. 16. *humii*, p. 293.

  B. No enlarged dorsal scales.

    1. Ventral scales keeled.

      a. No strongly enlarged scales below the ear.

Nuchal crest very low, reduced to a few widely separated, triangular lobes; dorsal crest indistinct.............. 17. *modestus*, p. 294.

Nuchal crest composed of lanceolate lobes;
dorsal crest distinct; dorsal scales
with strongly ascending keels; gular
scales keeled .... 18. *geelvinkianus*, p. 294.
A black spot on each side of the head,
involving the ear .... 19. *auritus*, p. 295.
Nuchal and dorsal crests composed of
lanceolate lobes; dorsal scales scarcely
keeled, slightly ascending; gular scales
nearly smooth .... 20. *bruijnii*, p. 295.

 b. Several strongly enlarged flat scales below the ear.

  a. Anterior border of gular pouch not distinctly toothed;
   spines of the dorsal crest not bony.

Crest uninterrupted, low; gular scales
small, keeled .... 21. *binotatus*, p. 295.
Crest subcontinuous, notched above
the shoulder; anterior gular scales
nearly smooth, some enlarged ...... 22. *godeffroyi*, p. 295.
Nuchal and dorsal crests separated;
gular scales keeled................ 23. *papuensis*, p. 297.

  β. Anterior border of gular
   pouch with strong serration
   formed by large com-
   pressed triangular scales;
   spines of the nuchal and
   dorsal crests bony...... 24. *boydii*, p. 297.

2. Ventral scales smooth .... 25. *grandis*, p. 298.

1. Gonyocephalus doriæ.

Gonyocephalus doriæ, *Peters, Mon. Berl. Ac.* 1871, p. 570, *and Ann. Mus. Genov.* iii. 1872, p. 28, pl. iii. fig. 1.

Head high, with strongly elevated projecting supraciliary borders forming an angle posteriorly; canthus rostralis strongly projecting; snout a little longer than the diameter of the orbit; tympanum much smaller than the eye-opening; upper head-scales small, subequal, smooth; no enlarged scales on the temple; twelve upper and as many lower labials. Gular sac well developed, with serrated anterior edge; gular scales much smaller than ventrals, smooth. Body very strongly compressed. Nuchal and dorsal crests continuous, the former scarcely higher than the latter, its height equalling nearly the diameter of the orbit; it begins on a line with the posterior border of the orbit; the scales of the crest are triangular, with a very regular basal series of large quadrangular ones. Dorsal scales small, equal, roundish or squarish, scarcely imbricate, smooth, without any enlarged ones; ventral scales large, perfectly smooth. Limbs above with equal smooth scales; third and fourth fingers equal; the adpressed hind limb reaches the posterior border of the orbit. Tail very strongly compressed, with serrated edge above; caudal scales perfectly smooth, except the two inferior series, which are enlarged and strongly keeled; length of the tail once and a half that of head and body, Olive-brown above, with a series of darker spots on each side of the back; tail with regular dark annuli.

Total length	331 millim.
Head	38 ,,
Width of head	26 ,,
Body	93 ,,
Fore limb	75 ,,
Hind limb	105 ,,
Tail	200 ,,

Borneo.

a. ♂.   Sarawak.   A. Everett, Esq. [C.].

2. Gonyocephalus chamæleontinus.

Gonyocephalus chameleontina, *Gray, Cat.* p. 238.
Iguana chamæleontina, *Laur. Syn. Rept.* p. 47.
Lacerta superciliosa, *Shaw, Zool.* iii. p. 220, pl. lxviii.
Agama gigantea, *Kuhl, Beitr. Zool. Vergl. Anat.* p. 106.
Lophyrus tigrinus, *Dum. & Bibr.* iv. p. 421, pl. xli. (part.); *Schleg. Bijdr. tot de Dierk.* i. p. 5, pl. i.
Galeotes lophyrus, *Schleg. Abbild.* p. 73, pl. xxiii.

Head high, with strongly elevated projecting supraciliary borders forming an angle posteriorly; snout as long as the diameter of the orbit, ending in a small rounded hump in the adult female; tympanum much smaller than the eye-opening; upper head-scales small, subequal, smooth; sides of the head with small subgranular scales, with a few enlarged subconical ones on the temple; ten to twelve upper, and eleven to fourteen lower labials. Gular sac well developed, with serrated anterior edge; gular scales much smaller than ventrals, smooth. Body very strongly compressed. Nuchal and dorsal crests continuous; the former begins on a line with the posterior border of the orbit, and is formed of long compressed spines with three rows of smaller ones at the base; its height in the adult equals the length of the snout; the dorsal crest rapidly decreases in height, and is soon reduced to a feebly serrated ridge. Dorsal scales very small and smooth, with the points directed upwards with a few scattered enlarged ones; ventral scales much larger smooth. Limbs above with subequal smooth or feebly keeled scales fourth finger a little longer than third; the adpressed hind limb reaches the posterior border of the orbit. Tail strongly compressed without crest, with keeled slightly serrated upper edge; upper caudal scales smooth or feebly keeled, lower larger and strongly keeled; length of the tail not quite twice that of head and body Olive or greenish above, with dark reticulation which may form more or less distinct bands across the back; tail with regular dark annuli.

Total length	356 millim.
Head	33 ,,
Width of head	26 ,,
Body	93 ,,
Fore limb	72 ,,
Hind limb	106 ,,
Tail	230 ,,

Java; Sumatra.

a. ♀.	Java.	J. E. Gray, Esq. [P.].
b. Hgr.	Java.	Leyden Museum.
c. ♂.	Bencoolen.	Gen. Hardwicke [P.].

3. Gonyocephalus kuhlii.

Lophyrus tigrinus, part., *Dum. & Bibr.* p. 421.
—— kuhlii, *Schleg. Bijdr. tot de Dierk.* i. p. 5, pl. ii.

This species differs from *G. chamæleontinus* in the following points:—Supraciliary border more projecting still, serrated posteriorly; supraorbital scales larger. Nuchal crest a little lower, beginning a little further back. Enlarged dorsal scales more numerous; ventral scales feebly keeled. Scales on upper surface of limbs of very unequal size; third and fourth fingers nearly equal; the adpressed hind limb reaches the eye.

Total length	243 millim.
Head	23 ,,
Width of head	17 ,,
Body	65 ,,
Fore limb	53 ,,
Hind limb	77 ,,
Tail	155 ,,

Java.

| a. ♀. | Java. |

4. Gonyocephalus sumatranus.

Lophyrus sumatranus, *Schleg. Bijdr. tot de Dierk.* i. p. 5, pl. iii. fig. 1.

Allied to *G. chamæleontinus*, from which it differs chiefly in the much higher nuchal crest, which is composed of very narrow lanccolate spines the length of which equals twice the diameter of the eye; this crest extends forwards to the interorbital region. Total length 450 millim.

Sumatra.

5. Gonyocephalus liogaster.

Tiaris liogaster, *Günth. Proc. Zool. Soc.* 1872, p. 592, pl. **xxxvi**.

Snout slightly longer than the diameter of the orbit; canthus rostralis and supraciliary edge projecting; tympanum not quite as large (in the female distinctly smaller) as the eye-opening; upper head-scales small, keeled, a little enlarged on the supraorbital region; a few enlarged conical scales on the temple; nine to eleven upper and nine to twelve lower labials. Gular sac moderately developed, without serrated anterior edge; gular scales smaller than ventrals, smooth. Nuchal and dorsal crests continuous, formed of lanceolate spines with smaller keeled ones at the base; in the male the crest is very high, measuring about the length of the snout on the middle of the back, higher still on the nape, where it does not quite reach the line of the posterior borders of the orbits; in the female and young the crest is much lower and disappears on the posterior half of the back. Dorsal scales very small, smooth or feebly keeled, with the points directed upwards and backwards, with irregularly scattered enlarged ones; ventral scales of moderate size, smooth. Limbs above with feebly keeled scales, which are subequal on the arms and legs, more or less unequal on the thighs; fourth finger slightly longer than third; the adpressed hind limb reaches the tip of the snout or beyond. Tail strongly compressed, with serrated upper edge; upper caudal scales smooth or feebly keeled, lower longer and strongly keeled; length of the tail a little more than twice that of head and body. Olive above, uniform or with more or less distinct dark cross bands; throat with black spots, more or less distinctly arranged in oblique series converging towards the median line of the pouch; tail with regular dark annuli.

Total length	443	millim.
Head	36	,,
Width of head	27	,,
Body	97	,,
Fore limb	83	,,
Hind limb	125	,,
Tail	310	,,

Borneo.

a. ♂.	Borneo.	Marquis G. Doria [P.].	⎫
b. ♂.	Borneo.	Dr. Bleeker.	⎬ (Types.)
c. Yg.	Borneo.	A. R. Wallace, Esq. [C.].	⎭
d, e. ♂ ♀.	Borneo.		

6. Gonyocephalus miotympanum.

Tiaris miotympanum, *Günth. Proc. Zool. Soc.* 1872, p. 592, pl. xxxvii. fig. B.

Very closely allied to *G. liogaster*, from which it differs in having the gular and ventral scales feebly keeled. Tympanum much smaller than the eye-opening. Third and fourth fingers equal. Scales on upper surface of limbs equal, strongly keeled. A series of enlarged scales along each side of the back; no other scattered large scales on the side. Olive-brown, tail with regular dark annuli; throat without spots.

Total length	419	millim.
Head	35	,,
Width of head	24	,,
Body	89	,,
Fore limb	75	,,
Hind limb	111	,,
Tail	295	,,

Borneo.

a, b. ♂.   Labuan.   L. L. Dillwyn, Esq. [P.]. (Types.)

7. Gonyocephalus borneensis.

Lophyrus bornensis, *Schleg. Bijdr. tot de Dierk.* i. p. 6, pl. iii. fig. 2.

Size and habit of *G. kuhlii*. Nuchal crest and anterior portion of the dorsal composed of very long and linear scales as in *G. sumatranus*, but not extending forwards beyond the posterior border of the occiput. Supraciliary edge not very projecting, not angular. Ventral scales keeled.

Borneo.

8. Gonyocephalus bellii.

Lophyrus bellii, *Dum. & Bibr.* iv. p. 416; *Peters, Mon. Berl. Ac.* 1867, p. 16.

Tympanum nearly entirely covered with scales. Upper head-scales small, keeled; two subconical tubercles on the occiput. Gular sac small, without serrated anterior edge; gular scales equal, smooth. Nuchal and dorsal crests continuous, commencing on the occiput; the crest is composed of lanceolate spines with two or three rows of smaller keeled spines on each side at its base, which gradually decrease in size on the back. Dorsal scales small, keeled, with the points turned upwards and backwards, intermixed with irregularly scattered enlarged ones; ventral scales strongly keeled. The adpressed hind limb reaches nearly the tip of the snout. Tail compressed, anteriorly with serrated upper edge; its length more than twice that of head and body. Brownish above, whitish-spotted; crest greenish grey; tail with regular dark annuli.

Total length	487	millim.
Head	45	,,
Body	107	,,
Fore limb	90	,,
Hind limb	135	,,
Tail	335	,,

Bengal.

9. Gonyocephalus sophiæ.

Tiaris bellii, (*non D. & B.*) *Gray, Cat.* p. 239.
Tiaris sophiæ, *Gray, l. c.* p. 240.
Tiaris petersii, *Günth. Zool. Rec.* iv. p. 136.
—— sophiæ, *Günth. Proc. Zool. Soc.* 1872, p. 593, pl. **xxx**vii. f. C.

Snout as long as, or slightly longer than, the diameter of the orbit; canthus rostralis and supraciliary edge projecting; tympanum smaller than the eye-opening; upper head-scales small, keeled, a little enlarged on the supraorbital region; a few enlarged conical scales on the temple; eight or nine upper and nine to eleven lower labials. Gular sac rather small, without serrated anterior edge; gular scales strongly keeled, smaller than ventrals, intermixed with a few enlarged ones. Nuchal and dorsal crests continuous, formed of lanceolate spines with smaller keeled ones at the base; in the male the crest is high, measuring the length of the snout on the nape, gradually decreasing in height on the back; the crest much lower in the female, almost absent in the young; anteriorly, the crest does not quite reach the line of the posterior borders of the orbits. Dorsal scales very small, smooth or feebly keeled, with the points directed upwards and backwards, with irregularly scattered enlarged ones; ventral scales of moderate size, strongly keeled. Limbs above with unequal strongly keeled scales; third and fourth fingers equal; the adpressed hind limb reaches the anterior border of the orbit or the nostril. Tail strongly compressed, with slightly serrated upper edge; caudal scales all keeled, largest inferiorly; length of the tail nearly twice that of head and body. Olive above, uniform or with dark cross bands; tail with regular dark annuli; throat without, or with rather indistinct, oblique blackish lines.

Total length	326	millim.
Head	33	,,
Width of head	22	,,
Body	78	,,
Fore limb	59	,,
Hind limb	90	,,
Tail	215	,,

Philippine Islands.

a, b–c. ♀ & yg.	Philippines.	H. Cuming, Esq. [C.].	(Types.)
d–f. ♂ ♀.	Philippines.	H. Cuming, Esq. [C.].	(Types of *T. petersii*.)

10. Gonyocephalus semperi.

Lophyrus (Tiaris) semperi, *Peters, Mon. Berl. Ac.* 1867, p. 16.

Nuchal crest high, not continuous with the dorsal, which is much lower. The enlarged dorsal scales forming on each side a regular series parallel with the dorsal crest. Ventral scales strongly keeled. Olive-brown or greenish-grey above, uniform in the adult, transversely barred with brown in the young; gular sac brownish, or marbled with bluish; borders of the eyelids and infraorbital border white; tail with broad brown annuli.

Philippine Islands.

11. Gonyocephalus interruptus. (PLATE XXI.)

Snout as long as the diameter of the orbit; canthus rostralis and supraciliary edge projecting; tympanum smaller than the eye-opening; upper head-scales small, keeled, slightly enlarged on the supraorbital region; scattered conical scales on the back of the head and on the temple; ten or eleven upper and eleven or thirteen lower labials. Gular sac small, without serrated anterior edge; gular scales strongly keeled, smaller than ventrals, intermixed with a few enlarged ones on the sides. Nuchal crest formed of lanceolate spines with smaller keeled ones at the base; its height equals two thirds the length of the snout; dorsal crest not continuous with, and lower than, the nuchal, gradually decreasing in height. Dorsal scales very small, smooth or indistinctly keeled, with the points directed upwards and backwards; enlarged scales on the flanks, forming irregular vertical series; numerous small tubercles on the nape; ventral scales of moderate size, keeled. Limbs with unequal keeled scales; third and fourth fingers equal; the adpressed hind limb reaches the nostril. Tail strongly compressed, with slightly serrated upper edge; caudal scales all keeled, largest inferiorly; length of the tail twice that of head and body. Pale olive above, the enlarged tubercles and two cross bands, one on the nape, the other between the shoulders, whitish; tail with regular dark-brown annuli; limbs with rather indistinct dark cross bands; throat with very indistinct dark lines.

Total length	280	millim.
Head	29	,,
Width of head	22	,,
Body	66	,,
Fore limb	53	,,
Hind limb	85	,,
Tail	185	,,

Philippine Islands.

a. ♂.   Mindanao.   Mr. G. Taylor [P.].

12. Gonyocephalus dilophus.

Tiaris megapogon (*Schleg.*), *Gray, Cat.* p. 239.
Lophyrus dilophus, *Dum. & Bibr.* iv. p. 419, pl. xlvi.
Gonyocephalus (Lophosaurus) dilophus, *Fitz. Syst. Rept.* p. 45.
Tiaris dilophus, *Doria, Ann. Mus. Genov.* vi. 1874, p. 345.
Gonyocephalus (Tiaris) dilophus, *Peters & Doria, Ann. Mus. Genov.* xiii. 1878, p. 376.

Upper head-scales small, strongly keeled; occipital enlarged; tympanum quite as large as the eye-opening; eleven upper and as many lower labials. Gular sac very large, with strongly serrated anterior edge; gular scales small, keeled, with scattered strongly enlarged ones. Nuchal crest not continuous with dorsal, its height greater than the length of the snout; it is composed of a few large lanceolate spines directed backwards, implanted on a strong dermal fold covered with small keeled scales; dorsal crest formed of spines similar to the nuchals. Dorsal scales small, keeled, with the points turned upwards and backwards, intermixed with scattered enlarged ones forming more or less regular transverse series; ventral scales twice as large as dorsals, keeled. Limbs above with subequal keeled scales. Tail strongly compressed, with a very high crest of lanceolate scales; caudal scales feebly keeled except inferiorly; tail once and a half the length of head and body. Reddish-brown above, with indistinct darker spots; dorsal crest olive.

Total length .... 550 millim.
Head .... 75 ,,
Body .... 145 ,,
Fore limb .... 105 ,,
Hind limb .... 193 ,,
Tail.................. 330

New Guinea.

13. Gonyocephalus tuberculatus.

Tiaris tuberculatus, *Günth. Proc. Zool. Soc.* 1872, p. 593, pl. xxxviii.

Snout as long as the diameter of the orbit; canthus rostralis and supraciliary edge sharp, projecting; tympanum as large as the eye-opening; occiput concave; upper head-scales small, strongly keeled; twelve upper and ten lower labials. Gular sac rather small, without serrated anterior edge; gular scales smooth, smaller than ventrals; a large subconical tubercle below the tympanum. Nuchal crest not continuous with dorsal, its greatest height somewhat exceeding the length of the snout; it is composed of lanceolate spines directed backwards, very small anteriorly, gradually becoming very large; these spines implanted on a dermal fold covered on each side with three or four rows of keeled pointed scales directed upwards; dorsal crest formed of large lanceolate spines like the posterior nuchals, with a basal series of smaller feebly keeled scales. Dorsal scales smooth or very feebly keeled, with the points directed upwards, intermixed with irregularly scattered enlarged ones; ventral scales moderately large, smooth. Limbs above with subequal feebly keeled scales; fourth finger slightly longer than third; the adpressed hind limb reaches the eye. Tail very strongly compressed, with an upper crest composed of triangular lobes gradually decreasing in size towards the end; caudal scales large, the upper feebly, the lower strongly keeled; length of the tail not twice that of head and body. Olive, with very indistinct brown spots; gular fold black.

Total length .... 375 millim.
Head .... 37 ,,
Width of head .... 24 ,,
Body .... 103 ,,

                Fore limb .... 77 millim.
                Hind limb 113   ,,
                Tail (end broken) ...... 235 ,,
East-Indian archipelago.

a. ♂.   East-Indian archipelago.   Dr. Bleeker. (Type.)

14. Gonyocephalus spinipes.

Lophyrus spinipes, *A. Dum. Cat. Méth. Rept.* p. 90, *and Arch. Mus.* viii. p. 568.

Snout slightly longer than the diameter of the orbit; canthus rostralis and supraciliary edge angular, projecting: tympanum as large as the eye-opening; occiput concave; upper head-scales small, subequal, keeled; fourteen or fifteen upper and as many lower labials. Gular sac rather small, with serrated anterior edge; gular scales strongly keeled, smaller than ventrals, intermixed with a few enlarged ones. Nuchal and dorsal crests subcontinuous; the former composed of triangular spines, the longest of which do not quite equal the vertical diameter of the tympanum, inserted on a fold of the skin; dorsal crest, a denticulation formed of triangular scales. Dorsal scales small, unequal, keeled, intermixed with irregularly scattered, enlarged, strongly keeled mucronate ones; the points of the scales directed upwards and backwards; ventral scales rather small, strongly keeled. Limbs above with unequal keeled scales, intermixed with strongly enlarged mucronate ones; fourth finger longer than third; the adpressed hind limb reaches the anterior border of the orbit or the tip of the snout. Tail compressed, at the base with a serrated upper edge; caudal scales strongly keeled, largest inferiorly; tail twice as long as head and body. Brown or dark olive above, with indistinct darker spots; a more or less distinct white line along the hind side of the thigh.

                Total length .... 368 millim.
                Head 39   ,,
                Width of head 27   ,,
                Body 84   ,,
                Fore limb 65   ,,
                Hind limb .... 105 ,,
                Tail .... 245 ,,
Queensland.

a. ♂.   Queensland.   H.M.S. 'Challenger.'
b-c. ♂.   Richmond River.   A. P. Godwin, Esq. [C.].

15. Gonyocephalus subcristatus.

Tiaris subcristata, *Blyth, Journ. As. Soc. Beng.* xxix. 1861, p. 109; *Stoliczka, Journ. As. Soc. Beng.* xxxix. 1870, p. 180, and xli. 1872, p. 116.

Coryphophylax maximiliani (*Fitz.*), *Steind. Novara, Rept.* p. 30, pl. ii. fig. 6.

Snout longer than the diameter of the orbit; canthus rostralis and supraciliary edge sharp; tympanum nearly as large as the eye-opening; upper head-scales of unequal size, strongly keeled; a few enlarged tubercles on the back of the head; seven or eight upper and as many lower labials. Gular sac very small; gular scales smaller than ventrals, keeled. Nuchal crest not continuous with dorsal, formed of triangular spines, the longest of which measure less than the diameter of the eye-opening, inserted on a slight fold of the skin; dorsal crest a serrated ridge. Dorsal scales small, keeled, with the points directed upwards and backwards, with a few irregularly scattered enlarged ones; ventral scales rather small, strongly keeled. Limbs above with subequal keeled scales; third and fourth fingers equal; the adpressed hind limb reaches the anterior border of the orbit, or between the latter and the nostril. Tail strongly compressed, with serrated upper edge, with keeled scales which are larger inferiorly; the length of the tail a little more than twice that of head and body. Brown or olive above, spotted or reticulated with black on the sides; sometimes a light, dark-edged band along each side of the back.

Total length	294	millim.
Head	26	,,
Width of head	16	,,
Body	58	,,
Fore limb	46	,,
Hind limb	82	,,
Tail	210	,,

Andaman and Nicobar Islands.

a-b. ♂.	Andaman Islands.	Dr. Anderson [P.].
c-e. Hgr.	Nicobar Islands.	Prof. Reinhardt [P.].
f-g. ♂ ♀.	Nicobar Islands.	

16. Gonyocephalus humii.

Tiaris humei, *Stoliczka, Journ. As. Soc. Beng.* xlii. 1873, p. 167.

Near *G. subcristatus*, but with the crest very much more developed; the nuchal part is considerably higher than the dorsal one. General coloration greenish olive, on the top of the head brownish; sides of the entire body more or less distinctly and rather densely reticulated and spotted with black and yellow; sides of head and neck and the gular sac tinged with purplish blue, labials spotted with blue; chin mostly yellow.

Head and body 110 millim.; tail 295 millim.

Tillingchang Island, Nicobars.

17. Gonyocephalus modestus.

Gonyocephalus (Hypsilurus) modestus, *Meyer, Mon. Berl. Ac.* 1874, p. 130.
—— (Arua) inornatus, *Doria, Ann. Mus. Genov.* vi. 1874, p. 345, pl. xi. fig. e.
—— (Arua) modestus, *Peters & Doria, Ann. Mus. Genov.* xiii. 1878, p. 380.

Snout rather pointed, a little longer than the diameter of the orbit; canthus rostralis and supraciliary edge not projecting; tympanum as large as the eye-opening; upper head-scales very small, subequal, sharply keeled; occipital enlarged; ten to twelve upper and seven or eight lower labials. Gular sac small, without serrated edge; gular scales very small, granular. Nuchal crest (in both sexes) reduced to a few widely separated, short, triangular spines, the anterior of which falls above the centre of the tympanum; no dorsal crest properly speaking, but a vertebral series of slightly enlarged keeled scales. Dorsal scales very small, equal, keeled, with the points directed upwards and backwards; ventral scales small, keeled. Limbs above with equal keeled scales; fourth finger a little longer than third; the adpressed hind limb reaches the tip of the snout. Tail feebly compressed, without serrated upper edge; caudal scales all keeled, largest inferiorly; length of the tail twice and two thirds that of head and body. Olive-brown; limbs and tail with darker cross bands.

Total length	335	millim.
Head	23	,,
Width of head	14	,,
Body	67	,,
Fore limb	54	,,
Hind limb	84	,,
Tail	245	,,

New Guinea.

a. ♂. Aru Island. Marquis G. Doria [P.].
(One of the types of *G. inornatus*.)

18. Gonyocephalus geelvinkianus.

Gonyocephalus (Arua) geelvinkianus, *Peters & Doria, Ann. Mus. Genov.* xiii. 1878, p. 381.

Closely allied to the preceding. Loreal scales smaller, in seven or eight series; eight or ten upper and eight or nine lower labials. Gular sac well developed; gular scales very small and keeled. Nuchal crest formed of lanceolate scales united at the base by a fold of the skin, decreasing abruptly in size between the shoulders; dorsal crest continuous with the nuchal, but much lower. Vinaceous brown or olive above, dotted or vermiculated with black; limbs and tail with dark cross bands.

Size of the preceding.

Northern New Guinea.

19. Gonyocephalus auritus.

Gonyocephalus (Hypsilurus) auritus, *Meyer, Mon. Berl. Ac.* 1874, p. 130.

—— (Arua) auritus, *Peters & Doria, Ann. Mus. Genov.* xiii. 1878, p. 382.

"Allied to *G. binotatus*, but much smaller. Crest hardly interrupted, low. A black spot on each side of the head, involving the ear. No large plates under the tympanum. Gular sac large, covered with small keeled scales."

New Guinea.

20. Gonyocephalus bruijnii.

Gonyocephalus (Hypsilurus) bruijnii, *Peters & Doria, Ann. Mus. Genov.* xiii. 1878, p. 379.

Upper head-scales strongly keeled; ten to twelve upper and nine or ten lower labials. Gular sac little developed (♀?); gular scales small, nearly smooth. Nuchal and dorsal crests subcontinuous, separated by a notch, the former slightly higher than the latter, composed of lanceolate lobes. Dorsal scales small, hardly keeled, forming slightly ascending transverse series; ventral scales much larger and strongly keeled. Brownish blue above, whitish beneath, with black spots at the base of the dorsal crest; tail with darker annuli.

Total length	580	millim.
Head	40	,,
Body	90	,,
Fore limb	70	,,
Hind limb	120	,,
Tail	450	,,

Mount Arfak, New Guinea.

21. Gonyocephalus binotatus.

Gonyocephalus (Hypsilurus) binotatus, *Meyer, Mon. Berl. Ac.* 1874, p. 130; *Doria, Ann. Mus. Genov.* vi. 1874, p. 345; *Peters & Doria, Ann. Mus. Genov.* xiii. 1878, p. 379.

"Crest uninterrupted, low, bluish. Gular sac moderately large, covered with small keeled scales. A large, strongly marked group of white plates below the tympanum. Beautifully marbled, especially on the nape. Two large, elongate, oval, black-edged, more strongly marked spots on the sides of the neck. Extremities and tail barred. Back black-dotted. Lower surface yellowish. During life with very fine pink and bluish tints."

New Guinea.

22. Gonyocephalus godeffroyi.

Lophura (Hypsilurus) godeffroyi, *Peters, Mon. Berl. Ac.* 1867, p. 707, pl. —. fig. 1.

Hypsilurus macrolepis, *Peters, Mon. Berl. Ac.* 1872, p. 775.
Tiaris longii, *Macleay, Proc. Linn. Soc. N. S. W.* ii. 1878, p. 103.

Snout longer than the diameter of the orbit; canthus rostralis and supraciliary edge angular, not projecting; occiput concave; tympanum a little larger than the eye-opening; cheeks strongly swollen under the ear in the males. Upper head-scales very small, equal, sharply keeled; occipital enlarged; ten to twelve upper and nine or ten lower labials. Gular sac moderately large, not serrated anteriorly; median gular scales enlarged and smooth, posterior very small and keeled; some strongly enlarged flat scales below the tympanum. Nuchal crest subcontinuous with the dorsal (a notch separating them); in the male it is formed of compressed lobes, the anterior the smallest, inserted on an elevated dermal fold covered with very small keeled scales turned upwards; the longest spines of the nuchal crest equal the greatest diameter of the tympanum; dorsal crest similar to the nuchal, but a little higher, its base supported by the greatly developed processes of the vertebræ. In the females and young the crests little developed. Dorsal scales small, equal, strongly keeled, with the points directed upwards and backwards; ventral scales moderate, very strongly keeled. Limbs above with equal strongly keeled scales; third and fourth fingers nearly equal; the adpressed hind limb reaches the anterior border of the orbit, or the tip of the snout. Tail strongly compressed, in the males with a high crest on its basal portion, covered with keeled scales which are larger inferiorly; its length more than three times that of head and body. Young greenish olive, uniform or with dark cross bars; adult dark brown, with more or less distinct yellowish cross bars.

Total length	1030	millim.
Head	65	,,
Width of head	55	,,
Body	170	,,
Fore limb	120	,,
Hind limb	195	,,
Tail	800	,,

New Ireland, New Britain, Solomon, and Fiji Islands; Northern Queensland.

a–d. ♂ ♀.	Duke of York Island.	Rev. G. Brown [C.].
e. ♂.	New Ireland.	
f–h. ♂ & hgr.	Solomon Islands.	G. Krefft, Esq.
i–k. Hgr. & yg.	Treasury Island.	H. B. Guppy, Esq. [P.].
l, m, n. Hgr. & yg.	Santa Anna Island.	H. B. Guppy, Esq. [P.].
o. Hgr. ♂.	San Christoval.	H.M.S. 'Herald.'
p. Hgr. ♂.	San Christoval.	Museum Economic Geology.
q. Yg.	San Christoval.	J. Brenchley, Esq. [P.].
r–t. Yg.	Fiji Islands.	
u. Yg.	——?	J. Macgillivray, Esq.

Gonyocephalus nigrigularis, Meyer, Mon. Berl. Ac. 1874, p. 129, said to be closely allied to the preceding, is not sufficiently characterized to be placed in the system.—New Guinea.

23. Gonyocephalus papuensis.

Tiaris papuensis, *Macleay, Proc. Linn. Soc. N. S. W.* ii. 1878, p. 101.
Gonyocephalus (Lophosteus) albertisii, *Peters & Doria, Ann. Mus. Genov.* xiii. 1878, p. 377.

Snout longer than the diameter of the orbit; canthus rostralis and superciliary edge angular, not projecting; occiput concave; tympanum much larger than the eye-opening; cheeks strongly swollen under the ear. Upper head-scales very small, equal, keeled; occipital enlarged; twelve upper and ten lower labials. Gular sac moderately large, not serrated anteriorly; gular scales feebly keeled, becoming very small posteriorly, the largest nearly as large as the ventrals; some strongly enlarged flat scales below the tympanum. Nuchal crest well separated from dorsal, formed of compressed lobes, the smallest being the anterior, inserted on an elevated dermal fold covered with very small feebly keeled scales turned upwards; the longest spines of the nuchal crest equal the greatest diameter of the tympanum; dorsal crest similar to, and not higher than, the nuchal, its base supported by the greatly developed spinose processes of the vertebræ. Dorsal scales small, equal, feebly keeled, the superior with the points directed upwards and backwards; ventral scales moderate, strongly keeled. Limbs above with equal keeled scales; third and fourth fingers equal; the adpressed hind limb reaches the eye. Tail strongly compressed, with a high upper crest on its basal portion, covered with keeled scales which are larger inferiorly; its length more than three times that of head and body. Whitish above, tinged with reddish, and with brown vermiculations; upper part of nape, and a spot on each side of the neck, brown; throat with wide-meshed brown reticulation.

Total length	680	millim.
Head	40	,,
Width of head	28	,,
Body	120	,,
Fore limb	85	,,
Hind limb	140	,,
Tail	520	,,

New Guinea.

a. ♂. Aleya, S.E. New Guinea.

24. Gonyocephalus boydii.

Tiaris boydii, *Macleay, Proc. Linn. Soc. N. S. W.* viii. 1884, p. 432.

Snout triangular and shelving; upper surface of head a little concave, covered with minute roughly keeled scales; scales largest below the mouth, expanding behind the ear into a few ivory-looking tubercles of larger size. Gular pouch covered with very minute pointed cales, with broad, pointed, compressed, triangular teeth along its median fold. A large compressed skinny nuchal crest, densely covered with very small smooth scales, and armed with three or four erect, broad, pointed, triangular, very compressed bony teeth; dorsal crest composed of acute, very much compressed, triangular bony spines. Dorsal scales small and more or less keeled; ventral scales larger and more strongly keeled. Scales on the legs keeled. Caudal scales very strongly keeled; occasionally bands of larger keeled scales across the tail at irregular intervals. Reddish brown, with seven or eight narrow dark-brown fasciæ on the body, and similar but indistinct fasciæ on the tail and feet; front of the head and nape greyish, sides of the head stone-blue; gular pouch and space between tympanum and eye yellow; underside of body greyish yellow.

Head and body 150 millim.; tail 300 millim.

Herbert River, Queensland.

25. Gonyocephalus grandis.

Dilophyrus grandis, Gray, Cat. p. 238.
Dilophyrus grandis, Cantor, Cat. Mal. Rept. p. 34, pl. xx.; *Günth. Rept. Brit. Ind.* p. 136.

Snout longer than the diameter of the orbit; canthus rostralis and supraciliary edge sharp, projecting; tympanum nearly as large as the eye-opening; upper head-scales very small, bluntly keeled, enlarged on the canthus rostralis and the supraciliary and supraorbital borders; one or two enlarged tubercles on each side behind the occiput; ten to twelve upper and as many lower labials. Gular sac moderately large, without serrated anterior edge; gular scales smaller than ventrals, smooth. Nuchal and dorsal crests subcontinuous, separated by a deep notch, composed of long lanceolate spines united together, free only at the tips, with smaller triangular smooth spines at the base; in the male, the height of the nuchal crest nearly equals the length of the snout, and the dorsal crest is a little lower; in the female, the former is scarcely developed and the latter is reduced to a slight serration. Dorsal scales very small, with the points directed upwards and backwards; ventral scales rather small, smooth. Limbs above with equal smooth or very feebly keeled scales; third and fourth fingers equal; the adpressed hind limb reaches between the eye and the tip of the snout. Tail strongly compressed, with sharp serrated upper edge; caudal scales smooth, with two rows of enlarged strongly keeled ones inferiorly; the length of the tail about twice and a half that of head and body. Brown or olive above, with or without darker cross bands; flanks with roundish yellow spots; throat sometimes with blue lines.

Total length	560	millim.
Head	44	,,
Width of head	27	,,
Body	111	,,

Fore limb .... 85 millim.
Hind limb .... 140 „
Tail .... 405 „

Birma; Malay peninsula; Sumatra; Borneo.

a, b. ♂.	Rangoon.	Dr. J. E. Gray [P.].	(Types.)
c. Skeleton.	Rangoon.	Dr. J. E. Gray [P.].	
d. ♂.	Penang.	Dr. Cantor.	
e-f. ♂.	Nias.	Hr. Sandemann [C.].	
g. ♂.	Matang.		
h, i. ♂ ♀.	Indian archipelago.	Dr. Bleeker.	

13. ACANTHOSAURA.

Acanthosaura, *Gray, Griff. Anim. Kingd.* ix. *Syn.* p. 56, *and Cat. Liz.* p. 240; *Fitzing. Syst. Rept.* p. 44; *Günth. Rept. Brit. Ind.* p. 147; *Stoliczka, Journ. As. Soc. Beng.* xli. 1872, p. 110.
Lophyrus, part., *Dum. & Bibr.* iv. p. 410.
Oriocalotes, *Günth. l. c.* p. 146.
Oriotiaris, *Günth. l. c.* p. 150.
Charasia, part., *Stoliczka, l. c.*

Tympanum distinct. Body generally compressed, limbs more or less elongate. Dorsal scales heterogeneous, small or moderate. A dorsal crest. No gular fold, but a more or less distinct oblique fold in front of the shoulder. No gular sac. No præanal or femoral pores.

South-Eastern Asia.

Synopsis of the Species.

I. Nuchal crest separated from dorsal.

  A. No spine on the side of the neck.

A long postorbital spine .... 1. *capra*, p. 300.

  B. A spine on each side of the neck.

Postorbital and nuchal spines nearly as
  long as the diameter of the orbit .... 2. *armata*, p. 301.
Postorbital spine half the diameter of
  the orbit .... 3. *crucigera*, p. 302.
Postorbital spine not half the diameter
  of the orbit .... 4. *lamnidentata*, p. 302.

II. Nuchal and dorsal crests continuous.

  A. A spine on each side of the neck.

The small dorsal scales much smaller
  than the ventrals .... 5. *coronata*, p. 303.
The smaller dorsal scales larger than the
  ventrals .... 6. *minor*, p. 304.

B. No spine on the side of the neck.

Dorsal scales not very unequal in size .. 7. *kakhienensis*, p. 305.
Dorsal scales very unequal in size .... 8. *major*, p. 306.
A serrated ridge on each side of the neck,
　parallel with the nuchal ridge ...... 9. *tricarinata*, p. 306.

1. Acanthosaura capra.

Acanthosaura capra, *Günth. Proc. Zool. Soc.* 1861, p. 188, *and Rept. Brit. Ind.* p. 148, pl. xiv. fig. F.

Snout as long as the diameter of the orbit; canthus rostralis and supraciliary edge angular; tympanum smaller than the eye-opening; upper head-scales small, subequal, keeled; a long spine, measuring three fourths the diameter of the orbit in the adult, half that diameter in the half-grown male, terminates the supraciliary edge; ten upper and twelve or thirteen lower labials; gular scales scarcely smaller than the ventrals, strongly keeled. A curved vertical fold on each side of the neck, in front of the shoulder. Nuchal crest not continuous with the dorsal, composed of large compressed spines which are a little longer than the postorbital spine, with the base concealed by two or three rows of smaller spines. Dorsal crest less developed than the nuchal, composed of triangular scales gradually decreasing in size behind. Dorsal scales very small, obtusely keeled, all with the points directed upwards; a few slightly enlarged ones are scattered among them; ventral scales larger, strongly keeled. Limbs above with subequal keeled scales; fourth finger a little longer than third; the hind limb stretched forwards reaches the temple or the eye. Tail compressed, once and one third to once and a half as long as head and body, covered with uniform strongly keeled scales, which are larger inferiorly. Greenish brown, with indistinct roundish light spots; throat of the adult blackish brown, with an orange-coloured streak along the middle.

The largest specimen, a skin, measures about 135 millim. from snout to vent; tail 175 millim.

The following measurements are taken from the smaller specimen:—

Total length	178	millim.
Head	20	,,
Width of head	15	,,
Body	53	,,
Fore limb	47	,,
Hind limb	60	,,
Tail	105	,,

Siam.

a–b. Ad. & hgr. ♂.	Chartaboum.	M. Mouhot [C.].	(Types.)
c. Ad., stuffed.	Chartaboum.	M. Mouhot [C.].	

2. Acanthosaura armata. (Plate XXII. fig. 1.)

Acanthosaura armata, *Gray, Cat.* p. 240.
Agama armata, *Gray, Zool. Journ.* iii. 1827, p. 216.
Calotes lepidogaster, *Cuv. R. A.* 2nd ed. ii. p. 39.
Acanthosaura armata, *Gray, Griff. A. K.* ix. *Syn.* p. 56; *Günth. Rept. Brit. Ind.* p. 148.
Lophyrus armatus, *Dum. & Bibr.* p. 413; *Cantor, Cat. Mal. Rept.* p. 32.
Gonyocephalus (Acanthosaura) armatus, *Fitz. Syst. Rept.* p. 44.

Snout as long as, or slightly shorter than, the diameter of the orbit; canthus rostralis and supraciliary edge angular; tympanum nearly as large as the eye-opening; upper head-scales keeled, much larger on the supraorbital region; a long spine, measuring from three fourths to once the diameter of the orbit, terminates the supraciliary edge; eleven to thirteen upper and thirteen to fifteen lower labials; gular scales strongly keeled, much smaller than the ventrals. An oblique fold on each side of the neck, in front of the shoulder; a large spine, of the same size as the postorbital, on each side of the nape above the tympanum. Nuchal crest not continuous with the dorsal, composed of large compressed spines, as long as the postorbital ones, with the base concealed by two or three rows of smaller spines. Dorsal crest anteriorly as high as the nuchal, becoming very low behind. Dorsal scales extremely small, all with the points directed upwards, intermixed with irregularly scattered, enlarged, rhomboidal strongly keeled scales; ventral scales a little larger than the enlarged dorsals, strongly keeled. Fore limb and tibia above with equal keeled scales, femur with unequal ones; fourth finger very slightly longer than third; the hind limb stretched forwards reaches between the temple and the nostril. Tail compressed, once and one fourth to once and a half as long as head and body, covered with uniform strongly keeled scales, which are larger inferiorly. Brownish above, with round lighter spots, or light with irregular dark spots; some dark lines radiate from the eye; an oblique dark-brown band down each side of the neck, from the interruption between the nuchal and dorsal crests, to the throat.

Total length	298	millim.
Head	33	,,
Width of head	26	,,
Body	105	,,
Fore limb	73	,,
Hind limb	101	,,
Tail	160	,,

Singapore; Penang; Tenasserim; Siam; Cochin China.

a, b. ♀ & hgr.	Singapore.	Gen. Hardwicke [P.]. (Types.)
c. ♀.	Singapore.	
d–e. ♂ ♀.	Penang.	Dr. Cantor.
f. ♂.	Chartaboum.	M. Mouhot [C.].

3. Acanthosaura crucigera. (Plate XXII. fig. 2.)

Acanthosaura armata, part., *Blanf. Journ. As. Soc. Beng.* xlviii. 1879, p. 130.

Snout as long as the diameter of the orbit; canthus rostralis and supraciliary edge angular; tympanum smaller than the eye-opening; upper head-scales rather feebly keeled, much larger on the supra-orbital region; a spine, measuring half the diameter of the orbit, terminates the supraciliary edge; nine or ten upper and as many lower labials; gular scales strongly keeled, much smaller than the ventrals. An oblique fold down each side of the neck, in front of the shoulder; a spine of the same size as the postorbital on each side of the nape, above the tympanum. Nuchal crest not continuous with the dorsal, composed of large compressed spines a little longer than the postorbital ones, with the base concealed by two or three rows of smaller spines. Dorsal crest much lower than the nuchal, gradually decreasing in size, reduced posteriorly to a very feeble denticulation. Dorsal scales extremely small, with the point directed upwards, intermixed with irregularly scattered, enlarged, rhomboidal strongly keeled scales; ventral scales larger than the enlarged dorsals, strongly keeled. Fore limb and tibia above with equal keeled scales, femur with unequal ones; fourth finger very slightly longer than third; the adpressed hind limb reaches the anterior border of the orbit in the female, the end of the snout in the male. Tail compressed, nearly twice as long as head and body, covered with uniform strongly keeled scales, which are larger inferiorly. Pinkish grey above, with brown spots and marblings enclosing roundish light spots; a large cruciform dark-brown marking on the nape, the lateral branches descending along the antehumeral fold to the throat; eyelids and a streak from the end of the snout to the tympanum, through the orbit, dark brown; limbs and tail with dark and light cross bands.

Total length	267	millim.
Head	26	,,
Width of head	19	,,
Body	71	,,
Fore limb	57	,,
Hind limb	84	,,
Tail	170	,,

Tenasserim.

a. ♂. Tavoy. W. Davison, Esq. [C.], W. T. Blanford, Esq. [P.].
b. ♀. Tavoy. W. Davison, Esq. [C.].

4. Acanthosaura lamnidentata. (Plate XXII. fig. 3.)

Snout as long as the diameter of the orbit; canthus rostralis and supraciliary edge angular; tympanum a little smaller than the eye-opening; upper head-scales keeled, larger on the supraorbital region; a spine, measuring two fifths the diameter of the orbit, terminates the supraciliary edge; ten or eleven upper and eleven or twelve lower labials; gular scales strongly keeled, smaller than the ventrals. An oblique fold down each side of the neck, in front of the shoulder; a spine a little shorter than the postorbital on each side of the nape above the tympanum. Nuchal crest not continuous with the dorsal, composed of compressed triangular scales, the longest of which equal the postorbital spine; the spines at the base of the nuchal crest small. Dorsal crest very low, formed of triangular scales, gradually decreasing in size posteriorly, where it is reduced to a slight denticulation. Dorsal scales extremely small, with the points directed upwards, intermixed with irregularly scattered, enlarged, rhomboidal strongly keeled scales; ventral scales larger than the enlarged dorsals, strongly keeled. Fore limb and tibia above with equal keeled scales, femur with unequal ones; fourth finger very slightly longer than third; the adpressed hind limb reaches the eye or the nostril. Tail compressed, once and two thirds or once and three fourths as long as head and body, covered with uniform strongly keeled scales, which are larger inferiorly. Brownish-olive above, with roundish lighter spots, or with a series of large dark-brown spots on each side of the back; a large dark-brown marking on the nape, widening anteriorly; hind limbs and tail with dark transverse bars.

Total length	280	millim.
Head	28	,,
Width of head	20	,,
Body	82	,,
Fore limb	57	,,
Hind limb	92	,,
Tail	170	,,

Pegu; Tenasserim.

| a. ♂. | Pegu. | W. Theobald, Esq. [C.] |
| b. ♀. | Tenasserim. | Col. Beddome [C.]. |

5. Acanthosaura coronata.

Acanthosaura coronata, *Günth. Proc. Zool. Soc.* 1861, p. 187, *and Rept. Brit. Ind.* p. 149, pl. xiv. fig. E.

Snout as long as the diameter of the orbit; canthus rostralis angular; supraciliary edge very prominent, serrated behind; tympanum smaller than the eye-opening; upper head-scales keeled, a little larger on the supraorbital region; a very small conical tubercle or spine behind the supraciliary edge; eleven upper and eleven or twelve lower labials; gular scales much smaller than the ventrals, strongly keeled. An oblique fold on each side in front of the shoulder. A small spine, shorter than the greatest diameter of the tympanum, on each side of the nape above the tympanum. Nuchal and dorsal crests continuous, the former composed of short compressed spines, shorter than the greatest diameter of the tympanum, the latter reduced to a denticulation. Dorsal scales very small, smooth or feebly keeled, with the points directed backwards or slightly upwards, intermixed with irregularly scattered, slightly enlarged, keeled ones; ventral scales larger than the enlarged dorsals, strongly keeled. Limbs, and especially digits, shorter than in the preceding species; scales on the fore limb and tibia subequal, strongly keeled, of femur more irregular; fourth finger slightly longer than third; the adpressed hind limb reaches the tympanum in the female, the eye in the male. Tail scarcely compressed, as long as, or a little longer than, head and body, covered with uniform strongly keeled scales, which are larger inferiorly. Ground-colour of the male grey, of the female brownish red; irregular dark-brown bands across the back and the legs; a yellowish-olive band edged with black across the crown, from one supraciliary edge to the other; an oblique short yellow band, broadly edged with brown, from below the orbit to the angle of the mouth; other dark-brown streaks radiating from the eye.

Total length	175	millim.
Head	24	,,
Width of head	20	,,
Body	66	,,
Fore limb	46	,,
Hind limb	65	,,
Tail	85	,,

Siam.

a–c. ♂ ♀. Chartaboum.   M. Mouhot [O.]. (Types.)

6. Acanthosaura minor. (PLATE XXIII. fig. 2.)

Calotes minor, *Gray, Cat.* p. 244.
Oriocalotes minor, *Günth. Rept. Brit. Ind.* p. 147.
Charasia minor, *Theob. Cat. Rept. Brit. Ind.* p. 113.

Snout a little longer than the diameter of the orbit; canthus rostralis and supraciliary edge angular; tympanum smaller than the eye-opening; upper head-scales keeled, of very irregular size, enlarged on the supraorbital region; a small spine, measuring a little less than the vertical diameter of the tympanum, behind the supraciliary edge; two other similar spines on each side of the back of the head, the foremost nearly equally distant from the tympanum and the nuchal crest, the other just above the tympanum; seven or eight upper and as many lower labials. Gular scales a little smaller than the ventrals. An oblique fold on each side of the neck, in front of the shoulder. Nuchal and dorsal crests continuous, low, reduced to a serrated ridge on the back. Dorsal scales rather large, strongly keeled, the upper with the points directed obliquely upwards and backwards; some enlarged scales are scattered on the sides; ventral scales very strongly keeled, smaller than dorsals. Limbs above with subequal strongly keeled scales; fourth finger very slightly longer than third; the adpressed hind limb reaches between the neck-fold and the temple. Tail scarcely compressed, nearly twice as long as head and body, covered with uniform strongly keeled scales, which are not larger inferiorly. Pale brownish-olive above, with irregular dark-brown spots or marblings, frequently forming irregular cross bands; several dark streaks radiate from the eye; throat with more or less distinct irregular dark transverse lines.

Total length	217 millim.
Head	21 ,,
Width of head	15 ,,
Body	51 ,,
Fore limb	33 ,,
Hind limb	45 ,,
Tail	145 ,,

Eastern Himalayas.

a–c. ♀ & yg.	[Afghanistan.]	East-India Co. [P.]. (Types.)
d. ♀.	Khasia.	East-India Co. [P.].
e–h. ♀ & yg.	Khasia.	T. C. Jerdon, Esq. [P.].
i. ♀.	Sikkim.	Messrs. v. Schlagintweit [C.].
k. ♀.	Sikkim.	T. C. Jerdon, Esq. [P.].
l. ♀.	——?	Dr. Griffith.

7. Acanthosaura kakhienensis.

Oriocalotes kakhienensis, *Anders. Zool. W. Yunnan*, p. 806, pl. lxxvi. fig. 1.

Canthus rostralis and supraciliary ridge not well defined; upper head-scales obtusely keeled, of different sizes; no spines on the head. A slight fold above and in front of the shoulder. Nuchal crest composed of six to eight triangular spines, disappearing a short way behind the shoulders. Dorsal scales of moderate size, keeled; those on the side of the back directed upwards and backwards, and those below downwards and backwards; a few large keeled scales scattered over the sides; scales of chest and belly of moderate size and strongly keeled. The hind limb extends to the angle of the jaw. Base of tail compressed, but thick and somewhat rounded. General colour above olive, irregularly variegated with brown and yellow, these colours having a tendency to arrange themselves in cross bands; a broad black band from the posterior margin of the eye to the tympanum; black streaks radiate from the eye; under surface olive-green.

Size of *A. major*.

Khasia Hills.

8. Acanthosaura major. (PLATE XXIII. fig. 3.)

Oriocalotes major, *Jerdon, Proc. As. Soc. Beng.* 1870, p. 77.
Charasia major, *Theob. Cat. Rept. Brit. Ind.* p. 113.

Snout longer than the diameter of the orbit; canthus rostralis and supraciliary edge angular; tympanum a little larger than the eye-opening; upper head-scales unequal, keeled, a series of larger ones bordering inwards the supraorbital region; no postorbital spine; a row of projecting triangular scales borders the head posteriorly; seven upper and as many lower labials; gular scales smaller than ventrals, feebly keeled. A rather indistinct oblique fold in front of the shoulder. Nuchal and dorsal crest continuous, reduced to a serrated ridge. Dorsal scales very irregular, some very small, others very large; all are distinctly keeled, and the upper have their points directed upwards and backwards; ventral scales distinctly keeled, smaller than the largest dorsals. Limbs above with unequal keeled scales; fourth finger very slightly longer than third; the adpressed hind limb reaches the tympanum. Tail compressed, once and two thirds as long as head and body, covered with unequal keeled scales. Olive above; back with six large angular blackish cross bands, with the apex directed backwards; flanks reticulated with blackish; head above with blackish cross streaks; an oblique black band from the eye to the tympanum, continued along the side of the neck; limbs and tail with dark cross bars.

Total length	248	millim.
Head	23	,,
Width of head	18	,,
Body	70	,,
Fore limb	42	,,
Hind limb	63	,,
Tail	155	,,

Western Himalaya.

a. ♀. Valley of the Sutlej, near Kotegurh. T. C. Jerdon, Esq. [P.]. (Type.)

9. Acanthosaura tricarinata.

Calotes tricarinatus, *Blyth, Journ. As. Soc. Beng.* xxii. 1854, p. 650.
Tiaris elliotti, *Günth. Proc. Zool. Soc.* 1860, p. 151, pl. xxv. fig. B.
Oriotiaris elliotti, *Günth. Rept. Brit. Ind.* p. 150; *Jerdon, Proc. As. Soc. Beng.* 1870, p. 77.
Oreotiaris tricarinata, *Anders. Proc. Zool. Soc.* 1871, p. 167.
Charasia (Oriotiaris) tricarinata, *Stoliczka, Journ. As. Soc. Beng.* xli. 1872, p. 112.

Snout slightly longer than the diameter of the orbit; canthus rostralis and supraciliary edge angular; tympanum nearly as large as the eye-opening; upper head-scales unequal, feebly keeled, a series of larger ones bordering inwards the supraorbital region; a conical tubercle behind the supraciliary edge; an oblique transverse row of conical ribbed tubercles on each side, bordering the head posteriorly; five or six upper and as many lower labials; gular scales smaller than ventrals, indistinctly keeled; a conical tubercle below the tympanum. Oblique fold in front of the shoulders very indistinct. Body not, or but very slightly, compressed. Nuchal and dorsal crests continuous, reduced to a series of enlarged strongly keeled scales; another parallel series of enlarged scales on each side of the nape, lost a short distance beyond the shoulder. Dorsal scales very irregular, scarcely imbricate, smooth or feebly keeled; strongly enlarged keeled scales are scattered on the flanks, and frequently others form angular series across the back; ventral scales as large as the enlarged dorsals, feebly keeled. Limbs above with unequal strongly keeled scales; fourth finger longer than third; the adpressed hind limb reaches the eye or the tip of the snout. Tail not compressed, twice to twice and a half as long as head and body, covered with rather unequal keeled scales. Brownish-grey or blue above, uniform or with angular brown markings on the back; lateral nuchal denticulation brown-edged; lower surfaces uniform whitish or with small black dots.

Total length	162	millim.
Head	14	,,
Width of head	10·5	,,
Body	38	,,
Fore limb	27	,,
Hind limb	46	,,
Tail	110	,,

Eastern Himalaya.

a. ♀.	Sikkim.	Messrs. v. Schlagintweit [C.].	(Types of
b–d. ♂ ♀.	Sikkim.	Sir J. Hooker [P.].	*Tiaris elliotti*.)
e–f. ♂ ♀.	Sikkim.	T. C. Jerdon, Esq. [P.].	
g–i. ♂ ♀.	Sikkim.	W. T. Blanford, Esq. [P.].	
k. ♂.	Darjeeling.	W. T. Blanford, Esq. [P.].	

14. JAPALURA.

Japalura, *Gray, Ann. & Mag. N. H.* (2) xii. 1853, p. 387; *Günth. Rept. Brit. Ind.* p. 132.
Biancia, *Gray, l. c.*
Diploderma, *Hallow, Proc. Ac. Philad.* 1860, p. 490.

Tympanum hidden. Body compressed. Dorsal scales heterogeneous. A dorsal crest. Gular pouch small or absent. An oblique fold in front of the shoulder; gular fold present or absent. Tail feebly compressed. No præanal or femoral pores.

East Indies; Southern China.

Synopsis of the Species.

I. Third and fourth fingers equal.

Infracaudal scales not larger than ventrals;
no transverse gular fold; tibia shorter
than the skull .... 1. *variegata*, p. 308.
Infracaudal scales (at the base of the tail)
larger than ventrals; no transverse gular
fold; tibia as long as the skull........ 2. *swinhonis*, p. 309.
Only seven upper labials .... 3. *polygonata*, p. 310.
Infracaudal scales larger than ventrals; a
transverse gular fold; tibia shorter than
the skull......................... 4. *yunnanensis*, p. 310.

II. Fourth finger longer than third.

No transverse gular fold; body subquad-
rangular........................... 5. *planidorsata*, p. 311.
A transverse gular fold; body strongly
compressed........................ 6. *nigrilabris*, p. 311.

1. Japalura variegata. (PLATE XXIV. fig. 1.)

Japalura variegata, *Gray, Ann. & Mag. N. H.* (2) xii. p. 388; *Günth. Rept. Brit. Ind.* p. 133; *Anders. Proc. Zool. Soc.* 1871, p. 164; *Stoliczka, Journ. As. Soc. Beng.* xli. 1872, p. 106.
Biancia niger, *Gray, l. c.* p. 387.
Japalura microlepis, *Jerdon, Proc. As. Soc. Beng.* 1870, p. 76; *Anders. Zool. W. Yunnan*, p. 804.

Snout a little longer than the diameter of the orbit; canthus rostralis sharp; upper head-scales irregular, sharply keeled · a small tubercular scale behind the supraciliary edge; other scattered conical scales on the hind part of the head; nine to eleven upper and as many lower labials. Male with a very small gular pouch. A well-developed nuchal crest. Body compressed, covered above and on the sides with small keeled scales intermixed with larger ones, all having their points obliquely directed upwards and backwards; dorsal crest, a slight serrated ridge; ventral scales as large as the largest dorsals, strongly keeled. Limbs above with scales of irregular size; third and fourth fingers nearly equal; fifth toe not quite as long as third; the adpressed hind limb reaches the eye; tibia shorter than the skull. Tail compressed, slender, once and a half to once and three fourths as long as head and body, covered above with unequal scales, inferiorly with equal larger ones, the size of which does not exceed that of the ventrals; all the caudal scales keeled. Olive-brown above, generally with alternating broad dark and narrower light chevron-shaped bands on the body; flanks reticulated; a more or less distinct light band on the upper lip and along each side of the neck; tail with dark annuli; middle of the throat generally black.

Total length	316 millim.
Head	29 ,,
Width of head	18 ,,
Body	82 ,,
Fore limb	53 ,,
Hind limb	90 ,,
Tail	205 ,,

Himalayas.

a–b. ♀ & hgr.	Sikkim.	Sir J. Hooker [P.]. (Types.)
c, d–e. ♂ ♀.	Sikkim.	Sir J. Hooker [P.]. (Types of *Biancia niger*.)
f. ♀.	Sikkim.	Messrs. v. Schlagintweit [C.].
g. ♀.	Sikkim.	T. C. Jerdon, Esq. [P.] (Type of *J. microlepis*.)
h–i. Yg.	Sikkim.	W. T. Blanford, Esq. [C.].
k–o. ♂.	Darjeeling.	
p–q. ♂ ♀.	Darjeeling.	T. C. Jerdon, Esq. [P.].

2. Japalura swinhonis.

Japalura swinhonis, *Günth. Rept. Brit. Ind.* p. 133, pl. xiv. fig. B.

Snout a little longer than the diameter of the orbit; canthus rostralis sharp; upper head-scales very small, irregular, keeled; a small tubercular scale behind the supraciliary edge, and others scattered on the hind part of the head; eight or nine upper and as many lower labials. Male with a very small gular pouch. A small nuchal crest. Body strongly compressed, covered above and on the sides with small keeled scales intermixed with larger ones; only the upper dorsal scales have their points directed upwards; dorsal crest, a very slightly serrated ridge; ventral scales only a little larger than the gulars, strongly keeled. Limbs above with scales of irregular size; third and fourth fingers nearly equal; fifth toe not quite as long as third; the adpressed hind limb reaches the anterior border of the orbit, or between the latter and the tip of the snout; tibia as long as the skull. Tail compressed, slender, twice and one third to twice and a half as long as head and body, covered above with unequal scales, inferiorly with equal large ones which are larger than the ventrals; all the caudal scales keeled. Olive-brown above, the back with dark chevron-shaped cross bands; a light longitudinal band on each side of the back; head generally dotted with blackish; sides reticulate with brown; throat greyish, with large white spots; tail with dark annuli.

Total length	284 millim.
Head	24 ,,
Width of head	16 ,,
Body	62 ,,
Fore limb	43 ,,
Hind limb	72 ,,
Tail	198 ,,

Formosa.

a. ♂.	Formosa.	R. Swinhoe, Esq. [C.]. (Type.)
b-c, d. ♂ & hgr.	Formosa.	R. Swinhoe, Esq. [C.].
e-f. ♂.	Formosa.	M. Dickson, Esq. [P.].

3. Japalura polygonata.

Diploderma polygonatum, *Hallow. Proc. Ac. Philad.* 1860, p. 490.
Japalura polygonata, *Günth. Rept. Brit. Ind.* p. 134.

Appears to be very closely allied to the preceding. Seven upper labials. Uniform greenish-olive above, somewhat deeper upon the back, much lighter beneath, with a marked tinge of yellow; eight dark bands upon the tail.

Amakarima Island, Loo Choo.

4. Japalura yunnanensis.

Japalura swinhoii (*non Gthr.*), *Swinhoe, Proc. Zool. Soc.* 1870, p. 411.
Japalura yunnanensis, *Anders. Zool. IV. Yunnan*, p. 803, pl. lxxvii. fig. 2.

Snout a little longer than the diameter of the orbit; canthus rostralis sharp; upper head-scales moderately large, irregular, keeled; a conical scale behind the supraciliary edge, and others scattered on the hind part of the head; eight upper and as many lower labials. No regular gular pouch; a well-marked transverse gular fold. A small nuchal crest. Body compressed, covered above and on the sides with small strongly keeled scales intermixed with larger ones which may form rather regular longitudinal series; upper dorsal scales with their points directed slightly upwards and backwards; dorsal crest a very slightly serrated ridge; dorsal scales a little larger than the gulars, strongly keeled. Limbs above with scales of unequal size; third and fourth fingers nearly equal; fifth toe nearly as long as third; the adpressed hind limb reaches the eye or the anterior border of the orbit; tibia shorter than the skull. Tail compressed, slender, a little more than twice as long as head and body, covered above with subequal scales; lower caudal scales larger, larger than the ventrals; all the caudal scales keeled. Olive above, head spotted with dark brown, the back with a series of dark spots or bands separated by narrow light interspaces; a more or less distinct light longitudinal band on each side of the back; sides reticulate with dark brown; limbs and tail with dark transverse bars.

Total length	288	millim.
Head	26	,,
Width of head	19	,,
Body	62	,,
Fore limb	46	,,
Hind limb	73	,,
Tail	200	,,

Western China.

a-c. ♂ ♀.   Szechuen.   R. Swinhoe, Esq. [C.].

5. Japalura planidorsata. (Plate XXIV. fig. 2.)

Japalura planidorsata, *Jerdon, Proc. As. Soc. Beng.* 1870, p. 70;
Stoliczka, Journ. As. Soc. Beng. xli. 1872, p. 106; *Anders. Zool.
W. Yunnan*, p. 804.

Snout not or scarcely longer than the diameter of the orbit; canthus rostralis sharp; upper head-scales irregular, sharply keeled, the largest on the supraorbital region; a few conical scales on the hinder border of the head; nine or ten upper and as many lower labials. Body subquadrangular, the back being plane, bordered by a slight ridge of enlarged scales on each side; dorsal and lateral scales very irregular, keeled, intermixed with enlarged ones which on the back form five angular transverse series, with the point turned backwards; a very slight serrated vertebral ridge; ventral scales as large as the largest dorsals, strongly keeled. Limbs above with scales of irregular size; fourth finger longer than third; fifth toe as long as third; the adpressed hind limb reaches the eye or a little beyond the tip of the snout; tibia as long as, or slightly shorter than, the skull. Tail slightly compressed, above with unequal scales, inferiorly with equal larger ones, the size of which does not exceed that of the ventrals; all the caudal scales keeled. Olive-brown above, flanks darker; a band on the upper lip, whitish.

Total length (tail broken).	95 millim.
Head	13 ,,
Width of head	10·5 ,,
Body	36 ,,
Fore limb	24 ,,
Hind limb	39 ,,

Khasia.

a–b. Hgr. (?).   Khasia.   T. C. Jerdon, Esq. [P.]. (Types.)

6. Japalura nigrilabris.

Otocryptis (Japalura) nigrilabris, *Peters, Mon. Berl. Ac.* 1864, p. 385.

Snout very short, shorter than the diameter of the orbit, with angular canthus rostralis; supraciliary ridge prominent, feebly serrated; upper head-scales irregular, sharply keeled; scattered enlarged tubercular scales on the posterior part of the head; thirteen upper and as many lower labials; gular scales small, spinose, with enlarged conical ones on the sides. Male with a small gular pouch. A strong transverse gular fold. Body compressed, covered above and on the sides with very small keeled scales intermixed with scattered enlarged ones, the points of which are directed upwards and backwards. Nuchal and dorsal crests a slight serrated ridge; ventral scales as large as the largest dorsals, strongly keeled. Limbs above with scales of irregular size; fourth finger longer than third; fifth toe as long as third; the adpressed hind limb reaches a little beyond the tip of the snout; tibia longer than the skull. Tail compressed, not twice as long as head and body, covered above with unequal scales, inferiorly with equal larger ones, the size of which exceeds that of the ventrals: all the caudal scales keeled. Light brown, with broad oblique blackish bands down each side; lips dark brown; some of the enlarged scales whitish.

Total length	132 millim.
Head	15 ,,
Width of head	11·5 ,,
Body	35 ,,
Fore limb	30 ,,
Hind limb	49 ,,
Tail	82 ,,

Borneo.

a. ♂. Mataug.

15. SALEA.

Salea, *Gray, Cat. Liz.* p. 242; *Günth. Rept. Brit. Ind.* p. 145.
Mecolepis, *A. Dum. Cat. Méth. Rept.* p. 87.
Lophosalea, *Beddome, Proc. Zool. Soc.* 1878, p. 153.

Tympanum distinct. Body compressed, covered with unequal-sized imbricate keeled scales. Male with a dorso-nuchal crest and a gular sac. No transverse gular fold. Tail compressed. No femoral or præanal pores.

Southern India.

1. Salea horsfieldii.

Salea horsfieldii, *Gray, Cat.* p. 242.
Salea jerdonii, *Gray, Ann. Mag. N. H.* xviii. 1846, p. 429; *Jerdon, Journ. As. Soc. Beng.* xxii. 1853, p. 473; *Kelaart, Prodr. Faun. Zeyl.* p. 167.
Mecolepis trispinosus, *A. Dum. Cat. Méth.* p. 88, *and Arch. Mus.* viii. p. 564, pl. xxiv. fig. 1.
—— hirsutus, *A. Dum. ll. cc.* pp. 88 & 566, fig. 2.
—— sulcatus, *A. Dum. ll. cc.* pp. 89 & 567, fig. 3.
Salea horsfieldii, *Günth. Rept. Brit. Ind.* p. 145.

Snout not more than once and a half as long as the diameter of the orbit, which equals about twice that of the tympanum; upper head-scales large, rugose, with a more or less marked curved series of regular ones bordering the supraorbital region internally; a row of three or four enlarged scales from the eye to above the tympanum. Gular scales lanceolate, keeled, ending in a spine as large as or a little larger than the ventrals. No fold in front of the shoulder. Nuchal crest, in the male composed of a few lanceolate spines directed backwards, the longest measuring nearly the length of the snout, with smaller ones at the base; in the female this crest is reduced to a double row of alternate oblique short spines; dorsal crest not continuous with nuchal, composed of similar slightly shorter lanceolate spines in the male, absent in the female. Dorsal scales large, rhomboidal, strongly keeled, pointing straight backwards; they are nearly always of unequal size, larger ones being scattered on the sides; ventral scales very strongly imbricate, strongly keeled and ending in a spine, nearly as large as the dorsals. Limbs moderately elongate, the adpressed hind limb reaching between the shoulder and the tympanum. Tail compressed, and with a small upper crest in the male, scarcely compressed, and without a crest in the female; caudal scales subequal, strongly keeled. Pale olive above, with irregular dark-brown cross bands, which may be broken up by a band of the light-brown colour running along each side of the back; the enlarged scales on the sides frequently white; a blackish band, edged below with white, extends from the eye to the fore limb, passing through the tympanum; tail usually with regular dark-brown annuli.

Total length	345 millim.
Head	27 ,,
Width of head	16 ,,
Body	68 ,,
Fore limb	44 ,,
Hind limb	61 ,,
Tail	250 ,,

Southern India.

a-b. ♀.	India.	East-India Company. (Types.)
c-d. ♂.	Madras.	T. C. Jerdon, Esq. [P.]. (Types of *Salea jerdonii*.)
e-f. ♀.	S. India.	T. C. Jerdon, Esq. [P.].
g. ♀.	Nilgherries.	Col. Beddome [C.].
h. Yg.	Malabar.	Col. Beddome [C.].

2. Salea anamallayana.

Lophosalea anamallayana, *Beddome, Proc. Zool. Soc.* 1878, p. 153, pl. xiv.

Snout long, measuring nearly twice the diameter of the orbit; tympanum once and two thirds or once and three fifths the diameter of the orbit; upper head-scales rather large, feebly keeled, with a regular curved series of enlarged scales bordering the supraorbital region internally; an enlarged tubercle behind the supraciliary edge and a few others scattered on the back of the head; a row of three or four enlarged scales from the eye to above the tympanum. Gular scales a little larger than ventrals, smooth or keeled. A well-marked curved fold on each side of the neck, in front of the shoulder. Dorso-nuchal crest continuous, composed of large lanceolate spines. Dorsal scales of unequal size, their arrangement varying considerably, strongly keeled, the upper ones pointing upwards and backwards, the others straight backwards or backwards and downwards; ventral scales very strongly imbricate, strongly keeled, and ending in a spine. The adpressed hind limb reaches the neck. Tail strongly compressed, in its anterior half with an upper crest nearly as much developed as the dorsal; caudal scales rather unequal in size, keeled. Pale olive above, with four broad angular dark-brown cross bands on the back, separated by narrow interspaces; head to the lip dark brown, with small light spots; limbs and tail with more or less regular dark-brown cross bars.

Total length	271	millim.
Head	33	,,
Width of head	19	,,
Body	78	,,
Fore limb	47	,,
Hind limb	68	,,
Tail (injured)	160	,,

Southern India.

a. ♂.   Anamallays, 6000 ft.   Col. Beddome [C.].   (Type.)
b-c. ♂.   Pulney hills.   Col. Beddome [C.].

16. CALOTES *.

Calotes, *Cuv. R. A.* ii. p. 35; *Fitzing. N. Classif. Rept.* p. 17; *Kaup, Isis,* 1827, p. 618; *Wagl. Syst. Amph.* p. 152; *Wiegm. Herp. Mex.* p. 14; *Dum. & Bibr.* iv. p. 391; *Fitzing. Syst. Rept.* p. 46; *Gray, Cat. Liz.* p. 242; *Günth. Rept. Brit. Ind.* p. 139.
Bronchocela, *Kaup, l. c.* p. 619; *Dum. & Bibr.* iv. p. 394; *Fitz. Syst. Rept.* p. 45; *Gray, Cat.* p. 240; *Günth. Rept. Brit. Ind.* p. 137.
Lophodeira, *Fitz. l. c.* p. 46.
Pseudocalotes, *Fitz. l. c.*

Tympanum distinct. Body compressed, covered with equal-sized scales. A dorso-nuchal crest. A more or less developed gular sac in the male; no transverse gular fold, or a very feebly marked one. Tail round or feebly compressed. No femoral or præanal pores.
Indian Region.

Synopsis of the Species.

I. No fold in front of the shoulder.

  A. Lateral scales pointing backwards and downwards.

    1. Ventral scales larger than latero-dorsals.

      a. Fourth finger nearly the same length as the fifth toe.

        a. Upper dorsal scales pointing upwards and backwards.

57 to 97 scales round the middle of the body; tympanum at least half the diameter of the orbit .... 1. *cristatellus*, p. 316.

---

* *Salea gularis*, Blyth, Journ. As. Soc. Beng. xxii. 1853, p. 473.—Mirzapore?

53 to 67 scales round the middle of the
body; tympanum hardly half the
diameter of the orbit............ 2. *celebensis*, p. 318.

    β. Only the scales of the row at the base of the dorsal crest point upwards.

61 to 75 scales round the middle of the
body; gular scales smaller than those
on upper surface of arm.......... 3. *marmoratus*, p. 318.
43 to 53 (65) scales round the middle
of the body; gular scales as large
as those on upper surface of arm .. 4. *jubatus*, p. 318.

    b. Fourth finger much longer than the fifth toe.

53 scales round the middle of the body 5. *smaragdinus*, p. 319.

    2. Ventral scales not larger than latero-dorsals.

43 to 53 scales round the middle of the
body .... 6. *tympanistriga*, p. 320.

  B. Lateral scales pointing backwards and upwards.

35 to 47 scales round the middle of
the body; tympanum at least half
the diameter of the orbit .... 7. *versicolor*, p. 321.
53 to 61 scales round the middle of the
body; tympanum not quite one
third the diameter of the orbit .... 8. *maria*, p. 322.

II. An oblique fold or pit covered with small granular scales in front of the shoulder, not extending across the throat.

  A. Dorsal scales larger than ventrals, keeled.

Two serrated parallel ridges on each
side of the back of the head; colour
green .... 9. *jerdonii*, p. 323.
A long spine behind the supraciliary
edge, and two others above the ear. 10. *emma*, p. 324.
No postorbital spine; a few small
spines above the ear .... 11. *mystaceus*, p. 325.

  B. Dorsal scales larger than ventrals, smooth.

Lateral scales pointing upwards and
backwards; 29 to 35 scales round
the middle of the body .... 12. *grandisquamis*, p. 325.
Lateral scales pointing upwards and
backwards; 37 to 43 scales round
the middle of the body .... 13. *nemoricola*, p. 326.
Lateral scales pointing downwards and
backwards .... 14. *liolepis*, p. 326.

C. Dorsal scales as large as or smaller than ventrals.

Lateral scales pointing upwards and backwards .... 15. *ophiomachus*, p. 327.
Lateral scales pointing downwards and backwards; ventrals much larger than dorsals .... 16. *nigrilabris*, p. 328.
Lateral scales pointing downwards and backwards; no spines whatever on the head..................... 17. *liocephalus*, p. 329.

III. An oblique fold in front of the shoulder, extending across the throat.

No spine at the posterior corner of the orbit....................... 18. *rouxii*, p. 330.
A small slender spine behind the supraciliary edge .... 19. *elliotti*, p. 330.

1. Calotes cristatellus.

Bronchocela cristatella, *Gray*, Cat. p. 241.
Agama cristatella, *Kuhl, Beitr. Zool. Vergl. Anat.* p. 108.
—— gutturosa, *Merr. Tent.* p. 51.
Calotes cristatellus, *Fitz. N. Class. Rept.* p. 49.
Agama moluccana, *Lesson, Coquille, Rept.* pl. i. fig. 2.
Calotes gutturosa, *Guérin, Icon. R. A., Rept.* pl. vii. fig. 3.
Agama vultuosa, *Gray, Griff. A. K.* ix. *Syn.* p. 56.
Bronchocela cristatella, *Kaup, Isis*, 1827, p. 619; *Dum. & Bibr.* iv. p. 395; *Cantor, Cat. Mal. Rept.* p. 30; *Girard, U.S. Explor. Exped., Herp.* p. 411; *Günth. Rept. Brit. Ind.* p. 138; *Peters, Mon. Berl. Ac.* 1867, p. 17; *Steind. Novara, Rept.* p. 27; *Stoliczka, Journ. As. Soc. Beng.* xxxix. 1870, p. 178.
—— (Lophodeira) cristatella, *Fitz. Syst. Rept.* p. 46.
—— moluccana, *Peters, Mon. Berl. Ac.* 1867, p. 17; *Stoliczka, l. c.* p. 179.
—— intermedia, *Peters & Doria, Ann. Mus. Genov.* xiii. 1878, p. 375.
—— burmana, *Blanf. Journ. As. Soc. Beng.* xlviii. 1878, p. 127.

Upper head-scales rather small, keeled, slightly enlarged on supraorbital region; a few more or less distinct, erect, compressed scales behind the supraciliary edge; tympanum half, or more than half, the diameter of the orbit. Gular pouch very small; gular scales keeled, smaller than ventrals, distinctly larger than latero-dorsals. No oblique fold in front of the shoulder. Nuchal crest composed of erect compressed spines, the longest never equalling the diameter of the orbit; dorsal crest a mere serrated ridge. 57 to 97 scales round the middle of the body; dorsal scales keeled, much smaller than ventrals, the upper ones directed upwards and backwards, the others downwards and backwards. The adpressed hind limb reaches between the eye and the tip of the snout; third and fourth fingers equal or nearly so, about as long as the fifth toe. Tail round, subtriangular at the base. Green, uniform or with reddish-white markings.

Total length	520	millim.
Head	37	,,
Width of head	22	,,
Body	83	,,
Fore limb	65	,,
Hind limb	105	,,
Tail	400	,,

Indo-Malayan peninsula and archipelago.

An examination of the series in the British Museum shows that a separation of *C. cristatellus* from *C. moluccanus* cannot be carried out, although indeed the extreme forms are rather different. The typical *C. cristatellus* has the rostral very small, separated from the nasal by three or four scales, the two first supralabials very small, the nasal being situated above the third labial, and eighty-one to ninety-seven scales round the middle of the body. *C. moluccanus* has the rostral larger, separated from the nasal by one or two scales, the nasal situated above the second labial, and fifty-nine to sixty-three scales. However, there are many exceptions, and as similar variations occur also, independently of localities, in the allied *C. jubatus*, I think a specific separation would not be justified.

We have specimens which, in the rostral scutellation, are typical *C. cristatellus*, but with seventy-five rows of scales; others with the same number of scales, but with the rostral scutellation of *C. moluccanus*; others have the nasal separated from the rostral by three scales and, nevertheless, situated above the second labials, &c.

That the number of scales does not exactly correspond with the localities is well shown by specimens *l-o*, which, of same size and similar in every other respect, vary in the number of scales round the middle of the body from seventy-five to ninety-one.

a, b. ♂ & hgr.	Penang.	Gen. Hardwicke [P.].
c-d. ♂.	Penang.	Dr. Cantor.
e-f. ♂.	Singapore.	
g-i. Hgr.	Singapore.	Sir E. Belcher [P.].
k. ♂.	Singapore.	Dr. Dennys [P.].
l-o. Hgr.	Nias.	Hr. Sandemann [C.].
p-q. ♂♀.	Borneo.	Sir E. Belcher [P.].
r-t, u-v. ♂ & hgr.	Borneo.	L. L. Dillwyn, Esq. [P.].
w. ♀.	Puerto Princesa.	A. Everett, Esq. [C.].
x-y. Hgr.	Dinagat Island.	A. Everett, Esq. [C.].
z-β. Hgr. & yg.	Celebes.	
γ. Hgr.	N. Ceram.	
δ-ζ. Hgr.	Mysol.	
η. Hgr.	Amboyna?	
θ. Hgr.	Timor Laut.	H. O. Forbes, Esq. [C.].
ι. Skeleton.	—— ?	

2. Calotes celebensis.

Bronchocela celebensis, *Gray, Cat.* p. 241.
Bronchocela celebensis, *Günth. Proc. Zool. Soc.* 1873, p. 168; *Peters & Doria, Ann. Mus. Genov.* xiii. 1878, p. 376.

Scarcely different from the preceding. The tympanum is smaller, the spines of the male's nuchal crest narrower, and the scales rather larger; 53 to 67 scales round the middle of the body.

Size and proportions as in the preceding.

Celebes.

a. ♂.	Celebes.	(Type.)
b–c. ♂ ♀.	N. Celebes.	Dr. A. B. Meyer [C.].
d–i. ♂, ♀, & hgr.	Manado.	Dr. A. B. Meyer [C.].

3. Calotes marmoratus.

Bronchocela marmorata, *Gray, Cat.* p. 242.
? *Hombr. & Jacq. Voy. Pôle Sud, Rept.* pl. iii.
? Lophyrus spinosus, *A. Dum. Cat. Méth. Rept.* p. 91.
Calotes (Bronchocela) philippinus, *Peters, Mon. Berl. Ac.* 1867, p. 16.
Bronchocela marmorata, *Günth. Proc. Zool. Soc.* 1873, p. 168.

Also closely allied to *C. cristatellus*, from which it differs in the following points:—The male's dorsal crest a little more developed; gular sac larger; gular scales smaller, scarcely larger than dorsolaterals; only the scales of the row contiguous to the dorsal crest point upwards. Upper and lateral head-scales very small; 61 to 75 scales round the middle of the body. Purplish (in spirit) with blue-green undulating lines which are more or less distinct.

Total length	512	millim.
Head	33	,,
Width of head	19	,,
Body	89	,,
Fore limb	68	,,
Hind limb	105	,,
Tail	390	,,

Philippine Islands.

a. ♂.	Philippines.	(Type.)
b. ♀.	Philippines.	C. Bowring, Esq. [P.].
c–d. ♂ ♀.	Luzon.	Dr. A. B. Meyer [C.].

4. Calotes jubatus.

Bronchocela gutturosa, *Gray, Cat.* p. 241.
Bronchocela gutturosa, (*non Merr.*) *Kaup, Isis*, 1827, p. 619.
—— jubata, *Dum. & Dibr.* p. 397; *Günth. Rept. Brit. Ind.* p. 139; *Steind. Novara, Rept.* p. 27; *Stoliczka, Journ. As. Soc. Beng.* xxxix. 1870, p. 179.

Upper head-scales moderate, keeled, not enlarged on supraorbital region; a few erect compressed scales behind the supraciliary edge; tympanum half or more than half the diameter of the orbit. Gular pouch well developed; gular scales keeled, smaller than the ventrals. No oblique fold in front of the shoulder. Nuchal crest large, composed of falciform spines directed backwards, the longest equalling or exceeding the diameter of the orbit, with several irregular rows of smaller spines at the base; dorsal crest well developed, though much less so than the nuchal, gradually decreasing in size towards the posterior part of the body. 43 to 53 (exceptionally 65) scales round the middle of the body; dorsal scales keeled, smaller than the ventrals; the scales of the row contiguous to the dorsal crest point upwards and backwards, those of the next row point straight backwards, all the others backwards and downwards. The adpressed hind limb reaches between the eye and the tip of the snout; third and fourth fingers equal, about as long as the fifth toe. Tail round, subtriangular, and with slight upper ridge at the base. Green, generally with yellow or reddish spots and short bands.

Total length	570	millim.
Head	34	,,
Width of head	21	,,
Body	96	,,
Fore limb	72	,,
Hind limb	107	,,
Tail	440	,,

East-Indian archipelago; Nicobars.

a, b. Several spec.: ♂, ♀, hgr., & yg.	Java.	C. Bowring, Esq. [P.].
c, d. ♂ ♀.	Java.	Dr. Ploem [C.].
e, f, g–h. ♂ ♀.	Java.	G. Lyon, Esq. [P.].
i–k. Hgr.	Java.	
l. ♂.	Batavia.	
m. ♂.	Soerabaya.	J. B. Jukes, Esq. [P.].
n. ♂.	Manado, Celebes.	Dr. A. B. Meyer [C.].
o. Yg.	E. Indies.	Dr. J. E. Gray [P.].
p. ♂.	E. Indies.	East India Comp.
q. Skeleton.	——?	

5. Calotes smaragdinus.

Bronchocela smaragdina, *Günth. Rept. Brit. Ind.* p. 138.

Upper head-scales small, keeled, slightly enlarged on supraorbital region; tympanum not half the diameter of the orbit. No gular pouch; anterior gular scales large, posterior smaller. No oblique fold in front of the shoulder. Nuchal and dorsal crests scarcely indicated (♀). 53 scales round the middle of the body; dorsal scales small, rather feebly keeled, the upper ones pointing straight backwards, the others backwards and downwards; ventral scales much larger than dorsals, strongly keeled. The adpressed hind limb reaches the tip of the snout or a little beyond; third and fourth fingers equal, very long, much longer than the fifth toe. Tail round. Green; a white line along the side of the body and tail and the hinder side of the thigh.

Total length	428 millim.
Head	23 ,,
Width of head	12 ,,
Body	80 ,,
Fore limb	66 ,,
Hind limb	98 ,,
Tail	325 ,,

Camboja.

a–b. ♀.   Camboja.   M. Mouhot [C.].   (Types.)

6. Calotes tympanistriga.

Bronchocela tympanistriga, *Gray, Cat.* p. 242.
Bronchocela tympanistriga (*Kuhl*), *Gray, Griff. A. K.* ix. *Syn.* p. 56; *Dum. & Bibr.* p. 399; *Steind. Novara, Rept.* p. 27.
—— (Pseudocalotes) tympanistriga, *Fitz. Syst. Rept.* p. 46.

Upper head-scales rather large, irregular, feebly keeled; a small tubercle behind the supraciliary edge, and a few others scattered on the back of the head; tympanum not quite half the diameter of the orbit. Gular pouch scarcely developed; gular scales feebly keeled, about the size of the ventrals. No oblique fold in front of the shoulder. Nuchal crest composed of very small spines; dorsal crest a slightly serrated ridge. 43 to 53 scales round the middle of the body. Dorsal scales smooth or very feebly keeled, as large as or slightly larger than ventrals, the uppermost pointing backwards and upwards, the others backwards and downwards; ventral scales strongly keeled. The adpressed hind limb reaches the neck or the tympanum; fourth finger slightly longer than third. Tail compressed, with a slight upper ridge at the base. Greenish, uniform or with more or less distinct dark markings.

Total length	264 millim.
Head	21 ,,
Width of head	12·5 ,,
Body	58 ,,
Fore limb	35 ,,
Hind limb	57 ,,
Tail	185 ,,

Java; Sumatra.

a, b. ♂ ♀.   —— ?
c. ♀.   Pajo, Sumatra.   Hr. Carl Bock [C.].

7. Calotes versicolor.

Calotes versicolor, *Gray, Cat.* p. 243.
Agama versicolor, *Daud. Rept.* iii. p. 395, pl. xliv.; *Kuhl, Beitr. Zool. Vergl. Anat.* p. 114.
—— tiedemanni, *Kuhl, l. c.* p. 109.
—— vultuosa, *Harl. Journ. Ac. Philad.* iv. 1825, p. 296, pl. xix.
Calotes tiedemanni, *Kaup, Isis,* 1827, p. 612, pl. viii.
Agama indica, *Gray, Zool. Journ.* iii. 1828, p. 217.
Calotes versicolor, *Fitz. N. Class. Rept.* p. 49; *Dum. & Bibr.* iv. p. 405; *Kelaart, Prodr. Faun. Zeyl.* p. 170; *Jerdon, Journ. As. Soc. Beng.* xxii. 1853, p. 470; *Blyth, ibid.* p. 649; *Günth. Rept. Brit. Ind.* p. 140; *Anders. Proc. Zool. Soc.* 1872, p. 381; *Blanf. E. Persia,* p. 313; *Anders. Zool. Yunnan,* p. 805.
—— cristatus, *Jacquemont, Voy. dans l'Inde, Rept.* pl. ii.
? Calotes viridis, *Gray, Ann. & Mag. N. H.* xviii. 1846, p. 429.

Upper head-scales rather large, smooth or feebly keeled, imbricate, more or less enlarged on supraorbital region; two well-separated spines (seldom absent or scarcely distinct) on each side of the back of the head, above the ear; tympanum half, or less than half, the diameter of the orbit. Gular pouch not developed; gular scales smooth or feebly keeled, as large as or larger than ventrals, largest and mucronate in the adult male. No oblique fold in front of the shoulder. Dorso-nuchal crest well developed in the male, composed of lanceolate spines gradually decreasing in size towards the posterior part of the back. Thirty-five to forty-seven scales round the middle of the body; dorsal scales more or less distinctly keeled, larger than ventrals, all directed upwards and backwards; ventral scales strongly keeled. The adpressed hind limb reaches the temple or the eye; fourth finger a little longer than third. Tail round. Light brownish or yellowish (in spirit), uniform or with dark transverse bands or spots, or dark olive-brown with light spots or longitudinal lines; belly sometimes with dark longitudinal lines.

Total length	405	millim.
Head	34	,,
Width of head	28	,,
Body	86	,,
Fore limb	62	,,
Hind limb	90	,,
Tail	285	,,

Ceylon, India, eastwards to Tenasserim, Cochin China, and Southern China, westwards to Afghanistan and Baluchistan.

a, b. ♂ ♀.	Ceylon.	
c–d. ♂ ♀.	Ceylon.	R. Templeton, Esq. [P.].
e–h, i. ♂, hgr., & yg.	Ceylon.	Messrs. v. Schlagintweit [C.].
k. Yg.	Ceylon.	W. H. Holdsworth, Esq. [C.].
l–m. Yg.	India.	Gen. Hardwicke [P.].
n–o, p. ♂, ♀, & hgr.	India.	W. Masters, Esq. [P.].

q. Hgr.	India.	C. Bowring, Esq. [P..
r-s. Hgr.	India.	T. C. Jerdon, Esq. [P.].
t, u, v-w. ♂, hgr., & yg.	India.	
x. Many specimens.	Malabar.	Col. Beddome [C.].
y, z-α. ♀ & yg.	Madras.	Major Beavan [P.].
β,γ-δ. ♂, hgr., & yg.	Madras.	J. E. Boileau, Esq. [P.].
ε. Hgr.	Godavery Valley.	W. T. Blanford, Esq. [P.].
ζ-κ, λ. ♂,♀, & hgr.	Deccan.	Col. Sykes [P.].
μ-ν. Hgr.	Ganjam.	F. Day, Esq. [P.].
ξ. ♂.	Patna.	W. Masters, Esq. [P.].
o-π. ♀ & yg.	Punjab?	Dr. Cantor.
ρ-φ, χ-bb.♂, ♀, & hgr.	Khasia.	T. C. Jerdon, Esq.
cc. ♂.	Jamu, Himalayas.	Messrs. v. Schlagintweit [C.].
dd, ee. Several specimens.	Nepal.	B. H. Hodgson, Esq. [P.].
ff. Hgr.	Sind.	Dr. Leith [P.].
gg, hh. ♀.	Afghanistan.	East India Comp.
ii. ♂.	Afghanistan.	
kk-ll. ♂ ♀.	Kalagan, Baluchistan.	W. T. Blanford, Esq. [C.].
mm. Yg.	Salween Valley, Burmah.	Lieut. R. C. Beavan [P.].
nn. Hgr.	Siam.	C. Bowring, Esq. [P.].
oo, pp-qq. Hgr. & yg.	Siam.	M. Mouhot [C.].
rr. ♂.	Pachebon.	M. Mouhot [C.].
ss-tt. Hgr.	Bangkok.	F. E. Lott, Esq. [P.].
uu. ♂.	Canton.	R. Swinhoe [C.].
vv-yy. ♂,♀, & yg.	Hainan.	R. Swinhoe [C.].
zz, aa, ββ-γγ, δδ-εε.♂,♀, hgr., & yg.	China.	J. R. Reeves, Esq. [P.].
ζζ. Skeleton.	Ceylon.	

8. Calotes maria.

Calotes maria, part., *Gray, Cat.* p. 243.
Calotes platyceps, *Blyth, in Kelaart, Prod. Faun. Zeyl., App.* p. 46.
—— maria, part., *Günth. Rept. Brit. Ind.* p. 144.
—— maria, *Jerdon, Proc. As. Soc. Beng.* 1870, p. 77; *Günth. Proc. Zool. Soc.* 1870, p. 778, pl. xlv. fig. 2; *Anders. Proc. Zool. Soc.* 1872, p. 382, *and Zool. Yunnan*, p. 806.

Upper head-scales rather large, smooth or feebly keeled, imbricate, very slightly enlarged on supraorbital region; two parallel longitudinal series of enlarged erect scales, terminating in long spines, above the temple, the lower series separated from the tympanum by several rows of scales; tympanum very small, measuring less than one third the diameter of the orbit. Gular pouch not developed; gular scales strongly keeled, larger than ventrals, not larger than dorsals. No oblique fold in front of the shoulder. Dorso-nuchal crest well developed anteriorly, the longest spines measuring nearly two thirds the diameter of the orbit, gradually decreasing in size and reduced to a feebly serrated edge on the posterior half of the back. Fifty-three to sixty-one scales round the middle of the body; dorsal scales feebly keeled, all directed upwards and backwards; ventral scales much smaller than dorsals, strongly keeled. The adpressed hind limb reaches the anterior border of the orbit, or between the latter and the nostril; third and fourth fingers nearly equal. Tail round. Green, with a few light (red) markings constantly present on the limb-joints.

Total length	488	millim.
Head	35	,,
Width of head	23	,,
Body	83	,,
Fore limb	65	,,
Hind limb	96	,,
Tail	370	,,

Khasia hills.

a. ♂.	[Afghanistan.]	E. India Compy. (Type.)
b. ♀.	[Afghanistan.]	
c-d, e-h. ♂ ♀.	Khasia.	T. C. Jerdon, Esq. [P.].

9. Calotes jerdonii.

Calotes maria, part., *Gray, Cat.* p. 243.
Calotes maria, part., *Günth. Rept. Brit. Ind.* p. 144.
—— platyceps (*non Blyth*), *Jerdon, Proc. As. Soc. Beng.* 1870, p. 77.
—— jerdonii, *Günth. Proc. Zool. Soc.* 1870, p. 778, pl. xlv. fig. 1; *Anders. Proc. Zool. Soc.* 1872, p. 382.

Upper head-scales rather large, smooth or feebly keeled, imbricate, very slightly enlarged on supraorbital region; two parallel longitudinal series of enlarged erect scales on the temple, forming serrated ridges, not terminating in spines; the lower serrated ridge separated from the tympanum by one or two rows of scales; tympanum not quite half the diameter of the orbit. Gular pouch not developed; gular scales strongly keeled, larger than dorsals. A strong oblique fold or pit in front of the shoulder. Dorso-nuchal crest very low, the longest spines (on the nape) equalling about the diameter of the tympanum. Forty-seven to fifty-five scales round the middle of the body; dorsal scales feebly keeled, all directed upwards and backwards; ventral scales much smaller than dorsals, strongly keeled. The adpressed hind limb reaches the eye, or between the latter and the nostril; fourth finger a little longer than third. Tail round. Green, uniform, or with light (red) markings, which may form a longitudino-lateral band; frequently black lines radiate from the eye: the fold on the side of the neck black.

Total length	362	millim.
Head	25	,,

Width of head	16 millim.
Body	67 ,,
Fore limb	53 ,,
Hind limb	72 ,,
Tail	270 ,,

Khasia hills; Himalayas.

a, b. Many spec.: ♂, ♀, & hgr.	Khasia.	T. C. Jerdon, Esq. [P.]. (Types.)
c-e. ♂ & yg.	Khasia.	East India Co. (Types of *C. maria*.)
f-g. ♂ & hgr.	Khasia.	Sir J. Hooker [P.].
h. ♀.	Jamu, Himalayas.	Messrs. v. Schlagintweit [C.].
i. ♂.	[Afghanistan.]	

10. Calotes emma. (Plate XXV. fig. 1.)

Calotes emma, *Gray, Cat.* p. 244.
Calotes emma, *Blyth, Journ. As. Soc. Beng.* xxii. 1853, p. 647; *Günth. Rept. Brit. Ind.* p. 144; *Anders. Zool. Yunnan*, p. 806.

Upper head-scales imbricate, keeled, strongly enlarged on supraorbital region; a long spine surrounded by small ones behind the supraciliary edge, and two others on each side between the tympanum and the nuchal crest; the posterior spine smallest, just above the tympanum; latter measuring half the diameter of the orbit. Gular pouch not developed; gular scales keeled, as large as, or a little larger than, ventrals. A very strong oblique curved fold in front of the shoulder. Nuchal crest composed of long lanceolate spines, soon decreasing in size, the dorsal crest being reduced to a feebly serrated ridge. Fifty-one to fifty-five scales round the middle of the body; dorsal scales keeled, all directed upwards and backwards; ventral scales smaller than dorsals, strongly keeled. The adpressed hind limb reaches the eye; third and fourth fingers nearly equal. Tail slightly compressed. Light brownish olive above, with transverse dark-brown bands on the back, which are interrupted by a more or less strongly defined light lateral band; black lines radiating from the eye; the fold in front of the arm black.

Total length	330 millim.
Head	25 ,,
Width of head	19 ,,
Body	65 ,,
Fore limb	47 ,,
Hind limb	72 ,,
Tail	240 ,,

Kakhyen and Khasia hills, Pegu, Mergui, and Tenasserim.

a. ♂.	[Afghanistan.]	East India Co. (Type.)
b. ♀.	Tavoy.	W. T. Blanford, Esq. [P.].
c. ♂.	India.	Prof. Oldham [P.].

11. Calotes mystaceus.

Calotes mystaceus, *Gray, Cat.* p. 245.
Calotes mystaceus, *Dum. & Bibr.* iv. p. 408; *Blyth, in Kelaart, Prodr. Faun. Zeyl., App.* p. 47; *Günth. Rept. Brit. Ind.* p. 141.
? Calotes gigas, *Blyth, Journ. As. Soc. Beng.* xxii. 1853, p. 648.

Upper head-scales smooth or feebly keeled, imbricate, scarcely enlarged on supraorbital region; a few small spines on each side of the head above the tympanum; latter measuring at least half the diameter of the orbit. Gular sac small; gular scales feebly keeled, as large as dorsals. An oblique fold in front of the shoulder. Dorso-nuchal crest well developed in the male, composed of falciform spines directed backwards, the longest measuring the diameter of the orbit; it gradually decreases in height on the back, being reduced to a mere denticulation on the sacrum. Forty-five to fifty-three scales round the middle of the body; dorsal scales keeled, nearly twice as large as ventrals, all directed upwards and backwards; ventral scales strongly keeled. The adpressed hind limb reaches the tympanum or the posterior border of the orbit; fourth finger slightly longer than third. Tail a little compressed, at the base with a slightly serrated upper ridge. Olive above, frequently with large transverse red spots on the back; lips yellowish.

Total length	380	millim.
Head	36	,,
Width of head	25	,,
Body	89	,,
Fore limb	57	,,
Hind limb	75	,,
Tail	255	,,

Arrakan; Pegu; Tenasserim; Siam; Nicobars; Ceylon.

a–c. ♂ ♀.	Pegu.	W. Theobald, Esq. [P.].
d. ♂.	Birma.	Lieut. R. C. Beavan [P.].
e. ♂.	Camboja.	M. Mouhot [C.].
f. ♀.	India.	Prof. Oldham [P.].

12. Calotes grandisquamis.

Calotes grandisquamis, *Günth. Proc. Zool. Soc.* 1875, p. 226, pl. xxx.

Upper head-scales smooth or very feebly keeled, imbricate, enlarged on supraorbital region; three or four small spines above the tympanum; latter measuring half the diameter of the orbit. A gular pouch; lateral gular scales larger than ventrals, smooth. A short oblique fold in front of the shoulder. Dorso-nuchal crest formed of large lanceolate spines, the longest of which (in the male) equals the diameter of the orbit, gradually decreasing in height on the back; in the female the dorsal crest is reduced to a very feebly serrated ridge. Twenty-nine to thirty-five scales round the middle of the body; dorsal scales nearly four times as large as ventrals, smooth, pointing backwards and upwards; ventral scales strongly keeled. The adpressed hind limb reaches the tympanum or slightly beyond; third and fourth fingers equal. Tail compressed. Green, uniform, or with broad black transverse bands on the back; in one of the specimens each scale within the black bands with a central orange-coloured spot.

Total length	463	millim.
Head	34	,,
Width of head	25	,,
Body	104	,,
Fore limb	68	,,
Hind limb	93	,,
Tail	325	,,

Malabar.

a-d. ♂ ♀.   Foot of Canoot Ghat.   Col. Beddome [C.].   (Types.)

13. Calotes nemoricola.

Calotes nemoricola, *Jerdon, Journ. As. Soc. Beng.* xxii. 1853, p. 471, and *Proc. As. Soc. Beng.* 1870, p. 78.

Upper head-scales feebly keeled, imbricate, enlarged on supraorbital region; three or four very small spines above the tympanum; latter measuring not quite half the diameter of the orbit. A gular pouch; gular scales larger than ventrals, very feebly keeled. A short oblique fold in front of the shoulder. Nuchal crest formed of large lanceolate spines, the largest of which measure three fourths the diameter of the orbit; dorsal crest a mere serrated ridge. Thirty-seven to forty-three scales round the middle of the body; dorsal scales three times as large as ventrals, squarish, smooth, pointing backwards and upwards; ventral scales strongly keeled. The adpressed hind limb reaches hardly the tympanum; third and fourth fingers equal. Tail compressed. Olive above, with indistinct darker markings; a black streak from the eye to above the tympanum; dark lines radiating from the eye; gular sac pink (in spirit).

Total length	470	millim.
Head	40	,,
Width of head	25	,,
Body	100	,,
Fore limb	62	,,
Hind limb	87	,,
Tail	330	,,

Malabar.

a-b. ♂ ♀.   Malabar.   • Col. Beddome [C.].

14. Calotes liolepis. (PLATE XXV. fig. 2.)

Calotes nemoricola (*non Jerd.*), *Günth. Proc. Zool. Soc.* 1869, p. 507.

Upper head-scales smooth, imbricate, strongly enlarged on supraorbital region; two distant spines on each side of the back of the head, between the ear and the nuchal crest; diameter of the tympanum nearly half that of the orbit. Gular sac not developed (♀); gular scales strongly keeled, as large as ventrals. A short oblique fold in front of the shoulder. Nuchal crest formed of narrow separated spines, the longest of which measure about the diameter of the tympanum; dorsal crest quite indistinct. Thirty-five or thirty-nine scales round the middle of the body; dorsal scales three times as large as ventrals, squarish, smooth, pointing backwards and downwards; ventral scales strongly keeled. The adpressed hind limb reaches hardly the tympanum; third and fourth fingers equal. Tail round. Pale olive, with indistinct brown transverse bands on the back; brown lines radiating from the eye, the lower reaching down to the lower lip.

Total length	256	millim.
Head	21	,,
Width of head	13	,,
Body	55	,,
Fore limb	36	,,
Hind limb	51	,,
Tail	180	,,

Ceylon.

a. ♀. Ceylon. B. H. Barnes, Esq. [P.].
b. Yg. Ceylon.

15. Calotes ophiomachus.

Calotes ophiomachus, *Gray, Cat.* p. 243.
Lacerta calotes, *Linn. S. N.* i. p. 367.
Iguana calotes, *Laur. Syn. Rept.* p. 49.
Agama calotes, *Daud. Rept.* iii. p. 361, pl. xlii.
—— ophiomachus, *Merr. Tent.* p. 51.
—— lineata, *Kuhl, Beitr. Zool. Vergl. Anat.* p. 108.
Calotes ophiomachus, *Kaup, Isis,* 1827, p. 621; *Dum. & Bibr.* iv. p. 403; *Kelaart, Prodr. Faun. Zeyl.* p. 169; *Günth. Rept. Brit. Ind.* p. 142.

Upper head-scales smooth, imbricate, enlarged on supraorbital region; two groups of spines on each side of the head, between the ear and the nuchal crest; tympanum nearly half the diameter of the orbit. Gular sac not developed; gular scales feebly keeled, nearly as large as ventrals. A short oblique fold in front of the shoulder. Dorso-nuchal crest composed of closely set lanceolate spines directed backwards, with smaller ones at the base; in the male the height of the crest on the nape equals or exceeds the diameter of the orbit, and decreases gradually on the back. Thirty-one to thirty-five scales round the middle of the body; dorsal scales very feebly keeled, or even quite smooth, with the points directed upwards and backwards; ventral scales larger than dorsals, very strongly keeled and mucronate. The adpressed hind limb reaches the anterior border of the orbit or a little beyond; third and fourth fingers nearly equal. Tail round. Green above, frequently with blackish-green cross bands, broader than the interspaces between them; young sometimes with a whitish longitudinal band on each side of the back.

Total length	650 millim.
Head	35 ,,
Width of head	25 ,,
Body	105 ,,
Fore limb	74 ,,
Hind limb	114 ,,
Tail	510 ,,

Ceylon; Southern India; Nicobars.

a, b-c. ♂,♀, & hgr.	Ceylon.	A. Paul, Esq. [P.].
d-g. ♂ & yg.	Ceylon.	R. Templeton, Esq. [P.].
h. ♀.	Ceylon.	Messrs. v. Schlagintweit [C.].
i. Hgr.	Ceylon.	Col. Beddome [C.].
k, l. Yg.	Ceylon.	
m, n, o. ♂, hgr., & yg.	——?	

16. Calotes nigrilabris.

Calotes rouxii (*non* D. & B.), *Blyth, Journ. As. Soc. Beng.,* xxii. 1853, p. 647.
—— (Bronchocele) nigrilabris, *Peters, Mon. Berl. Ac.* 1860, p. 183.
—— nigrilabris, *Günth. Rept. Brit. Ind.* p. 143.

Upper head-scales smooth, imbricate, not or but slightly enlarged on supraorbital region; a short series of three to six small spines above and behind the posterior part of the tympanum; latter measuring half, or more than half, the diameter of the orbit. Gular pouch not developed; gular scales more or less distinctly keeled, as large as ventrals. A strong oblique fold or pit in front of the shoulder. Dorso-nuchal crest composed of small lanceolate spines, the longest of which, on the nape, measure about two thirds the diameter of the orbit; on the back the crest is very distinct, but the spines gradually decrease in size. Forty-three to forty-seven scales round the middle of the body; dorsal scales rather feebly keeled, the upper pointing straight backwards, the others downwards and backwards; ventral scales much larger than dorsals, very strongly keeled, mucronate. The appressed hind limb reaches the eye; third and fourth fingers equal. Tail round; in the males the scales on the basal part of the tail very large and hard, those of the median upper row forming a slightly serrated ridge. Green, uniform on the back and tail, or with angular whitish black-edged angular cross bars or ocelli, with or without a reddish-brown vertebral band; lips and sides of head with a broad black band or variegated with black in the male; female usually with a white, black-edged horizontal streak below the eye.

Total length	410	millim.
Head	35	,,
Width of head	25	,,
Body	65	,,
Fore limb	56	,,
Hind limb	85	,,
Tail	310	,,

Ceylon.

a. ♂.	Ceylon.	Dr. Kelaart.
b, c, d. ♂ ♀.	Ceylon.	
e. ♂.	Ceylon.	B. H. Barnes, Esq. [P.].
f. Yg.	Ceylon.	G. H. K. Thwaites, Esq. [P.].
g, h, i–k. ♀ & yg.	Ceylon.	W. H. Holdsworth, Esq. [C.].

17. Calotes liocephalus. (Plate XXVI.)

Calotes liocephalus, *Günth. Ann. & Mag. N. H.* (4) ix. 1872, p. 86.

Upper head-scales smooth, imbricate, enlarged on supraorbital region; no spines whatever on the side of the head; tympanum measuring a little more than half the diameter of the orbit. Gular pouch not developed; gular scales much larger than ventrals, rather feebly keeled. An oblique fold in front of the shoulder. Nuchal crest composed of narrow spines the length of which equals the diameter of the tympanum; dorsal crest a mere serrated ridge. Forty-five scales round the middle of the body; dorsal scales feebly keeled, the upper ones pointing straight backwards or slightly upwards, the others downwards and backwards; ventral scales strongly keeled, nearly the same size as the dorsals. The adpressed hind limb reaches the eye; third and fourth fingers equal. Tail round, the scales of its basal part very large and hard, those of the median upper row forming a serrated edge. Pale olive-green, above with transverse bands of a darker green; four angular reddish-brown cross bands on the back; a dark streak from the eye to above the tympanum; limbs and tail with alternate lighter and darker cross bands.

Total length	360	millim.
Head	30	,,
Width of head	22	,,
Body	60	,,
Fore limb	49	,,
Hind limb	75	,,
Tail	270	,,

Ceylon.

a. ♂.	Ceylon.	G. H. K. Thwaites, Esq. [P.]. (Type.)

18. Calotes rouxii.

Calotes rouxii, *Gray, Cat.* p. 245.
Calotes rouxii, *Dum. & Bibr.* iv. p. 407; *Günth. Rept. Brit. Ind.* p. 142, *and Proc. Zool. Soc.* 1869, p. 501.
—— elliotti, (*non Gthr.*) *Stoliczka, Journ. As. Soc. Beng.* xli. 1872, p. 113.

Upper head-scales feebly keeled, imbricate, strongly enlarged on supraorbital region ; two small slender spines on each side of the back of the head, the anterior midway between the nuchal crest and the tympanum, the posterior just above the latter, which measures half the diameter of the orbit. Gular sac small ; gular scales strongly keeled, nearly as large as the ventrals. A strong oblique fold or pit in front of the shoulder, and a transverse gular fold. Nuchal crest composed of a few slender erect spines, the longest of which measures the diameter of the tympanum ; dorsal crest a mere denticulation. Fifty-five or fifty-seven scales round the middle of the body ; dorsal scales of nearly the same size as the ventrals, keeled, those on the upper half of the body directed upwards and backwards, the others downwards and backwards ; ventral scales strongly keeled and mucronate. The adpressed hind limb reaches the posterior border of the orbit ; fourth finger longer than third. Tail slightly compressed, the scales on its basal part large and hard, those of the median upper row strongly enlarged and forming a serrated ridge. Olive-brown above, vertebral zone rather lighter ; a dark band from the eye, through the tympanum, along each side of the neck ; the oblique fold in front of the shoulder black ; dark lines radiating from the eye.

Total length	241	millim.
Head	21	,,
Width of head	15	,,
Body	55	,,
Fore limb	37	,,
Hind limb	60	,,
Tail	165	,,

India.
a-b. ♂.   Matheran, Bombay.   Dr. Leith [P.].

19. Calotes elliotti. (Plate XXV. fig. 3.)

Calotes rouxii, (*non D. & B.*) *Jerdon, Journ. As. Soc. Beng.* xxii. 1853, p. 471.
—— elliotti, *Günth. Rept. Brit. Ind.* p. 142 ; *Jerdon, Proc. As. Soc. Beng.* 1870, p. 77 ; *Stoliczka, Journ. As. Soc. Beng.* xli. 1872, p. 113.
Bronchocela indica, *Theob. Cat. Rept. Brit. Ind.* p. 105.

Upper head-scales feebly keeled, imbricate, strongly enlarged on supraorbital region ; a small spine behind the supraciliary edge, and two others on each side, the anterior midway between the nuchal crest and the tympanum, the posterior just above the latter, which measures nearly half the diameter of the orbit. Gular sac not developed; gular scales strongly keeled, smaller than the ventrals. A strong oblique fold or pit in front of the shoulder, and a transverse gular fold. Nuchal crest composed of a few widely separated slender spines, the longest of which measures about two thirds the diameter of the orbit; dorsal crest a mere denticulation. Fifty-three to sixty-one scales round the middle of the body; dorsal scales of nearly the same size as the ventrals, keeled, the uppermost with the points directed straight backwards, the others directed downwards and backwards; ventral scales strongly keeled. The adpressed hind limb reaches the anterior border of the orbit, or the tip of the snout; fourth finger longer than third. Tail scarcely compressed. Olive above, with more or less distinct angular darkbrown cross bands on the body; an angular black mark on each side of the neck; a white spot below the orbit; dark lines radiating from the eye.

Total length	282 millim.
Head	20 ,,
Width of head	12 ,,
Body	52 ,,
Fore limb	38 ,,
Hind limb	60 ,,
Tail	210 ,,

Southern India.

a–d. ♂ ♀.	Anamallays, 6000 ft.	Col. Beddome [C.].
e. ♂.	Anamallays.	Col. Beddome [C.].
f–h. Hgr.	Top of Sivagiri Ghat.	Col. Beddome [C.].
i. ♂.	Tinnevelly.	Col. Beddome [C.].
k–n. ♂ & hgr.	Malabar.	Col. Beddome [C.].

17. CHELOSANIA.

Chelosania, *Gray, Cat. Liz.* p. 245.

Tympanum distinct. Body compressed, covered with equal, very small keeled scales; a very feeble dorsal crest. No transverse gular fold; male with a gular sac. Tail compressed. No præanal or femoral pores.

Australia.

1. Chelosania brunnea. (PLATE XXVII.)

Chelosania brunnea, *Gray, Cat.* p. 245.

Head large, swollen at the cheeks, covered above with equal small polygonal rough tubercles scarcely larger than the scales on the body; no canthus rostralis; nostril equally distant from the eye and the tip of the snout; labials very small; the diameter of the tympanum equals two thirds that of the orbit. Scales on throat and body very small, equal, feebly imbricate, those on the body forming regular transverse series; gular and dorsal scales with a short keel or tubercle, ventrals strongly keeled along the whole length; a slight indication of a dorso-nuchal crest. Limbs and digits rather short; the adpressed hind limb reaches the axilla. Tail compressed, once and a half as long as head and body, ending obtusely; caudal scales slightly larger than dorsals, equal, strongly keeled; a slight serrated upper ridge. Pale brown.

Total length	263 millim.
Head	31 ,,
Width of head	23 ,,
Body	72 ,,
Fore limb	46 ,,
Hind limb	55 ,,
Tail	160 ,,

West Australia.

a. ♂.   W. Australia.   B. Bynoe, Esq. [C.].   (Type.)

18. CHARASIA.

Agama, part., *Dum. & Bibr.* iv. p. 481.
Psammophilus, (non *Dahlb.*) *Fitzing. Syst. Rept.* p. 79.
Charasia, *Gray, Cat. Liz.* p. 246; *Günth. Rept. Brit. Ind.* p. 156.
Brachysaura, *Blyth, Journ. As. Soc. Beng.* xxv. 1856, p. 448.
Charasia, part., *Stoliczka, Journ. As. Soc. Beng.* xli. 1872, p. 109.

Tympanum distinct. Body slightly depressed, covered with uniform keeled scales. A very feeble dorsal crest. A pit on each side of the throat, usually connected by a transverse fold. No gular sac. Tail round or slightly compressed. No præanal or femoral pores.

India.

Synopsis of the Species.

Tail longer than head and body; 115 to
  140 scales round the middle of the body . 1. *dorsalis*, p. 332.
Tail longer than head and body; 80 to
  100 scales round the middle of the body 2. *blanfordiana*, p. 333.
Tail shorter than head and body; gular
  scales larger than ventrals .... 3. *ornata*, p. 334.

1. Charasia dorsalis.

Charasia dorsalis, *Gray, Cat.* p. 246.
Agama dorsalis, *Gray, Griff. A. K.* ix. *Syn.* p. 56; *Dum. & Bibr.* iv. p. 486; *Jerdon, Journ. As. Soc. Beng.* xxii. 1853, p. 475.
Phrynopsis (Psammophilus) dorsalis, *Fitzing. l. c.* p. 80.
Charasia dorsalis, *Günth. Rept. Brit. Ind.* p. 156.

Head rather elongate in the adult, covered with small irregular smooth or feebly keeled scales, which are a little enlarged on the supraorbital region; two very small spines usually present above the tympanum; latter larger than the eye-opening; cheeks strongly swollen in the adult male. A strongly marked transverse gular fold; gular scales very small, subimbricate, smooth, very feebly keeled in the young. Scales on the body very small (115 to 140 round the middle), subimbricate, and smooth or indistinctly keeled in the adult, more distinctly keeled and imbricate in the young, the keels of the dorsal scales directed obliquely towards the vertebral line. Dorso-nuchal crest scarcely indicated by a slight ridge. Limbs covered with small scales slightly larger than those on the body; digits moderately elongate; the adpressed hind limb reaches between the tympanum and the posterior border of the orbit in the adult, beyond the latter point in the young. Tail round, slightly compressed and much swollen at the base in the adult male, covered with small, more or less distinctly keeled scales which are a little enlarged inferiorly and on the upper median series; its length equals nearly once and three fourths that of head and body. Young olive-brown, spotted or marbled with brown, and with a series of rounded white spots along each side of the back; this coloration more or less distinctly retained in the female; male yellow or red, with a broad black lateral band commencing from the eye.

Total length	339	millim.
Head	38	,,
Width of head	32	,,
Body	86	,,
Fore limb	60	,,
Hind limb	87	,,
Tail	215	,,

Southern India.

a, b, c. ♂, ♀, & hgr.	India.	East India Company [P.].
d. ♂.	India.	Gen. Hardwicke [P.].
e. ♀.	Madras.	T. C. Jerdon, Esq. [P.].
f. ♂.	Malabar.	Col. Beddome [C.].
g. Yg.	Nilgherries.	F. Day, Esq. [P.].
h. Hgr.	—?	

2. Charasia blanfordiana.

Charasia dorsalis, (non Gray) Blanf. Journ. As. Soc. Beng. xxxix. 1870, p. 368; Anders. Proc. Zool. Soc. 1872, p. 382.
—— blanfordana, Stoliczka, Journ. As. Soc. Beng. xli. 1872, p. 110.

This species is very closely allied to the preceding, but remains smaller; the scales are a little larger (80 to 100 round the middle of the body) and always distinctly keeled and imbricate; and the hind limb stretched forwards extends at least to the posterior border of the orbit. The nuchal crest is a little more developed. Frequently a small spine behind the supraciliary edge, and a few scattered slightly enlarged scales on the flanks. Young olive-brown above, spotted or marbled with brown, and with a series of more or less distinct large lozenge-shaped brown spots on the back; these markings persist in the adult female, but are absent in the male, which has a more or less distinct dark lateral band as in C. dorsalis.

Total length	255 millim.
Head	26 ,,
Width of head	21 ,,
Body	59 ,,
Fore limb	47 ,,
Hind limb	72 ,,
Tail	170 ,,

Central India.

a, b-c, d, e-i. ♂,♀, & hgr.	Godavery Valley.	W. T. Blanford, Esq. [C.].
k-m. ♂ ♀.	Jeypore.	Col. Beddome [C.].
n-p. ♂, ♀, & hgr.	Near Ranchi, Nagpur.	W. T. Blanford, Esq. [C.].

3. Charasia ornata.

Brachyura ornata, *Blyth, Journ. As. Soc. Beng.* xxv. 1856, p. 448; *Stoliczka, Proc. As. Soc. Beng.* 1872, p. 77; *Cockburn, Journ. As. Soc. Beng.* li. 1882, p. 50.

Habit stout; head short and convex. Upper head-scales subequal, tubercular, obtusely keeled; two small spines surrounded by still smaller ones on each side of the back of the head, above the tympanum: latter not quite as large as the eye-opening. Gular fold not well marked; gular scales larger than ventrals, feebly keeled. Fifty-seven scales round the middle of the body. Dorsal scales much larger than ventrals, rhomboidal, imbricate, keeled, the keels converging obliquely towards the vertebral line; ventral scales feebly keeled. Nuchal and dorsal crests not continuous, reduced to a mere serrated ridge. Limbs, and especially digits, rather short; the adpressed hind limb reaches the neck; scales on the limbs much smaller than those on the back. Tail round, not quite as long as head and body, covered with equal, strongly keeled scales. Pale brown above, with three rows of darker, light-edged spots on the back, the median row most distinct and formed of rhomboidal spots; limbs and tail with dark cross bars.

Total length	94 millim.
Head	16 ,,
Width of head	13 ,,
Body	34 ,,
Fore limb	24 ,,
Hind limb	32 ,,
Tail	44 ,,

Central and Northern India.

a. Hgr.	Cutch.	F. Stoliczka [C.].

19. AGAMA.

Agama, part., *Daud. Hist. Rept.* iii. p. 333; *Cuvier, Règne An.* ii. p. 33; *Merr. Tent. Syst. Amph.* p. 50; *Fitzing. Neue Classif. Rept.* p. 17; *Dum. & Bibr. Erp. Gén.* iv. p. 481.

Stellio, part., *Daud. l. c.* iv. p. 5; *Cuv. l. c.* p. 31; *Fitz. l. c.*
Trapelus, *Cuvier, l. c.* p. 35; *Kaup, Isis,* 1827, p. 614; *Wagler, Syst. Amph.* p. 144; *Gray, Cat. Liz.* p. 258.
Tapaya, *Fitzing, l. c.*
Stellio, *Wagl. l. c.* p. 145; *Dum. & Bibr. l. c.* p. 526; *Gray, l. c.* p. 254.
Phrynopsis, *Fitzing. Syst. Rept.* p. 80.
Podorrhoa, p. 80.
Pseudotrapelus, p. 81.
Planodes, p. 81.
Trapeloides, p. 81.
Psammorrhoa, p. 81.
Eremioplanis, p. 82.
Acanthocercus, p. 84.
Laudakia, *Gray, l. c.* p. 254.
Agama, *Gray, l. c.* p. 256; *Peters, Reise n. Mossamb.* iii. p. 38.
Plocederma, *Blyth, Journ. As. Soc. Beng.* xxiii. 1854, p. 737.
Barycephalus, *Günth. Proc. Zool. Soc.* 1860, p. 150.

Tympanum distinct. Body more or less depressed. Dorsal crest absent or feebly developed. A pit on each side of the throat and a transverse gular fold. Gular sac present or absent. Tail round or feebly compressed. No femoral pores; males with callose præanal scales.

Africa; Southern Asia; South-eastern Europe.

Synopsis of the Species.

I. Occipital scale not enlarged; caudal scales not forming annuli.

   A. Third toe the longest; two much enlarged scales at the base of the claws.

Median dorsal scales smooth, or nearly so, not larger than the ventrals .... 1. *mutabilis*, p. 338.
Median dorsal scales keeled, larger than the ventrals .... 2. *sinaita*, p. 339.

   B. Fourth toe the longest.

     1. Dorsal scales equal.

       a. Ear-opening larger than the eye-opening; third and fourth toes nearly equally long .... 3. *hartmanni*, p. 340.

       b. Ear-opening not larger than the eye-opening.

         α. Tail compressed; ventral scales strongly keeled .... 4. *tournevillii*, p. 340.

         β. Tail rounded.

            * Dorsal scales much larger than the ventrals.

Nostril pierced just below the canthus rostralis; tibia as long as the skull (to occiput); ventral scales smooth or nearly so .... 5. *agilis*, p. 341.

Nostril pierced just above the canthus rostralis; tibia a little longer than the skull; ventral scales feebly keeled (or smooth) .... 6. *isolepis*, p. 342.
Nostril pierced just above the canthus rostralis; dorsal scales spinose, ventrals strongly keeled .... 7. *sanguinolenta*, [p. 343.

    ** Dorsal scales scarcely larger than ventrals; latter smooth. 8. *latastii*, p. 344.

  2. Dorsal scales unequal.

    a. Scales on the upper surface of the hind limb equal or nearly so.

      a. Claws of the fingers not much larger than those of the toes.

        * Males with a large gular pouch; on the middle of the back the larger scales are not twice as large as the smaller.

Dorsal scales strongly keeled .... 10. *persica*, p. 345.
Dorsal scales feebly keeled .... 11. *leucostigma*, p. 346.

        ** Males without a large gular pouch.

Dorsal scales irregular .... 9. *inermis*, p. 344.
Dorsal scales small, equal, intermixed with enlarged ones..................... 12. *rubrigularis*, p. 346.

      β. Claws of the fingers much larger than those of the toes 13. *megalonyx*, p. 347.

    b. Scales of the upper surface of the hind limb intermixed with strongly enlarged ones.

Nostril pierced just below the canthus rostralis; all the dorsal scales more or less distinctly keeled .... 14. *ruderata*, p. 348.
Nostril pierced just above the canthus rostralis; the smaller dorsal scales smooth or indistinctly keeled................ 15. *pallida*, p. 348.

II. Occipital scale enlarged §.

  A. Dorsal scales unequal.

    1. Fifth toe not extending as far as first; ear-opening smaller than the eye-opening.

Third toe the longest; ventral scales keeled 16. *hispida*, p. 349.
Fourth toe slightly longer than third; ventrals smooth.................... 17. *brachyura*, p. 350.

---

§ Slightly in *A. atra.*

2. Fifth toe extending as far as first; ear-opening larger than the eye-opening.

Third toe slightly longer than fourth; ventral scales smooth .... 18. *aculeata*, p. 351.
Third toe slightly longer than fourth; ventrals keeled .... 19. *armata*, p. 352.
Fourth toe slightly longer than third; ventrals smooth .... 20. *atra*, p. 352.

B. Dorsal scales equal.

1. A slight denticulation or crest on the vertebral line.

Dorsal scales considerably larger than ventrals .... 21. *mossambica*, p. 353.
Dorsal scales scarcely larger than ventrals . 22. *kirkii*, p. 354.

2. No dorsal denticulation or crest.

a. Body moderately depressed; dorsal scales mucronate.

α. The longest spines near the ear and on the sides of the neck at least two thirds as long as the diameter of the ear-opening.

Dorsal scales much smaller than the caudals, 40 to 50 from the origin of the fore limbs to the origin of the hind limbs .... 23. *spinosa*, p. 355.
Dorsal scales nearly as large as the caudals, 26 or 27 from the origin of the fore limbs to the origin of the hind limbs .... 24. *rueppellii*, p. 355.

β. The longest spines near the ear and on the sides of the neck not two thirds the diameter of the ear-opening.

Four or five labials anterior to the front border of the eye; fourth toe as long as the distance from the posterior border of the eye to the end of the snout .... 25. *colonorum*, p. 356.
Seven or eight labials anterior to the front border of the eye; fourth toe as long as the distance from the posterior border of the eye to the nostril .... 26. *bibronii*, p. 357.

b. Body much depressed; dorsal scales not or but very slightly mucronate .... 27. *planiceps*, p. 358.

III. Occipital scale not enlarged; caudal scales forming more or less distinct annuli; ventral scales smooth; digits compressed.

A. The caudal segments, if at all distinct, composed of more than two whorls of scales.

1. Scales on the vertebral region unequal, irregular.

Nostril pierced on the canthus rostralis .. 28. *atricollis*, p. 358.
Nostril pierced below the canthus rostralis 29. *cyanogaster*, p. 359.

2. Scales on the vertebral region equal, or forming longitudinal series.
   a. Ventral scales not much smaller than the largest dorsals.
     α. Nostril pierced on the canthus rostralis; tail compressed .. 30. *annectens*, p. 360.
     β. Nostril pierced below the canthus rostralis; tail not compressed.

The largest caudal scales scarcely larger than the largest dorsals .... 31. *stoliczkana*, p. 360.
Dorsal scales keeled; upper head-scales smooth or feebly keeled .... 32. *tuberculata*, p. 361.
Dorsal scales keeled; upper head-scales sharply keeled .... 33. *dayana*, p. 362.
Dorsal scales smooth or very feebly keeled 34. *himalayana, p. 362.

   b. Ventral scales not more than half the size of the largest dorsals.

Upper head-scales keeled; dorsal scales sharply keeled; flanks with enlarged scales .... 35. *agrorensis*, p. 363.
Median dorsal scales of unequal width, forming regular longitudinal series .... 36. *melanura*, p. 363.
Median dorsal scales equal, forming regular longitudinal series; no enlarged scales on the flanks .... 37. *lirata*, p. 364.
Dorsal scales large, equal, mucronate, the keels forming oblique lines .... 38. *nupta*, p. 365.

B. Tail divided into distinct segments, each composed of two whorls of scales.

Scales on upper surface of tail not larger than those on upper surface of tibia; more than 200 scales round the middle of the body; gular scales smooth ...... 39. *microlepis*, p. 366.
Scales on upper surface of tail not larger than those on upper surface of tibia; not more than 160 scales round the middle of the body; gular scales smooth .... 40. *caucasica*, p. 367.
Scales on upper surface of tail larger than those on the tibia, strongly spinose; gular scales strongly keeled .... 41. *stellio*, p. 368.

1. Agama mutabilis.

Trapelus sinaitus (*non Heyd.*), *Gray*, p. 259.
Geoffroy, Descr. Egypte, p. 127, pl. v. figs. 3, 4.
Agama mutabilis, *Merr. Tent.* p. 50.
Trapelus mutabilis, *Kaup, Isis,* 1827, p. 617.
Agama arenaria, *Heyden, in Rüpp. Atlas N. Afr., Rept.* p. 12.
Eremioplanis ægyptiaca, *Fitzing. Syst. Rept.* p. 82.

Head convex, subcordiform. Nostril tubular, turned upwards and backwards, pierced on the canthus rostralis in a convex nasal. Upper head-scales smooth; occipital not enlarged; a spine-like scale in front of the ear, but no others on the head; ear entirely exposed, quite as large as the eye-opening; male without gular pouch. Body rounded, a little depressed; dorsal scales very small, juxtaposed or subimbricate, equal, smooth or feebly keeled, and not larger than the ventrals, which are smooth. Limbs long and thin, digits short; tibia longer than the skull; third finger longer than fourth, fifth not extending as far as second; third toe a little longer than fourth, fifth extending as far as first. Tail more than twice as long as the distance from gular fold to vent, compressed, covered with keeled scales which are much larger than those on the body. Males with a row of six large anal pores. Colour very variable; head sometimes blue.

Total length	239 millim.
Head	27 ,,
Width of head	23 ,,
Body	67 ,,
Fore limb	50 ,,
Hind limb	81 ,,
Tail	145 ,,

Egypt.

a–b, c. ♂.   Egypt.   J. Burton, Esq. [P.].

2. Agama sinaita.

Agama sinaita, *Heyden, in Rüpp. Atlas N. Afr., Rept.* p. 10, pl. iii.; *Dum. & Bibr.* iv. p. 509.
Podorrhoa (Pseudotrapelus) sinaita, *Fitzing. Syst. Rept.* p. 81.
Trapelus sinaita, *Günth. Proc. Zool. Soc.* 1864, p. 489; *Tristr. Faun. Palest.* pl. xvi. fig. 3.

Head convex, subcordiform. Nostril tubular, turned upwards and backwards, pierced on the canthus rostralis in a convex nasal. Upper head-scales smooth; occipital not enlarged; a small spine-like scale in front of the ear, but no others on the hinder part of the head; ear entirely exposed, larger than the eye-opening; male without gular pouch. Body rounded, moderately depressed; dorsal scales small, equal, rhomboidal, imbricate, keeled; lateral and ventral scales very small, smooth. Limbs very long and thin; digits short; tibia much longer than the skull; third finger longer than fourth, fifth not extending as far as second; third toe longer than fourth, fifth extending as far as or a little further than first. Tail more than twice as long as the distance from gular fold to vent, compressed, covered with keeled scales which are larger than the dorsals. Males with a row of four to eight large anal pores. Olive or grey above, with a few more or less distinct dark cross bands; head blue; collar-pit black; belly dirty white.

Total length	223	millim.
Head	24	,,
Width of head	21	,,
Body	64	,,
Fore limb	50	,,
Hind limb	77	,,
Tail	135	,,

North Arabia; Syria.

a–b. ♀ & hgr.	Mount Sinai.	
c. ♂.	Sinaitic Peninsula.	H. C. Hart, Esq. [C.].
d. ♂.	Dead Sea.	Canon Tristram [C.].

3. Agama hartmanni.

Agama hartmanni, *Peters, Mon. Berl. Ac.* 1869, p. 65.

General proportions and scaling as in *A. agilis* and *isolepis*, *i. e.* with equal keeled dorsal scales and moderately long limbs. The tympanum much larger, entirely exposed. Third and fourth toes nearly equal in length. Ventral scales quite smooth. Olive-brown above, with a yellow vertebral line from occiput to base of tail, irregular small dark-brown spots on the back, and on each of the sides two rather indistinct rows of large yellow, black-edged spots; lower surfaces ochraceous.

Dongola.

4. Agama tournevillii.

Agama tournevillei, *Lataste, Le Naturaliste*, 1880, p. 325.

Habit rather slender. Head oval, snout sloping gradually. Nostril not tubular, directed upwards and backwards, pierced on the canthus rostralis in the posterior part of a small flat nasal. Upper head-scales convex; occipital not enlarged; no spinose scales; a very slight fringe of pointed scales on the upper border of the ear, which is smaller than the eye-opening; male with a large gular pouch. Body round, not at all depressed, covered above with rhomboidal, imbricate, strongly keeled, not mucronate scales; lateral and ventral scales a little smaller, strongly keeled. Limbs rather slender; tibia a little shorter than the skull (to occiput); third finger slightly shorter than fourth, fifth not extending as far as second; third toe much shorter than fourth, fifth not extending quite as far as first. Tail twice and a half as long as the distance from gular fold to vent, distinctly compressed, covered with keeled scales. Male with a row of anal pores. Sandy-coloured above; a transverse brown band between the eyes, two longitudinal ones from the occiput along the nape, and two others on each side of the head; back with regular longitudinal series of quadrangular transverse brown spots separated by rather indistinct longitudinal light lines; these spots form annuli on the tail; lower surfaces whitish; the gular pouch grey.

Total length	256	millim.
Head	24	,,
Width of head	19	,,
Body	70	,,
Fore limb	45	,,
Hind limb	62	,,
Tail	162	,,

Algerian Desert.

a. ♂. Sahara.

5. Agama agilis.

Agama agilis, *Gray, Cat.* p. 257.
Agama agilis, *Olivier, Voy. Emp. Ottoman*, ii. p. 438, pl. xxix. fig. 2; *Dum. & Bibr.* iv. p. 496.
Podorrhoa (Planodes) agilis, part., *Fitzing. Syst. Rept.* p. 81.
Agama agilis, part., *Blanf. Proc. Zool. Soc.* 1881, p. 672.

Head moderate, convex, subcordiform. Nostril not tubular, lateral, pierced just below the canthus rostralis in the posterior part of a flat nasal. Upper head-scales convex, smooth; occipital not enlarged; a few short spines on the post-temporal region, and two or three on the upper edge of the ear, which is not larger than the eye-opening, and exposed; males with a very slight indication of a gular pouch. Body roundish, moderately depressed, covered above with equal, rhomboidal, imbricate, feebly keeled, shortly mucronate scales; lateral and ventral scales much smaller, smooth, or indistinctly keeled. Limbs moderate; tibia as long as the skull (to occiput); third finger slightly shorter than fourth, fifth not extending as far as second; third toe much shorter than fourth, fifth not extending quite as far as first. Tail about once and three fourths as long as the distance from gular fold to vent, rounded, covered with keeled scales. Male with a single or double row of anal pores. Sandy-coloured above, with more or less distinct broad transverse dark bars, enclosing an elongate narrow lighter spot on the vertebral line; the collar-pit blackish; lower surfaces cream-coloured; the male's throat lined with brown.

Total length	200	millim.
Head	22	,,
Width of head	20	,,
Body	68	,,
Fore limb	43	,,
Hind limb	63	,,
Tail	110	,,

Bagdad; Southern Persia.

a. ♀. Kazeroon, S. Persia.
b–c. ♂ ♀. Abadeh, S. Persia.

6. Agama isolepis.

Agama agilis, (*non Oliv.*) *Aud. Desc. Egypte, Rept. Suppl.* p. 169,
pl. i. fig. 5; *Anders. Proc. Zool. Soc.* 1872, p. 384; *Blanf. E. Pers.*
p. 314, *and Journ. As. Soc. Beng.* xlv. 1876, p. 22, and xlviii. 1879,
p. 129; *Murray, Zool. Sind,* p. 371.
—— savignii, (*non Aud.*) *Dum. & Bibr.* iv. p. 508; *Peters, Mon.
Berl. Ac.* 1869, p. 66.
Phrynopsis savignyi, *Fitzing. Syst. Rept.* p. 80.
Trapelus, sp., *Jerdon, Proc. As. Soc. Beng.* 1870, p. 78.
—— megalonyx, (*non Gthr.*) *Stoliczka, Proc. As. Soc. Beng.* 1872,
p. 128; *Murray, l. c.* p. 370.
Agama agilis, part., *Blanf. Proc. Zool. Soc.* 1881, p. 672.

Head moderate, convex, subcordiform. Nostril not tubular, superior, pierced just above the canthus rostralis in the posterior part of a flat nasal. Upper head-scales convex, smooth or feebly keeled; occipital not enlarged; generally a few short spines on the post-temporal region and a few on the upper edge of the ear, which is exposed and not larger than the eye-opening; males with a small gular pouch. Body roundish, moderately depressed, covered above with equal, rhomboidal, imbricate, keeled, more or less distinctly mucronate scales; lateral and ventral scales much smaller, feebly keeled, exceptionally smooth. Limbs moderate; tibia a little longer than the skull (to occiput); third finger shorter than fourth, fifth not extending as far as second; third toe much shorter than fourth, fifth not extending quite as far as first. Tail twice and one fourth to twice and two thirds as long as the distance from gular fold to vent, rounded, covered with keeled scales. Male with a single or double row of anal pores. Sandy grey or brown above, with or without dark transverse bands and light rhomboidal vertebral spots; the collar-pit blackish; lower surfaces white, often with dark longitudinal lines more or less distinctly marked; males during the breeding-season with the sides of the body and the throat deep blue.

Total length	280	millim.
Head	27	,,
Width of head	23	,,
Body	78	,,
Fore limb	53	,,
Hind limb	78	,,
Tail	175	,,

From Egypt to Sind.

a-c. ♂ & yg.	S. Persia.	W. T. Blanford, Esq. [C.].
d-f. ♂ ♀.	Dehbid, north of Shiraz.	
g-i. ♀ & yg.	Persia.	Gen. Goldsmid [P.].
k. ♂.	Near Rigan, Narmashir, S.E. Persia.	W. T. Blanford, Esq. [C.].
l. ♂.	Between Magas and Bampur, Baluchistan.	W. T. Blanford, Esq. [C.].
m-o. ♂ & yg.	Istandak, Baluchistan.	W. T. Blanford, Esq. [C.].
p-q. ♀ & bgr.	Bahu Kalat, Baluchistan.	W. T. Blanford, Esq. [C.].

r. Yg.	Mand, Baluchistan.	W. T. Blanford, Esq. [C.].
s. Yg.	Sind.	W. T. Blanford, Esq. [C.].
t. Yg.	Sind.	W. T. Blanford, Esq. [C.].
u. Hgr.	Alpine Punjab.	T. C. Jerdon, Esq. [P.].

7. Agama sanguinolenta.

Trapelus? sanguinolentus, *Gray, Cat.* p. 259.
Lacerta sanguinolenta, *Pall. Zoogr. Ross.-As.* iii. p. 23.
Agama aralensis, *Lichstenst. in Eversm. Reise,* p. 144, *and Verz. Doubl. Mus. Berl.* p. 101; *Eichw. Zool. Spec.* iii. p. 185.
—— oxiana, *Eichw. l. c.*
Trapelus aralensis, *Eversm. Nouv. Mém. Ac. Mosc.* iii. p. 366.
—— sanguinolentus, *Eichw. Faun. Casp. Cauc.* p. 89, pl. xiv. figs. 3, 4.
Podorrhoa (Trapeloidis) sanguinolenta, *Fitz. Syst. Rept.* p. 81.
Agama sanguinolenta, *A. Dum. Cat. Méth. Rept.* p. 102, *and Arch. Mus.* viii. p. 576; *Schreib. Herp. Eur.* p. 465.
—— agilis, part., *Blanf. Proc. Zool. Soc.* 1881, p. 674.

Head moderate, convex, subcordiform. Nostril not tubular, lateral, pierced just below the canthus rostralis in the posterior part of a flat nasal. Upper head-scales convex; occipital not enlarged; a few spine-like scales on the post-temporal region; ear-opening as large as the eye-opening, partly concealed by a strong fringe of spinose scales superiorly; males with a very small gular pouch. Body roundish, moderately depressed, covered above with equal, rhomboidal, imbricate, very strongly keeled scales, ending in a strong, spinose, raised mucro; lateral and ventral scales much smaller, distinctly keeled. Limbs moderate; tibia nearly as long as the skull (to occiput); third finger a little shorter than fourth, fifth not extending as far as second; third toe much shorter than fourth, fifth not extending quite as far as first. Tail a little more than twice as long as the distance from gular fold to vent, rounded, covered with strongly keeled scales. Males with a double or triple row of anal pores. Pale olive or sandy above, with or without darker transverse bars interrupted by a light vertebral spot; lower surfaces paler, the throat dark blue in the breeding male.

Total length	282	millim.
Head	29	,,
Width of head	25	,,
Body	83	,,
Fore limb	46	,,
Hind limb	72	,,
Tail	170	,,

Borders of the Caspian Sea; Central Asia; Turkestan.

a. ♂.	Mangyschlak.	St. Petersburg Museum [E.].
b. ♂.	West Goloduaja.	St. Petersburg Museum [E.].
c, d. ♂.	Syr Darya.	M. Severzow [C.].

8. Agama latastii.

Head short, convex, subcordiform. Nostril not tubular, superior, pierced in the posterior part of a flat nasal just above the canthus rostralis. Upper head-scales smooth; occipital not enlarged; no spinose scales on the back of the head; a fringe of three or four pointed scales on the upper edge of the ear, which is smaller than the eye-opening. Body depressed, covered above with equal, rhomboidal, imbricate, smooth or indistinctly keeled scales, the size of which scarcely exceeds that of the ventrals; latter smooth or indistinctly keeled. Limbs moderate; tibia slightly shorter than the skull (to occiput); third finger shorter than fourth, fifth not extending as far as second; third toe much shorter than fourth, fifth not extending quite as far as first. Tail not quite twice as long as the distance from gular fold to vent, rounded, covered with equal keeled scales. Back with four pairs of large quadrangular dark spots.

Total length	161	millim.
Head	18	,,
Width of head	17	,,
Body	56	,,
Fore limb	36	,,
Hind limb	50	,,
Tail	87	,,

Egypt.

a. ♀. Egypt. Sir J. Wilkinson [P.].

9. Agama inermis.

Trapelus savignii, (*non* D. & B.) *Gray, Cat.* p. 258.
? Agama deserti, *Lichtenst. Verz. Doubl. Mus. Berl.* p. 101.
Agama inermis, *Reuss, Mus. Senck.* p. 33.
—— agilis, (*non. Oliv.*) *Strauch, Erp. Alg.* p. 28.
—— mutabilis, (*non Merr.*) *Lataste, Le Nat.* 1880, p. 325.

Head short, convex, subcordiform. Nostril not tubular, directed upwards and backwards, pierced on the canthus rostralis in the posterior part of a flat nasal. Upper head-scales convex, smooth, or very slightly keeled; occipital not enlarged; a few scattered small spinose scales on the back of the head; a fringe of small spinose scales on the upper edge of the ear, which is smaller than the eye-opening; male without, or with only an indication of, a gular pouch. Body depressed, covered above with unequal, rhomboidal, imbricate, keeled, more or less distinctly mucronate scales; ventral scales smooth or indistinctly keeled. Limbs moderate; tibia as long as, or a little shorter than, the skull (to occiput); third finger shorter than fourth, fifth not extending as far as second; third toe much shorter than fourth, fifth not extending as far as first. Tail about twice as long as the distance from gular fold to vent, rounded, covered with equal keeled scales. Males with a double row of anal pores. Grey-brown or sandy-coloured above, with more or less distinct quadrangular brown or reddish spots arranged symmetrically on the back; sometimes some of the larger scales lighter; lower surfaces dirty white; the breeding male's throat with blue longitudinal lines.

Total length	197 millim.
Head	20 ,,
Width of head	19 ,,
Body	65 ,,
Fore limb	38 ,,
Hind limb	56 ,,
Tail	112 ,,

Egypt to Algeria.

a-b. ♂.	Egypt.	J. Burton, Esq. [P.].
c-d. ♂ ♀.	Tunis.	Mr. Fraser [C.].

10. Agama persica.

Agama persica, *Blanf. Proc. Zool. Soc.* 1881, p. 674, pl. lix.

Head moderate, convex, subcordiform. Nostril not tubular, directed upwards and backwards, pierced on the canthus rostralis in the posterior part of a flat nasal. Upper head-scales all more or less distinctly keeled; occipital not enlarged; a few short spines on the post-temporal region, and a few on the upper edge of the ear, which is exposed and not larger than the eye-opening; males with a large gular pouch. Body roundish, moderately depressed, covered above with unequal, rhomboidal, imbricate, keeled, shortly mucronate scales, the largest not being twice as large as the smallest; lateral and ventral scales smaller, more or less distinctly keeled. Limbs moderate; tibia nearly as long as the skull (to occiput); third finger shorter than fourth, fifth not extending as far as second; third toe much shorter than fourth, fifth not extending as far as first. Tail once and three fourths to twice and a half as long as the distance from gular fold to vent, rounded, covered with equal keeled scales. Males with a double row of anal pores. Grey-brown or sandy-coloured above, with more or less distinct regular series of large quadrangular dark spots on the back, and dark annuli on the tail; some of the enlarged scales frequently white; the collar-pit black; lower surfaces whitish; the breeding male's throat dark blue.

Total length	240 millim.
Head	23 ,,
Width of head	20 ,,
Body	70 ,,
Fore limb	45 ,,
Hind limb	67 ,,
Tail	147 ,,

Persia; Euphrates.

a-d. ♂, ♀, & hgr.	Dehbid, Persia.	(Types.)
e. Hgr.	Kazeroon, Persia.	
f. ♂.	Euphrates Expedition.	

11. Agama leucostigma.

Agama leucostygma, *Reuss, Mus. Senck.* p. 44.
Trapelus flavimaculatus, *Rüppell, N. Wirbelth. Faun. Abyss., Rept.* p. 12, pl. vi. fig. 1.

Head short, very convex. Nostril not tubular, superior, pierced in the posterior part of a flat nasal above the canthus rostralis. Upper head-scales smooth; occipital not enlarged; no distinct spines on the hinder part of the head; a denticulation formed by three pointed scales on the upper edge of the ear, which is not larger than the eye-opening. [Males with a large gular pouch.] Body depressed, covered above with unequal, rhomboidal, imbricate, feebly keeled, not mucronate scales, the largest of which are not twice as large as the smallest; ventral scales smaller, feebly keeled. Limbs moderate; tibia shorter than the skull (to occiput); digits short; third finger shorter than fourth, fifth not extending as far as second; third toe shorter than fourth, fifth not extending as far as first. Tail nearly twice as long as the distance between gular fold and vent, rounded, covered with equal, keeled scales. Sandy-coloured above, the back with five dark transverse bars each enclosing a light vertebral elongate rhomboidal spot; lower surfaces white, throat with brown lines.

Total length	136	millim.
Head	17	,,
Width of head	16	,,
Body	43	,,
Fore limb	32	,,
Hind limb	41	,,
Tail	76	,,

Arabia; Egypt.

a. Hgr. Arabia.

12. Agama rubrigularis.

Trapelus rubrigularis, *Blanf. Proc. As. Soc. Beng.* 1875, p. 233, *and Journ. As. Soc. Beng.* xlv. 1876, p. 23, pl. i. fig. 1; *Murray, Zool. Sind,* p. 371.

Head short, convex, subcordiform. Nostril not tubular, pierced in the posterior part of a slightly swollen nasal just above the canthus rostralis. Upper head-scales keeled; occipital not enlarged; no distinct spines on the hinder part of the head; a very slight fringe of pointed scales on the upper edge of the ear, which is hardly as large as the eye-opening. Male without gular pouch. Body depressed, covered above with small rhomboidal, imbricate, smooth scales intermixed with much larger, flat, feebly keeled ones forming more or less irregular transverse series; no enlarged scales on the limbs; ventral scales smooth. Limbs moderate; tibia not quite as long as the skull (to occiput); third finger shorter than fourth, fifth not extending as far as second; third toe much shorter than fourth, fifth not extending as far as first. Tail once and two thirds to once and three fourths as long as the distance from gular fold to vent, rounded, rather depressed at the base, covered with keeled scales. Male with a row of anal pores. "Colour olive-brown to grey, spotted with pale yellow, each enlarged scale of the back being in the middle of a pale spot; a dusky longitudinal line on each side of the neck and three or four pairs of dark spots on the back; a bright red patch beneath the throat in living specimens of both sexes."

Total length	175	millim.
Head	20	„
Width of head	19	„
Body	60	„
Fore limb	38	„
Hind limb	53	„
Tail	95	„

Sind; Baluchistan; Persia.

a-b. ♂ & yg.　　Upper Sind.　　W. T. Blanford, Esq. [P.].　(Types.)

13. Agama megalonyx.

Trapelus megalonyx, *Günth. Rept. Brit. Ind.* p. 159, pl. xiv. fig. C.

Head short, convex, subcordiform. Nostril not tubular, superior, pierced in the posterior part of a small flat nasal just above the canthus rostralis. Upper head-scales keeled; occipital not enlarged; no distinct spines on the hinder part of the head; a feeble fringe of pointed scales on the upper edge of the ear, which is smaller than the eye-opening. Body depressed, covered above with unequal rhomboidal imbricate scales, the smaller being smooth or indistinctly keeled, the larger being more than twice as large and distinctly keeled; ventral scales smooth. Limbs moderate, the fore pair relatively more developed and armed with very long claws; tibia as long as the skull (to occiput); third finger shorter than fourth, fifth not extending as far as second; third toe much shorter than fourth, fifth not extending as far as first. Tail not quite twice as long as the distance from gular fold to vent, rounded, covered with equal keeled scales. Greyish above, with darker transverse bands on the back, and a vertebral series of dark-edged light spots; a dark longitudinal band on each side of the neck; collar-pit black; lower surfaces whitish.

Total length	140	millim.
Head	18	„
Width of head	16	„
Body	47	„
Fore limb	36	„
Hind limb	46	„
Tail	75	„

Afghanistan?

a. ♀.　　　　Afghanistan?　　　Dr. Griffith [C.].　(Type.)

14. Agama ruderata.

Agama ruderata, *Olivier, Voy. Emp. Ottom.* ii. p. 429, pl. xxix. fig. 3.
—— loricata, *Reuss, Mus. Senckenb.* p. 40.
—— mutabilis, part., *Dum. & Bibr.* iv. p. 505.
Eremioplanis ruderata, *Fitzing. Syst. Rept.* p. 82.
Agama lessonæ, *De Filippi, Viag. Pers.* p. 353.
Trapelus ruderatus, *Anders. Proc. Zool. Soc.* 1872, p. 384; *Blanf. E. Persia*, p. 315; *Murray, Zool. Sind*, p. 370.

Head short and very convex. Nostril not tubular, lateral, pierced in the posterior part of a small nasal just below the canthus rostralis. Upper head-scales keeled or striated; occipital not enlarged; no regular spines on the back of the head; a small denticulation partly covering the ear, which is smaller than the eye-opening. Males without gular pouch. Body short, depressed, covered above with small, irregular, more or less distinctly keeled, imbricate scales, intermixed with scattered large, nail-like, pointed keeled scales; ventral scales smooth. Limbs moderate, with heterogeneous scales; tibia as long as the skull (to occiput); third finger a little shorter than fourth, fifth not extending as far as second; third toe much shorter than fourth. Tail once and a half to twice as long as the distance from gular fold to vent, rounded, with strongly keeled scales intermixed with enlarged ones at the base. Males with a double row of anal pores. Greyish or sandy-coloured above, variegated with darker or with dark spots arranged in pairs along the back; tail with dark cross bars; lower surfaces whitish, the male's throat variegated with bluish grey.

Total length	167	millim.
Head	19	,,
Width of head	17	,,
Body	60	,,
Fore limb	35	,,
Hind limb	51	,,
Tail	88	,,

Syria to Sind.

a–b. ♂ ♀.	Shiraz.	W. T. Blanford, Esq. [C.].
c. ♂.	Road to Euphrates.	C. G. Danford, Esq. [P.].
d. ♀.	Between Baalbeck and Shtora.	Dr. Anderson [P.].

15. Agama pallida.

Trapelus ruderatus, (*non Oliv.*) *Gray, Cat.* p. 258.
Agama ruderata, (*non Oliv.*) *Aud. Descr. Égypte, Rept. Suppl.* p. 169, pl. i. fig. 6.
—— pallida, *Reuss, Mus. Senckenb.* p. 38, pl. iii. fig. 3.
—— nigrofasciata, *Reuss, l. c.* p. 42.
—— mutabilis, part., *Dum. & Bibr.* iv. p. 505.
Trapelus ægyptius, *Duvern. R. A., Rept.* pl. xiv. fig. 2.
—— mutabilis, (*non Merr.*) *Blanf. E. Persia*, p. 316.

Head short, thick, very convex. Nostril not tubular, superior, pierced in the posterior part of a slightly swollen nasal just above the canthus rostralis. Upper head-scales convex, smooth; occipital not enlarged; no distinct spines on the hinder part of the head; a fringe of small pointed scales on the upper edge of the ear, which is hardly as large as the eye-opening. Male without gular pouch. Body short, depressed, covered above with small irregular, imbricate, smooth or indistinctly keeled scales intermixed with irregularly scattered much larger ones, each of which bears a short keel or spine; limbs also with scattered enlarged scales; ventral scales smooth. Limbs moderate; tibia nearly as long as the skull (to occiput); third finger shorter than fourth, fifth not extending as far as second; third toe much shorter than fourth, fifth not extending as far as first. Tail once and a half to twice as long as the distance from gular fold to vent, rounded, with feebly keeled scales intermixed with enlarged ones at the base. Males with a double row of anal pores. Grey or sandy-coloured above, with paired dark markings, one on the neck and one in front of the sacrum being more constant; tail with dark cross bars; the breeding male's head bluish, and the throat with bluish-grey lines; lower surfaces whitish.

Total length	174	millim.
Head	20	,,
Width of head	19	,,
Body	50	,,
Fore limb	37	,,
Hind limb	54	,,
Tail	104	,,

Egypt; North Arabia.

a–b. ♀.	Egypt.	Sir J. Wilkinson [P.].
c, d. ♂ ♀.	Egypt.	
e–f. ♀.	Nile.	Mr. Petherick [C.].
g–i. ♂ & hgr.	Cairo.	Sir R. Owen [P.].
k. Hgr.	Cairo.	R. MacAndrew, Esq. [P.].
l. ♂.	Suez.	R. MacAndrew, Esq. [P.].
m. ♂.	Sinaitic Peninsula.	H. C. Hart, Esq. [C.].
n–o. ♂ & hgr.	Mount Sinai.	

16. Agama hispida.

Agama hispida, *Gray, Cat.* p. 257.
Lacerta hispida, *Linn. Mus. Ad. Fred.* p. 44, and *S. N.* (ed. 10) i. p. 205.
—— orbicularis, part., *Linn. S. N.* (ed. 12), i. p. 265.
Agama aspera, part., *Daud. Rept.* iii. p. 402.
—— orbicularis, part., *Daud. l. c.* p. 406.
—— gemmata, *Daud. l. c.* p. 402.
Trapelus hispidus, *Kaup, Isis*, 1827, p. 616, pl. vii.; *Gravenh. Nova Acta Ac. Leop. Carol.* xvi. 1833, p. 917, pl. lxiv. figs. 1–8.
Agama spinosa, (non *Gray*) *Dum. & Bibr.* iv. p. 502, pl. xli. bis. fig. 2.
—— hispida, part., *Peters, Mon. Berl. Ac.* 1870, p. 112.

Head short, convex, subcordiform. Nostril not tubular, lateral, pierced just below the canthus rostralis in a convex nasal. Upper head-scales unequal, rough, keeled, or subconical; forehead convex, with enlarged subconical scales; occipital enlarged; back of head with large conical spinose scales; ear-opening small, not quite as large as the eye-opening. Male without a regular gular pouch. Body depressed, covered with moderately large, imbricate, strongly keeled scales intermixed with strongly enlarged, spinose, raised ones; a well-marked dorsal and nuchal crest; ventral scales more or less strongly keeled and mucronate. Limbs moderate; tibia as long as the skull (to occiput); fingers short, third longest; toes short, third longer than fourth, fifth not extending as far as first. Tail once and one fourth to once and a half as long as the distance from gular fold to vent, rounded, covered with strongly keeled scales, the basal portion intermixed with enlarged spinose scales. Males with a row of anal pores. Brown or olive, spotted with darker; lower surfaces dirty white, in the young with wide-meshed grey reticulation; throat of the breeding male blue.

Total length	187	millim.
Head	23	,,
Width of head	24	,,
Body	67	,,
Fore limb	49	,,
Hind limb	62	,,
Tail	97	,,

South Africa.

a, b, c, d, e, f. Ad., stuffed.	Cape of Good Hope.	
g. ♂.	Cape of Good Hope.	Prof. Busk [P.].
h. ♀.	Cape of Good Hope.	Sir E. Belcher [P.].
i. ♂.	Cape of Good Hope.	Dr. Lea [P.].
k. ♂.	Cape of Good Hope.	— Townsend, Esq. [P.].
l, m, n-o. Hgr. & *yg.*	Cape of Good Hope.	
p. ♂.	Africa.	

17. Agama brachyura. (PLATE XXVIII. fig. 1.)

Head short, convex, subcordiform. Nostril not tubular, lateral, pierced just below the canthus rostralis in a convex nasal. Upper head-scales smooth; forehead convex; occipital enlarged; sides of head posteriorly and neck with small spines; ear-opening a little smaller than the eye-opening. Body depressed, covered with moderately large imbricate strongly keeled scales intermixed with strongly enlarged spinose ones; ventral scales smooth. Limbs moderate; tibia as long as the skull (to occiput); fingers short, third longest; fourth toe very slightly longer than third, fifth not extending as far as first. Tail shorter than head and body, a little longer than the distance from gular fold to vent, rounded, covered with strongly keeled scales. Olive-brown above, with a double series of large dark spots on the back, each pair separated on the vertebral line by a ⊃⊂-shaped whitish marking; lower surfaces whitish, with wide-meshed blackish network.

Total length	134	millim.
Head	19	,,
Width of head	20	,,
Body	52	,,
Fore limb	38	,,
Hind limb	51	,,
Tail	63	,,

South Africa.

a. ♀. Cape of Good Hope. Sir A. Smith [P.].

18. Agama aculeata.

Agama aculeata, *Gray, Cat.* p. 257.
Agama aculeata, *Merr. Tent.* p. 53; *Dum. & Bibr.* iv. p. 499.
Trapelus (Psammorrhoa) bibronii, *Fitzing. Syst. Rept.* p. 81.
Agama hispida, part., *Peters, Mon. Berl. Ac.* 1870, p. 112.

Head short, convex, subcordiform. Nostril slightly tubular, pierced on the canthus rostralis or just below, in a convex nasal. Upper head-scales smooth or feebly keeled; forehead slightly convex; occipital enlarged; sides of head posteriorly and neck with spinose scales; ear-opening larger than the eye-opening. Male without a regular gular pouch. Body depressed, covered with moderately large, imbricate, strongly keeled scales intermixed with strongly enlarged ones; a feeble but very distinct dorsal and nuchal crest; ventral scales perfectly smooth. Limbs moderate; tibia as long as the skull (to occiput); third finger very slightly longer than fourth; fourth toe slightly longer than third, fifth extending as far as first. Tail nearly twice as long as the distance from gular fold to vent, rounded in the female, a little compressed in the male, covered with strongly keeled scales. Male with a row of præanal pores. Yellowish or olive-brown above, with a double series of large brown spots on the back; lower surfaces whitish, the throat with longitudinal brown lines in the female, blackish blue in the breeding male.

Total length	252	millim.
Head	24	,,
Width of head	23	,,
Body	81	,,
Fore limb	48	,,
Hind limb	69	,,
Tail	147	,,

South Africa.

a. ♂. S. Africa.
b–d. ♀ & hgr. Damara Land.
e–f. ♀. ——?
g. Skeleton. S. Africa. Sir A. Smith [P.].

19. Agama armata.

Agama aculeata, (*non Merr.*) *Bianc. Spec. Zool. Mossamb.* p. 27, pl. i. fig. 2.
—— armata, *Peters, Mon. Berl. Ac.* 1854, p. 616, *and Reise n. Mossamb.* iii. p. 42, pl. vii. fig. 2.
—— infralineata, *Peters, Mon. Berl. Ac.* 1877, p. 613.

Head moderately long, convex. Nostril slightly tubular, lateral, pierced in a small nasal just below the canthus rostralis. Upper head-scales keeled; occipital enlarged; back of head with scattered short spines; ear-opening larger than the eye-opening. Body depressed, covered with moderately large, imbricate, strongly keeled scales intermixed with enlarged ones, forming longitudinal series; a slight dorsal and nuchal crest; ventral scales strongly keeled and mucronato. Limbs rather short; tibia shorter than the skull (to occiput); third finger slightly longer than fourth; third toe slightly longer than fourth, fifth extending beyond first. Tail once and two thirds as long as the distance from gular fold to vent, rounded, covered with strongly keeled scales. Olive-brown above, the enlarged scales lighter; a double series of darker spots along the back; lower surfaces lighter, the throat with darker longitudinal lines.

Total length	167	millim.
Head	20	,,
Width of head	17	,,
Body	57	,,
Fore limb	36	,,
Hind limb	44	,,
Tail	90	,,

South Africa.

a. ♀.   Natal.   J. H. Gurney, Esq. [P.].

20. Agama atra.

Agama atra, *Gray, Cat.* p. 256.
Agama atra, *Daud. Rept.* iii. p. 349; *Dum. & Bibr.* iv. p. 493.
—— subspinosa, *Gray, Ann. Philos.* 1827, p. 214.
Trapelus subhispidus, *Kaup, Isis,* 1827, p. 616.
Phrynopsis atra, *Fitzing. Syst. Rept.* p. 80.

Head moderate, rather depressed. Nostril tubular, pierced in a small convex nasal upon the canthus rostralis. Upper head-scales smooth or feebly keeled; occipital slightly enlarged; sides of head posteriorly and neck with groups of small spines; ear-opening larger than the eye-opening. Male without a regular gular pouch. Body much depressed, covered with small imbricate keeled scales, largest on the vertebral line; these scales sometimes intermixed with scattered enlarged ones; a more or less distinct curved fold along each side of the back; a short nuchal crest; no distinct dorsal crest; ventral scales small and smooth. Limbs moderate;

tibia as long as the skull (to occiput); third finger very slightly shorter than fourth; fourth toe slightly longer than third, fifth extending beyond first. Tail nearly twice as long as the distance from gular fold to vent, roundish in the female, strongly compressed and crested in the male, covered with equal strongly keeled scales forming more or less distinct annuli. Male with a row of anal pores. Olive-brown above, spotted or speckled with black, sometimes nearly black; a lighter, yellowish-brown vertebral band; lower surfaces with more or less distinct dark network or lines, blackish blue in the breeding male.

Total length	203 millim.
Head	20 ,,
Width of head	18 ,,
Body	66 ,,
Fore limb	43 ,,
Hind limb	60 ,,
Tail	117 ,,

South Africa.

a, b. Ad., stuffed.	Cape of Good Hope.	
c, d-e. ♂ ♀.	Cape of Good Hope.	
f. ♂.	Cape of Good Hope.	J. Ford, Esq. [P.].
g. ♂.	Cape Recife.	A. E. Craven, Esq. [P.].
h. ♀.	Port Elizabeth.	
i. ♂.	S. Africa.	Haslar Collection.
k. ♀.	S. Africa.	
l-m, n, o. Hgr. & yg.	S. Africa.	
p. Skeleton.	S. Africa.	Sir A. Smith [P.].
q. Skeleton.	Cape of Good Hope.	

21. Agama mossambica.

Agama mossambica, *Peters, Mon. Berl. Ac.* 1854, p. 616, *and Reise n. Mossamb.* iii. p. 38, pl. vii. fig. 1; *Fischer, Jahresb. Naturh. Mus. Hamb.* f. 1883, p. 21, pl. ii. fig. 6.

—— cariniventris, *Peters, Mon. Berl. Ac.* 1874, p. 159.

Head moderate, convex. Nostril tubular, directed upwards and backwards, pierced in the posterior part of a small nasal on the canthus rostralis. Upper head-scales feebly keeled; occipital enlarged; sides of head near the ear, and neck with groups of short spines; ear-opening large, entirely exposed, much larger than the eye-opening; no regular gular pouch. Body moderately depressed, covered above with rhomboidal, mucronate, imbricate scales with strong keels converging towards the vertebral line; a slight nuchal and dorsal crest; ventral scales keeled; seventy-seven to eighty-one scales round the middle of the body. Limbs moderate; tibia as long as the skull (to occiput); third and fourth fingers equal, or fourth very slightly longer; fourth toe slightly longer than third, fifth extending beyond first. Tail twice to twice and a half as long as the distance from gular fold to vent, slightly compressed, covered with strongly keeled scales. Olive-brown above, with a row of more or less distinct large brown spots on each side of the vertebral line; lower surfaces lighter; throat dotted with blackish.

Total length	278 millim.
Head	24 ,,
Width of head	23 ,,
Body	84 ,,
Fore limb	50 ,,
Hind limb	70 ,,
Tail	170 ,,

East Africa.

. ♀. Quellimane. Sir J. Kirk [P.].
b. Hgr. Zambesi Exped. Sir J. Kirk [P.].

22. Agama kirkii. (Plate XXVIII. fig. 2.)

Head moderate, convex. Nostril tubular, directed upwards, pierced in the posterior part of a small nasal on the canthus rostralis. Upper head-scales smooth or feebly keeled; occipital enlarged: sides of head near the ear and neck, with groups of very small spinose scales; ear-opening entirely exposed, larger than the eye-opening; no regular gular pouch. Body moderately depressed, covered above with small rhomboidal, not or but very shortly mucronate, imbricate scales with strong keels converging towards the vertebral line; a low but very distinct nuchal and dorsal crest; ventral scales scarcely smaller than dorsals, smooth; ninety-nine scales round the middle of the body. Limbs moderate; tibia a little longer than the skull (to occiput); fourth finger a little longer than third; fourth toe a little longer than third, fifth extending beyond first. Tail nearly twice and a half as long as the distance from gular fold to vent, compressed and keeled above, covered with strongly keeled scales forming rather distinct annuli. Male with a row of anal pores. Olive above, with dark network enclosing light ocelli; collar-pit black; lower surfaces whitish, with indistinct darker lines.

Total length	209 millim.
Head	19 ,,
Width of head	19 ,,
Body	60 ,,
Fore limb	42 ,,
Hind limb	61 ,,
Tail	130 ,,

East Africa.

a. ♂. Zambesi Expedition. Sir J. Kirk [P.].

23. Agama spinosa.

Agama colonorum (*non D. & B.*), *Gray, Cat.* p. 256.
Agama spinosa, *Gray, Griff. A. K.* ix., *Syn.* p. 57.
—— colonorum, *Rüppell, N. Wirbelth. Faun. Abyss.* p. 14, pl. iv; *Blanf. Zool. Abyss.* p. 449.

Head small, rather depressed. Nostril tubular, directed upwards and backwards, pierced in the posterior part of a small nasal on the canthus rostralis. Upper head-scales smooth; generally no elongate scale on the snout; occipital enlarged; sides of head near the ear, and neck with groups of rather long spines, the largest of which measure more than half the diameter of the ear-opening; latter large, exposed, larger than the eye-opening. Throat much plicate; no regular gular pouch. Body not much depressed, covered above with rather large rhomboidal mucronate imbricate scales, with strong keels converging towards the vertebral line; these dorsal scales, smaller than the caudals, number forty to fifty from the origin of the fore limbs to the origin of the hind limbs (counting on the vertebral line); a distinct nuchal crest composed of rather long spines; no dorsal crest; ventral scales much smaller than the dorsals, smooth; sixty to seventy-four scales round the middle of the body. Limbs long and strong, digits elongate; tibia longer than the skull (to occiput); third and fourth fingers equal; fourth toe slightly longer than third, fifth extending beyond first. Tail twice and a half as long as the distance from gular fold to vent, roundish in the female, compressed and keeled above in the male, covered with large strongly keeled scales forming rather distinct annuli. Male with a row of anal pores. Olive-brown above; lower surfaces yellowish, with a well-marked black collar; the middle of the throat orange-red in the male.

Total length	310	millim.
Head	23	,,
Width of head	22	,,
Body	87	,,
Fore limb	54	,,
Hind limb	75	,,
Tail	200	,,

North-east Africa; Arabia.

a. ♂.	Egypt.	Sir J. G. Wilkinson [P.]
b, c. ♂ ♀.	Egypt.	J. Burton, Esq. [P.].
d. ♂.	Rairo Habab, Abyssinia.	W. T. Blanford, Esq. [P.].
e–f. ♂.	Mt. Sinai.	

24. Agama rueppellii.

Agama ruppelli, *Vaillant, in Révoil, Faun. Somalis, Rept.* p. 6, pl. i.

Distinguished from *A. colonorum* by the shorter, more globulose head, the longer spines on the back of the head and neck, which partly conceal the ear-opening, and the larger dorsal scales, the size of which nearly equals that of the caudals; twenty-six or twenty-seven dorsal scales from the origin of the fore limbs to the origin of the hind limbs.

Somali Land.

25. Agama colonorum.

Agama occipitalis, *Gray, Cat.* p. 256.
Agama colonorum, part., *Daud. Rept.* iii. p. 356.
—— occipitalis, *Gray, Griff. A. K.* ix., *Syn.* p. 56.
—— colonorum, *Dum. & Bibr.* iv. p. 489.
Podorrhoa colonorum, *Fitzing. Syst. Rept.* p. 80.
Agama colonorum, var. congica, *Peters, Mon. Berl. Ac.* 1877, p. 612.
—— picticauda, *Peters, Mon. Berl. Ac.* 1877, p. 612.

Head small, rather depressed. Nostril tubular, directed upwards and backwards, pierced in the posterior part of a small nasal on the canthus rostralis. Upper head-scales smooth or feebly keeled; generally a narrow elongate scale on the middle of the snout; occipital enlarged; nine or ten upper labials; sides of head, near the ear, and neck with groups of short spines, the longest of which are considerably shorter than the diameter of the ear-opening; latter large, entirely exposed, larger than the eye-opening. Throat much plicate; no regular gular pouch. Body not much depressed, the back slightly keeled, covered with rather large, rhomboidal, mucronate imbricate scales with strong keels converging towards the vertebral line; these dorsal scales smaller than the caudals, numbering forty to sixty from the origin of the fore limbs to the origin of the hind limbs (counting on the vertebral line); a distinct nuchal but no dorsal crest; ventral scales much smaller than the dorsals, smooth or feebly keeled; sixty to eighty scales round the middle of the body. Limbs long and strong, digits elongate; tibia longer than the skull (to occiput); third and fourth fingers equal, or fourth slightly longer; fourth toe a little longer than third, fifth extending beyond first. Tail about twice and a half as long as the distance from gular fold to vent, roundish in the female, compressed and keeled above in the male, covered with large strongly keeled scales forming rather distinct annuli. Male with a row of anal pores. Olive or brown above, the young with light spots and lines arranged symmetrically; lower surfaces lighter, uniform or marbled with grey, blackish in the breeding male.

Total length	347	millim.
Head	27	,,
Width of head	23	,,
Body	100	,,
Fore limb	63	,,
Hind limb	92	,,
Tail	220	,,

West Africa.

a–d, e. ♂.	W. Africa.	Dr. Stanger [P.].
f, g, h. ♂ & hgr.	W. Africa.	W. Raddon, Esq. [P.].
i–k. Yg.	W. Africa.	Mr. Fraser [C.].

l-p, q, r. ♂ & yg.	W. Africa.	
s. ♀.	Senegal.	
t. ♂.	Gambia.	
u. Hgr.	Sierra Leone.	
v. Several spec.: ♂, ♀, & hgr.	Ancober River, Gold Coast.	Major Burton and Capt. Cameron [P.].
w. Yg.	Ambriz, Angola.	
x. ♂.	Carangigo.	Dr. Welwitsch [P.].
y, z. ♂ ♀.	Benguela.	J. J. Monteiro, Esq. [C.].
a–δ. ♂ & hgr.	S. Africa.	Sir A. Smith [P.].
ε. Skeleton.	——?	

26. Agama bibronii *.

Agama colonorum, (non D. & B.) Guich. Expl. Sc. Alg., Rept. p. 7; Strauch, Erp. Alg. p. 27.
—— bibronii, A. Dum. Cat. Méth. Rept. p. 101, and Arch. du Mus. viii. p. 574; Boettg. Abh. Senck. Ges. xiii. 1883, p. 127.
—— colonorum, var. impalearis, Boettg. Abh. Senck. Ges. ix. 1874, p. 132.

Head moderate, somewhat less depressed than in *A. colonorum*. Nostril tubular, directed laterally and posteriorly, pierced just below the canthus rostralis. Upper head-scales smooth or indistinctly keeled; a very slightly elongate scale on the middle of the snout; occipital enlarged; labials very small, eleven to fourteen upper and as many lower; sides of head and neck with groups of spines which are slightly longer than in *A. colonorum*; ear-opening entirely exposed, larger than the eye-opening. Throat strongly plicate; no regular gular pouch. Body depressed, covered above with large rhomboidal, mucronate, imbricate scales with strong keels converging towards the vertebral line; these scales scarcely smaller than the caudals, numbering about forty from the origin of the fore limbs to the origin of the hind limbs (counting on the vertebral line); a slight nuchal crest; no dorsal crest; ventral scales much smaller than the dorsals, smooth; sixty-two or sixty-four scales round the middle of the body. Limbs shorter than in *A. colonorum*, digits also shorter; tibia as long as the skull; third and fourth fingers equal; third and fourth toes nearly equal, measuring the distance from the posterior border of the eye to the nostril; fifth toe extending beyond first. Tail about twice as long as the distance from gular fold to vent, round in the female, feebly compressed in the male; scaling as in *A. colonorum*. Male with one [to three] row of anal pores. Coloration as in *A. colonorum*.

Total length ..	250 millim.		Fore limb	49 millim.
Head	27 ,,		Hind limb	73 ,,
Width of head	25 ,,		Tail	150 ,,
Body	83 ,,			

Morocco; Algeria; Tunis.

---

* Described from Algerian specimens in M. Lataste's collection.

27. Agama planiceps.

Agama planiceps, *Peters, Mon. Berl. Ac.* 1862, p. 15.

Head small, much depressed. Nostril tubular, directed upwards and backwards, pierced in the posterior part of a small nasal on the canthus rostralis. Upper head-scales smooth or feebly keeled; occipital enlarged; sides of head near the ear, and neck, with groups of short spines; ear-opening large, entirely exposed, much larger than the eye-opening. Throat much plicate; no regular gular pouch. Body much depressed, covered above with moderate-sized, roundish, not or but very slightly mucronate, imbricate scales, with more or less strong keels converging towards the vertebral line; these dorsal scales much smaller than the caudals, numbering fifty to sixty-five from the origin of the fore limbs to the origin of the hind limbs (counting on the vertebral line); a slight nuchal but no dorsal crest; ventral scales much smaller than the dorsals, smooth; sixty-four to eighty-six scales round the middle of the body. Limbs long and strong, digits elongate; tibia longer than the skull (to occiput); third and fourth fingers equal; fourth toe slightly longer than third, fifth extending beyond first. Tail about twice and a half as long as the distance from gular fold to vent, roundish in the female, compressed and keeled above in the male, covered with large strongly keeled scales forming rather distinct annuli. Male with a row of small pores. Olive-brown above, with or without yellow spots and bands; lower surfaces yellowish, darker in the male.

Total length (tail broken)	239 millim.	Body	96 millim.
Head	26 ,,	Fore limb	63 ,,
Width of head	23 ,,	Hind limb	86 ,,

South-west Africa.

a. ♂.	Duque de Bragance, Angola.	Prof. J. du Bocage [P.].
b. ♀.	Carangigo.	Dr. Welwitsch [C.].
c-d. ♀ & hgr.	Damara Land.	
e, f. ♂ & yg.	——?	Berlin Museum.

28. Agama atricollis.

Agama atricollis, *Smith, Ill. S. Afr., Rept. App.* p. 14.
Stellio capensis, *A. Dum. Cat. Méth. Rept.* p. 106, and *Arch. Mus.* viii. p. 579.

Head large (especially in the male, which has the cheeks strongly swollen), convex, with very prominent supraciliary ridges; nostril lateral, slightly tubular, pierced on the canthus rostralis; upper head-scales mostly feebly keeled; occipital not enlarged; small scattered spinose scales on the cheeks; ear exposed, larger than the eye-opening; male without a regular gular pouch. Body moderately depressed; dorsal scales small, rhomboidal, keeled, larger on the vertebral region, everywhere intermixed with enlarged, strongly keeled, pointed scales, which sometimes form longitudinal series on the vertebral region; a slight nuchal denticulation; a more or less marked curved fold from the neck to the middle of the side; ventral scales smooth. Limbs moderately elongate, with compressed digits; the scales on the upper surface of the limbs irregular in size, keeled; third and fourth fingers nearly equal; fourth toe very slightly longer than third, fifth extending beyond first. Tail very slightly compressed, about twice as long as the distance from gular fold to vent; the scales strongly keeled, and forming rather regular annuli, those on the upper surface much enlarged, with denticulated edge. Male with a double row of anal pores. Olive or brown above, marbled or reticulated with blackish; young with ✗✗-shaped black markings across the back; a large black spot in front of the shoulder; lower surfaces lighter; throat with dark network, blue in the male.

Total length ..	360 millim.	Fore limb	70 millim.
Head	38 „	Hind limb	95 „
Width of head	37 „	Tail	220 „
Body	102 „		

South Africa.

a-c. ♂, ♀, & hgr.	Natal.	Sir A. Smith [P.]. (Types.)
d-e. ♂ ♀.	Natal.	J. Ayres, Esq. [C.].
f, g. ♀ & yg.	Natal.	
h. ♂.	Lamo.	Sir J. Kirk [P.].
i. ♂.	S.E. Africa.	Sir J. Kirk [P.].
k. ♂.	Duque de Bragance, Angola.	Prof. B. du Bocage [P.].
l. ♀.	S. Africa.	J. H. Gurney, Esq. [P.].

29. Agama cyanogaster.

Stellio cyanogaster, *Gray, Cat.* p. 255.
? Agama gularis, *Reuss, Mus. Senck.* p. 36.
Stellio cyanogaster, *Rüpp. N. Wirbelth. Faun. Abyss., Rept.* p. 10, pl. v.; *Dum. & Bibr.* iv. p. 532.
Acanthocercus cyanogaster, *Fitzing. Syst. Rept.* p. 84.
Agama (Stellio) cyanogaster, *Blanf. Zool. Abyss.* p. 446.

Head moderately depressed; nostril lateral, not tubular, below the canthus rostralis; upper head-scales smooth; occipital not enlarged; small conical spinose scales on the sides of the head near the ear; latter exposed, larger than the eye-opening; throat strongly plicate, no gular pouch. Body depressed, with a distinct fold on each side of the back; dorsal scales rather large, irregular, feebly keeled on the vertebral region, minute and intermixed with enlarged ones on the sides; a series of small spinose scales on the dorsal folds; ventral scales smooth, smaller than the largest dorsals. Limbs moderately elongate, with compressed digits; the scales on the upper surface of the limbs irregular in size, keeled; fourth finger very slightly longer than third; fourth toe very slightly longer than third, fifth extending beyond first. Tail round, nearly twice as long as the distance from gular fold to vent; the scales strongly keeled, shortly mucronate, forming regular annuli. Male with a patch of anal pores. Olive-brown above, spotted or reticulated with blackish; the vertebral region usually lighter; lower surfaces whitish in the female, with bluish network on the throat, dark blue in the breeding male.

Total length..	265 millim.	Fore limb	60 millim.
Head	32 ,,	Hind limb......	89 ,,
Width of head	29 ,,	Tail	143 ,,
Body........	90 ,,		

Abyssinia; Arabia.

a. ♂. Abyssinia. W. Jesse, Esq. [C.].
b–d. ♀ & yg. Tigré, Abyssinia. W. T. Blanford, Esq. [P.].

30. Agama annectens.

Agama annectens, *Blanf. Zool. Abyss.* p. 446, fig.

Head depressed; snout as long as the diameter of the orbit; nostril lateral, slightly tubular, on the canthus rostralis. Upper head-scales smooth; occipital not enlarged; a few small spinose scales on the side of the head near the ear; latter entirely exposed, much larger than the eye-opening. Throat much plicate; no gular pouch. Body depressed; dorsal scales small, rhomboidal, imbricate, very slightly keeled, largest on the vertebral region, very small on the sides; scattered small spines on the nuchal region; ventral scales perfectly smooth, nearly as large as the median dorsals. Limbs long and strong; digits compressed; fourth finger very slightly longer than third; fourth toe a little longer than third, fifth extending beyond first. Tail compressed, covered with moderately large keeled scales forming annuli in front, but becoming gradually imbricate towards the tip. Male with a double row of præanal pores. Rufous-brown above, with small irregular blackish rings and a lighter vertebral band.

Total length (tail broken)	285 millim.	Body........	102 millim.
Head.................	33 ,,	Fore limb....	71 ,,
Width of head	29 ,,	Hind limb....	105 ,,

Abyssinia.

a. ♂. Sooroo Pass. W. T. Blanford, Esq. [P.]. (One of the types.)

31. Agama stoliczkana.

Stellio stoliczkanus, *Blanf. Journ. As. Soc. Beng.* xliv. 1875, p. 191, *and 2nd Yark. Miss., Rept.* p. 3, pl. i. figs. 1, 2.

Head depressed, its length considerably exceeding its breadth; nostrils directed backwards, situated below the canthus rostralis. Upper head-scales smooth; occipital not enlarged; some spinose scales round the tympanum; groups of spinose scales scattered over the sides and back of the neck, the former being the larger. Throat strongly plicate; no gular pouch. Scales on the back of the neck granular, passing gradually into the bluntly keeled scales of the middle of the back; these are considerably larger than the scales of the sides, being about twice as broad; latter distinctly keeled, with a few conspicuously larger scales scattered amongst them; ventral scales smooth, about the same size as those of the middle of the back. Limbs long and strong, covered above with strongly keeled scales; digits as in the preceding. Tail rounded, covered with keeled scales ending in a short spine posteriorly, all in verticils, the largest scarcely larger than the largest dorsal scales. Males with two or three rows of anal pores. General colour pale yellowish, mixed with dusky black; chin and throat dusky, more or less mottled or speckled with pale yellow.

Total length...... 374 millim. Fore limb...... 70 millim.
From snout to vent 140 ,, Hind limb...... 100 ,,
Eastern Turkestan.

32. Agama tuberculata.

Laudakia tuberculata, *Gray, Cat.* p. 254.
Agama tuberculata, *Gray, Zool. Journ.* iii. 1828, p. 218, *and Ill. Ind. Zool.*; *Dum. & Bibr.* iv. p. 488.
Podorrhoa tuberculata, *Fitz. Syst. Rept.* p. 80.
Stellio indicus, *Blyth, Journ. As. Soc. Beng.* xxii. 1853, p. 646.
Barycephalus sykesii, *Günth. Proc. Zool. Soc.* 1860, p. 150, pl. xxv. fig. A.
Stellio tuberculatus, *Günth. Rept. Brit. Ind.* p. 157; *Steind. Novara, Rept.* p. 22; *Stoliczka, Journ. As. Soc. Beng.* xli. 1872, p. 115, pl. iii. fig. 3.

Head much depressed; snout longer than the diameter of the orbit; nostril lateral, below the canthus rostralis, slightly tubular. Upper head-scales smooth or feebly keeled; occipital not enlarged; small closely set spinose scales on the sides of the head near the ear, and on the neck; ear entirely exposed, larger than the eye-opening. Throat strongly plicate; no gular pouch. Body depressed, with a more or less distinct fold on each side of the back; scales on the neck and sides minute, almost granular, keeled, uniform or intermixed with scattered enlarged scales; those on the vertebral region enlarged, equal, rhomboidal, imbricate, strongly keeled; a very slight indication of a nuchal denticulation; ventral scales smooth, nearly as large as the enlarged dorsals. Limbs strong, with compressed digits; the scales on the upper surface of the limbs strongly enlarged and very strongly keeled; third and fourth fingers equal, or fourth very slightly longer; fourth toe slightly longer than third, fifth extending beyond first. Tail rounded, much depressed at the base, covered with moderate-sized strongly keeled scales arranged in rings; its length equals twice and a half to three times the distance from gular fold to vent. Males with a large patch of anal pores and a similar patch of pores on the middle of the belly. Olive-brown above, spotted or speckled with blackish, sometimes with small yellowish spots; the breeding male's throat blue, with light spots; sometimes a light vertebral band.

Total length..	315 millim.	Fore limb	52 millim.
Head	27 ,,	Hind limb......	77 ,,
Width of head	22 ,,	Tail	205 ,,
Body........	83 ,,		

Himalayas.

a. ♂.	Bengal (?).	Gen. Hardwicke [P.]. (Type.)
b. ♀.	Garhval.	Messrs. v. Schlagintweit [C.]. ⎫ (Types of
c. ♀.	Simla.	Messrs. v. Schlagintweit [C.]. ⎬ *Baryceph.*
d. Hgr.	Ladak.	Messrs. v. Schlagintweit [C.]. ⎭ *sykesii.*)
e–g. Yg.	Tibet.	T. C. Jerdon, Esq. [P.].
h–k. ♂ & hgr.	Kashmere.	T. C. Jerdon, Esq. [P.]
m. ♀.	Marri.	W. T. Blanford, Esq. [P.].
n–o, p. ♂ & hgr.	—— ?	

33. Agama dayana.

Stellio dayanus, *Stoliczka, Journ. As. Soc. Beng.* xli. 1872, p. 113, pl. iii. fig. 4.

Differs from *A. tuberculata* in having the head-scales sharply keeled, and the enlarged dorsal scales distinctly continuous, though considerably smaller, on the neck and up to the occiput.

Hardwar, N. India.

34. Agama himalayana.

Stellio himalayanus, *Steind. Novara, Rept.* p. 22, pl. i. fig. 8; *Stoliczka, Journ. As. Soc. Beng.* xli. 1872, p. 113; *Blanf. 2nd Yark. Miss., Rept.* p. 3.

Head much depressed; snout slightly longer than the diameter of the orbit; nostril lateral, below the canthus rostralis, slightly tubular. Upper head-scales smooth; occipital not enlarged; small closely set spinose scales on the head near the ear, and on the neck; ear entirely exposed, larger than the eye-opening. Throat strongly plicate; no gular pouch. Body depressed, with a more or less distinct fold on each side of the back; scales on the neck and sides small, smooth or very feebly keeled, uniform, those on the vertebral region enlarged, equal, roundish-hexagonal, imbricate, smooth or very feebly keeled; ventral scales smooth, a little smaller than the enlarged dorsals. Limbs strong, with compressed digits; the scales on the upper surface large and strongly keeled; fourth finger slightly longer than third; fourth toe considerably longer than third, the extremity of the claw of the latter not reaching the base of the claw of the former; fifth toe extending beyond first. Tail rounded, much depressed at the base, covered with moderate-sized strongly keeled scales arranged in rings; its length equals twice and a half to three times the distance from gular fold to vent. Males with a double or triple row of anal pores. Olive above, marbled with black, and generally with round light spots producing a network; sometimes the black spots forming a festooned band on each side of the vertebral line; the male's throat marbled with blackish.

Total length..	212 millim.	Fore limb	40 millim.
Head	19 ,,	Hind limb	63 ,,
Width of head	17 ,,	Tail	134 ,,
Body	59 ,,		

Himalayas.

a–d. ♂, ♀, & yg.	Tibet.	T. C. Jerdon, Esq. [P.].
e. Yg.	Balti, Tibet.	Messrs. v. Schlagintweit [C.].

35. Agama agrorensis.

Stellio agrorensis, *Stoliczka, Proc. As. Soc. Beng.* 1872, p. 128; *Blanf. 2nd Yark. Miss., Rept.* pl. i. fig. 3.

Head much depressed; snout longer than the diameter of the orbit; nostril lateral, below the canthus rostralis. Upper head-scales keeled; occipital not enlarged; small spinose scales on the sides of the head near the ear, and on the neck; ear entirely exposed, larger than the eye-opening. Throat strongly plicate; no gular pouch. Body depressed, with a slight fold on each side of the back; scales on the neck and sides of the back very small, almost granular, keeled, those on the vertebral region and on the flanks enlarged, rhomboidal, imbricate, strongly keeled; a very slight indication of nuchal denticulation; ventral scales smooth, much smaller than the enlarged dorsals. Limbs and tail as in *A. tuberculata*. Male with a patch of anal pores and a double longitudinal row of pores on the middle of the belly. Olive above, spotted with yellow (sometimes with dark and light longitudinal bands); the male's throat reticulated with grey.

From snout to vent 92 millim.

Punjab and Kachmere.

a. ♂.	Agror Valley.	W. T. Blanford, Esq. [P.].

36. Agama melanura.

Laudakia (Plocederma) melanura, *Blyth, Journ. As. Soc. Beng.* xxiii. 1854, p. 737.

—— tuberculata (*non Gray*), *Theob. Cat. Rept. As. Soc. Mus.* 1868, p. 38.

Stellio melanurus, *Anders. Proc. As. Soc. Beng.* 1871, p. 189; *Stoliczka, Journ. As. Soc. Beng.* xli. 1872, p. 113, *and Proc. As. Soc. Beng.* 1872, p. 129.

Head depressed, triangular, as broad as long; snout as long as the diameter of the orbit; nostril lateral, just below the canthus rostralis, slightly tubular. Upper head-scales smooth; occipital not enlarged; a group of small spines in front of the ear, and other scattered ones below; ear-opening twice as large as the eye-opening. Throat strongly plicate; no gular pouch. Body depressed, with a slight fold on each side of the back; scales on the neck and sides very small, on the latter region mucronate and feebly keeled: median dorsal scales enlarged, rhomboidal, imbricate, strongly keeled, forming regular longitudinal series, some of which are larger than the others; a very slight indication of a nuchal denticulation; ventral scales smooth, smaller than the largest dorsals. Limbs strong, with compressed digits; the scales on the upper surface of the limbs large and very strongly keeled and spinose; third and fourth fingers nearly equal; fourth toe very slightly longer than third, fifth extending beyond first. Tail rounded, depressed at the base, covered with large strongly keeled spinose scales forming annuli; it is four times as long as the distance from gular fold to vent. Males with a patch of anal pores, and another patch of pores on the middle of the belly. Head and neck yellow, the rest of the body blackish brown.

Total length	438 millim.
Head	28 ,,
Width of head	28 ,,
Body	90 ,,
Fore limb	63 ,,
Hind limb	99 ,,
Tail	320 ,,

N.W. India.

a. ♂.   Khirthar Range, Sind.   W. T. Blanford, Esq. [P.].

37. Agama lirata.

Stellio liratus, *Blanf. Ann. & Mag. N. H.* (4) xiii. 1874, p. 453, *and E. Persia*, p. 320, pl. xx. fig. 2.

Head rather flat, longer than broad. Nostril in the middle of the canthus rostralis. Scales of the upper part of the head transversely keeled on the occiput, smooth in the supraorbital regions, convex or bluntly keeled longitudinally on the snout; occipital not enlarged; a group of spinose scales in front of the large tympanum, and a very few scattered spines below and behind it. Two or three folds across the throat. Body depressed; scales of the back of the neck very small; a rudimentary nuchal crest; a fold from the side of the neck over the shoulder, running back towards the groin; in the middle of the back are six or seven rows of large keeled scales, the keels forming continuous parallel lines: lateral scales small, keeled; ventral scales smooth, much larger than the lateral ones, but not half as large as the median dorsals; no enlarged scales on the sides. Limbs strong, with compressed digits; the scales on the upper surface of the limbs sharply keeled and pointed; fourth finger very slightly longer than third, fourth toe slightly longer than third. Tail much longer than the body, covered with keeled pointed scales, which do not form distinct rings; the scales at the base of the tail smooth. Dusky above, with imperfect blackish transverse markings; a small blackish pit before each shoulder.

From snout to vent	100	millim.
Head	25	,,
Fore limb	50	,,
Hind limb	75	,,

Baluchistan.

38. Agama nupta.

Agama nupta, *De Fil. Giorn. Ist. Lomb.* vi. 1843, p. 407.
Stellio carinatus, *A. Dum. Cat. Méth. Rept.* p. 107, *and Arch. Mus.* viii. p. 580.
—— nuptus, *De Fil. Viagg. Persia,* p. 352; *Blanf. E. Pers.* p. 317, pl. xix. fig. 1.

Head depressed, triangular; nostril rather large, slightly tubular, pierced on the canthus rostralis; upper head-scales smooth; occipital not enlarged; borders of the tympanum and sides of neck with groups of well-developed spines; ear-opening much larger than the eye-opening. Throat strongly plicate; no gular pouch; a transverse fold on the neck. Body depressed; a distinct curved fold from the neck to the sides; dorsal scales rather large, rhomboidal, strongly imbricate, shortly mucronate, keeled, the keels forming oblique lines converging towards the vertebral line; flanks with minute scales; ventral scales smooth, much smaller than dorsals and much larger than laterals. Limbs strong, digits compressed; scales on upper surface of limbs as large as, or a little smaller than, the dorsals; third and fourth fingers nearly equal; fourth toe slightly longer than third, fifth extending beyond first. Tail round or slightly compressed, depressed at the base, nearly three times as long as the distance from gular fold to vent; caudal scales larger than dorsals, mucronate, feebly keeled, forming regular annuli. Male with a patch of anal pores and another of ventral pores. Yellowish-brown above, uniform or with a darker network or dots; fore limbs dusky or blackish; posterior half of tail black, or annulate with black; gular region marbled yellow and dark blue.

Total length	456	millim.
Head	41	,,
Width of head	37	,,
Body	115	,,
Fore limb	76	,,
Hind limb	113	,,
Tail	300	,,

Persia.

a. Hgr.	Persia.	Marquis G. Doria [P.].
b-d. ♂, ♀, & hgr.	S.E. of Rigan.	W. T. Blanford, Esq. [C.].
e-f, g-h, i. ♂, ♀, & yg.	Kazeroon.	
k. ♂.	Shiraz.	India Museum.
l. Ad., stuffed.	Dizful.	W. K. Loftus, Esq. [P.].

Var. fusca.

Stellio nuptus, var. fuscus, *Blanf. E. Persia*, p. 319.

The transverse nuchal fold less distinct, and the spines on the sides of head and neck rather longer. General colour blackish-brown, throat yellowish.

Baluchistan, Persia.

a. ♂.	Near Jalk, Baluchistan.	W. T. Blanford, Esq. [C.].
β-γ. ♂.	Shiraz, Persia.	

39. Agama microlepis.

Stellio microlepis, *Blanf. Ann. & Mag. N. H.* (4) xiii. 1874, p. 453, and *E. Persia*, p. 326, pl. xix. fig. 2.

Head much depressed; snout not longer than the diameter of the orbit: nostril lateral, below the canthus rostralis, slightly tubular. Upper head-scales smooth or feebly keeled; occipital not enlarged; very small conical spinose scales on the sides of the head near the ear and on the neck; ear entirely exposed, larger than the eye-opening. Throat strongly plicate; no gular pouch. Body much depressed, with a very indistinct lateral fold; nuchal and dorsal scales minute, granular; a narrow zone on the vertebral line covered with flat, smooth or feebly keeled, rather irregular, enlarged scales about the size of those on the snout; on the flanks are scattered some feebly enlarged, conical spinose scales; no nuchal denticulation; ventral scales smooth, not quite as large as the enlarged dorsals. 210 to 220 scales round the middle of the body. Limbs strong, with compressed digits; the scales on the upper surface of the limbs strongly enlarged, strongly keeled and spinose; fourth finger very slightly longer than third; fourth toe a little longer than third, fifth extending beyond first. Tail rounded, depressed at the base, covered with rather large spinose scales arranged in rings, two rings forming a distinct segment; the length of the tail equals once and a half to once and three fourths the distance from gular fold to vent. Male with a large patch of anal pores and an enormous patch of ventral pores. Olivaceous above, dotted or reticulated with black, and with yellowish spots; throat dark-marbled.

Total length	301	millim.
Head	31	,,
Width of head	29	,,
Body	110	,,
Fore limb	63	,,
Hind limb	88	,,
Tail	160	,,

Persia.

a-d. ♂.	Kushkizard, N. of Shiraz.	W. T. Blanford, Esq. [C.]. (Types.)
e. ♂.	——?	

40. Agama caucasica.

Stellio caucasicus, *Gray, Cat.* p. 255.
Lacerta muricata (*non Shaw*), *Pall. Zoogr. Ross.-As.* iii. p. 20.
—— stellio (*non L.*), *Pall. l. c.* p. 24.
Stellio vulgaris (*non Latr.*), *Ménétr. Cat. Rais.* p. 64.
—— caucasicus, *Eichw. Zool. Spec.* p. 187, *and Faun. Casp.-Cauc.* p. 80, pl. xiii. figs. 1-8; *A. Dum. Cat. Méth. Rept.* p. 105, *and Arch. Mus.* viii. p. 578; *De Fil. Viagg. Pers.* p. 352; *Blanf. E. Persia*, p. 322, pl. xx. fig. 1.
—— persicus, *Anders. Proc. Zool. Soc.* 1872, p. 382, fig.

Head much depressed; snout not longer than the diameter of the orbit; nostril lateral, below the canthus rostralis, slightly tubular. Upper head-scales smooth; occipital not enlarged; small conical spinose scales on the side of the head near the ear, and on the neck: ear entirely exposed, larger than the eye-opening. Throat strongly plicate; no gular pouch. Body much depressed, with a very indistinct lateral fold; nuchal and latero-dorsal scales very small, granular; vertebral region with enlarged flat, feebly keeled, rather irregular scales; flanks with enlarged, strongly keeled or spinose scales; no nuchal denticulation; ventral scales smooth, distinctly smaller than the enlarged dorsals. 150 to 160 scales round the middle of the body. Limbs strong, with compressed digits; the scales on the upper surface of limbs much enlarged, strongly keeled, generally spinose; fourth finger slightly longer than third; fourth toe a little longer than third, fifth extending beyond first. Tail rounded, depressed at the base. covered with rather large spinose scales arranged in rings, two rings forming a distinct segment; the length of the tail does not equal quite twice the distance from gular fold to vent. Male with a large patch of anal pores and an enormous patch of ventral pores. Olivaceous above, with round yellowish black-edged spots, the black frequently forming a network; vertebral region yellowish; limbs with more or less distinct yellowish cross bars; lower surfaces yellowish in the female, blackish in the breeding male.

Total length	312	millim.
Head	33	,,
Width of head	30	,,
Body	99	,,
Fore limb	64	,,
Hind limb	97	,,
Tail	180	,,

Caucasus to Persia.

a. Hgr. ♂. Elizabethpol.
b-c. ♂ & hgr. Persia. Marquis G. Doria [P.].
d-h. ♂, ♀, & yg. Kohrud, N. of Isfahan. W. T. Blanford, Esq. [C.].

41. Agama stellio.

Stellio cordylina, *Gray, Cat.* p. 255.
Lacerta stellio, *Linn. S. N.* i. p. 361; *Hasselq. Reise Paläst.* p. 351.
Iguana cordylina, *Laur. Syst. Rept.* p. 47.
Cordylus stellio, *Laur. l. c.* p. 52.
Stellio vulgaris, *Latr. Rept.* ii. p. 22; *Daud. Rept.* iv. p. 16; *Geoffr. Descr. Egypte, Rept.* p. 127, pl. ii. fig. 3; *Heyd. in Rüpp. Atlas N. Afr., Rept.* p. 6. pl. ii.; *Guérin, Icon. R. A.* pl. vi. fig. 2; *Bibr. in Bory, Expéd. Morée, Rept.* p. 68, pl. xi. fig. 1 & pl. xiii. fig. 1; *Dum. & Bibr.* iv. p. 528; *Duvern. R. A., Rept.* pl. xiii. fig. 1; *Schreib. Herp. Eur.* p. 469; *Boettg. Ber. Senck. Ges.* 1879-80, p. 196.
Agama sebæ, *Merr. Tent.* p. 55.
—— cordylea, *Merr. l. c.*
Stellio antiquorum, *Eichw. Zool. Spec.* p. 117.
—— cyprius, *Fitzing. Syst. Rept.* p. 85.

Head moderately depressed, with swollen cheeks in the male; nostril lateral, below the canthus rostralis; upper head-scales smooth or feebly keeled; occipital not enlarged; sides of head and neck with numerous short spine-like scales; ear exposed, a little larger than the eye-opening. Throat with strong transverse folds; no gular pouch. Body much depressed, with a distinct dorso-lateral fold, and vertical folds on the sides; an irregular transverse fold on the neck; vertebral region with irregular keeled scales intermixed with strongly enlarged, feebly keeled ones; the rest of the back with minute scales intermixed with scattered enlarged ones, forming more or less regular transverse series; on the dorso-lateral fold and on the flanks the enlarged scales conical, spinose; ventral scales smooth; gulars strongly keeled, spinose posteriorly. Limbs strong, with compressed digits; scales on upper surface of limbs large, strongly keeled, spinose; third and fourth fingers equal, or fourth very slightly longer; fourth toe a little longer than third, fifth extending beyond first. Tail round, depressed at the base, about twice as long as the distance from gular fold to vent; the scales large, strongly spinose, arranged in rings, two of which form a well-marked segment. Male with three to five rows of anal pores, and a double series of pores on the middle of the belly. Olive above, spotted with black, generally with a vertebral series of large yellow or orange spots; the male's throat reticulated with bluish grey.

Total length	284	millim.
Head	31	,,
Width of head	30	,,
Body	83	,,
Fore limb	55	,,
Hind limb	85	,,
Tail	170	,,

S.E. Europe; Asia Minor; Syria; N. Arabia; Egypt.

a, b–e. ♂ ♀.	Zebil, Bulgar Dagh, Cilician Taurus, 4000 ft.	C. G. Danford, Esq. [P.].
f. ♂.	Biredijk, Euphates.	C. G. Danford, Esq. [P.].
g. ♂.	Rumkaleh, Euphrates.	C. G. Danford, Esq. [P.].
h. ♀.	Asia Minor.	C. Fellows, Esq. [P.].
i. ♂.	Asia Minor.	A. Christy, Esq. [P.].
k–n. ♂ ♀.	Asia Minor.	Sir J. Richardson [P.].
o, p. ♀ & yg.	Xanthus.	Haslar Collection.
q–s. ♂ ♀.	Lake of Galilee.	Canon Tristram [C.].
t. Hgr.	Damascus.	Dr. Anderson [P.].
u–v. Hgr. & yg.	Ramleh, Judea.	Dr. Anderson [P.].
w. ♀.	Jerusalem.	Dr. Anderson [P.].
x. ♀.	Mt. Sinai.	
y. ♀.	Egypt.	M. Lefebvre [P.].
z, a. Hgr. & yg.	Egypt.	Rev. W. Hennah [P.].
β. Skeleton.	——?	

20. PHRYNOCEPHALUS*.

Phrynocephalus, *Kaup, Isis*, 1826, p. 591; *Fitzing. N. Classif. Rept.* p. 15; *Wagl. Syst. Amph.* p. 144; *Wiegm. Herp. Mex.* p. 17; *Dum. & Bibr. Erp. Gén.* iv. p. 512; *Gray, Cat. Liz.* p. 259; *Günth. Rept. Brit. Ind.* p. 160.
Megalochilus, *Eichw. Zool. Spec. Ross. Pol.* iii. p. 185; *Gray, l. c.* p. 261.

Tympanum hidden. Body depressed. No dorsal crest. A transverse gular fold; no gular sac. Tail round. No femoral or præanal pores.

South-eastern Europe; Central Asia.

Synopsis of the Species.

I. No lobe of skin at the angle of the mouth.

A. Dorsal lepidosis heterogeneous.

Nasals generally in contact; toes feebly fringed; tail with deep black bars inferiorly .... 1. *olivieri*, p. 370.
Nasals separated; toes strongly fringed .. 2. *helioscopus*, p. 371.

B. Dorsal lepidosis homogeneous, the scales on the vertebral region considerably larger than those on the flanks; no spinose scales on the sides of the head and neck.

1. Pectoral scales smooth or very feebly keeled; tail not twice as long as the distance from gular fold to vent.

Pectoral scales quite smooth; nostrils widely separated, directed outwards .... 3. *vlangalii*, p. 372.

---

* 1. *P. varius*, Eichw. Zool. Spec. iii. p. 186.—S. Siberia.
2. *P. melanurus*, Eichw. l. c.—S. Siberia.
3. *P. nigricans*, Eichw. l. c.—S. Siberia.

Pectoral scales quite smooth; fourth toe
feebly fringed .... 4. *theobaldi*, p. 373.
Pectoral scales slightly keeled; fourth toe
strongly fringed .... 5. *versicolor*, p. 374.

 2. Pectoral scales strongly keeled.

  a. Fourth toe strongly fringed.

Posterior ventral scales smooth; tibia as
long as the skull .... 6. *frontalis*, p. 375.
Posterior ventral scales smooth; tibia longer   [p. 375.
than the skull .... 7. *caudivolvulus*,
Posterior ventral scales keeled .... 8. *przewalskii*, p. 377.

  b. Fourth toe feebly fringed.

Ventral scales feebly keeled .... 9. *affinis*, p. 377.

 3. Pectoral scales smooth; tail nearly twice as long as the distance from gular fold to vent.

End of tail black inferiorly .... 10. *maculatus*, p. 377.
End of tail not black .... 11. *axillaris*, p. 378.

 C. Dorsal lepidosis homogeneous; sides of head and neck with projecting spine-like scales.

Limbs very long..................... 12. *interscapularis*,
                 [p. 378.

II. A large, folded, fringed lobe of the skin at the angle of the mouth.

Digits very strongly fringed .... 13. *mystaceus*, p. 379.

1. Phrynocephalus olivieri.

Phrynocephalus olivieri, *Gray, Cat.* p. 260.
Phrynocephalus tickelii, *Gray, l. c.*
Olivier, *Voy. Emp. Ottom.* pl. xlii. fig. 1.
Phrynocephalus olivieri, *Dum. & Bibr.* iv. p. 517; *Anders. Proc. Zool. Soc.* 1872, p. 386; *Blanf. Zool. E. Pers.* p. 327.
—— tickelii, *Günth. Rept. Brit. Ind.* p. 160.

Snout extremely short, vertically truncate; upper head-scales large, smallest on the middle of the supraorbital region, where they are about the size of the median dorsals, largest and convex behind the nostrils and on the occiput; nostril directed forwards and upwards; nasal large, in contact with its fellow (quite exceptionally separated by a series of scales). Dorsal scales flat, smooth, juxtaposed or slightly imbricate, with scattered enlarged nail-like scales; sides more or less folded, with small granular scales intermixed with scattered enlarged ones. Gular, pectoral, and ventral scales smooth, the former very small but not granular; ventrals square, forming slightly oblique transverse series. Scales on upper surface of limbs large, smooth or feebly keeled. Outer edge of third and fourth toes feebly denticulated. Tibia longer than the skull. Tail round, gradually tapering to a fine point, covered with smooth or slightly keeled scales; on the basal portion of the tail they are intermixed with enlarged conical scales; the length of the tail equals once and two thirds to a little more than twice the distance from gular fold to vent. Grey above, marbled with darker and guttate with lighter; sometimes a blackish cross band behind the scapular region, and another in front of the sacral region; sometimes a large zone on the middle of the back uniform light grey; limbs and tail with blackish cross bars, on the latter forming annuli which are deep black on the lower surface; body and limbs inferiorly white.

Total length	105	millim.
Head	11·5	,,
Width of head	12	,,
Body	31·5	,,
Fore limb	27	,,
Hind limb	39	,,
Tail	62	,,

Persia; Baluchistan.

a-d. ♀ & hgr.	Persia.	Marquis G. Doria [P.].
e-g. ♀.	Rayen, S.E. of Karman.	W. T. Blanford, Esq. [C.].
h-k. ♂ ♀.	Near Rigan, Narmashir, S.E. Persia.	W. T. Blanford, Esq. [C.].
l. ♂.	Between Tehran and Kazvin.	W. T. Blanford, Esq. [C.].
m-p. ♂ ♀.	Ghistigan, Bampusht, Baluchistan.	W. T. Blanford, Esq. [C.].
q. ♂.	Near Bampur, Baluchistan.	W. T. Blanford, Esq. [C.].

2. Phrynocephalus helioscopus.

Phrynocephalus helioscopus, *Gray, Cat.* p. 260.
Lacerta helioscopa, *Pall. Reise,* i. p. 457.
—— uralensis, *Gmel. S. N.* p. 1073.
Stellio helioscopa, *Latr. Rept.* ii. p. 30.
—— uralensis, *Latr. l. c.* p. 39.
Agama helioscopa, *Daud. Rept.* iii. p. 419; *Licht. in Eversm. Reise,* p. 144.
—— uralensis, *Daud. l. c.* p. 422.
Phrynocephalus uralensis, *Fitz. N. Class. Rept.* p. 47.
—— helioscopus, *Wagl. Syst. Amph.* p. 144; *Eichw. Zool. Spec.* iii. p. 186; *Gravenh. N. Acta Ac. Leop.-Carol.* xvi. 1833, ii. p. 934, pl. lxiv. figs. 9, 14; *Eversm. N. Mém. Soc. Moscou,* iii. 1834, p. 364, pl. xxxii. fig. 3; *Dum. & Bibr.* iv. p. 519.
—— persicus, *De Filippi, Arch. p. la Zool.* ii. p. 387, *and Viagg. Pers.* p. 353; *Anders. Proc. Zool. Soc.* 1872, p. 388, fig.; *Blanf. Zool. E. Pers.* p. 329.

Forehead convex, with enlarged convex scales; occipital scales enlarged; median supraorbital scales as large as, or rather smaller than, the median dorsals; nostril directed forwards; nasal separated from its fellow by three to five (exceptionally one) longitudinal series of scales. Dorsal scales irregular, on the vertebral region enlarged, more or less imbricate, smooth or very feebly keeled; on the sides granular, intermixed with irregularly arranged enlarged nail-like or pointed scales. Gular, pectoral, and ventral scales smooth, none mucronate. Scales on upper surface of limbs smooth or feebly keeled, intermixed with enlarged ones. Outer and inner edge of third and fourth toes with a fringe of acute lobules. Tibia longer than the skull. Tail round, depressed, and more or less swollen at the base, ending in an obtuse point, covered with more or less distinctly keeled scales, on the basal portion with scattered enlarged projecting tubercles; the length of the tail equals once and one fourth to once and two thirds the distance from gular fold to vent. Grey above, with darker spots sometimes forming interrupted transverse bars; tail above with dark cross bars, which are either absent or of a pale grey on the lower surface. Lower surfaces white; throat and breast generally marbled with grey.

Total length	116	millim.
Head	13	,,
Width of head	14	,,
Body	45	,,
Fore limb	31	,,
Hind limb	44	,,
Tail	58	,,

South-western Siberia; Turkestan; South-eastern Russia; Persia.

a. ♂.	Kirghiz Steppe, Astrachan.	Dr. A. Günther [P.].
b. ♂.	Astrachan.	
c–d. ♂ ♀.	Siberia.	Berlin Museum.
e. ♀.	Siberia.	Dr. J. E. Gray [P.].
f–h. ♂ & hgr	Sysfyk-Kul.	Geographical Society of Bremen.
i–k. ♂ ♀.	West Golodnaja.	St. Petersburg Museum.
l–q. ♂, ♀, & hgr.	Mangyschlak.	St. Petersburg Museum.
r–s. ♂ ♀.	Turkestan.	M. Severzow [C.].
t. ♀.	Between Tehran and Kazvin.	W. T. Blanford, Esq. [C.].
u–v, w–y. ♂ ♀.	Between Shiraz and Isfahan.	W. T. Blanford, Esq. [C.].

3. Phrynocephalus vlangalii.

Phrynocephalus vlangalii, *Strauch, Voy. Przewalski, Rept.* p. 20, pl. iii. fig. 3.

Forehead convex, with enlarged convex scales; nostril anterior, directed not upwards but outwards, pierced under the canthus rostralis; nasal very concave, separated from its fellow by six or seven longitudinal series of scales. Back with very distinct, mostly transverse folds; scales homogeneous, slightly imbricate, enlarged on the vertebral region, smaller on the sides, all more or less convex but not keeled. Gular, pectoral, and ventral scales quite smooth, the former rather convex. Outer edge of third and fourth toes distinctly fringed. Tail short and thick, suddenly enlarged and much depressed at the base, then roundish, ending obtusely; caudal scales smooth, except posteriorly, where the upper are slightly and the lower distinctly keeled. Dirty grey above, yellowish inferiorly; an immaculate vertebral band, on each side of which irregular black and yellowish-orange spots; throat and chest blackish; a large elongate deep black spot on the middle of the belly, and another on the lower surface of the end of the tail.

Total length	114 millim.
Head and neck	18 ,,
Body	39 ,,
Tail	57 ,,

Kuku Lake.

4. Phrynocephalus theobaldi.

Phrynocephalus tickelii, part., *Günth. Proc. Zool. Soc.* 1860, p. 161.
—— olivieri, (*non D. & B.*) *Theob. Journ. As. Soc. Beng.* xxxi. 1862, p. 518.
—— theobaldi, *Blyth, tom. cit.* xxxii. 1863, p. 90; *Blanf. tom. cit.* xliv. 1875, p. 192, *and 2nd Yark. Miss., Rept.* p. 6.
—— caudivolvulus, part., *Günth. Rept. Brit. Ind.* p. 161.
—— stoliczkai, *Steind. Novara, Rept.* p. 23, pl. i. figs. 6, 7.
—— caudivolvulus, *Anders. Proc. Zool. Soc.* 1872, p. 387.

Forehead very convex, with enlarged convex scales; supraorbital scales about as large as the largest dorsals; occipital scales much larger; nostril directed forwards; nasal separated from its fellow by one or three longitudinal rows of scales. Dorsal scales homogeneous, small, granular on the sides, enlarged, flat, and subimbricate on the vertebral region, all perfectly smooth; pectoral and ventrals small, perfectly smooth, not mucronate; scales on upper surface of limbs smooth. Outer side of third and fourth toes feebly but distinctly denticulated. Tibia as long as the skull. Tail roundish, thickened and depressed at the base, covered with smooth scales, ending obtusely; its length is once and a half to once and two thirds the distance from gular fold to vent. Grey above, more or less spotted with blackish or with light, dark-edged ocelli; lower surfaces white; male with the middle of the throat and of the belly, and the end of the tail inferiorly, black.

Total length	110 millim.
Head	11 ,,
Width of head	11 ,,
Body	38 ,,
Fore limb	26 ,,
Hind limb	39 ,,
Tail	61 ,,

Tibet; Eastern Turkestan.

a. ♂.	Tsomoriri, Tibet.	Messrs. v. Schlagintweit [C.].
b–c. ♀ & hgr.	Turkestan.	M. Severzow [C.].

Var. forsythii.

Phrynocephalus forsythii, *Anders. Proc. Zool. Soc.* 1872, p. 390, fig.

—— theobaldi, var. forsythii, *Blanf. 2nd Yark. Miss., Rept.* p. 8.

Very nearly allied to *P. theobaldi*, but distinguished in the following characters:—The toes are less distinctly denticulated laterally, the middle ones being almost deprived of any such denticulation; all the scales, and especially those on the upper surface of the head, smaller; scales on the limbs keeled; tail longer, nearly twice as long as the distance from gular fold to vent, tapering to a fine point. Light grey above, dotted with darker; two series of small dark spots along the back; end of tail black inferiorly; lower surfaces white, with a faint blackish line along the middle of the belly.

Total length	112	millim.
Head	11	,,
Width of head	11	,,
Body	37	,,
Fore limb	24	,,
Hind limb	35	,,
Tail	64	,,

Yarkand.

a. ♂.   Yarkand.   India Museum, Calcutta. (One of the types.)

5. Phrynocephalus versicolor.

Phrynocephalus versicolor, *Strauch, Voy. Przewalski, Rept.* p. 18, pl. iii. fig. 2.

Forehead slightly convex and sloping, with enlarged convex scales; occipital scales enlarged; median supraorbitals smaller than median dorsals; nostril directed forwards and slightly upwards; nasal separated from its fellow by three longitudinal series of scales. Dorsal scales homogeneous, granular on the sides, enlarged, flat, smooth, and imbricate on the vertebral region. Gular and abdominal scales smooth, pectorals very slightly keeled. Third and fourth toes on both sides with a strong fringe of acute lobules. Tibia as long as the skull. Tail roundish, tapering to a rather fine point, covered with keeled scales: its length equals once and a third to once and three fourths the distance from gular fold to vent. Greyish above, spotted and vermiculated with darker; limbs and tail with dark cross bars, which form three or four complete annuli, black inferiorily, at the end of the latter; lower surfaces white.

Total length	108	millim.
Head	12	,,
Width of head	11	,,
Body	42	,,

                Fore limb .... 24 millim.
                Hind limb 36   ,,
                Tail 54   ,,

The younger specimen measures—from snout to vent 37 millim., tail 47 millim.
Mongolia.

a–b. ♀ & hgr.   Desert Gobi.   J. Brenchley, Esq. [P.].

6. Phrynocephalus frontalis *.

Phrynocephalus frontalis, *Strauch, Voy. Przewalski, Rept.* p. 15, pl. iii. fig. 1.

Forehead sloping, subconvex, with enlarged convex scales; nostril anterior, scarcely turned upwards; nasal separated from its fellow by three or four longitudinal series of scales. Dorsal scales homogeneous, subimbricate, larger and more or less distinctly keeled on the vertebral region, decreasing in size to the sides, where they are convex or tectiform. Gular and abdominal scales smooth, pectorals strongly keeled. Third and fourth toes distinctly fringed on the outer side. Tibia as long as the skull. Tail long, little thickened at the base, gradually tapering, everywhere slightly depressed, the end curling upwards, covered with keeled scales. Brown above, irregularly guttate with yellow, uniform yellowish inferiorly; a slightly arched præocular transverse blackish band; two series of blackish spots on the back, sometimes confluent and forming transverse bands; end of tail black inferiorly.

                Total length .... 116 millim.
                Head and neck 16   ,,
                Body   38   ,,
                Tail 62   ,,

Chinese Province Ordos.

7. Phrynocephalus caudivolvulus.

Phrynocephalus caudivolvulus, *Gray, Cat.* p. 260.
Lacerta caudivolvula, *Pall. Zoogr. Ross.-As.* iii. p. 27.
Agama guttata, *Daud. Rept.* iii. p. 426.
—— caudivolvula, *Lichtenst. in Eversm. Reise*, p. 143.
—— ocellata, *Lichtenst. l. c.*
Phrynocephalus caudivolvulus, *Fitz. N. Classif. Rept.* p. 48; *Eichw. Zool. Spec.* p. 186, *and Faun. Casp. Cauc.* p. 107, pl. xii. figs. 6, 7, & pl. xiii. figs. 9–14; *Dum. & Bibr.* iv. p. 522.
—— ocellatus, *Eichw. Zool. Spec.* p. 186.
—— reticulatus, *Eichw. l. c.*
—— tickelii, part., *Günth. Proc. Zool. Soc.* 1860, p. 161.
—— caudivolvulus, part., *Günth. Rept. Brit. Ind.* p. 161.

Forehead convex and slightly sloping, with moderately enlarged convex scales; occipital scales either small, scarcely larger than

---

* I have examined two specimens, belonging to M. Lataste, collected by Abbé David in Mongolia.

the median dorsals, or more enlarged; supraorbital scales very small, granular, or nearly as large as median dorsals; nostril directed forwards and slightly upwards; nasal separated from its fellow by three to five longitudinal series of scales. Dorsal scales homogeneous, small, granular on the sides, enlarged, flat and subimbricate, smooth or feebly keeled on the vertebral region. Gular scales smooth or indistinctly keeled; pectoral and anterior ventral scales strongly keeled and mucronate, posterior ventrals smooth. Scales on upper surface of limbs keeled. Both sides of third and fourth toes with a strong fringe of acute lobules. Tibia longer than the skull. Tail swollen and much depressed at the base, then roundish depressed, tapering to a rather fine point, the end curling upwards; except on the basal portion of the tail, the scales are keeled. Greyish above, spotted, marbled, or vermiculated with blackish; sometimes with whitish, dark-edged spots forming ocelli on the back; tail inferiorly with black bars alternating with white ones of nearly equal width; lower surfaces white.

The specimens fall into two distinct forms, which are perhaps entitled to specific separation:—

A. (*P. caudivolvulus.*) Occipital scales much larger than dorsals. Tail not quite twice as long as the distance from gular fold to vent. Three to five black spots under the tail.

Total length	114 millim.
Head	11 ,,
Width of head	11 ,,
Body	38 ,,
Fore limb	23 ,,
Hind limb	40 ,,
Tail	65 ,,

Shores of the Caspian Sea; Turkestan; Tibet.

a, b. ♀.	Astrachan.	
c. ♂.	Turkestan.	M. Severzow [C.].
d. ♂.	Ladak.	Messrs. v. Schlagintweit [C.].

B. (*P. ocellatus.*) Occipital scales not larger than the largest dorsals. Tail more than twice as long as the distance from gular fold to vent. Six to nine black spots under the tail.

Total length	125 millim.
Head	12 ,,
Width of head	12 ,,
Body	38 ,,
Fore limb	24 ,,
Hind limb	43 ,,
Tail	75 ,,

Eastern Central Asia.

a–b. ♀.	Shore of Lake Alakul.	St. Petersburg Museum.
c–e. ♂.	West Kharn-Kum.	St. Petersburg Museum.
f–g. Yg.	Lepsinskaja Staniza.	St. Petersburg Museum.

8. Phrynocephalus przewalskii.

Phrynocephalus przewalskii, *Strauch, Voy. Przewalski, Rept.* p. 10, pl. ii. fig. 1.

Forehead convex, with large keeled or tectiform scales; nostril anterior, turned upwards; nasal separated from its fellow by three longitudinal series of scales. Dorsal scales homogeneous, subimbricate, large, plane, and very distinctly keeled on the vertebral region, smaller, very convex, tectiform or spiniform on the sides. Gular, pectoral, and abdominal scales imbricate and distinctly keeled. Third and fourth toes with a fringe of long acute lobules on the outer side. Tail long, thickened and scarcely depressed at the base, then rounded, the end curling upwards, and covered with imbricate keeled scales. Yellowish-orange above, yellowish inferiorly; a broad sinuous black vertebral band, ending on the sacral region; sides black-dotted.

Total length	193	millim.
Head and neck	27	,,
Body	54	,,
Tail	112	,,

Alashan Desert.

9. Phrynocephalus affinis.

Phrynocephalus affinis, *Strauch, Voy. Przewalski, Rept.* p. 13, pl. ii. fig. 2.

Forehead convex, with large convex scales; nostril anterior, turned upwards; nasal separated from its fellow by three or four longitudinal series of scales. Dorsal scales homogeneous, subimbricate, larger and slightly keeled on the vertebral region, smaller, slightly tectiform on the sides. Gular and ventral scales hardly keeled, pectorals very distinctly keeled. Third and fourth toes with a fringe of short acute lobules on the outer side. Tail long, thickened, and depressed at the base, then roundish, slightly compressed, the end curling upwards, with imbricate keeled scales. Dirty yellowish grey above, the sides of the neck and body frequently reddish; two series of blackish spots, sometimes confluent on the back; pectoral and abdominal region, and sometimes also the throat, more or less blackish.

Total length	174	millim.
Head and neck	23	,,
Body	50	,,
Tail	101	,,

Chinese Province Ordos.

10. Phrynocephalus maculatus.

Phrynocephalus maculatus, *Anders. Proc. Zool. Soc.* 1872, p. 389, fig.; *Blanf. Zool. E. Pers.* p. 331.

Forehead convex, with slightly enlarged scales; median supraorbital scales as large as or smaller than the median dorsals; occipital scales slightly enlarged; nostril directed upwards; nasal separated from its fellow by one scale. Dorsal scales homogeneous, smaller on the sides, rhomboidal, imbricate, smooth. Gular, pectoral, and abdominal scales smooth. Scales on the limbs smooth or feebly keeled. Third and fourth toes with feeble lateral denticulation. Tibia longer than the skull. Tail depressed, especially at the base, tapering to a point, covered with keeled scales; its length equals nearly twice the distance from gular fold to vent. Greyish above, speckled or guttate with whitish, and with more or less distinct darker dots, spots, or cross bands; lower surfaces white; end of tail black inferiorly.

Total length	166	millim.
Head	18	,,
Width of head	16	,,
Body	47	,,
Fore limb	36	,,
Hind limb	58	,,
Tail	101	,,

Persia.

a. ♀.	Near Bam, S.E. Persia.	W. T. Blanford, Esq.	[C.].
b. Hgr.	Karman.	W. T. Blanford, Esq.	[O.].
c. ♂.	Salt Marsh, Sar-i-jum.	W. T. Blanford, Esq.	[C.].
d. Hgr.	Between Karman and Shiraz.	W. T. Blanford, Esq.	[C.].

11. Phrynocephalus axillaris.

Phrynocephalus axillaris, *Blanf. Journ. As. Soc. Beng.* xliv. 1875, p. 192, *and 2nd Yark. Miss., Rept.* p. 9, pl. i. fig. 4.

Very closely allied to the preceding, from which it differs in rather shorter limbs, and the outer edge of the fourth toe more strongly fringed. A red spot behind the shoulder; end of the tail never black.

Total length about 15 centim., in which the tail counts for three fifths.

Yarkand.

12. Phrynocephalus interscapularis.

Phrynocephalus interscapularis, *Lichtenst. Nomencl. Rept. Mus. Berol.* p. 12.

Head much depressed, snout sloping; upper head-scales small, keeled, not enlarged on occipital region; nostril turned upwards; nasal separated from its fellow by one scale. Dorsal scales small, more or less distinctly keeled, not enlarged on vertebral region; small projecting spinose scales on the sides of the head and neck, and sometimes on the flanks; a fold along the flanks. Gular scales pointed, smooth or feebly keeled; pectoral and ventral scales mucronate, the former smooth or feebly keeled, the latter smooth. Scales on limbs smooth or keeled; a more or less distinct fringe of pointed scales borders the thighs posteriorly. Tibia longer than the skull. Toes very long, the third and fourth on both sides with a very strong fringe of acute lobules. Tail depressed, tapering to a fine point, covered with keeled scales; its length equals once and a half to once and two thirds the distance from gular fold to vent. Sandy-grey above, dotted and vermiculated with blackish, and dotted with white; lower surfaces white; end of tail inferiorly black, with two to four black spots in front.

Total length	77 millim.
Head	8 ,,
Width of head	8 ,,
Body	28 ,,
Fore limb	19 ,,
Hind limb	34 ,,
Tail	41 ,,

Eastern borders of the Caspian Sea.

a-c. ♂ ♀.	Tschelekän Island.	St. Petersburg Museum.
d. Many spec.	Krasnowodsk.	St. Petersburg Museum.

13. Phrynocephalus mystaceus.

Megalochilus auritus, *Gray, Cat.* p. 261.
Lacerta mystacea, *Pallas, Reise*, iii. p. 702.
—— aurita, *Pall. Zoogr. Ross.-As.* iii. p. 21.
—— lobata, *Shaw, Zool.* iii. p. 244.
Gecko auritus, *Latr. Rept.* ii. p. 61.
Agama aurita, *Daud. Rept.* iii. p. 429, pl. xlv. fig. 2; *Licht. in Eversm. Reise,* p. 142.
—— mystacea, *Merr. Tent.* p. 53.
Phrynocephalus auritus, *Fitz. N. Class.* p. 47; *Eversm. Nouv. Mém. Soc. Mosc.* iii. 1834, p. 360; *Dum. & Bibr.* iv. p. 524, pl. xl. fig. 1; *Schreib. Herp. Eur.* p. 460.
—— mystaceus, *Kaup, Isis,* 1827, p. 614.
Megalochilus auritus, *Eichw. Zool. Spec.* iii. p. 185.

Forehead convex, with enlarged tectiform or striated scales; occipital scales rather small; supraorbital scales small, keeled; nostril directed forwards and upwards; nasal separated from its fellow by one or three series of scales; sides of the head with spine-like projecting scales; a large, fringed, folded lobe of skin at the angle of the mouth. Dorsal scales small, not enlarged on the vertebral region, rhomboidal, imbricate, keeled, intermixed with more or less scantily scattered enlarged scales of similar shape. Gular scales keeled; pectoral and ventral scales mucronate, the former strongly, the latter feebly keeled. Scales on upper surface of limbs keeled. Tibia longer than the skull. Third and fourth digits of both pair of limbs very strongly fringed on each side.

Tail depressed, tapering to a rather blunt point, covered with strongly keeled scales; its length equals nearly once and a half the distance from gular fold to vent. Greyish above, reticulated with blackish and spotted with whitish; lower surfaces white; generally a black spot on the chest; end of tail black inferiorly.

Total length	184	millim.
Head	22	,,
Width of head	22	,,
Body	71	,,
Fore limb	49	,,
Hind limb	73	,,
Tail	91	,,

Borders of the Caspian Sea; Southern Russia; Turkestan.

a, b. ♂ & hgr.	Russia.	Dr. J. E. Gray [P.].
c. ♀.	Island of Tschelekän.	St. Petersburg Museum.
d. ♂.	Turkestan.	M. Severzow [C.].

21. AMPHIBOLURUS *.

Grammatophora, (*non Steph.*) *Kaup, Isis*, 1827, p. 621; *Dum. & Bibr.* iv. p. 468; *Fitzing. Syst. Rept.* p. 84; *Gray, Cat. Liz.* p. 251.
Amphibolurus, *Wagl. Syst. Rept.* p. 145; *Wiegm. Herp. Mex.* p. 17; *Fitzing. l. c.* p. 83.
Homalonotus, *Fitzing. l. c.*
Ctenophorus, *Fitzing. l. c.*

Tympanum distinct. Body more or less depressed. Dorsal crest absent or feebly developed. No gular sac; a strong transverse gular fold. Tail round or feebly compressed. Præanal and femoral pores.

Australia.

Synopsis of the Species.

I. The hind limb stretched forwards reaches beyond the orbit.

 A. Ventral scales keeled; tail not compressed.

No nuchal crest; nostril below the canthus rostralis .... 1. *maculatus*, p. 381.
A low nuchal crest; nostril upon the canthus rostralis .... 2. *imbricatus*, p. 382.

 B. Ventral scales smooth or very feebly keeled.

No nuchal crest; tail not compressed .. 3. *ornatus*, p. 382.
A nuchal crest; tail a little compressed; dorsal scales intermixed with enlarged ones .... 4. *cristatus*, p. 383.

---

* *Grammatophora jugularis*, Macleay, Proc. Linn. Soc. N. S. W. ii. 1877, p. 104.—Cape Grenville.

A nuchal crest; tail a little compressed;
dorsal scales uniform .... 5. *caudicinctus*, p. 384.

II. The hind limb stretched forwards does not reach beyond the orbit.

A. Præanal and femoral pores 20 or more altogether, the latter forming a long series on the whole length, or the proximal two thirds, of the thighs.

1. Ventral scales smooth.

Dorsal scales uniform, lateral scales intermixed with very small tubercles .... 6. *decresii*, p. 385.
Dorsal and lateral scales uniform ...... 7. *pictus*, p. 385.
Dorsal scales heterogeneous .... 8. *reticulatus*, p. 386.

2. Ventral scales keeled.

Gular scales keeled; body with longitudinal series of angular black spots . 9. *adelaidensis*, p. 387.
Gular scales smooth or very feebly keeled 10. *pulcherrimus*, p. 388.
Gular scales keeled; upper surfaces uniform pale brown.................. 11. *pallidus*, p. 388.

B. Præanal and femoral pores 22 or less altogether, the latter extending only on the proximal half of the thighs.

The adpressed hind limb reaches the tympanum or between the latter and the orbit; no dorsal crest or serrated ridge .... 12. *angulifer*, p. 389.
The adpressed hind limb reaches the eye or between the latter and the tympanum; a small vertebral crest or serrated ridge .... 13. *muricatus*, p. 390.
The adpressed hind limb reaches the axilla or the shoulder; no dorsal crest or ridge .... 14. *barbatus*, p. 391.

1. Amphibolurus maculatus.

Grammatophora maculata, *Gray, Cat.* p. 253.
Uromastyx maculatus, *Gray, Griff. A. K.* ix., *Syn.* p. 62.
Grammatophora gaimardii, *Dum. & Bibr.* p. 470.
Homalonotus gaimardii, *Fitz. Syst. Rept.* p. 83.
Grammatophora isolepis, *Fischer, Arch. f. Nat.* 1881, p. 232, pl. xii. figs. 10-12.

Habit slender. Head moderately large; snout as long as the diameter of the orbit, with angular canthus rostralis; nostril a little nearer the eye than the end of the snout, situated below the canthus rostralis; tympanum large, measuring a little more than half the diameter of the orbit; upper head-scales subequal, strongly keeled, smallest on supraorbital region. Sides of neck rather strongly plicate. Gular scales a little smaller than ventrals, keeled. Body moderately depressed, covered with small, uniform, strongly keeled scales, smallest on the sides; on the back the keels slightly converge towards the vertebral line; no dorsal crest. Limbs and digits very long, the adpressed hind limb reaching beyond the tip of the snout; the length of the foot equals that of the fore limb. An uninterrupted series of about fifty femoral and præanal pores, extending along the whole length of the thighs and forming an angle in front of the præanal region. Tail slender, round, depressed at the base, twice and a half as long as head and body; caudal scales equal, strongly keeled. Grey above; with a longitudinal white band on each side from the neck to the base of the tail, dividing regular transverse black spots; a large Λ-shaped blackish mark on the throat, and a transverse bar of the same colour on the chest; a black longitudinal streak along each side of the tail.

Total length	185	millim.
Head	14	,,
Width of head	10	,,
Body	38	,,
Fore limb	23	,,
Hind limb	50	,,
Tail	133	,,

Western Australia.

a. ♂. Champion Bay, N.W. Australia. Mr. Duboulay [C.].

2. Amphibolurus imbricatus.

Amphibolurus imbricatus, *Peters, Mon. Berl. Ac.* 1876, p. 520.

Differs from the preceding as follows:—Nostril pierced in the canthus rostralis, directed more upwards than sidewards. Dorsal scales increasing strongly in size towards the vertebral line and the posterior part of the trunk, where they are as large as those on the outer side of the limbs or on the tail. A low nuchal crest. Only twenty-two femoral and præanal pores. Yellowish brown above, with black spots and dots which on the sides of the neck and body are confluent into two or three longitudinal lines, and white spots confluent into transverse bands; tail with alternate broad brown and narrow white cross bars, and a median white longitudinal line; throat with bluish-black network.

From snout to vent 46 millim., tail 102 millim.

Southern Australia.

3. Amphibolurus ornatus.

Grammatophora ornata, *Gray, Cat.* p. 253.
Grammatophora ornata, *Gray, Zool. Erebus & Terror, Rept.* pl. xviii. fig. 4.

Habit slender. Head moderately large; snout slightly longer than the diameter of the orbit; canthus rostralis swollen, not angular; nostril distinctly tubular, directed slightly upwards, much nearer the eye than the end of the snout; tympanum large, three fifths the diameter of the orbit; upper head-scales tubercular, rough, smallest on supraorbital region. Sides of neck rather strongly plicate; a distinct dorso-lateral fold. Gular scales minute, smooth. Body much depressed, covered above with small keeled scales, largest and uniform on the vertebral region, minute and intermixed with widely scattered slightly enlarged ones on the sides; no dorsal crest; ventral scales small, smooth. Limbs and digits long, the adpressed hind limb reaching the tip of the snout; the scales on the limbs strongly keeled, those on the arm and tibia much enlarged. A series of sixty pores extending along the whole length of the thighs, slightly interrupted on the præanal region. Tail slender, round, depressed at the base; twice as long as head and body; caudal scales equal, much larger than dorsals, strongly keeled. Black above; a series of large irregular yellowish spots along the vertebral line and a few very small ones scattered on the sides; limbs and tail with yellowish cross bars; throat punctate with blackish; a large black spot covers the chest.

Total length	264	millim.
Head	23	,,
Width of head	18	,,
Body	63	,,
Fore limb	39	,,
Hind limb	77	,,
Tail	178	,,

West Australia.

a. ♂.   West Australia.   (Type.)

4. Amphibolurus cristatus. (PLATE XXIX. fig. 1.)

Grammatophora cristata, *Gray, Cat.* p. 251.
Grammatophora cristata, *Gray, in Grey's Trav. Austr.* ii. p. 439.

Habit slender. Head moderately large; snout as long as the diameter of the orbit, with subangular canthus rostralis; nostril a little nearer the eye than the end of the snout; tympanum large, measuring three fourths the diameter of the orbit; upper head-scales subequal, strongly keeled; a few small spines on the back of the head. Sides of neck strongly plicate; a distinct dorso-lateral fold. Gular scales smaller than ventrals, very feebly keeled. Body much depressed, covered above with small keeled scales, largest and more strongly keeled on the middle of the back, minute on the sides, all intermixed with scattered enlarged scales, which are spine-like on the dorso-lateral fold; a well-developed nuchal crest composed of a few widely separated compressed spines; a serrated ridge, or crest, along the back; ventral scales small, smooth or very feebly keeled. Limbs and digits very long, the adpressed hind limb reaching beyond the tip of the snout; the length of the foot equals that of the fore limb; scales similar to those on the back. A series of fifty pores extending along the whole length of the thighs, scarcely interrupted on the præanal region. Tail slender, slightly compressed, twice as long as head and body, in its basal portion with a slight upper ridge; caudal scales equal (scattered, slightly enlarged flat ones at the base of the tail), keeled. Dark olive above, head and neck varied with darker and lighter: tail with regular dark and light annuli; middle of throat, chest, and lower surface of limbs black.

Total length	287	millim.
Head	24	,,
Width of head	18	,,
Body	68	,,
Fore limb	44	,,
Hind limb	96	,,
Tail	195	,,

West Australia.

a. ♂.   West Australia.   Mr. Gilbert [C.]. (Type.)

5. Amphibolurus caudicinctus. (PLATE XXIX. fig. 2.)

Grammatophora caudicincta, *Günth. Zool. Erebus & Terror, Rept.* p. 19.

Habit moderate. Head rather large; snout as long as the diameter of the orbit, with swollen canthus rostralis; nostril directed slightly upwards, much nearer the eye than the tip of the snout; tympanum large, two thirds the diameter of the orbit; upper head-scales subequal, keeled, smallest on supraorbital region. Sides of neck feebly plicate; no dorso-lateral fold. Gular scales smaller than ventral, smooth. Body moderately depressed, covered with small uniform keeled scales, smallest on the back; on the back the keels slightly converge towards the vertebral line; a very small nuchal crest and a slight ridge along the middle of the back; ventral scales small, smooth. Limbs and digits long, the adpressed hind limb reaching nearly the tip of the snout; scales on upper surface of limbs a little larger than dorsals. A series of thirty-three pores extending along the whole length of the thighs, interrupted on the præanal region. Tail compressed, slightly keeled above, once and a half as long as head and body: caudal scales equal, keeled. Uniform pale brown above; tail with regular black rings, narrower than the interspaces between them; throat slightly mottled with greyish; chest black.

Total length	203	millim.
Head	24	,,

Width of head	19 millim.
Body	54 „
Fore limb	37 „
Hind limb	72 „
Tail	25 „

North-west Australia.

a. ♂.   Nicol Bay.   Mr. Duboulay [C.]. (Type.)

6. Amphibolurus decresii.

Grammatophora decresii, *Gray, Cat.* p. 253.
Grammatophora decresii, *Dum. & Bibr.* iv. p. 472, pl. xli. *bis.* fig. 1.
Ctenophorus decresii, *Fitz. Syst. Rept.* p. 83.
Amphibolurus decresii, *Peters, Mon. Berl. Ac.* 1863, p. 229.

Habit stout. Head short; nostril equally distant from the eye and the end of the snout; tympanum rather large; small spinose tubercles near the ear; upper head-scales subequal, feebly keeled, slightly larger on supraorbital region. Dorsal scales small, equal, keeled; a very small nuchal crest and a slight ridge along the middle of the back; lateral scales very small, smooth, intermixed with very small tubercles; ventral scales smooth. The adpressed hind limb reaches the eye. A series of fifty pores extending along the whole length of the thighs, interrupted on the præanal region. Tail round, nearly twice as long as head and body, covered with equal keeled scales. Olive-brown above, sides spotted or striped with fulvous; throat vermiculate with brown; chest sometimes black.

Total length	230 millim.
Head	26 „
Body	54 „
Fore limb	33 „
Hind limb	65 „
Tail	150 „

South Australia.

7. Amphibolurus pictus.

Amphibolurus ornatus, (*non* Gray) *Peters, Mon. Berl. Ac.* 1863, p. 230.
—— pictus, *Peters, Mon. Berl. Ac.* 1866, p. 88.
Grammatophora picta, *Günth. Zool. Ereb. & Terr., Rept.* p. 18.

Habit stout. Head very short, snout shorter than the diameter of the orbit; nostril equally distant from the eye and the tip of the snout; tympanum large, nearly two thirds the diameter of the orbit; upper head-scales subequal, tubercular, smallest on supraorbital region: a series of enlarged scales from the nostril to above the tympanum, passing below the eye. Sides of neck strongly plicate; no dorso-lateral fold. Gular scales smaller than ventrals, smooth.

Body much depressed, covered with very small uniform feebly keeled scales smallest on the sides; a slight ridge along the middle of the back ; ventral scales smooth. Limbs and digits rather short, the adpressed hind limb reaching the tympanum or between the latter and the orbit; scales on upper surface of limbs small, equal, keeled. A series of thirty-two to forty-five pores extending along the whole length of the thighs, continuous or interrupted on the præanal region. Tail round, a little depressed at the base, not twice as long as head and body, covered with equal feebly keeled scales. Grey-brown above, with small darker and lighter spots; a series of transverse black spots on the back separated or connected by a black vertebral line ; throat and chest mottled with blackish.

Total length	150	millim.
Head	15	,,
Width of head	13	,,
Body	42	,,
Fore limb	23	,,
Hind limb	42	,,
Tail	93	,,

South Australia.

a–b. ♀ & yg. ——? G. Krefft, Esq. [P.].

8. Amphibolurus reticulatus. (PLATE XXX. fig. 1.)

Grammatophora reticulata, *Gray, Cat.* p. 252.
Grammatophora decresii, (*non D. & B.*) *Gray, in Grey's Trav. Austr.* ii. p. 439.
—— lævis, *Günth. Ann. & Mag. N. H.* (3) xx. 1867, p. 52.

Habit stout. Head very short, snout a little shorter than the diameter of the orbit; nostril nearly equally distant from the eye and the tip of the snout; tympanum half the diameter of the orbit; upper head-scales subequal, subtubercular, obtusely keeled, smallest on supraorbital region; small conical tubercles or spines near the tympanum and on the neck. Sides of neck rather strongly plicate; no dorso-lateral fold. Gular scales smaller than ventrals, smooth. Body much depressed, covered above with small smooth scales intermixed with enlarged, flat, obtusely keeled ones; a regular series of the latter along the vertebral line; ventral scales smooth. Limbs and digits short, the adpressed hind limb reaching the shoulder in the female, between the latter and the tympanum in the male; scales on the limbs keeled. A series of thirty-five to forty-five pores extending along the whole length of the thighs, slightly interrupted on the præanal region. Tail round, a little depressed at the base, not more than once and a half the length of head and body; scales strongly keeled, on the upper surface of the basal part of the tail intermixed with enlarged flat scales. Pale olive or yellowish above, spotted or reticulated with black, the black reticulation generally enclosing small round light spots; tail with rather irregular dark annuli; sometimes a regular series of large brown spots on each side of the vertebral region; throat reticulate with grey; chest frequently brownish in the male.

Total length	195	millim.
Head	20	,,
Width of head	19	,,
Body	60	,,
Fore limb	32	,,
Hind limb	51	,,
Tail	115	,,

West Australia.

a-b. ♂ ♀.	W. Australia.	Mr. Gilbert [C.]. (Types.)
c-d. ♂.	W. Australia.	Mr. Duboulay [C.].
e-g. ♀ & hgr.	Perth, W. Australia.	Mr. Duboulay [C.].
h. Yg.	Nicol Bay, N.W. Australia.	Mr. Duboulay [C.].
i-m. ♂, ♀, & yg.	Champion Bay, N.W. Australia.	Mr. Duboulay [C.]. (Types of *Grammatophora lævis*.)

9. Amphibolurus adelaidensis.

Grammatophora angulifera, var. 2, *Gray, Cat.* p. 253.
Grammatophora muricata, var. adelaidensis, *Gray, in Grey, Austr.* ii. p. 430.
—— adelaidensis, *Gray, Zool. Erebus & Terror, Rept.* pl. xviii. fig. 2.

Habit stout. Head short; snout nearly as long as the diameter of the orbit; nostril equally distant from the eye and the end of the snout; tympanum scarcely half the diameter of the orbit; upper head-scales strongly keeled; small spinose tubercles on the back of the head; sides of neck strongly plicate; a more or less distinct dorso-lateral fold. Gular scales smaller than ventrals, keeled. Body much depressed, covered with irregular strongly keeled scales, largest on the vertebral region, intermixed with enlarged trihedral spinose scales forming very irregular longitudinal series; a more or less regular vertebral series of enlarged scales; ventral scales keeled. Limbs short, the adpressed hind limb reaching the shoulder or the neck in females, the tympanum or a little beyond in males; scales on upper surface of limbs unequal, strongly keeled. A series of twenty to thirty pores extending on more than the proximal half of the thighs, continuous or interrupted on the præanal region. Tail round, depressed at the base, not once and two thirds the length of head and body; scales strongly keeled at the base with four or five longitudinal series of enlarged ones, the outer series, on the side, composed of large trihedral tubercles. Pale olive-grey above, with a regular series of angular dark brown, white-edged spots on each side of the vertebral region, and another more or less regular along each side; head with symmetrical dark markings; limbs with irregular dark cross bars; tail with two series of dark spots: lower parts white, the throat marbled with black in the male, less distinctly with grey in the female; in the male an elongate black spot on the chest and blackish variegations on the chest and belly.

Total length	126 millim.
Head	13 ,,
Width of head	11 ,,
Body	35 ,,
Fore limb	21 ,,
Hind limb	33 ,,
Tail	78 ,,

South Australia.

a. Many spec.: ♂ ♀.   Swan River.   (Types.)
b-c. ♂.   Swan River.
d. ♀.   Australia.   Lord Derby [P.].

Var. tasmaniensis.

Distinguished by smaller dorsal scales and different markings on the lower surfaces. Two ⋀-shaped black streaks on the throat, less marked in the female; in the male a large elongate black spot on the chest, and three more or less distinct longitudinal black lines on the belly.

Tasmania.—Perhaps a distinct species.

a. Several spec.: ♂, ♀, & yg.   Tasmania.

10. Amphibolurus pulcherrimus. (Plate XXX. fig. 2.)

Closely allied to the preceding, from which it differs in having the gular scales smooth or very feebly keeled, and in coloration. Grey above, with four longitudinal rows of angular black spots on the body; head with symmetrical markings; throat with two ⋀-shaped black streaks; a black spot on the chest and three longitudinal black bands, uniting in front and behind, on the belly.

Total length 100 millim.

West Australia.

a-d. ♂.   West Australia.   Mr. Duboulay [C.].

11. Amphibolurus pallidus. (Plate XXX. fig. 3.)

Grammatophora angulifera, var. 3, *Gray, Cat.* p. 253.

Also very closely allied to *A. adelaidensis*. Enlarged dorsal scales forming very regular longitudinal series. Uniform pale brown above, without any spots; lower surfaces white, immaculate in the female; male with the throat black-marbled, and a black streak along the middle of the belly.

Total length 113 millim.

West Australia.

a–e. ♂, ♀, & yg.	West Australia.		
f. ♀.	Perth, West Australia.	Mr. Duboulay [C.].	

12. Amphibolurus angulifer.

Grammatophora angulifera, var. 1, *Gray, Cat.* p. 252.
Grammatophora muricata, var. diemensis, *Gray, Grey's Austr.* ii. p. 439.
Agama cælaticeps, *Smith, Ill. S. Afr., Rept.* pl. lxxiv.
Grammatophora angulifera, *Gray, Zool. Erebus & Terror,* pl. xviii. fig. 3.

Habit stout. Head short; snout as long as the diameter of the orbit; nostril equally distant from the eye and the tip of the snout; tympanum measuring nearly half the diameter of the orbit; upper head-scales rough, strongly keeled. Sides of neck strongly plicate and studded with small spines; a distinct dorso-lateral fold. Gular scales a little smaller than ventrals, keeled. Body much depressed, covered above with very irregular strongly keeled scales intermixed with enlarged spinose ones; the latter form a zigzag series on each side of the vertebral region, the scales of which are not enlarged, and a longitudinal series following the dorso-lateral fold; they are irregularly scattered on the flanks; ventral scales strongly keeled and mucronate. Limbs and digits short; the adpressed hind limb reaches the tympanum or between the latter and the orbit; spinose scales scattered on the limbs. Femoral pores four to six on each side, not extending beyond the basal half of the thighs; præanal pores two to five on each side. Tail round, depressed at the base, once and two thirds to once and three fourths as long as head and body, above with five longitudinal series of strongly enlarged spinose scales. Brown above, sides darker; a festooned dark brown, black-edged band along the back; lower surfaces pale brown, usually dotted or reticulated with darker.

Total length	199	millim.
Head	20	,,
Width of head	17	,,
Body	56	,,
Fore limb	32	,,
Hind limb	47	,,
Tail	123	,,

Tasmania; South-east Australia.

a. Several spec.: ♂ & yg.	Tasmania.		
b–c. ♂ ♀.	Tasmania.	C. Darwin, Esq. [P.].	} (Types.)
d. ♀.	Tasmania.	Lord Derby [P.].	
e, f, g–h. ♂ ♀.	Tasmania.	R. Gunn, Esq. [P.].	
i–n. ♂, ♀, & yg.	Tasmania.	Dr. Millinger [P.].	
o. ♀.	Tasmania.	G. Krefft, Esq. [P.].	
p. ♀.	Port Denison.	G. Krefft, Esq. [P.].	
q–s. ♂.	Sydney.	J. B. Jukes, Esq. [P.].	
t. ♂.	Australia.	Sir J. Richardson [P.].	

u. ♂.   Australia.   G. Krefft, Esq.
v. Several spec., ♀ & yg.   —— ?
w–x. ♀.   —— ?   Sir A. Smith [P.].
(Types of *Agama cælaticeps*.)

13. Amphibolurus muricatus.

Grammatophora muricata, *Gray, Cat.* p. 251.
Lacerta muricata, *White, Journ. N.S. Wales, App.* p. 244, pl. xxxi. fig. 1; *Shaw, Zool.* iii. p. 211, pl. lxv. fig. 2.
Agama muricata, *Daud. Rept.* iii. p. 391.
—— jacksoniensis, *Kuhl, Beitr. Zool. Vergl. Anat.* p. 113; *Guérin, Icon. R. A., Rept.* pl. iii.
Grammatophora muricata, *Kaup, Isis*, 1827, p. 621; *Dum. & Bibr.* iv. p. 475.
Amphibolurus muricatus, *Wiegm. Herp. Mex.* p. 17; *Girard, U. S. Explor. Exped., Herp.* p. 414.
—— maculiferus, *Girard, Proc. Ac. Philad.* 1857, p. 198, and *U. S. Explor. Exped., Herp.* p. 417.

Habit moderate. Head rather elongate, snout longer than the diameter of the orbit; canthus rostralis angular: nostril equally distant from the eye and the end of the snout; tympanum measuring nearly half the diameter of the orbit; upper head-scales strongly keeled: back of head and borders of the tympanum with small spines. Sides of neck strongly plicate; a more or less distinct dorso-lateral fold, frequently disappearing altogether in the adult. Gular scales a little smaller than ventrals, feebly keeled. Body moderately depressed, covered above with very irregular small keeled scales intermixed with very numerous, enlarged, strongly keeled, spinose scales, some of which form regular series along the back; a low serrated vertebral ridge or crest; ventral scales feebly keeled, shortly mucronate. Limbs moderately elongate, the adpressed hind limb reaching the eye or between the latter and the tympanum; limbs with strongly keeled scales of unequal size. Femoral pores three or four on each side, not extending beyond the proximal half of the thigh; præanal pores two on each side. Tail round, twice or more than twice as long as head and body, covered above with strongly keeled scales of unequal size. Brown above, with a series of angular darker spots along the middle of the back; sometimes a lighter band along each side of the latter; lower surfaces lighter brown, uniform or indistinctly spotted with darker.

Total length	307	millim.
Head	29	,,
Width of head	24	,,
Body	73	,,
Fore limb	41	,,
Hind limb	76	,,
Tail	205	,,

Australia and Tasmania.

a. ♂.   Australia.   Capt. White [P.].   (Type.)
b. ♀.   Australia.   Lord Derby [P.].

c. Hgr.	Australia.	J. Hunter, Esq. [P.].
d-f. Yg.	Australia.	P. L. Sclater, Esq. [P.].
g. ♂.	Australia.	G. Krefft, Esq.
h, i. ♂ ♀.	Australia.	
k-o. ♂, ♀, & hgr.	West Australia.	Mr. Gilbert [C.].
p-t. ♂ ♀.	Sydney.	J. B. Jukes, Esq. [P.].
u. ♂.	Sydney.	G. Krefft, Esq.
v. ♂.	Melbourne.	G. Krefft, Esq.
w. Hgr.	Tasmania.	C. Darwin, Esq. [P.].

14. Amphibolurus barbatus.

Grammatophora barbata, *Gray, Cat.* p. 252.
Agama barbata, *Cuv. R. A.* 2nd ed. ii. p. 35; *Duvern. R. A., Rept.* pl. xiv. fig. 1.
Grammatophora barbata, *Kaup, Isis,* 1827, p. 621; *Dum. & Bibr.* iv. p. 478; *Gray, Zool. Erebus & Terror, Rept.* pl. xviii. fig. 1.
Amphibolurus barbatus, *Wiegm. Herp. Mex.* p. 7.

Habit stout. Head large, swollen at the sides; snout a little longer than the diameter of the orbit, with angular canthus rostralis; nostril large, directed backwards, nearly equally distant from the eye and the end of the snout; tympanum nearly half the diameter of the orbit; upper head-scales keeled, largest on the snout; a transverse series of larger scales borders the head posteriorly, forming a right angle with another series above the ear. Sides of neck with group of spines; no distinct dorso-lateral fold. Gular scales as large as ventrals, feebly keeled, more or less strongly mucronate, sometimes produced into spines. Body much depressed; scales on the middle of the back largest, unequal, keeled, the enlarged ones sometimes forming transverse series; on the sides, the scales almost granular and intermixed with numerous erect conical spines; ventral scales feebly keeled. Limbs, and especially digits, short; the adpressed hind limb reaches the axilla or the shoulder; four or five femoral and two or three præanal pores on each side. Tail round, depressed at the base, once and a half to twice as long as head and body, above with large unequal strongly keeled or spinose scales forming more or less regular cross series. Brown above, uniform or with symmetrical darker markings; usually a black spot on each side of the neck; lower surfaces brown or brownish, uniform or with lighter or darker spots; the throat blackish in the adult male.

Total length	530 millim.
Head	67 ,,
Width of head	65 ,,
Body	163 ,,
Fore limb	92 ,,
Hind limb	123 ,,
Tail	300 ,,

Australia.

a. Hgr.	Queensland.	
b. ♂.	Sydney.	Hr. R. Schütte [C.].
c. Hgr.	Sydney.	J. B. Jukes, Esq. [P.].

d. ♂.	Cook River, near Sydney.	G. Krefft, Esq.
e. ♀.	West Australia.	G. F. Moore, Esq. [P.].
f-g, h-i, k-l. Hgr. & yg.	West Australia.	
m-o. Hgr. ♂ ♀.	Swan River.	Capt. G. Grey [P.].
p-r, s-u, v. Hgr. & yg.	Swan River.	
w. ♀.	Houtman's Abrolhos.	Mr. Gilbert [C.].
x. Hgr. ♂.	Houtman's Abrolhos.	J. Gould, Esq. [C.].
y. ♀.	Nicol Bay, North-west Australia.	Mr. Duboulay [C.].
z. Hgr.	Champion Bay, N.W. Australia.	Mr. Duboulay [C.].
α. ♂.	N.W. Australia.	Haslar Collection.
β. Yg.	N.W. Australia.	
γ. Hgr. ♂.	Australia.	G. Krefft, Esq.
δ-ζ. Yg.	Australia.	Lord Derby [P.].
η. Yg.	Australia.	Capt. Stokes [C.].
θ, ι, κ-λ. ♂ & hgr.	Australia.	
μ. Ad., stuffed.	Australia.	A. Cunningham, Esq. [P.].
υ. Ad., stuffed.	Australia.	
ξ. Skeleton.	Australia.	Sir A. Smith [P.].

22. TYMPANOCRYPTIS.

Tympanocryptis, *Peters, Mon. Berl. Ac.* 1863, p. 230.

Tympanum hidden. Body depressed, covered above with heterogeneous scales. No dorsal crest. No gular sac; a strong transverse gular fold. Tail round. A præanal pore on each side, sometimes absent in the female; no femoral pores.

Australia.

1. Tympanocryptis lineata.

Tympanocryptis lineata, *Peters, l. c.*

Habit very stout. Head short; nostril nearer the eye than the tip of the snout; upper head-scales moderately large, very strongly keeled, with slightly enlarged ones on the occiput. Dorsal scales very strongly keeled, the enlarged ones nail-shaped, raised, not or scarcely mucronate; gular and ventral scales indistinctly keeled. The adpressed hind limb reaches the shoulder or the neck. Tail rather slender, covered with very strongly keeled scales, not more than once and a half the length of head and body. Brownish above, with regular darker transverse spots, and five interrupted longitudinal light lines, three on the back and one on each side; limbs and tail with dark bars.

Total length	122	millim.
Head	15	,,
Width of head	14	,,
Body	43	,,

Fore limb	23	millim.
Hind limb	33	,,
Tail	64	,,

South Australia.

| a–b. ♂ ♀. | South Australia. | G. Krefft, Esq. [P.]. |
| c. ♂. | Kangaroo Island. | |

2. Tympanocryptis cephalus. (Plate XXXI. fig. 1.)

Tympanocryptis cephalus, *Günth. Ann. & Mag. N. H.* (3) xx. 1867, p. 52.

Differs from the preceding in the following characters:—Nostril much nearer the eye than the tip of the snout; upper head-scales larger, less strongly keeled; large flat tubercles on the occiput. Dorsal scales rather feebly keeled, the enlarged ones strongly mucronate and spinose. Brown above, with more or less indistinct darker cross bands on the body and limbs, the one across the neck being darkest; tail with dark bars separated by narrow light interspaces.

Total length	133	millim.
Head	17	,,
Width of head	15	,,
Body	45	,,
Fore limb	29	,,
Hind limb	40	,,
Tail	71	,,

North-west Australia.

| a–b. ♀ & yg. | Nicol Bay. | Mr. Duboulay [C.]. | (Types.) |

23. DIPOROPHORA.

Diporophora, *Gray, Zool. Misc.* p. 53, and *Cat. Liz.* p. 250.
Gindalia, *Gray, Cat.* p. 246.
Calotella, *Steind. Novara, Rept.* p. 28.

Tympanum distinct. Body slightly depressed. No dorsal crest. No gular sac; gular fold present or absent. Tail round. One or two præanal pores on each side, sometimes absent in the female; no femoral pores.

Australia.

Synopsis of the Species.

No transverse gular fold	1. *bilineata*, p. 394.
A gular fold; tail twice as long as head and body	2. *australis*, p. 394.
A gular fold; tail slightly longer than head and body	3. *bennettii*, p. 395.

1. Diporophora bilineata.

Diporophora bilineata, *Gray, Cat.* p. 250.
Diporophora bilineata, *Gray, Zool. Misc.* p. 54, *and Zool. Erebus & Terr., Rept.* pl. xix. fig. 1.
Grammatophora calotella, *Günth. Ann. & Mag. N. H.* (3) xx. 1867, p. 52.
Amphibolurus (Diporophora) bilineatus, *Peters & Doria, Ann. Mus. Genov.* xiii. 1878, p. 384.

Head moderately large, with distinct canthus rostralis, above with equal sharply keeled scales; nostril nearly equally distant from the eye and tip of the snout: tympanum rather large. A slight oblique fold on each side of the neck, but no transverse gular fold. Dorsal scales strongly keeled, the keels forming regular straight longitudinal series, uniform or more usually with longitudinal series of enlarged ones forming ridges along the back; gular and ventral scales strongly keeled. Limbs and digits rather elongate; the adpressed hind limb reaches the tympanum or the eye. Tail twice to twice and two thirds the length of head and body. Colour as variable as the dorsal lepidosis; upper parts greyish, brownish, or reddish, uniform or with more or less distinct angular darker spots; frequently two or three light bands along the back.

Total length	260	millim.
Head	20	,,
Width of head	14	,,
Body	50	,,
Fore limb	31	,,
Hind limb	51	,,
Tail	190	,,

North Australia.

a–b. ♂ ♀.	Cape York.	
c. ♂.	Port Darwin.	R. G. S. Buckland, Esq. [C.].
d–e. ♂.	Port Essington.	Mr. Gilbert [C.]. (Types.)
f–g. ♂.	Port Essington.	Mr. Gilbert [C.].
h. ♀.	Port Essington.	B. Bynoe, Esq. [P.].
i. ♀.	Port Essington.	Lord Derby [P.].
k–l. ♂ & hgr.	N. Australia.	J. R. Elsey, Esq. [P.].
m. Hgr.	Nicol Bay, N.W. Australia.	Mr. Duboulay [C.].
n. ♂.	Australia.	Sir J. Richardson [P.].
o–p, q. ♂ ♀.	Australia.	Haslar Collection.
r, s. ♂.	Australia.	

2. Diporophora australis.

Calotella australis, *Steind. Novara, Rept.* p. 29, pl. i. fig. 9.
Grammatophora macrolepis, *Günth. Ann. & Mag. N. H.* (3) xx. 1867, p. 51.

Very closely allied to the preceding, and apparently equally variable in the dorsal lepidosis, and to be distinguished only by the presence of a gular fold. The dorsal scales are subequal in specimen *a*;

ridges of enlarged scales are present in specimen *b*. Brownish or pale olive above, with two series of dark spots along the back and tail; specimen *a* with a large black spot on each side of the neck. North Australia.

a. ♀.   Australia.   G. Krefft, Esq. [P.]. (Type of *Grammatophora macrolepis*.)
b. ♂.   [New Guinea and Islands.]

3. Diporophora bennettii. (Plate XXXI. fig. 2.)

Gindalia bennettii, *Gray, Cat.* p. 247.

Head moderately large, with obtuse canthus rostralis, covered above with equal sharply keeled scales; nostril equally distant from the eye and the tip of the snout; tympanum half the diameter of the orbit. A transverse gular fold. Dorsal scales strongly keeled, equal, about as large as the head-scales, the keels forming pretty regular longitudinal series; gular and ventral scales strongly keeled, smaller than dorsals. Limbs moderate, digits short; the adpressed hind limb reaches the neck. Tail a little longer than head and body. Olive-brown above, with very indistinct darker cross bars on the limbs and tail.

Total length	119	millim.
Head	15	,,
Width of head	12	,,
Body	36	,,
Fore limb	20	,,
Hind limb	31	,,
Tail	68	,,

North-west Australia.

a. ♀.   N.W. Australia.   Sir J. Richardson [P.]. (Type.)

24. PHYSIGNATHUS [*].

Physignathus, *Cuv. R. A.* 2nd ed. ii. p. 41; *Wagl. Syst. Amph.* p. 151; *Wiegm. Herp. Mex.* p. 14; *Fitzing. Syst. Rept.* p. 49; *Gray, Cat.* p. 248; *Günth. Rept. Brit. Ind.* p. 152.
Istiurus, part., *Dum. & Bibr.* iv. p. 376.
Lophognathus, *Gray, Zool. Misc.* p. 53, and *Cat.* p. 250.
Istiurus, *Fitzing. Syst. Rept.* p. 49.
Redtenbacheria, *Steind. Novara, Rept.* p. 31.

Tympanum distinct. Body more or less compressed. Nuchal and dorsal crests present. No gular sac; a strong transverse gular fold. Tail round or more or less compressed. Toes not lobate. Femoral pores present, at least in the male.

Australia and Papuasia; Siam and Cochinchina.

---

[*] *Lophognathus lateralis*, Macleay, Proc. Linn. Soc. N. S. W. ii. 1877, p. 103.—Katow, New Guinea.

Synopsis of the Species.

I. Tail round or feebly compressed, not crested.

   A. Keels of the upper dorsal scales forming parallel lines with the dorsal crest.

Nostril a little nearer the orbit than the tip of the snout .... 1. *gilberti*, p. 396.

   B. Keels of the upper dorsal scales obliquely directed towards the vertebral line.

Nostril a little nearer the orbit than the tip of the snout; gular scales very feebly keeled .... 2. *longirostris*, p. 397.

Nostril a little nearer the tip of the snout than the orbit; gular scales very feebly keeled .... 3. *temporalis*, p. 397.

Nostril equally distant from the orbit and the tip of the snout; gular scales strongly keeled .... 4. *maculilabris*, p. 398.

II. Tail strongly compressed, crested.

Dorsal scales intermixed with enlarged tubercles .... 5. *lesueurii*, p. 398.

Dorsal scales equal; a series of six or seven shields along each side of the throat......................... 6. *cochinchinensis*, p. 399.

Dorsal scales equal; a series of eleven shields along each side of the throat.. 7. *mentager*, p. 400.

1. Physignathus gilberti.

Lophognathus gilberti, part., *Gray, Cat.* p. 250.
Lophognathus gilberti, *Gray, Zool. Misc.* p. 53, *and Zool. Erebus & Terr., Rept.* pl. xix. fig. 2; *Günth. Ann. & Mag. N. H.* (3) xx. 1867, p. 81; *Bouleng. Ann. & Mag. N. H.* (5) xii. 1883, p. 225.
Redtenbacheria fasciata, *Steind. Novara, Rept.* p. 31.
Grammatophora temporalis, part., *Günth. l. c.* p. 52.

Head rather elongate; snout not longer than the distance between the orbit and the posterior border of the tympanum; nostril a little nearer the orbit than the tip of the snout; tympanum half the diameter of the orbit; cheeks swollen, with a few small erect spines; upper head-scales strongly keeled, rather large on the snout and between the orbits, minute on the back of the head. Gular scales equal, imbricate, feebly keeled, slightly enlarged on the sides of the jaw. Dorsal scales imbricate, keeled, the median ones largest, with the keels forming straight continuous series parallel with the vertebral line; the scales on the sides directed upwards and backwards; nuchal and dorsal crests a small serrated ridge; ventral scales strongly keeled, a little smaller than the largest dorsals. Limbs long; the adpressed hind limb reaches the tip of the snout or between the latter point and the orbit. Two to four femoral and two or three præanal pores on each side. Tail slightly compressed in the adult male, not crested, the scales on the lower surface not larger than those on the upper. Olive-brown above, head and sides darker; a broad light band along upper and lower lip; a light band along each side of the back; young with dark cross bars.

Total length	468 millim.
Head	39 ,,
Width of head	29 ,,
Body	84 ,,
Fore limb	50 ,,
Hind limb	106 ,,
Tail	345 ,,

Australia.

a. ♂.	Port Essington.	Sir J. Richardson [P.]. (Type.)
b. Yg.	Nicol Bay.	Mr. Duboulay [C.]. (One of the types of *Grammatophora temporalis*.)
c. Yg.	N. Australia.	J. R. Elsey, Esq. [P.].
d–f. ♂.	Swan River.	Mr. Dring [C.].

2. Physignathus longirostris. (PLATE XXXI. fig. 3.)

Lophognathus longirostris, *Bouleng. Ann. & Mag. N. H.* (5) xii. 1883, p. 225.

Differs from the preceding in the following points:—Snout longer than the distance between the orbit and the posterior border of the tympanum. Dorsal scales all obliquely directed upwards; ventral scales as large as dorsals. Five to eight femoral and two or three præanal pores on each side. Olive or reddish brown above; a light band bordering the lower lip, and another along each side of the back.

Total length	372 millim.
Head	27 ,,
Width of head	15 ,,
Body	70 ,,
Fore limb	40 ,,
Hind limb	80 ,,
Tail	275 ,,

North-west Australia.

a. ♀.	Champion Bay.	Mr. Duboulay [C.]. } (Types.)
b–c. Hgr.	Nicol Bay.	Mr. Duboulay [C.].

3. Physignathus temporalis. (PLATE XXXI. fig. 4.)

Lophognathus gilberti, part., *Gray, Cat.* p. 250.
Grammatophora temporalis, part., *Günth. Ann. & Mag. N. H.* (3) xx. 1867, p. 52.
Lophognathus labialis, *Bouleng. Ann. & Mag. N. H.* (5) xii. 1883, p. 225.

Differs from *P. gilberti* in the following points:—Nostril a little nearer the tip of the snout than the orbit. Dorsal scales all obliquely directed upwards; ventral scales larger than dorsals. Two femoral and two præanal pores on each side. Reddish brown above, head darker; more or less distinct dark cross bands on the back; a light band bordering the upper lip and another along each side of the back.

```
Total length (tail broken) 215 millim.
Head ................   25    ,,
Width of head ........   17    ,,
Body ................   66    ,,
Fore limb ............   42    ,,
Hind limb ...........   84    ,,
```

North Australia.

a. ♀.   Port Essington. Lord Derby [P.]. (Type.)
b. ♂.   Port Essington. Sir J. Richardson [P.]. } Types of
c. ♂, bad state. Port Essington. Mr. Gilbert [C.]. } *Lophognathus labialis.*

4. Physignathus maculilabris.

Lophognathus maculilabris, *Bouleng. Ann. & Mag. N. H.* (5) xii. 1883, p. 226, *and Proc. Zool. Soc.* 1883, p. 386, pl. xli.

Differs from *P. gilberti* in the following points:—Nostril equally distant from the eye and the tip of the snout. Gular scales strongly keeled. Dorsal scales all obliquely directed upwards; ventral scales slightly larger than dorsals. No pores (♀). Upper surfaces olive, with blackish transverse markings across the back, tail, and limbs; a broad blackish band from orbit to tympanum, bordered inferiorly by a light band extending to above the fore limb; lips light-coloured, variegated with blackish; lower surfaces whitish, dotted all over with blackish.

```
Total length .... ....  388 millim.
Head ................   22    ,,
Width of head ........   17    ,,
Body ................   76    ,,
Fore limb ............   46    ,,
Hind limb ............   94    ,,
Tail ................  290    ,,
```

Timor Laut.

a–b. ♀.   Timor Laut.   H. O. Forbes, Esq. [C.]. (Types.)

5. Physignathus lesueurii.

Physignathus lesueurii, *Gray, Cat.* p. 248.
Lophura lesueurii, *Gray, Griff. A. K.* ix., *Syn.* p. 60.
Istiurus lesueurii, *Dum. & Bibr.* p. 384, pl. xl.

Amphibolurus heterurus, *Peters, Mon. Berl. Ac.* 1866, p. 86.
Physignathus lesueurii, *Günth. Ann. & Mag. N. H.* (3) xx. 1867, p. 51.

Head moderately elongate, large and thick in the male; snout slightly longer than the diameter of the orbit; nostril nearer the end of the snout than the orbit; canthus rostralis, supraciliary and supraorbital borders forming slight ridges; tympanum half the diameter of the orbit; upper head-scales very small, very strongly keeled; occiput and temple with numerous conical and compressed tubercles. Gular scales subimbricate, indistinctly keeled, intermixed on the sides with enlarged suboval tubercles forming irregular longitudinal series; some of the hindermost of these tubercles conical; a row of slightly enlarged shields on each side, parallel with the infralabials. Nuchal crest composed of a few triangular compressed spines; dorsal crest a serrated ridge. Dorsal scales minute, granular or subimbricate, keeled, intermixed with enlarged, roundish, keeled tubercles forming irregular transverse series; ventral scales larger than dorsals, imbricate, keeled. Limbs long, scaled like the back; the adpressed hind limb reaches between the eye and the end of the snout. Sixteen to twenty-two femoral pores on each side. Tail strongly compressed, crested like the back, twice and a half as long as the body; superolateral scales very small, intermixed at the base of the tail with enlarged tubercles; lower scales larger. Dark olive above, with darker and lighter cross bands; a broad black band from the eye to above the shoulder, involving the tympanum; belly pale olive, dotted with black; throat with black longitudinal lines in the young.

Total length	466	millim.
Head	46	,,
Width of head	39	,,
Body	120	,,
Fore limb	80	,,
Hind limb	150	,,
Tail (reproduced)	300	,,

Queensland.

a, b. Adult, stuffed.	Australia.	
c–d, e. ♀ & yg.	Australia.	G. Krefft, Esq.
f, g. ♂ ♀.	Australia.	

6. Physignathus cochinchinensis.

Physignathus concinnus, *Gray, Cat.* p. 248.
Physignathus concinnus, *Cuv. R. A.* 2nd ed. ii. p. 41.
Istiurus cochinchinensis, *Guérin, Icon. R. A., Rept.* pl. ix. fig. 2.
Lophura concinna, *Gray, Griff. A. K.* ix., Syn. p. 60.
—— cuvieri, *Gray, l. c.*
Istiurus physignathus, *Dum. & Bibr.* iv. p. 387.
Physignathus cochinchinensis, *Günth. Rept. Brit. Ind.* p. 153.

The chief difference between this species and the following appears to be the greater number of lateral or molar teeth (eighteen on each side in each jaw) and the presence of only six or seven shields along each side of the throat. In other respects, both forms are very similar.

Cochinchina.

7. Physignathus mentager.

Dilophyrus mentager, *Günth. Proc. Zool. Soc.* 1861, p. 188.
Physignathus mentager, *Günth. Rept. B. Ind.* p. 153, pl. xv.

Head moderately elongate; snout slightly longer than the diameter of the orbit; canthus rostralis well marked; nostril nearer the end of the snout than the orbit; tympanum one third the diameter of the orbit; upper head-scales minute, granular, keeled, a little enlarged on canthus rostralis and supraorbital border. Eleven molar teeth on each side of the upper jaw, and twelve in the lower. Gular scales, large juxtaposed granules, unequal on the sides; three or four of the posterior lateral ones largest and conical; a series of eleven large shields on each side, parallel with the infralabials. Nuchal crest composed of lanceolate spines, which are very long in the adult male, short in the half-grown, supported by a fold of the skin covered with granular scales; dorsal crest formed of similar spines, a little lowered above the shoulders, but continuous with the nuchal. Dorsal scales equal, minute, granular, feebly keeled; ventral scales larger, imbricate, smooth. Limbs long, covered with minute, slightly imbricate, keeled scales; the adpressed hind limb reaches between the orbit and the tip of the snout. Five pores under the basal part of each thigh. Tail strongly compressed, about twice and a half the length of head and body, crested above, the crest as much developed as, but not continuous with, the dorsal; lower caudal scales enlarged and strongly keeled. Green, a little lighter inferiorly; rather indistinct, narrow oblique lighter bands down each side of the body; enlarged gular shields and tubercles whitish; tail with dark-brown annuli.

The largest specimen measures 23 centim. from snout to vent, and the tail measures 55 centim. The following measurements are taken from the smaller specimen:—

Total length	550	millim.
Head	38	,,
Width of head	27	,.
Body	112	,,
Fore limb	70	,,
Hind limb	130	,,
Tail	400	,,

Siam.

a. Adult ♂, stuffed.   Chartaboum.   M. Mouhot [C.].   (Type.)
b. Hgr. ♂.   Pachebone.   M. Mouhot [C.].

25. CHLAMYDOSAURUS.

Chlamydosaurus, *Gray, in King's Voy. Austr.* ii. p. 424; *Wagl. Syst. Amph.* p. 151; *Wiegm. Herp. Mex.* p. 14; *Dum. & Bibr.* iv. p. 440; *Fitzing. Syst. Rept.* p. 48; *Gray, Cat. Liz.* p. 249.

Tympanum distinct. Body slightly compressed. No dorsal crest. Neck with a large frill-like dermal expansion on each side, confluent on the throat. Tail round or slightly compressed. Præanal and femoral pores.
Australia.

1. Chlamydosaurus kingii.

Chlamydosaurus kingii, *Gray, Cat.* p. 249.
Chlamydosaurus kingii, *Gray, in King's Voy. Austr.* ii. p. 425, pl. A, and *Zool. Misc.* p. 53; *Dum. & Bibr.* iv. p. 441, pl. xlv.; *Krefft, Austr. Vert.* 1871, p. 47; *De Vis, Proc. Linn. Soc. N. S. W.* viii. 1883, p. 1, pls. xiv.-xvi.

Snout rather pointed, with angular canthus rostralis, its upper surface slightly concave; upper head-scales small, keeled; tympanum more than half the diameter of the orbit. The frill-like expansion much more developed in the adult than in the young, covered with keeled scales larger than the dorsals; its upper edge strongly dentate. Scales on the body keeled, very small on the sides, larger on the vertebral region and belly. The adpressed hind limb reaches the tympanum or the eye. Three to seven femoral and one or two præcanal pores on each side. Tail round, slightly compressed in the adult male, nearly twice as long as head and body, covered with rather large, equal, strongly keeled scales. Pale brown above, uniform or varied with dark brown, or blackish varied with yellow.

Total length	810	millim.
Head	70	,,
Width of head	55	,,
Body	190	,,
Fore limb	117	,,
Hind limb	200	,,
Tail	550	,,

Queensland; Northern and North-western Australia.

a. ♂.	Islands of Torres Straits.	Rev. S. Macfarlane [C.].
b. Yg.	[New Guinea.]	Mrs. Stanley [P.].
c-d, e, f. ♀ & yg.	Port Essington.	J. Gould, Esq. [C.].
g. Hgr.	Port Essington.	J. B. Jukes, Esq. [P.].
h. ♂.	N. Australia.	Dr. J. Elsey [P.].
i. Hgr.	Nicol Bay.	Mr. Duboulay [C.].
k-l. ♂ & yg.	N.W. Australia.	Haslar Collection.
m, n-p. ♂, hgr., & yg.	Australia.	Haslar Collection.

q. Ad., stuffed.   Port Essington.   J. Gould, Esq. [C.].
r. Hgr., stuffed. Australia.   Allan Cunningham, Esq. [P.].
s. Ad., skeleton. Australia.   Haslar Collection.

26. LOPHURA.

Lophura, *Gray, Phil. Mag.* ii. 1827, p. 56; *Wagl. Syst. Amph.* p. 151; *Wiegm. Herp. Mex.* p. 14; *Fitzing. Syst. Rept.* p. 49; *Gray, Cat. Liz.* p. 247.
Istiurus, *Cuv. R. A.* 2nd ed. ii. p. 41.
Hydrosaurus, *Kaup, Isis*, 1828, p. 1147.
Istiurus, part., *Dum. & Bibr.* iv. p. 376.

Tympanum distinct. Body compressed. A dorsal crest. Throat longitudinally plicate; a transverse gular fold. Toes covered inferiorly with small granular scales, with a lateral fringe of large united scales most developed on the outer side. Tail strongly compressed, in the adult with a very high crest supported by the enormously developed spinose processes of the caudal vertebræ. Femoral pores present.

East Indies.

1. Lophura amboinensis.

Lophura amboinensis, *Gray, Cat.* p. 247.
Lophura shawii, *Gray, l. c.*
Lacerta amboinensis, *Schlosser, Epist.*; *Hornst. Abh. Acad. Stockh.* vi. 1785, p. 130, pl. v. fig. 1.
Iguana amboinensis, *Latr. Rept.* i. p. 271.
Basiliscus amboinensis, *Daud. Rept.* iii. p. 322.
Lophura amboinensis, *Gray, Phil. Mag.* ii. p. 56; *Wagl. Icon. Amph.* pl. xxviii.; *Günth. Proc. Zool. Soc.* 1873, p. 168; *Peters & Doria, Ann. Mus. Gen.* xiii. 1878, p. 383.
Hydrosaurus amboinensis, *Kaup, Isis*, 1828, p. 1147.
Istiurus amboinensis, *Cuv. R. A.* 2nd ed. ii. p. 41; *Dum. & Bibr.* iv. p. 380; *Duvern. R. A., Rept.* pl. xv. fig. 1.
Histiurus pustulosus, *Eschsch. Zool. Atlas*, pl. vii.; *Lichtenst. Nomencl. Rept. Mus. Berol.* p. 10.
Lophura pustulosa, *Fitz. Syst. Rept.* p. 49.
Istiurus microlophus, *Bleek. Natuurk. Tijdschr. Ned. Ind.* xxii. 1860, p. 80.
Lophura celebensis, *Peters, Mon. Berl. Ac.* 1872, p. 581.

Head rather small; snout elongate, with a small longitudinal crest of enlarged scales in the male; tympanum large; upper head-scales small, strongly keeled. Gular scales small, granular, of unequal size; a row of enlarged shields on each side, parallel with the infralabials, commencing from the very large mental. Dorsal and nuchal crests continuous, composed of compressed lanceolate spines. Dorsal scales small, imbricate, keeled, the keels directed upwards and backwards, intermixed with scattered, enlarged, roundish, shortly keeled scales, varying considerably in size. Ventral scales larger than dorsals, subquadrangular, smooth, forming transverse series; enlarged scales on the sides of the chest. Limbs long; the adpressed hind limb reaches the eye or between the latter and the tip of the snout; scales enlarged on anterior face of fore limb. Femoral pores seven to sixteen on each side. Tail covered with minute quadrangular keeled scales above and on the sides, with much larger ones inferiorly; caudal crest very high in the males, with feebly denticulated border, present only on the anterior part of the tail; length of tail more than twice that of head and body. Olive above. spotted or vermiculated with black; oblique fold in front of the shoulder black.

Total length	108	centim.
Head	7	,,
Width of head	4·5	,,
Body	26	,,
Fore limb	15	,,
Hind limb	25	,,
Tail	75	,,

Philippines; Java; Celebes; Moluccas.

a. Ad., stuffed.	Amboyna.	Leyden Museum.
b. Hgr.	Celebes.	Leyden Museum.
c-d. Hgr. & yg.	Philippines.	H. Cuming, Esq. [C.].
e. Hgr.	Philippines.	F. Veitch, Esq. [P.].
f. ♂.	Placer, Mindanao.	A. Everett, Esq. [C.].
g. ♀.	S. Negros.	A. Everett, Esq. [C.].
h. Ad., stuffed.	S. Negros.	A. Everett, Esq. [C.].
i. Hgr.	Surigao.	A. Everett, Esq. [C.].
k. ♂.	Dinagat Island.	A. Everett, Esq. [C.].
l. ♂.	Luzon.	Dr. A. B. Meyer [C.].
m. ♂.	Zebu.	H.M.S. 'Challenger.'
n, o. ♀ & yg.	——?	Dr. Bleeker.
p. ♂.	——?	Dr. Rüppell.
q, r, s, t, u. ♂, hgr., & yg.	——?	
v. ♀.	——?	(Type of *L. shawii*.)

27. LIOLEPIS.

Leiolepis, *Cuv. R. A.* 2nd ed. ii. p. 37; *Wiegm. Herp. Mex.* p. 17; *Dum. & Bibr.* iv. p. 463; *Fitzing. Syst. Rept.* p. 86; *Gray, Cat.* p. 262; *Günth. Rept. Brit. Ind.* p. 153.

Tympanum distinct. Body depressed; no crest; scales very small. No gular pouch; a strong transverse gular fold. Tail long, round, feebly depressed. Femoral pores.

South-eastern Asia.

1. Liolepis bellii.

Leiolepis bellii, *Gray, Cat.* p. 262.
Leiolepis reevesii, *Gray, l. c.* p. 263.
Uromastyx bellii, *Gray, Zool. Journ.* iii. 1827, p. 216, *and Ill. Ind. Zool.*
Leiolepis guttatus, *Cuv. R. A.* 2nd ed. ii. p. 37; *Guér. Icon. R. A., Rept.* pl. vii. fig. 2; *Dum. & Bibr.* iv. p. 465, pl. xliii.

Uromastyx reevesii, *Gray, Griff. A. K.* ix., *Syn.* p. 62.
Liolepis bellii, *Cantor, Cat. Mal. Rept.* p. 41; *Günth. Rept. Brit. Ind.* p. 154; *Swinhoe, Proc. Zool. Soc.* 1870, p. 240; *Sanders, Proc. Zool. Soc.* 1872, p. 154.
Leiolepis reevesii, *Theob. Journ. Linn. Soc.* x. 1868, p. 34.

Head rather small; snout with strongly curved profile, as long as or little longer than the diameter of the orbit; nostril large, directed backwards; tympanum large, vertically oval; upper head-scales small and strongly keeled on the snout and interorbital region, minute and granular on the supraorbital region and the occiput; no occipital. Gular scales small, granular; a series of chin-shields on each side, parallel with the infralabials, the first shield in contact with the pentagonal mental and the first infralabial. Side of neck plicate; a more or less distinct fold along the side of the body. Dorsal scales minute, granular, feebly keeled, uniform; ventral scales larger, subimbricate, smooth. Limbs rather long, with long slender digits and long claws; the adpressed hind limb reaches the neck or the tympanum; fourteen to twenty-one femoral pores on each side. Tail about twice as long as head and body, round, depressed at the base, tapering to a fine point, covered with small, equal, keeled scales, largest inferiorly. Grey, brownish or blackish above, with whitish black-edged spots which may form ocelli or be confluent into longitudinal bands; sides with black and white (orange) vertical bars: lower surfaces white (orange), uniform or variegated with black (or blue).

Total length	500	millim.
Head	35	,,
Width of head	28	,,
Body	135	,,
Fore limb	58	,,
Hind limb	103	,,
Tail	330	,,

Southern China; Siam; Arracan; Birma; Malay peninsula; Southern India.

a. ♂.	China.	J. Reeves, Esq. [P.]. (Type of *L. reevesii*.)
b-c. ♂.	China.	
d-e. ♂.	Hainan.	R. Swinhoe, Esq. [C.].
f-g. ♀.	Mergui.	Prof. Oldham [P.].
h-i. ♂.	Camboja.	M. Mouhot [C.].
k. ♂.	Tavoy.	W. T. Blanford, Esq. [P.].
l. ♀.	Birma.	Col. Beddome [C.].
m-n, o. ♂♀.	Penang.	Dr. Cantor.
p. ♂.	Penang.	Capt. Stafford [P.].
q. Hgr.	S. Canara.	Col. Beddome [C.].

28. UROMASTIX.

Uromastix, *Merrem, Tent. Syst. Amph.* p. 56; *Fitzing. N. Classif. Rept.* p. 17; *Wagl. Syst. Amph.* p. 145; *Wiegm. Herp. Mex.* p. 17; *Dum. & Bibr.* iv. p. 533; *Fitzing. Syst. Rept.* p. 86; *Gray, Cat.* p. 261; *Günth. Rept. Brit. Ind.* p. 155, and *Phil. Trans.*
Mastigura, *Flem. Phil. Zool.* ii. p. 277.
Centrocercus, *Fitzing. Syst. Rept.* p. 85.
Saara, *Gray, Cat.* p. 85.
Centrotrachelus, *Strauch, Bull. Ac. St. Pétersb.* vi. 1863, p. 479.

Tympanum distinct. Incisors large, uniting in the adult into one or two cutting-teeth, separated from the molars by a toothless interspace. Body depressed, without crest. No gular pouch; a transverse gular fold. Tail short, depressed, covered with whorls of large spinose scales. Præanal and femoral pores present.

Arid tracts of North Africa and Southern Asia.

The species resembling one another so much in general structure, the following description will suffice for all :—

Head small, feebly depressed, with short snout and obtuse canthus rostralis; nostril large, directed backwards, nearer the end of the snout than the eye; tympanum large, vertically elliptic; upper head-scales smooth, much larger than those on the body, smallest on supraorbital region; occipital not enlarged; labials small and numerous. Neck strongly plicate. Limbs short and thick; hind limb with spinose conical tubercles; digits short and armed with strong claws. Scales on upper surface of body very small, on belly larger, flat, smooth, juxtaposed or subimbricate.

Synopsis of the Species.

I. The whorls of spinose scales on the upper surface of the tail not separated by small scales.

   A. The caudal scales form regular annuli, those on the lower surface being as long as those on the upper .... 1. *ornatus*, p. 406.

   B. Two or more transverse rows of scales on the lower surface of the tail correspond to one on the upper.

Scales larger (about 90 from the gular fold to the præanal pores); no enlarged scales on the flanks .... 2. *acanthinurus*, p. 406.
Scales on the body very minute; flanks with scattered slightly enlarged scales ...... 3. *spinipes*, p. 407.
Scales very minute; no enlarged ones on the flanks .... 4. *microlepis*, p. 407.

II. The whorls of spinose scales on the upper surface of the tail separated by small scales.

Caudal spines small, the lateral ones largest; no strongly enlarged tubercles on the

back: a large black spot on the anterior
face of the thigh .... 5. *hardwickii*, p. 408.
Caudal spines large, longer than the interspaces between them, subequal; enlarged
tubercles on the back; 20 transverse rows
of scales on the middle of the belly, on a
space corresponding to the length of the
head .... 6. *asmussii*, p. 409.
Caudal spines large, subequal; enlarged
tubercles on the back; 30 to 40 transverse
rows of scales on the middle of the belly,
on a space corresponding to the length of
the head .... 7. *loricatus*, p. 409.

1. Uromastix ornatus.

Uromastix ornatus, *Gray, Cat.* p. 261.
Uromastix ornatus, *Rüpp. Atlas N. Afr., Rept.* p. 1, pl. i.; *Dum. & Bibr.* iv. p. 538.

Anterior border of ear not denticulated; no spines on the back of the head and neck. Scales small, equal. Nine or ten femoral and four præanal pores on each side. Tail strongly depressed, with complete annuli of large scales which are not separated by smaller ones; lateral caudal scales strongly spinose; inferior caudal scales as long as but narrower than the upper. Sand-coloured above, uniform or with brown marblings or cross bands of orange brown-edged round spots.

Total length	298	millim.
Head	33	,,
Width of head	29	,,
Body	130	,,
Fore limb	65	,,
Hind limb	82	,,
Tail	135	,,

Egypt; Syria.

a, b. ♀ & yg.   Egypt.   J. Burton, Esq. [P.].

2. Uromastix acanthinurus.

Uromastix acanthinurus, *Gray, Cat.* p. 262.
Uromastix acanthinurus, *Bell, Zool. Journ.* i. 1825, p. 457, pl. xvii.; *Dum. & Bibr.* iv. p. 543; *Guichen. Expl. Sc. Alg., Rept.* p. 8; *Strauch, Erp. Alg.* p. 30.
—— dispar, *Rüpp. Atlas N. Afr.* p. 5.
—— temporalis, *Valenc. C.R. Ac. Paris*, xxxix. 1854, p. 89.

Anterior border of ear denticulated. Scales small, much larger than in the following species, there being about ninety on the middle line of the belly, from the gular fold to the præanal pores; no enlarged tubercles on the flanks. Nine to eleven femoral and four or five præanal pores on each side. Tail strongly depressed, above with whorls of large spinose scales not separated by smaller ones, inferiorly with smaller scales, two or three whorls corresponding to one of the upper surface. Greyish above, dotted or vermiculated with blackish; lower surfaces lighter, uniform or marbled with blackish.

Total length	339	millim.
Head	47	,,
Width of head	41	,,
Body	167	,,
Fore limb	70	,,
Hind limb	97	,,
Tail	125	,,

North Africa.

a. Stuffed.	N. Africa.	
b, c. ♂.	N. Africa.	Zoological Society.
d. ♂.	Biskra, Algeria.	J. Brenchley, Esq. [P.].
e, f. Hgr., skeletons.	Algeria.	

3. Uromastix spinipes.

Uromastix spinipes, *Gray, Cat.* p. 261.
Stellio spinipes, *Daud. Rept.* iv. p. 31; *Geoffr. Descr. Egypte*, i. p. 125, pl. ii. fig. 2.
Uromastix spinipes, *Merr. Tent.* p. 56; *Dum. & Bibr.* iv. p. 541; *Duvern. R. A., Rept.* pl. xiii. fig. 2; *Schreib. Herp. Eur.* p. 474.
Mastigura spinipes, *Flem. Phil. Zool.* ii. p. 277.

Distinguished from the preceding by the much smaller scales and the presence of scattered small tubercles on the flanks. Sandy-grey or greenish above, uniform or clouded with brown.

Total length	452	millim.
Head	46	,,
Width of head	46	,,
Body	216	,,
Fore limb	112	,,
Hind limb	140	,,
Tail	190	,,

Egypt; Crete; Arabia.

a. Ad., stuffed.	Egypt.	
b, c. ♂ & hgr.	Egypt.	J. Burton, Esq. [P.].
d, e. ♀.	Egypt.	
f. Skeleton.	Egypt.	Dr. Rüppell.
g, h. Skeletons.	Egypt.	
i. Hgr.	Arabia.	Capt. Burton [P.].

4. Uromastix microlepis.

Uromastix microlepis, *Blanf. Proc. Zool. Soc.* 1874, p. 656, pl. liii., and *E. Persia*, p. 334.

Distinguished from the preceding by the still smaller scales and the absence of enlarged tubercles on the flanks. Upper surfaces greyish olive, with rather indistinct darker dots.

Total length	540 millim.
Head	47 ,,
Width of head	47 ,,
Body	273 ,,
Fore limb	121 ,,
Hind limb	170 ,,
Tail	220 ,,

Persia.

a, b. ♂.   Basrah.   Zoological Society [P.]. (Types.)

5. Uromastix hardwickii.

Saara hardwickii, *Gray, Cat.* p. 262.
Uromastix hardwickii, *Gray, Zool. Journ.* iii. 1827, p. 219, *and Ill. Ind. Zool.*; *Dum. & Bibr.* iv. p. 546; *Günth. Rept. Brit. Ind.* p. 155; *Anders. Proc. Zool. Soc.* 1871, p. 167; *Murray, Zool. Sind.* p. 372.
—— reticulatus, *Cuv. R. A.* 2nd ed. ii. p. 34; *Guér. Icon. R. A.* pl. vii. fig. 4.
? Uromastix griseus, *Cuv. l. c.*; *Dum. & Bibr.* iv. p. 548.
Centrocercus hardwickii, *Fitz. Syst. Rept.* p. 86.
—— similis, *Fitz. l. c.*

Anterior border of ear very slightly denticulated. Scales very small, uniform or with slightly enlarged ones forming irregular cross series on the back. Nine to eleven femoral and five or six præanal pores on each side. Tail above with cross series of enlarged spinose scales, largest on the side, gradually decreasing in size towards the median line; these rows of spines separated from one another by two to four rows of small keeled scales; lower surface of the tail covered with small scales scarcely larger than the abdominals. Sand-coloured above, uniform or with darker dots or vermiculations; lower surfaces whitish; a large black spot on the anterior face of the thigh.

Total length	279 millim.
Head	34 ,,
Width of head	30 ,,
Body	86 ,,
Fore limb	62 ,,
Hind limb	95 ,,
Tail	165 ,,

Northern India; Baluchistan.

a. ♂.   Plains of Kanouge, Hindustan.   Gen. Hardwicke [P.]. (Type.)
b. ♂.   Hindustan.
c. Hgr.   Goojerat.   Dr. Cantor.

d. ♂.	Sind.	W. T. Blanford, Esq. [P.].
e. ♀.	Kurrachee.	Messrs. v. Schlagintweit [C.].
f. ♂.	N. India.	Messrs. v. Schlagintweit [C.].
g, h. Ad., stuffed.	India.	
i. Yg.	S.E. coast of Arabia (?). H. J. Carter, Esq. [P.].	

6. Uromastix asmussii.

Centrotrachelus asmussi, *Strauch*, *Bull. Ac. St. Pétersb.* vi. 1863, p. 479; *Blanf. E. Persia*, p. 337, pl. xxi.

Anterior border of ear very slightly denticulated. Back of head and neck with spinose tubercles. Scales of back small, rhomboidal, subimbricate, intermixed with enlarged subconical tubercles with the points directed backwards, forming equidistant interrupted transverse series; ventral scales rather large, there being about twenty transverse rows on the middle of the belly on a space corresponding to the length of the head. Six or seven femoral and two or three præanal pores on each side. Tail feebly depressed, above with transverse series of large spinose conical tubercles, the series separated from one another by several rows of small scales; inferiorly with small scales also forming transverse series, there being three or four series in one annulus. Head, limbs, and tail blackish above, body buff, the larger tubercles scarlet in life; sometimes uniform yellowish olivaceous; belly yellowish; throat, chest, and lower surface of limbs marbled with black.

Total length	460	millim.
Head	45	,,
Width of head	42	,,
Body	205	,,
Fore limb	90	,,
Hind limb	130	,,
Tail	210	,,

Eastern Persia; Baluchistan.

a. ♂.	Near Rigan, Narmashir, S.E. Persia.	W. T. Blanford, Esq. [C.].

7. Uromastix loricatus. (PLATE XXXII.)

Centrotrachelus loricatus, *Blanf. Proc. Zool. Soc.* 1874, p. 660, and *E. Persia*, p. 340, and *Proc. Zool. Soc.* 1881, p. 677.

Closely allied to the preceding; differing in the following points:—No trace of a denticulation in front of the ear. Scales smaller, the enlarged tubercles not forming so regular transverse series; thirty to forty transverse rows of scales on the middle of the belly on a space corresponding to the length of the head. Twelve or thirteen femoral and three or four præanal pores on each side. Yellowish grey or cream-colour above, with small brown spots, sometimes with transverse series of round yellow brown-edged spots; lower surfaces yellowish white.

Total length	420 millim.
Head	43 ,,
Width of head	40 ,,
Body	197 ,,
Fore limb	82 ,,
Hind limb	116 ,,
Tail	180 ,,

Is said to reach a larger size than the preceding.
Persia.

a. ♀. Bushire. P. L. Sclater, Esq. [P.]. (Type.)
b, c. Hgr. & yg. Ghorak.

29. APOROSCELIS.

Tympanum distinct. Incisors large, uniting in the adult, separated from the molars by a toothless interspace. Body depressed, without crest. No gular pouch; a transverse gular fold. Tail very short, much depressed, covered above with series of large spines. Præanal and femoral pores absent; males with callose præanal scales.

East Africa.

1. Aporoscelis princeps.

Uromastix princeps, *O'Shaughn. Proc. Zool. Soc.* 1880, p. 445, pl. xliii.; *Vaillant, Miss. Révoil, Rept.* p. 9.

Head thick, moderately large, with the cheeks strongly swollen; snout very short, without canthus rostralis; nostril large, directed slightly backwards; ear-opening a long vertical slit; upper head-scales smooth, irregular, largest on the snout; two enlarged scales between the nostril and the orbit; a row of enlarged infraorbitals; labials very small. Body covered above with juxtaposed, flat, smooth uniform granules; ventral scales larger, smooth, subquadrangular, slightly imbricate, arranged in transverse series. Limbs rather short; hind limb reaching the axil; digits short; scales smooth, enlarged on the upper and anterior face of the thighs and on the tibial side of the leg; a few enlarged conical tubercles near the knee. Tail much flattened, not quite half the length of head and body, nearly as broad as the trunk, subelliptical in shape; its upper surface covered with enormous spines arranged in six longitudinal and eleven transverse series, those of the outer series the largest and hooked; on the lower surface the scales are small, arranged in regular transverse series, two of which correspond to one on the upper surface. Olive-grey; tail and upper surface of thighs reddish-brown.

Total length	187 millim.
Head	30 ,,
Width of head	29 ,,
Body	96 ,,

Fore limb .... 52 millim.
Hind limb .... 67 ,,
Tail .... 61 ,,

Zanzibar; Somali Land; Aden?

a. ♂.   Zanzibar Island.   Sir J. Kirk [C.]. (Type.)

2. Aporoscelis batilliferus.

Uromastix batilliferus, *Vaillant, Miss. Révoil, Rept.* p. 10, pl. ii.

Snout more elongate; ear-opening large, round. Dorsal scales small, rhomboidal, subimbricate, intermixed with enlarged ones forming irregular longitudinal series. One or two series of large trihedral spines on the thigh, extending to the leg. Tail very short, discoid at the base, attenuate posteriorly, racket-shaped; its upper surface with large spinose scales, arranged on the upper dilated portion in eight or ten longitudinal and seven transverse series; the lower surface with small imbricate scales similar in shape and arrangement to the ventrals. Greyish above; females with black lines and spots or with large ocelli on the back; males with the tail yellow above and the throat and chest dark blue.

Somali Land.

30. MOLOCH.

Moloch, *Gray, in Grey's Trav. Austr.* ii. p. 440, *and Cat. Liz.* p. 263; *A. Dum. Cat. Méth. Rept.* p. 109.

Mouth small; lateral teeth in the upper jaw implanted horizontally, directed inwards. Tympanum distinct. Body depressed. Tail short, rounded. Upper and lower parts covered with small scales or tubercles intermixed with large spinose tubercles. Nape with a large roundish protuberance. No femoral or præanal pores.

Australia.

1. Moloch horridus.

Moloch horridus, *Gray, Cat.* p. 263.
Moloch horridus, *Gray, Grey's Trav. Austr.* ii. p. 441, pl. ii.; *A. Dum. Cat. Méth.* p. 110, *and Arch. Mus.* viii. p. 582; *Gerv. Ann. Sc. Nat.* (4) xv. 1861, p. 86; *Wilson, Journ. Linn. Soc.* x. 1868, p. 69; *Gerv. Journ. Zool.* ii. 1873, p. 451.

Head very small; snout extremely short, the nostrils pierced on its upper surface; eye small, surmounted by a large horn curved outwards and backwards; tympanum a little smaller than the eye; upper surface of head with small, polygonal, granulate, juxtaposed tubercles; a spinose conical tubercle behind the nostril, another behind the supraorbital horn, another larger one in front of the tympanum, and a pair on the occiput; no distinct labials. Gular scales small, unequal, the largest conical and spinose; a pair of large tubercles on each side, behind the angle of the mouth. Neck as long as the head, from which it is separated by a fold; a transverse row of seven large spines, followed by a globular protuberance covered with enlarged scales, and a pair of large conical spines pointing outwards and backwards. Dorsal scales polygonal, juxtaposed, rough with small granules, some indistinctly keeled; these scales slightly enlarged round the base of the large spinose tubercles; latter conical, ribbed or keeled, forming ten more or less regular longitudinal series, those of the outer series the largest. Ventral scales unequal, rough, feebly keeled, subimbricate, intermixed with enlarged roundish, keeled, and spinose tubercles. Limbs rather strong, covered with polygonal scales and large spinose tubercles; digits extremely short, thick, armed with strong claws. Tail shorter than head and body, cylindrical, ending obtusely, covered with scales and tubercles as on the body. Yellowish, with large chestnut-brown or reddish brown, darker-edged, symmetrical markings.

Total length	212	millim.
Head	18	,,
Width of head	18	,,
Body	104	,,
Fore limb	51	,,
Hind limb	58	,,
Tail	90	,,

Southern and Western Australia.

a. Ad.	W. Australia.	
b. Hgr.	W. Australia.	Lady Harvey [P.].
c. Ad.	Swan River.	J. B. Jukes, Esq. [P.].
d. Ad.	Swan River.	C. W. Smith, Esq. [P.].
e. Ad.	Australia.	J. B. Jukes, Esq.
f–g. Ad.	Australia.	Capt. Stokes [C.].
h. Ad.	Australia.	Mr. Bland [C.].
i. Ad. skeleton.	W. Australia.	Sir G. Grey [P.].

The following genus was established upon a single female specimen, and its place in the system can hardly be assigned unless it be known whether or not præanal or femoral pores exist in the male:—

OREODEIRA, Girard, Proc. Ac. Philad. 1857, p. 199,
and U. S. Explor. Exped., Herp. p. 419.

"Head depressed, rather broad and short, covered with small, subtuberculous and subimbricated plates. Occipital plate rather large. Nostrils lateral, situated within the extension of the supraciliary ridge along the snout. Supralabial plates and temporal scales carinated. Tongue fleshy, anteriorly notched or bifid. Compressed and cutting-teeth on the sides of the jaws, and subconical incisors anteriorly. Auricular aperture moderate, subserrated; tympanum situated near its surface. A gular and a pectoral cross fold. Sides of the neck variously folded or rumpled. Nape with a small crest, vanishing away along the back. Scales on the back moderate, homogeneous, keeled, disposed upon oblique series; on the abdomen small, subequal, and keeled also; on the occiput and neck very small, subgranular. Limbs slender and elongated, terminated by slender, compressed, unequal, clawed fingers and toes. Tail slender, subconical and tapering. Femoral and præanal pores wanting."

OREODEIRA GRACILIPES, Gir. *ll. cc.*

"Three complete series of supralabials, smaller than the labials; infralabials exiguous, constituting likewise three series. Mental shield quite small, scarcely distinguishable from the infralabials. Posterior aspect of thighs scaly. Caudal scales conspicuously larger than the dorsal and abdominal ones. Brownish olive above, anteriorly maculated; beneath yellowish olive; chin obsoletely spotted."
New South Wales.

ADDENDA.

Add :—	Page 119. **Hemidactylus reticulatus.**	
h. ♀.	High Wavy, Madura.	Col. Beddome [C.].
Add :—	Page 121. **Hemidactylus frenatus.**	
ff. ♀.	Gilolo.	Dr. Platen [C.].
Add :—	Page 185. **Gecko stentor.**	
i-k. ♂.	Nias.	Hr. Sandemann [C.].

Page 196. The Synopsis of the species of *Tarentola* is to be modified as follows:—

I. A supraorbital bone.

  A. Tail rounded on the sides, elliptical in section.

Dorsal tubercles strongly keeled; scales on the throat much smaller than those on the occiput .... 1. *mauritanica*.

Dorsal tubercles smooth or feebly keeled; scales
on the throat nearly as large as those on the
occiput .... 2. *ephippiata.*

B. Tail perfectly flat inferiorly, with sharpish lateral edge.

Anterior border of ear denticulated .... 3. *annularis.*
Anterior border of ear not denticulated ...... 4. *senegalensis.*

II. No supraorbital bone.

    5. *delalandii.*   6. *gigas.*

Some specimens of this genus show a minute, almost imperceptible claw on the second and fifth toes.

Page 198. Add a new species:—

2*a*. **Tarentola senegalensis.**

Platydactylus ægyptiacus, (*non Cuv.*) *Steindachn. Sitzb. Ak. Wien,*
lxii. 1870, i. p. 327.
Tarentola delalandii, (*non D. & B.*) *Boettg. Abh. Senck. Ges.* xii. 1881,
p. 406.

Very closely allied to *T. annularis*, from which it differs in the absence of a denticulation in front of the ear. Even in coloration the two forms are perfectly similar.

Senegambia.

a. Ad.   Goree.   Baron v. Maltzan [C.].

Add:—   Page 198. **Tarentola ephippiata.**

c–f. Ad.   McCarthy Isd., Senegambia.   Capt. Moloney [P.].

These specimens show that this species reaches the same size as *T. annularis, senegalensis,* and *gigas.* In one of the specimens the three chin-shields on each side are in contact with the infralabials. The species is readily distinguishable from its nearest allies *T. annularis* and *senegalensis* by the larger gular scales and the shape of the tail, which is elliptical in section.

Page 199. **Tarentola delalandii.**

Erase spec. *m.*

ALPHABETICAL INDEX.

abbreviatus (Draco), 265.
acanthinurus (Uromastix), 406.
Acanthocercus, 335.
Acanthosaura, 299.
Acrodonta, 3.
Acrodontes, 250.
aculeata (Agama), 351, 352.
aculeatus (Gecko), 122.
adelaidensis (Amphibolurus), 387.
adelaidensis (Grammatophora), 387.
ægyptiaca (Eremioplanis), 338.
ægyptiaca (Tarentola), 197.
ægyptiacus (Platydactylus), 197, 414.
ægyptius (Trapelus), 348.
Æluronyx, 193.
Ælurosaurus, 73.
afer (Rhoptropus), 217.
affinis (Ceramodactylus), 14.
affinis (Cyrtodactylus), 42.
affinis (Gymnodactylus), 42.
affinis (Hemidactylus), 128.
affinis (Phrynocephalus), 377.
affinis (Phyllodactylus), 89.
Agama, 332, 335.
Agamidæ, 250.
Agamoidea, 250.
Agamura, 50.
agilis (Agama), 341, 342, 343, 344.
agilis (Podorrhoa), 341.
agrorensis (Agama), 363.
agrorensis (Stellio), 363.
albertisii (Gonyocephalus), 297.
albigularis (Gonatodes), 59.

albofasciatus (Gymnodactylus), 37.
albofasciolatus (Gecko), 184.
albogularis (Gonatodes), 59.
albogularis (Gymnodactylus), 59, 60.
alopex (Sphærodactylus), 222.
Alsophylax, 19.
amboinensis (Basiliscus), 402.
amboinensis (Draco), 264.
amboinensis (Hydrosaurus), 402.
amboinensis(Iguana),402.
amboinensis (Istiurus), 402.
amboinensis (Lacerta), 402.
amboinensis (Lophura), 402.
americana (Tarentola), 195.
americanus (Platydactylus), 195.
Amphibolurus, 380.
Amphisbæna, 3.
Amphisbænoidea, 3.
Amydosaurus, 162.
anamallayana (Lophosalea), 313.
anamallayana (Salea), 313.
anamallensis (Gecko), 175.
anamallensis (Hoplodactylus), 175.
andamanense(Phelsuma), 212.
androyensis (Phyllodactylus), 76.
angulatus (Hemidactylus), 113.
angulifer (Amphibolurus), 389.
angulifer (Chondrodactylus), 11.

angulifera (Grammatophora), 387, 388, 389.
annectens (Agama), 360.
annularis (Gecko), 197.
annularis(Tarentola),197.
Annulati, 3.
annulatus (Diplodactylus), 97.
annulatus (Gecko), 183.
Anomalurus, 22.
anomalus (Phyllodactylus), 75.
Anoplopus, 200, 209.
anthracinus (Sphærodactylus), 225.
antiquorum(Stellio),368.
Aphaniotis, 274.
Aporoscelis, 410.
Aprasia, 245.
Aprasiadæ, 239.
aralensis (Agama), 343.
aralensis (Trapelus), 343.
arenaria (Agama), 338.
arfakianus (Gymnodactylus), 40.
argentii (Nubilia), 134.
argus (Sphærodactylus), 223.
Aristelliger, 146.
armata (Acanthosaura), 301, 302.
armata(Agama),301,352.
armatus (Gecko), 122.
armatus (Gonyocephalus), 301.
armatus (Hemidactylus), 122.
armatus (Lophyrus), 301.
arnouxii (Gymnodactylus), 39, 40.
Arpephorus, 279.
articulatus (Perochirus), 156.
Arua, 282.
Ascalabotæ, 3, 250.
Ascalabotes, 3, 16, 195.
ascalabotes (Gecko), 110.
Ascalabotoidea, 3.
asmussii (Centrotrachelus), 409.

ALPHABETICAL INDEX.

asmussii (Uromastix), 409.
asper (Nephrurus), 9.
aspera (Agama), 349.
aspera (Ceratophora), 278.
ateles (Hemidactylus), 154, 155, 156.
ateles (Perochirus), 154.
atra (Agama), 352.
atra (Phrynopsis), 352.
atricollis (Agama), 358.
atropunctatus (Gymnodactylus), 22.
aubryanus (Rhacodactylus), 177.
aurantiacus (Hemidactylus), 164.
aurantiacus (Lepidodactylus), 164.
aureus (Calodactylus), 108.
auriculatus (Ceratolophus), 179.
auriculatus (Platydactylus), 179.
auriculatus (Rhacodactylus), 179.
aurita (Agama), 379.
aurita (Lacerta), 379.
auritus (Gecko), 379.
auritus (Gonyocephalus), 295.
auritus (Megalochilus), 379.
auritus (Phrynocephalus), 379.
australis (Calotella), 394.
australis (Diporophora), 394.
australis (Gehyra), 152.
australis (Goniodactylus), 72.
australis (Hoplodactylus), 75.
australis (Thecadactylus), 112.
Autarchoglossæ, 239.
axillaris (Phrynocephalus), 377.

baliola (Gehyra), 150.
baliolus (Hemidactylus), 150.
barbata (Agama), 391.
barbatus (Amphibolurus), 391.
Barycephalus, 335.
batilliferus (Aporoscelis), 411.
batilliferus (Uromastix), 411.
bavayi (Hemidactylus), 167.
beccarii (Draco), 264.
beddomii (Gymnodactylus), 67.
beddomii (Otocryptis), 272.
bellii (Gonyocephalus), 288.
bellii (Hemidactylus), 136.
bellii (Liolepis), 403.
bellii (Lophyrus), 288.
bellii (Tiaris), 288.
bellii (Uromastyx), 403.
bengalensis (Hemidactylus), 137.
bennettii (Diporophora), 395.
bennettii (Gindalia), 395.
berdmorei (Doryura), 139.
berdmorei (Hemidactylus), 139.
berdmorei (Leiurus), 139.
bergii (Pachydactylus), 205.
Biancia, 307.
bibronii (Agama), 357.
bibronii (Homodactylus), 201.
bibronii (Pachydactylus), 201.
bibronii (Tarentola), 201.
bibronii (Trapelus), 351.
bicatenata (Lialis), 247.
bilineata (Diporophora), 394.
bilineatus (Amphibolurus), 394.
bilineatus (Diplodactylus), 93.
bimaculatus (Draco), 263.
binoei (Heteronota), 74, 75.
binotatus (Gonyocephalus), 295.
bivittata (Otocryptis), 271.
bivittatus (Gecko), 186.
bivittatus (Platydactylus), 186.
bivittis (Scalabotes), 159.
blanfordiana (Charasia), 333.
blanfordii (Draco), 267.
blanfordii (Hemidactylus), 141.
bocagii (Hemidactylus), 125.
boettgeri (Ebenavia), 96.
boettgeri (Uroplates), 238.
boiei (Gonatodes), 72.
boiei (Goniodactyus), 72.
Boltalia, 113.
borneensis (Gonyocephalus), 288.
borneensis (Lophyrus), 288.
borneensis (Pentadactylus), 73.
borneensis (Tarentola), 199.
bouvieri (Hemidactylus), 118.
bowringii (Doryura), 139.
bowringii (Hemidactylus), 139.
boydii (Gonyocephalus), 297.
boydii (Tiaris), 297.
Brachydactylus, 234.
Brachysaura, 332.
brachyura (Agama), 350.
braconnieri (Dactychilikion), 209.
braconnieri (Goniodactylus), 60.
brevicaudis (Peripia), 147.
brevipalmata (Gehyra), 150.
brevipalmatus (Hemidactylus), 150.
brevipes (Gymnodactylus), 28.
Bronchocela, 314.
brookii (Hemidactylus), 128.
bruijnii (Gonyocephalus), 295.
brunnea (Chelosania), 331.
brunneus (Ælurosaurus), 74.
brunneus (Pentadactylus), 74.
buckleyi (Goniodactylus), 61.
Bunopus, 19.
burmana (Bronchocela) 316.
burtonii (Lialis), 247.

ALPHABETICAL INDEX. 417

cælaticeps (Agama), 389.
Calodactylus, 108.
Calotes, 250.
Calotella, 393.
calotella (Grammatophora), 394.
Calotes, 314.
calotes (Agama), 327.
calotes (Iguana), 327.
calotes (Lacerta), 327.
candeloti (Gymnodactylus), 165.
cantoris (Peripia), 165.
capensis (Hemidactylus), 160, 161.
capensis (Liurus), 161.
capensis (Lygodactylus), 160.
capensis (Pachydactylus), 202.
capensis (Stellio), 358.
capensis (Tarentola), 202.
capra (Acanthosaura), 300.
caracal (Gecko), 120.
carinatus (Stellio), 365.
cariniventris (Agama), 353.
carteri (Pristurus), 55.
carteri (Spatalura), 55.
casiculus (Sphærodactylus), 222.
caspius (Gymnodactylus), 26.
caucasica (Agama), 367.
caucasicus (Stellio), 367.
caudicincta (Grammatophora), 384.
caudicinctus (Amphibolurus), 384.
caudicinctus (Psilodactylus), 230.
caudicinctus (Stenodactylus), 230.
caudiscutatus (Gonatodes), 61.
caudiscutatus (Goniodactylus), 61.
caudiscutatus (Gymnodactylus), 61.
Caudiverbera, 236.
caudiverbera (Crossurus), 113.
caudiverbera (Gecko), 236.
caudiverbera (Lacerta), 236.
caudivolvula (Agama), 375.

caudivolvula (Lacerta), 375.
caudivolvulus (Phrynocephalus), 373, 375.
celebensis (Bronchocela), 318.
celebensis (Calotes), 318.
celebensis (Lophura), 402.
Centrocercus, 405.
Centrotrachelus, 405.
cepedeanus (Anoplopus), 211.
cepedianum (Phelsuma), 211.
cepedianus (Gecko), 211.
cepedianus (Pachydactylus), 214, 215.
cepedianus (Platydactylus), 211, 214.
cephalus (Tympanocryptis), 393.
Ceramodactylus, 13.
Ceratolophus, 176.
Ceratophora, 277.
cessacii (Hemidactylus), 118.
ceylanica (Cophotis), 275.
ceylonensis (Lepidodactylus), 164.
chahoua (Chameleonurus), 178.
chahoua (Platydactylus), 177, 178.
chahoua (Rhacodactylus), 177.
chamæleontina (Iguana), 285.
chamæleontinus (Gonyocephalus), 285.
chameleon (Gecko), 212.
Chameleonurus, 176.
Charasia, 299, 332.
chaus (Gecko), 120.
Chelosania, 331.
cheverti (Gymnodactylus), 41.
chinensis (Gecko), 188.
Chiroperus, 236.
Chlamydosaurus, 401.
Chondrodactylus, 10.
ciliaris (Diplodactylus), 98.
ciliatus (Correlophus), 180.
ciliatus (Rhacodactylus), 180.
cinereus (Sphærodactylus), 220.
Cionocrania, 3.
clamosa (Iguana), 281.
clypeata (Tarentola), 195.

cochinchinensis (Istiurus), 399.
cochinchinensis (Physignathus), 399.
coctæi (Hemidactylus), 136, 137, 138, 139.
cocteaui (Hoplopodion), 137.
Coleonyx, 234.
coleonyx (Gymnodactylus), 235.
collaris (Pristurus), 55.
collaris (Spatalura), 55.
collegalensis (Gymnodactylus), 34.
colonorum (Agama), 355, 356, 357.
colonorum (Podorrhoa), 356.
Colopus, 208.
concinna (Lophura), 399.
concinnatus (Gonatodes), 61.
concinnatus (Goniodactylus), 61.
concinnus (Physignathus), 399.
congica (Agama), 356.
consobrinus (Gymnodactylus), 47.
Cophotis, 275.
copii (Sphærodactylus), 225.
cordylea (Agama), 368.
cordylina (Iguana), 368.
cordylina (Stellio), 368.
cornutus (Draco), 258.
coronata (Acanthosaura), 303.
Correlophus, 176.
Coryphophylax, 282.
Cosymbotus, 114.
cowanii (Microscalabotes), 158.
crepuscularis (Lepidodactylus), 163.
crepuscularis (Platydactylus), 163, 165.
cristata (Grammatophora), 383.
cristatella (Agama), 316.
cristatella (Bronchocela), 316.
cristatellus (Calotes), 316.
cristatellus (Draco), 266.
cristatus (Amphibolurus), 383.
cristatus (Calotes), 321.

2 E

cristatus (Gecko), 236.
Crossurus, 113.
crucifer (Gymnodactylus), 55.
crucifer (Pristurus). 55.
cruciger (Gecko), 122.
crucigera (Acanthosaura), 302.
cruralis (Agamura), 50.
Cryptodelma, 242.
Ctenophorus, 380.
cubana (Tarentola), 195.
cubanus (Platydactylus), 195.
Cubina, 23.
cumingii (Luperosaurus), 181.
cuvieri (Lophura), 399.
cuvieri (Phyllurus), 49.
cyanodactylus (Gecus), 126.
cyanodactylus (Hemidactylus), 127, 128.
cyanogaster (Acanthocercus), 359.
cyanogaster (Agama), 359.
cyanogaster (Stellio), 359.
Cyclosaura, 3.
cyclura (Hemidactylus), 168.
cyclura (Peripia), 167.
cyclurus (Lepidodactylus), 167.
cyprius (Stellio), 368.
Cyrtodactylus, 22.
Cyrtopodion, 23.

Dactychilikion, 209.
Dactylocnemis, 171.
Dactyloperus, 147.
darwinii (Cubina), 39.
darwinii (Homonota), 21.
Dasyderma, 23.
daudinii (Draco), 256.
dayana (Agama), 362.
dayanus (Stellio), 362.
deccanensis (Gymnodactylus), 36.
deccanensis (Sitana), 270.
decresii (Amphibolurus), 385.
decresii (Ctenophorus), 385.
decresii (Grammatophora), 385, 386.
dekkanensis (Gymnodactylus), 36.
delalandii (Ascalabotes), 199.

delalandii (Platydactylus), 199.
delalandii (Tarentola), 199, 414.
Delma, 243.
Dendrobatæ, 250.
depressus (Hemidactylus), 134.
depressus (Perochirus), 155.
derbiana (Heteronota), 75.
deserti (Agama), 344.
diemensis (Grammatophora), 389.
dilophus (Gonyocephalus), 290.
dilophus (Lophyrus), 290.
dilophus (Tiaris), 290.
Dilophyrus, 282.
Diplodactylus, 76, 97.
Diploderma, 307.
Diporophora, 393.
Discodactylus, 76.
discosura (Agama), 49.
dispar (Uromastix), 406.
dorbignii (Cubina), 33.
dorbignii (Gymnodactylus), 33.
doriæ (Ceramodactylus), 13.
doriæ (Gonyocephalus), 284.
doriæ (Phyllodactylus), 91.
dorsalis (Ælurosaurus), 74.
dorsalis (Agama), 332.
dorsalis (Charasia), 332, 333.
dorsalis (Pentadactylus), 74.
dorsalis (Phrynopsis), 332.
Doryura, 114.
dovii (Eublepharis), 233.
Draco, 253.
Dracocella, 253.
Dracones, 250.
Draconoidea, 250.
Dracontoidis, 253.
Dracunculus, 253.
dubia (Peripia), 147.
dubius (Pachydactylus), 215.
dumerilii (Discodactylus), 100.
dunstervilloi (Stenodactylus), 16.

dussumieri (Draco), 268.
dussumieri (Dracocella), 268.
duvaucelii (Hoplodactylus), 172.
duvaucelii (Pentadactylus), 172.
duvaucelii (Platydactylus), 172, 178.
ebenaui (Uroplates), 238.
Ebenavia, 96.
eboracensis (Heteronota), 76.
echinatus (Hemidactylus), 123.
elegans (Coleonyx), 235.
elegans (Gymnodactylus), 169.
elegans (Naultinus), 168.
elegans (Pachydactylus), 202.
elegans (Sphærodactylus), 220.
elegans (Stenodactylus), 17.
elliotti (Calotes), 330.
elliotti (Tiaris), 306.
elongatus (Gymnodactylus), 30.
emma (Calotes), 324.
Emphyodontes, 250.
ephippiata (Tarentola), 198, 414.
Eremioplanis, 335.
Eublepharidæ, 229.
Eublepharis, 230.
Euleptes, 76.
europæus (Phyllodactylus), 90.
Eurydactylus, 192.
everetti (Draco), 258.
eversmanni (Stenodactylus), 19.

facetanus (Platydactylus), 196.
fantasticus (Sphærodactylus), 221, 223, 224.
fasciata (Cubina), 31.
fasciata (Heteronota), 41, 145.
fasciata (Rodtenbacheria), 396.
fasciata (Teratolepis), 145.
fasciatus (Eublepharis), 234.
fasciatus (Gymnodactylus), 31.

ALPHABETICAL INDEX. 419

fasciatus (Hemidactylus), 124.
fasciatus (Uromastix), 26.
fascicularis (Gecko), 196.
fascicularis (Platydactylus), 196.
fasciolatus (Eublepharis), 232.
fasciolatus (Gymnodactylus), 44.
fasciolatus (Naultinus), 44.
felinus (Ælurosaurus), 73.
felinus (Pentadactylus), 73.
feuillæi (Oiacurus), 236.
feuillæi (Ptyodactylus), 236.
fimbriatus (Draco), 265.
fimbriatus (Gecko), 237.
fimbriatus (Ptyodactylus), 237.
fimbriatus (Stellio), 237.
fimbriatus (Uroplates), 237.
flavimaculatus (Trapelus), 346.
flavipunctatus (Gymnodactylus), 52.
flavipunctatus (Pristurus), 52.
flavipunctatus (Saurodactylus), 52.
flaviviridis (Hemidactylus), 113.
formosus (Hemidactylus), 124.
formosus (Pachydactylus), 203.
forsythii (Phrynocephalus), 374.
fraseri (Delma), 243.
frenatus (Gymnodactylus), 34, 42.
frenatus (Hemidactylus), 120, 122, 413.
frontalis (Phrynocephalus), 375.
furcosus (Diplodactylus), 100.
fusca (Agama), 366.
fusca (Aphaniotis), 274.
fusca (Otocryptis), 274.
fuscus (Draco), 256.
fuscus (Gymnodactylus), 59.
fuscus (Stellio), 366.
fuscus (Stenodactylus), 59.

gaimardii (Grammatophora), 381.
gaimardii (Homalonotus), 381.
galapagoensis (Phyllodactylus), 82.
garnotii (Doryura), 141.
garnotii (Hemidactylus), 141.
garnotii (Hoplopodion), 141.
garrulus (Ptenopus), 15.
garrulus (Stenodactylus), 15.
gaudama (Doryura), 142.
gaudichaudi (Homonota), 21.
gaudichaudii (Gymnodactylus), 21, 63.
Gecconidæ, 3.
Gecko, 182.
gecko (Lacerta), 183.
gecko (Stellio), 110, 183.
Geckoella, 23.
geckoides (Gymnodactylus), 26, 27, 29, 39.
Geckolepis, 192.
Geckonidæ, 3, 229, 236.
Geckotidæ, 3.
Geckotiens, 3.
geelvinkianus (Gonyocephalus), 294.
Gehyra, 147.
Geissosaura, 3.
Gekko, 182.
gemmata (Agama), 349.
georgeensis (Idiodactylus), 146.
gerrhopygus (Diplodactylus), 95.
gerrhopygus (Phyllodactylus), 95.
gigantea (Agama), 285.
giganteus (Hemidactylus), 138.
gigas (Ascalabotes), 200.
gigas (Calotes), 325.
gigas (Tarentola), 200.
gilberti (Lophognathus), 396, 397.
gilberti (Physignathus), 396.
gillii (Gonatodes), 60.
gilvitorques (Sphærodactylus), 227.
Gindalia, 393.
girardi (Gymnodactylus), 39.
glaucus (Sphærodactylus), 221.

gleadovii (Hemidactylus), 129.
godeffroyi (Gonyocephalus), 295.
godeffroyi (Lophura), 295.
Gonatodes, 56.
Goniodactylus, 56.
Gonyocephali, 250.
Gonyocephalus, 282.
Gonyodactylus, 22.
goyazensis (Phyllopezus), 145.
gracilis (Gonatodes), 70.
gracilis (Gymnodactylus), 70.
gracilis (Hemidactylus), 119.
gracilis (Pletholax), 245.
gracilis (Pygopus), 245.
gracilipes (Oreodeira), 413.
Grammatophora, 380.
grandis (Dilophyrus), 298.
grandis (Gonyocephalus), 298.
grandis (Phelsuma), 214.
grandisquamis (Calotes), 325.
granosus (Hemidactylus), 126.
granulatus (Hoplodactylus), 174.
granulatus (Naultinus), 174.
grayi (Gecko), 152.
grayii (Delma), 243.
grayii (Naultinus), 168.
grayii (Nisara), 243.
griseus (Uromastix), 408.
guentheri (Draco), 257.
guentheri (Perochirus), 155.
guentheri (Phelsuma), 213.
guentheri (Phyllodactylus), 90.
guineensis (Hemidactylus), 128, 131.
gularis (Agama), 359.
gularis (Otocryptis), 273.
gularis (Ptyctolæmus), 273.
gularis (Salea), 314.
guppyi (Lepidodactylus), 166.
guttata (Agama), 375.
guttatus (Gecko), 183.
guttatus (Liolepis), 403.

2 E 2

guttatus (Platydactylus), 183.
guttatus (Ptyodactylus), 110.
guttatus (Stenodactylus), 17, 29.
gutturalis (Hemidactylus), 161.
gutturalis (Lygodactylus), 161.
gutturosa (Agama), 316.
gutturosa (Bronchocela), 318.
gutturosa (Calotes), 316.
Gymnodactylus, 16, 22, 52, 56, 230.
Gymnophthalmi, 239.
Gymnophthalmoidea, 239.
gymnopygus (Phyllodactylus), 95.

haematopogon (Draco), 267.
haematopogon (Dracocella), 267.
hardwickii (Centrocercus), 408.
hardwickii (Eublepharis), 231, 232.
hardwickii (Saara), 408.
hardwickii (Uromastix), 408.
Harpesaurus, 279.
harrieti (Gecko), 148.
hartmanni (Agama), 340.
hasselquistii (Ptyodactylus), 110.
helioscopa (Agama), 371.
helioscopa (Lacerta), 371.
helioscopa (Stellio), 371.
helioscopus (Phrynocephalus), 371.
Hemidactylus, 113, 147.
heterocercus (Gymnodactylus), 30.
Heteronota, 56, 74.
heteronotus (Gymnodactylus), 41.
Heteropholis, 169.
heterurus (Amphibolurus), 399.
hexaceros (Ceratolophus), 179.
hexaspis (Hemidactylus), 122.
hildebrandti (Scalabotes), 159.

himalayana (Agama), 362.
himalayanum (Nycteridium), 143.
himalayanus (Stellio), 362.
hirsutus (Mecolepis), 312.
hispida (Agama), 349, 351.
hispida (Lacerta), 349.
hispidus (Trapelus), 349.
homalocephala (Lacerta), 190.
homalocephalum (Ptychozoon), 190.
homalocephalus (Gecko), 190.
homalocephalus (Platydactylus), 190.
Homalonotus, 380.
Homodactylus, 200.
homoeolepis (Hemidactylus), 117.
homolepis (Ptyodactylus), 111.
Homonota, 21.
Homopholis, 191.
Hoplodactylus, 171.
Hoplopodion, 113.
horridus (Moloch), 411.
horsfieldii (Pteropleura), 190.
horsfieldii (Salea), 312.
humei (Gymnodactylus), 69.
humeralis (Gonatodes), 62.
humeralis (Gymnodactylus), 62.
humii (Gonyocephalus), 293.
humii (Tiaris), 293.
Humivagae, 250.
Hydrosaurus, 402.
hypoxantha (Rhacoëssa), 237.
Hypsilurus, 282.
Hysteropus, 240.

Idiodactylus, 146.
Iguaniens, 250.
imbricatus (Amphibolurus), 382.
imbricatus (Sphaerodactylus), 226.
impalearis (Agama), 357.
impar (Delma), 244.

impar (Pseudodelma), 244.
inaequalis (Phyllodactylus), 83.
incanesceus (Gecko), 122.
incertus (Gymnodactylus), 62.
indica (Agama), 321.
indica (Bronchocela), 330.
indicus (Gekko), 183.
indicus (Gonatodes), 64.
indicus (Goniodactylus), 64.
indicus (Gymnodactylus), 64.
indicus (Stellio), 361.
inermis (Agama), 344.
inermis (Phyllurus), 49.
infralineata (Agama), 352.
inornatus (Gonyocephalus), 294.
inornatus (Hemidactylus), 120.
inornatus (Sphaerodactylus), 221.
insignis (Pristurus), 54.
insulensis (Dactyloperus), 150.
insulensis (Gehyra), 150.
intermedia (Bronchocela), 316.
interruptus (Gonyocephalus), 290.
interruptus (Lophocalotes), 275.
interscapularis (Phrynocephalus), 378.
inunguis (Ebenavia), 96.
inunguis (Gecko), 205.
isolepis (Agama), 342.
isolepis (Grammatophora), 381.
Istiurus, 395, 402.

jacksoniensis (Agama), 390.
jamori (Platydactylus), 188.
Japalura, 307.
japonicus (Gecko), 188.
japonicus (Platydactylus), 188.
javanicus (Hemidactylus), 120.
jerdoni (Gymnodactylus), 71.
jerdonii (Calotes), 323.

ALPHABETICAL INDEX. 421

jerdonii (Gonatodes), 71.
jerdonii (Salea), 312.
jeyporensis (Gymnodactylus), 36.
jubata (Bronchocela), 318.
jubatus (Calotes), 318.
jugularis (Grammatophora), 380.
kachhensis (Gymnodactylus), 29.
kakhienensis (Acanthosaura), 305.
kakhienensis (Oriocalotes), 305.
kandianus (Gonatodes), 68.
kandianus (Gymnodactylus), 68.
karachiensis (Hemidactylus), 127.
karenorum (Doryura), 140.
karenorum (Hemidactylus), 140.
kelaartii (Hemidactylus), 136.
kendallii (Gonatodes), 63.
kendallii (Heteronota), 63.
keyserlingii (Teratoscincus), 12.
khasiensis (Gymnodactylus), 44.
khasiensis (Pentadactylus), 44.
kingii (Chlamydosaurus), 401.
Kionocrania, 3.
kirkii (Agama), 354.
kotschyi (Gymnodactylus), 29.
kuhlii (Gonyocephalus), 286.
kuhlii (Lophyrus), 286.
kushmorensis (Hemidactylus), 135.

labialis (Gecko), 166.
labialis (Lepidodactylus), 166.
labialis (Lophognathus), 397.
Lacertilia vera, 3.
lævis (Gecko), 111.
lævis (Grammatophora), 386.
lævis (Thecadactylus), 111.

lamnidentata (Acanthosaura), 302.
lar (Aristelliger), 147.
latastii (Agama), 344.
lateralis (Lophognathus), 395.
laticauda (Pachydactylus), 215.
laticauda (Phelsuma), 215.
Laudakia, 335.
lawderanus (Gymnodactylus), 32.
leachianus (Hoplodactylus), 176.
leachianus (Platydactylus), 176.
leachianus (Rhacodactylus), 176.
Leiolepides, 250.
Leiurus, 114.
Lepidodactylus, 162.
lepidogaster (Calotes), 301.
lepidopodus (Bipes), 240.
lepidopus (Pygopus), 240.
lepidopygus (Diplodactylus), 95.
Leptoglossi, 3.
leptorhyncha (Lialis), 247.
leschenaultii (Hemidactylus), 136.
lessonæ (Agama), 348.
lesueurii (Diplodactylus), 107.
lesueurii (Istiurus), 398.
lesueurii (Lophura), 398.
lesueurii (Œdura), 107.
lesueurii (Phyllodactylus), 107.
lesueurii (Physignathus), 398.
leucostigma (Agama), 346.
Lialis, 246.
Lialisidæ, 239.
lineata (Agama), 327.
lineata (Tympanocryptis), 392.
lineatum (Phelsuma), 212, 216.
lineatus (Diplodactylus), 92.
lineatus (Draco), 264.
linentus (Dracontoidis), 264.
lineatus (Dracunculus), 264.
lineatus (Naultinus), 109.

lineatus (Oiacurus), 238.
lineatus (Phyllodactylus), 92.
lineatus (Ptyodactylus), 238.
lineatus (Uroplates), 238.
lineolatus (Sphærodactylus), 221.
liocephalus (Calotes), 329.
liogaster (Gonyocephalus), 286.
liogaster (Tiaris), 286.
Liolepis, 403.
liolepis (Calotes), 326.
lirata (Agama), 364.
liratus (Stellio), 364.
littoralis (Gonatodes), 71.
littoralis (Gymnodactylus), 71.
lobata (Lacerta), 379.
lobatus (Gecko), 110.
lobatus (Ptyodactylus), 110.
Lonchurus, 236.
longicauda (Peripia), 147.
longicephalus (Hemidactylus), 125.
longiceps (Hemidactylus), 120, 125.
longii (Tiaris), 296.
longipes (Pristurus), 55.
longirostris (Lophognathus), 397.
longirostris (Physignathus), 397.
Lophocalotes, 274.
Lophodeira, 314.
Lophognathus, 395.
Lophosalea, 312.
Lophosaurus, 282.
Lophosteus, 282.
Lophura, 402.
Lophuræ, 250.
Lophyrus, 282, 299.
lophyrus (Galeotes), 285.
loricata (Agama), 348.
loricatus (Centrotrachelus), 409.
loricatus (Uromastix), 409.
ludekingii (Hemidactylus), 142.
lugubris (Amydosaurus), 165.
lugubris (Lepidodactylus), 165.
lugubris (Peripia), 165.
lugubris (Platydactylus), 165.
lunatus (Gymnodactylus), 231.

Luperosaurus, 181.
Lygodactylus, 158.
Lyriocephalus, 281.

mabouia (Gecko), 122.
mabouia (Hemidactylus), 122.
mabuya (Hemidactylus), 122.
macgregorii (Lyriocephalus), 281.
macrodactylus (Phyllodactylus), 89.
macrolepis (Grammatophora), 394.
macrolepis (Hypsilurus), 296.
macrolepis (Sphærodactylus), 226.
macularius (Cyrtodactylus), 232.
macularius (Eublepharis), 232.
maculata (Geckolepis), 192.
maculata (Grammatophora), 381.
maculatus (Amphibolurus), 381.
maculatus (Draco), 262.
maculatus (Dracunculus), 262.
maculatus (Gymnodactylus), 59, 66.
maculatus (Hemidactylus), 129, 132.
maculatus (Hoplodactylus), 171, 173.
maculatus (Pachydactylus), 206.
maculatus (Phrynocephalus,) 377.
maculatus (Ptenopus), 15.
maculatus (Uromastyx), 381.
maculiferus (Amphibolurus), 390.
maculilabris (Lophognathus), 398.
maculilabris (Physignathus), 398.
madagascariense (Phelsuma), 214.
madagascariensis (Lygodactylus), 160.
madagascariensis (Scalabotes), 160.
major (Charasia), 306.
major (Draco), 256, 267.
major (Oriocalotes), 306.

malabaricus (Gymnodactylus), 70.
mandellianus (Hemidactylus), 142.
margaritaceus (Lyriocephalus), 281.
marginatus (Gecko), 143.
marginatus (Hemidactylus), 143.
maria (Calotes), 322, 323.
mariquensis (Pachydactylus), 207.
marmorata (Bronchocela), 318.
marmorata (Heteronota), 41.
marmorata (Œdura), 101, 104.
marmorata (Peripia), 147.
marmoratus (Calotes), 318.
marmoratus (Cyrtodactylus), 46.
marmoratus (Diplodactylus), 88.
marmoratus (Gonatodes), 67.
marmoratus (Gonyodactylus), 44.
marmoratus (Gymnodactylus), 44, 67.
marmoratus (Hemidactylus), 113, 136.
marmoratus (Phyllodactylus), 88.
marmoratus (Phyllurus), 44.
Mastigura, 405.
mauritanica (Lacerta), 196.
mauritanica (Tarentola), 196.
mauritanicus (Ascalabotes), 196.
mauritanicus (Gecko), 196.
mauritanicus (Goniodactylus), 33.
mauritanicus (Platydactylus), 196.
mauritanicus (Saurodactylus), 33.
mauritanicus (Stellio), 196.
mauritanicus (Stenodactylus), 17.
maximiliani (Coryphophylax), 293.
Mecolepis, 312.
Megalochilus, 369.
megalonyx (Agama), 347.

megalonyx (Trapelus), 342, 347.
megapogon (Tiaris), 290.
melanospilos (Sphærodactylus), 224.
melanura (Agama), 363.
melanura(Laudakia),363.
melanurus (Phrynocephalus), 369.
melanurus (Stellio), 363.
mentager (Dilophyrus), 400.
mentager (Physignathus), 400.
mentomarginatus (Pachydactylus), 207.
mercatorius (Hemidactylus), 122.
meridionalis (Gecko), 126.
meyeri (Hemidactylus), 165.
meyeri (Peripia), 165.
Microdactylus, 114.
microlepis (Agama), 366.
microlepis (Japalura), 308.
microlepis (Sphærodactylus), 224.
microlepis (Stellio), 366.
microlepis (Uromastix), 407.
microlophus (Istiurus), 402.
microphyllus (Phyllodactylus), 84.
Microscalabotes, 157.
microtis (Gymnodactylus), 19.
milbertii (Platydactylus), 195.
miliusii (Gymnodactylus), 48.
miliusii (Phyllurus), 48.
millepunctatus (Sphærodactylus), 221.
minor (Acanthosaura), 304.
minor (Calotes), 304.
minor (Charasia), 304.
minor (Draco), 256.
minor (Oriocalotes), 304.
minor (Sitana), 270.
miotympanum (Gonyocephalus), 287.
miotympanum (Tiaris), 287.
mitratus (Brachydactylus), 235.
modestus (Gonyocephalus), 294.

ALPHABETICAL INDEX. 423

mœstus (Gecko), 165.
mölleri (Delma), 243.
Moloch, 411.
moluccana (Agama), 316.
moluccana (Bronchocela), 316.
monarchus (Gecko), 187.
monarchus (Platydactylus), 187.
mortoni (Hemidactylus), 113.
mossambica (Agama), 353.
multicarinatus (Gymnodactylus), 40.
muralis (Platydactylus), 196.
muricata (Agama), 390.
muricata (Grammatophora), 387, 389, 390.
muricata (Lacerta), 367, 390.
muricatus (Amphibolurus), 390.
muricatus (Gekko), 196.
muriceus (Hemidactylus), 123.
mutabilis (Agama), 338, 344, 348.
mutabilis (Trapelus), 338, 348.
mutilata (Gehyra), 148.
mutilata (Peripia), 149.
mutilatus (Hemidactylus), 148.
mutilatus (Peropus), 148.
mutilatus (Spathodactylus), 157.
mutilatus (Spathoscalabotes), 157.
mysorensis (Peripia), 165.
mysoriensis (Gonatodes), 68.
mysoriensis (Gymnodactylus), 68.
mystacea (Agama), 379.
mystacea (Lacerta), 379.
mystaceus (Calotes), 325.
mystaceus (Phrynocephalus), 379.

nanus (Hemidactylus), 188.
Naultinus, 168, 171.
nebulosus (Gymnodactylus), 34.
neglecta (Gehyra), 150.
neglectus (Peropus), 150.
nemoricola (Calotes), 326.
neocaledonicus (Lepidodactylus), 167.

Nephrurus, 9.
newtonii (Gecko), 228.
newtonii (Phelsuma), 212.
niger (Biancia), 308.
nigricans (Phrynocephalus), 369.
nigriceps (Cryptodelma), 242.
nigrigularis (Gonyocephalus), 296.
nigrilabris (Calotes), 328.
nigrilabris (Japalura), 311.
nigrilabris (Otocryptis), 311.
nigrofasciata (Agama), 348.
nigrofasciatus (Phyllodactylus), 82.
nigropunctatus (Sphærodactylus), 220.
nilii (Cyrtodactylus), 48.
Nisara, 243.
notatus (Gymnodactylus), 59.
notatus (Sphærodactylus), 226.
novæ-hollandiæ (Hysteropus), 240.
novæ-hollandiæ (Sheltopusik), 240.
nupta (Agama), 365.
nuptus (Stellio), 365.
Nycteridium, 114.
Nyctisaura, 3.

occipitalis (Agama), 353, 356.
oceanica (Gehyra), 152.
oceanicus (Gecko), 152.
ocellata (Agama), 375.
ocellata (Œdura), 105.
ocellatus (Cyrtodactylus), 60.
ocellatus (Diplodactylus), 93.
ocellatus (Gecko), 205.
ocellatus (Gonatodes), 60.
ocellatus (Goniodactylus), 60.
ocellatus (Pachydactylus), 205.
ocellatus (Phrynocephalus), 375.
ocellatus (Phyllodactylus), 93.
ocellatus (Platydactylus), 205.
octolineata (Aprasia), 246.
Œdura, 104.

Oiacurus, 236.
oldhami (Gymnodactylus), 38.
olivieri (Phrynocephalus), 370, 373.
Onychopus, 114.
ophiomachus (Agama), 327.
ophiomachus (Calotes), 327.
Ophiosauri, 3.
orbicularis (Agama), 349.
orbicularis (Lacerta), 349.
Oreodeira, 412.
orientalis (Cryptodelma), 242.
orientalis (Delma), 242.
orientalis (Stenodactylus), 16.
Oriocalotes, 299.
Oriotiaris, 299.
ornata (Brachysaura), 334.
ornata (Charasia), 334.
ornata (Grammatophora), 382.
ornata (Peripia), 147.
ornata (Rhynchœdura), 12.
ornatum (Phelsuma), 211.
ornatus (Amphibolurus), 382, 385.
ornatus (Diplodactylus), 100.
ornatus (Draco), 259.
ornatus (Gonatodes), 66.
ornatus (Gymnodactylus), 66.
ornatus (Leiurus), 124.
ornatus (Uromastix), 406.
oshaughnessyi (Pachydactylus), 204.
Otocryptæ, 250.
Otocryptis, 271.
oualensis (Hemidactylus), 152.
oualensis (Peropus), 152.
oudrii (Ptyodactylus), 110.
oviceps (Phyllodactylus), 85.
oxiana (Agama), 343.
oxyrrhinus (Sphærodactylus), 222.

Pachydactylus, 200.
Pachyglossa, 3.
Pachyglossæ, 250.
Pachyglossi, 3.
Pachyurus, 104.
pacificus (Dactylocnemis), 173.

pacificus (Hoplodactylus), 173.
pacificus (Naultinus), 171.
pacificus (Platydactylus), 173.
packardii (Peropus), 149.
pallida (Agama), 348.
pallidus (Amphibolurus), 388.
papuana (Gehyra), 152.
papuensis (Gonyocephalus), 297.
papuensis (Peripia), 147.
papuensis (Tiaris), 297.
pardus (Gecko), 148.
Parœdura, 76.
pelagicus (Gymnodactylus), 40.
Pentadactylus, 171.
perfoliatus (Stellio), 111.
Peripia, 147, 162.
Perochirus, 154.
Perodactylus, 147.
peronii (Hemidactylus), 148.
peronii (Peripia), 148.
peronii (Peropus), 148.
peronii (Phyllodactylus), 88.
Peropus, 147.
persica (Agama), 345.
persica (Agamura), 51.
persicus (Gymnodactylus), 51.
persicus (Hemidactylus), 131.
persicus (Phrynocephalus), 371.
persicus (Stellio), 367.
personatus (Dracontoidis), 260.
peruviana (Caudiverbera), 236.
peruvianum (Hoplopodion), 141.
peruvianus (Hemidactylus), 141.
petersii (Stenodactylus), 18.
petersii (Tiaris), 288.
petrensis (Gymnodactylus), 29.
phacophorus (Discodactylus), 84.
phacophorus (Phyllodactylus), 84.
Phelsuma, 209.
philippinicus (Gymnodactylus), 46.

philippinus (Calotes), 318.
Phoxophrys, 280.
Phrynocephali, 250.
Phrynocephalus, 369.
Phrynopsis, 335.
Phyllodactylus, 76, 97, 104.
Phyllopezus, 145.
Phyllurus, 22.
phyllurus (Gymnodactylus), 49.
phyllurus (Stellio), 49.
Phyria, 228.
Physignathus, 395.
physignathus (Istiurus), 399.
picta (Grammatophora), 385.
picticauda (Agama), 356.
picturatus (Hemidactylus), 161.
picturatus (Lygodactylus), 161.
pictus (Amphibolurus), 385.
pictus (Diplodactylus), 91.
pictus (Phyllodactylus), 91.
pictus (Scalabotes), 158.
pieresii (Hemidactylus), 134.
pipiens (Alsophylax), 19.
pipiens (Ascalabotes), 19.
pipiens (Eublepharis), 19.
pipiens (Gymnodactylus), 19.
pipiens (Lacerta), 19.
pipiens (Stenodactylus), 19.
planiceps (Agama), 358.
planidorsata (Japalura), 311.
planipes (Gymnodactylus), 71.
Planodes, 335.
platura (Cyrtodactylus), 49.
platura (Lacerta), 49.
platurus (Gonyodactylus), 49.
platurus (Gymnodactylus), 49.
platurus (Phyllurus), 49.
platurus (Stellio), 49.
platycephalus (Hemidactylus), 122.
platyceps (Calotes), 322, 323.

Platydactyles, 195, 200, 209.
Platydactylus, 111, 162, 171, 176, 182, 189, 195, 200, 209.
platyura (Agama), 49.
platyurum (Hoplopodion), 143.
Platyurus, 114.
platyurus (Crossurus), 143.
platyurus (Gecko), 143.
platyurus (Gymnodactylus), 49.
platyurus (Hemidactylus), 143.
platyurus (Nycteridium), 143.
platyurus (Stellio), 143.
Pletholax, 245.
Pleurodonta, 3.
Pleuropterus, 253.
Plocederma, 335.
Pneustoidea, 250.
Pnoëpus, 114.
Podorrhoa, 335.
pollicaris (Thecadactylus), 122.
polygonata (Diploderma), 310.
polygonata (Japalura), 310.
polyophthalmus (Diplodactylus), 101.
pomarii (Hoplodactylus), 173.
ponticeriana (Sitana), 270.
ponticerianus (Semiophorus), 270.
porphyreus (Gecko), 87.
porphyreus (Phyllodactylus), 87, 88.
porphyreus (Sphærodactylus), 87.
præpos (Draco), 256.
præsignis (Aristelliger), 146.
præsignis (Hemidactylus), 146.
princeps (Aporoscelis), 410.
princeps (Uromastix), 410.
Pristurus, 52.
przewalskii (Phrynocephalus), 377.
Psammophilus, 332.
Psamorrhoa, 335.
Pseudocalotes, 314.
Pseudodelma, 243.

ALPHABETICAL INDEX. 425

Pseudotrapelus, 335.
Psilodactylus, 229.
Ptenopus, 15.
Pteropleura, 189.
Pterosaurus, 253.
Ptychozoon, 189.
Ptyctolæmus, 273.
Ptyodactylus, 109, 236.
Puellula, 23.
pulchella (Aprasia), 245.
pulchellus (Cyrtodactylus), 46.
pulchellus (Gonyodactylus), 46.
pulchellus (Gymnodactylus), 46.
pulcher (Diplodactylus), 102.
pulcher (Discodactylus), 80.
pulcher (Lepidodactylus), 166.
pulcher (Phyllodactylus), 80.
pulcher (Stenodactylopsis), 102.
pulcherrimus (Amphibolurus), 388.
pumilus (Hemidactylus), 120.
punctata (Geckoella), 38.
punctatissimus (Sphærodactylus), 220.
punctatus (Hemidactylus), 120.
punctatus (Naultinus), 168.
punctatus (Pachydactylus), 206.
punctulata (Lialis), 247.
punctulata (Phyria), 228.
pusillus (Lepidodactylus), 167.
pusillus (Peropus), 167.
pustulosa (Lophura), 402.
pustulosus (Hemidactylus), 136.
pustulosus (Histiurus), 402.
Pygopidæ, 239.
Pygopodidæ, 239.
Pygopus, 240.

quadriocellatus (Pachydactylus), 216.
Quedenfeldtia, 23.
quinquefasciatus (Draco), 269.
quinquefasciatus (Dracunculus), 269.

rapicauda (Gecko), 111.
rapicauda (Lacerta), 111.
rapicaudus (Thecadactylus), 111.
Redtenbacheria, 395.
reevesii (Gecko), 183.
reevesii (Leiolepis), 403.
reevesii (Platydactylus), 183.
reevesii (Uromastyx), 403.
reissii (Phyllodactylus), 80.
reticulata (Grammatophora), 386.
reticulatus (Amphibolurus), 386.
reticulatus (Draco), 257.
reticulatus (Hemidactylus), 118, 413.
reticulatus (Phrynocephalus), 375.
reticulatus (Uromastix), 408.
Rhacodactylus, 176.
Rhacodracon, 253.
Rhacoëssa, 236.
rhombifer (Œdura), 107.
Rhoptropus, 217.
Rhynchœdura, 11.
richardsonii (Hemidactylus), 143.
richardsonii (Sphærodactylus), 227.
richardsonii (Velernesia), 143.
riebeckii (Diplodactylus), 94.
riebeckii (Phyllodactylus), 94.
robusta (Œdura), 106.
robustus (Hemidactylus), 126.
roseus (Peropus), 162.
rostratus (Draco), 261.
rouxii (Calotes), 328, 330.
rubida (Puellula), 45.
rubidus (Cyrtodactylus), 46.
rubidus (Gymnodactylus), 45.
rubrigularis (Agama), 346.
rubrigularis (Trapelus), 346.
ruderata (Agama), 348.
ruderata (Eremioplanis), 348.
ruderatus (Trapelus), 348.

rudis (Heteropholis), 170.
rudis (Naultinus), 170.
rueppellii (Agama), 355.
rugosus (Pachydactylus), 204.
rupestris (Pristurus), 53.

Saara, 405.
sakalava (Hemidactylus), 113.
Salea, 312.
sancti-johannis (Parœdura), 86.
sancti-johannis (Phyllodactylus), 86.
sanguinolenta (Agama), 343.
sanguinolenta (Lacerta), 343.
sanguinolenta (Podorrhoa), 343.
sanguinolentus (Trapelus), 343.
sarroubea (Gecko), 237.
Sarruba, 236.
sarrube (Ptyodactylus), 237.
Saurodactylus, 23.
sauragii (Lepidodactylus), 168.
savignii (Agama), 342.
savignyi (Gecko), 197.
savignyi (Phrynopsis), 342.
saviguyi (Trapelus), 17.
scaber (Gonyodactylus), 27.
scaber (Gymnodactylus), 27, 29.
scaber (Stenodactylus), 27.
Scalabotes, 158.
scapularis (Gymnodactylus), 235.
Scelotretus, 182.
schneideri (Nycteridium), 143.
schneideriana (Lacerta), 143.
schneiderianus (Platyurus), 143.
Scincoidiens, 239.
scincus (Stenodactylus), 12.
scincus (Teratoscincus), 12.
scutata (Agama), 281.
scutata (Iguana), 281.
scutata (Lacerta), 281.
scutatus (Lyriocephalus), 281.

scutellatus (Perochirus), 156.
sebæ (Agama), 368.
sebæ (Hemidactylus), 113.
Semiophori, 250.
Semiophorus, 270.
semperi (Gonyocephalus), 289.
senegalensis (Tarentola), 414.
seychellensis (Æluronyx), 193.
seychellensis (Platydactylus), 193.
seychellensis (Thecodactylus), 193.
seychellensis (Theconyx), 193.
shawii (Lophura), 402.
Sheltopusik, 240.
silvestris (Naultinus), 169.
similis (Centrocercus), 408.
sinaita (Agama), 339.
sinaita (Podorrhoa), 339.
sinaitus (Hemidactylus), 126.
sinaitus (Trapelus), 338, 339.
sisparensis (Gonatodes), 66.
sisparensis (Gymnodactylus), 66.
Sitana, 270.
smaragdina (Bronchocela), 319.
smaragdinus (Calotes), 319.
smithii (Gecko), 184.
sophiæ (Gonyocephalus), 288.
sophiæ (Tiaris), 288.
Spathodactylus, 156.
Spathoscalabotes, 156.
spatulatus (Phyllodactylus), 81.
speciosus (Gymnodactylus), 31.
Sphærodactylus, 76, 217.
spilonotus (Draco), 265.
spilopterus (Draco), 260.
spilopterus (Dracunculus), 260.
spinigera (Strophura), 99.
spinigerus (Diplodactylus), 99.
spinigerus (Phyllodactylus), 99.
spinipes (Gonyocephalus), 292.

spinipes (Lophyrus), 292.
spinipes (Mastigura), 407.
spinipes (Stellio), 407.
spinipes (Uromastix), 407.
spinosa (Agama), 349, 355.
spinosus (Lophyrus), 318.
spinulosus (Gonyodactylus), 39.
spixii (Cyrtodactylus), 39.
sputator (Anolis), 219, 220.
sputator (Gecko), 219.
sputator (Lacerta), 219.
sputator (Sphærodactylus), 219, 220.
squamiceps (Pygopus), 240.
steindachneri (Diplodactylus), 102.
stellatus (Hemidactylus), 130.
Stellio, 335.
stellio (Agama), 368.
stellio (Cordylus), 368.
stellio (Gecko), 196.
stellio (Lacerta), 367, 368.
stellio (Tarentola), 196.
Stelliones, 250.
Stellionidæ, 250.
Stenodactylopsis, 97.
Stenodactylus, 16, 22.
stentor (Gecko), 184, 413.
stentor (Platydactylus), 184.
steudneri (Gymnodactylus), 34.
sthenodactylus (Ascalabotes), 17.
stoddartii (Ceratophora), 277.
stoliczkæ (Gymnodactylus), 31.
stoliczkai (Phrynocephalus), 373.
stoliczkana (Agama), 360.
stoliczkanus (Stellio), 360.
strigatus (Lygodactylus), 160.
Strobilosaura, 3.
Strophura, 97.
Strophurus, 97.
strophurus (Diplodactylus), 100.

strophurus (Phyllodactylus), 100.
stumpffi (Phyllodactylus), 86.
subcristata (Tiaris), 292.
subcristatus (Gonyocephalus), 292.
subhispidus (Trapelus), 352.
sublævis (Boltalia), 137.
subpalmatus (Gecko), 189.
subspinosa (Agama), 352.
subtriedrus (Hemidactylus), 134.
sulcatus (Goniodactylus), 62.
sulcatus (Mccolepis), 312.
sulphureus (Naultinus), 169.
sumatrana (Cophotis), 276.
sumatranus (Gonyocephalus), 286.
sumatranus (Lophyrus), 286.
superciliosa (Lacerta), 285.
surinamensis (Gecko), 111.
swinhoii (Japalura), 310.
swinhonis (Gecko), 189.
swinhonis (Japalura), 309.
sykesii (Barycephalus), 361.
sykesii (Hemidactylus), 132.

Tachybates, 114.
tæniopterus (Draco), 269.
Tapaya, 335.
Tarentola, 195, 413.
tasmaniensis (Amphibolurus), 388.
temporalis (Grammatophora), 396, 397.
temporalis (Physignathus), 397.
temporalis (Uromastix), 406.
tennentii (Ceratophora), 278.
tenuis (Gymnodactylus), 22.
Teratolepis, 144.
Teratoscincus, 12.
teres (Gekko), 183.
tessellatus (Diplodactylus), 103.

ALPHABETICAL INDEX. 427

tessellatus (Stenodactylopsis), 103.
tetradactylus (Gecko), 237.
tetradactylus (Stellio), 237.
Thecadactylus, 111, 236.
Thecodactylus, 111.
theconyx (Platydactylus), 111.
theobaldi (Phrynocephalus), 373.
thomensis (Lygodactylus), 161.
thomensis (Scalabotes), 161.
Tiaris, 282.
tickelii (Phrynocephalus), 370, 373, 375.
tiedemanni (Agama), 321.
tiedemanni (Calotes), 321.
tigrinus (Lophyrus), 285, 286.
tigris (Gecko), 46.
timorensis (Draco), 261.
timorensis (Gonatodes), 63.
timorensis (Goniodactylus), 72.
timoriensis (Gymnodactylus), 63.
tolampyæ (Hemidactylus), 113.
Tolarenta, 16.
torresiana (Peripia), 151.
tournevillii (Agama), 340.
trachyblepharus (Gymnodactylus), 34.
trachyblepharus (Saurodactylus), 34.
trachycephalus (Chameleonurus), 178.
trachygaster (Æluronyx), 194.
trachygaster (Platydactylus), 194.
trachylæmus (Gecko), 186.
trachyrhynchus (Rhacodactylus), 178.
Trapeli, 250.
Trapeloides, 335.
Trapelus, 335.
tricarinata (Acanthosaura), 306.
tricarinata (Charasia), 306.
tricarinata (Oriotiaris), 306.

tricarinatus (Calotes), 306.
tricinctus (Arpephorus), 279.
tricinctus (Harpesaurus), 279.
triedrus (Gecko), 80, 133.
triedrus (Gymnodactylus), 38.
triedrus (Hemidactylus), 127, 133.
trihedrus (Hemidactylus), 133.
trilineatum (Phelsuma), 212.
tripolitanus (Tropiocolotes), 19.
tripolitanus (Stenodactylus), 19.
trispinosus (Mecolepis), 312.
tristis (Hemidactylus), 113.
Tropiocolotes, 16.
tuberculata (Agama), 361, 363.
tuberculata (Laudakia), 361, 363.
tuberculata (Phoxophrys), 280.
tuberculata (Podorrhoa), 361.
tuberculatus (Alsophylax), 20.
tuberculatus (Bunopus), 20.
tuberculatus (Gonyocephalus), 293.
tuberculatus (Stellio), 361.
tuberculatus (Tiaris), 293.
tuberculosus (Discodactylus), 79.
tuberculosus (Gecko), 122.
tuberculosus (Hemidactylus), 122.
tuberculosus (Phyllodactylus), 79.
turcica (Lacerta), 126.
turcicus (Hemidactylus), 126.
turneri (Homodactylus), 201.
tympanistriga (Bronchocela), 320.
tympanistriga (Calotes), 320.
Tympanocryptis, 392.
typica (Geckolepis), 192.
tytleri (Gecko), 129.

unctus (Diplodactylus), 94.
unctus (Phyllodactylus), 94.
unistriata (Lacerta), 185.
uralensis (Agama), 371.
uralensis (Lacerta), 371.
uralensis (Phrynocephalus), 371.
uralensis (Stellio), 371.
Uromasticidæ, 250.
Uromastix, 405.
Uroplates, 236.
Uroplatidæ, 236.
Uroplatus, 236.

variegata (Gehyra), 151.
variegata (Japalura), 308.
variegata (Peripia), 151.
variegatus (Coleonyx), 233.
variegatus (Eublepharis), 233.
variegatus (Gymnodactylus), 43.
variegatus (Hemidactylus), 151, 161.
variegatus (Naultinus), 43.
variegatus (Peropus), 151.
variegatus (Stenodactylus), 233.
varius (Phrynocephalus), 369.
Velernesia, 114.
ventralis (Phyllodactylus), 80.
verreauxi (Gecko), 184.
verrillii (Œdura), 108.
verrucosus (Hemidactylus), 126.
verruculatus (Gecko), 126.
verruculatus (Hemidactylus), 127, 128.
versicolor (Agama), 321.
versicolor (Calotes), 321.
versicolor (Phrynocephalus), 374.
verticillatus (Gecko), 183.
verus (Gecko), 183.
vieillardi (Platydactylus), 192.
vieillardii (Eurydactylus), 192.
viridis (Calotes), 321.
viridis (Draco), 256.
vittatus (Diplodactylus), 100.
vittatus (Gecko), 185.

vittatus (Gonatodes), 60.
vittatus (Gymnodactylus), 60.
vittatus (Hemidactylus), 120.
vittatus (Phyllodactylus), 100.
vittatus (Platydactylus), 185.
vlangalii (Phrynocephalus), 372.
volans (Draco), 256.
vorax (Gehyra), 153.
vulgaris (Stellio), 367, 368.
vulpecula (Doryura), 142.

vultuosa (Agama), 316, 321.

wagleri (Phyllodactylus), 91.
wahlbergii (Colopus), 208.
wahlbergii (Gecko), 191.
wahlbergii (Homopholis), 191.
whitii (Homonota), 22.
wicksii (Gymnodactylus), 69.
wiegmanni (Otocryptis), 271.
wilkinsonii (Stenodactylus), 18.

wilkinsonii (Tolarenta), 18.
wynaadensis (Gymnodactylus), 65.
wynadensis (Gonatodes), 65.
wynadensis (Goniodactylus), 65.

xanti (Phyllodactylus), 79.

yarkandensis (Cyrtodactylus), 31.
yunnanensis (Japalura), 310.

LIST OF PLATES.

Plate I.

Nephrurus asper, Gthr., p. 9.
  a. Upper view.
  b. Lower view.
  c. Lateral view of head.
  d. Lower view of foot, × 2.
  e. Lower view of hand, × 2.
  f. Lower view of finger, × 4.
  g. Group of dorsal tubercles, × 4.

Plate II.

Fig. 1. *Rhynchœdura ornata*, Gthr., p. 12.
  1 *a.* —— ——, lateral view of head, × 2.
  1 *b.* —— ——, lower view of fourth toe, × 6.
2. *Ptenopus garrulus*, Smith, p. 15.
  2 *a.* —— ——, lower view of hand, × 3.
  2 *b.* —— ——, lower view of foot, × 3.
  2 *c.* —— ——, lower view of fourth finger, × 5.
  2 *d.* —— ——, lower view of fourth toe, × 5.
3. *Teratoscincus scincus*, Schleg., p. 12, lower view of fourth toe, × 4.
4. *Ceramodactylus doriæ*, Blanf., p. 13, lower view of fourth toe, × 5.
5. *Chondrodactylus angulifer*, Ptrs., p. 11, lower view of hand, × 2.
  5 *a.* —— ——, lower view of fourth finger, × 4.

Plate III.

Fig. 1. *Stenodactylus orientalis*, Blanf., p. 16.
- 1 a. —— ——, lower view of fourth toe, ×4.
- 2. —— *guttatus*, Cuv., lower view of fourth toe, ×4.
- 3. —— *wilkinsonii*, Gray, p. 18, lateral view of head, ×2.
- 4. —— *petersii*, Blgr., p. 18.
- 4 a. —— ——, dorsal scales, ×4.
- 5. *Alsophylax pipiens*, Pall., p. 19, lower view of fourth toe, ×4.
- 6. *Homonota whitii*, Blgr., p. 22.
- 6 a. —— ——, lateral view of head, ×2.
- 7. —— *darwinii*, Blgr., p. 21, lateral view of head, ×2.
- 7 a. —— ——, lower view of fourth toe, ×4.
- 7 b. —— ——, dorsal scales, ×6.
- 8. *Ælurosaurus felinus*, Gthr., p. 73, lower view of fourth toe, ×4.
- 8 a. —— ——, lateral view of fourth toe, ×4.

Plate IV.

Fig. 1. *Gymnodactylus nebulosus*, Bedd., p. 34, var. B.
- 1 a. —— ——, var. A.
- 2. —— *jeyporensis*, Bedd., p. 36.
- 3. —— *albofasciatus*, Blgr., p. 37.

Plate V.

Fig. 1. *Gonatodes ocellatus*, Gray, p. 60.
- 1 a. —— ——, lower view of fourth toe, ×4.
- 2. —— *caudiscutatus*, Gthr., p. 61, male.
- 2 a. —— ——, female.
- 3. —— *humeralis*, Guich., p. 62, lower view of fourth toe, ×4.
- 4. —— *kendallii*, Gray, p. 63.
- 4 a. —— ——, lateral view of head, ×2.

Plate VI.

Fig. 1. *Gonatodes indicus*, Gray, p 64.
    1 a. —— ——, dorsal scales, strongly enlarged.
    2. —— *wynadensis*, Bedd., p. 65.
    2 a. —— ——, dorsal scales, strongly enlarged.
    3. —— *ornatus*, Bedd., p. 66.
    4. —— *marmoratus*, Bedd., p. 67.
    4 a. —— ——, lower view of fourth toe, ×3.
    5. —— *gracilis*, Bedd., p. 70.
    6. —— *littoralis*, Jerd., p. 71, lower view of fourth toe, ×4.

Plate VII.

Fig. 1. *Phyllodactylus sancti-johannis*, Gthr., p. 86.
    1 a. —— ——, lower view of fourth toe, ×4.
    2. —— *macrodactylus*, Blgr., p. 89.
    2 a. —— ——, lower view of fourth toe, ×4.
    2 b. —— ——, upper view of digital expansion, ×4.
    3. —— *guentheri*, Blgr., p. 90.
    3 a. —— ——, lower view of fourth toe, ×4.
    4. —— *affinis*, Blgr., p. 89, lateral view of snout, ×4.
    4 a. —— ——, lower view of fourth toe, ×4.
    5. —— *porphyreus*, Daud., p. 87, lateral view of snout, ×4.
    6. —— *marmoratus*, Gray, p. 88, lateral view of snout, ×4.

Plate VIII.

Fig. 1. *Ebenavia boettgeri*, Blgr., p. 96.
    2. *Diplodactylus ciliaris*, Blgr., p. 98.
    2 a. —— ——, lower view of foot, ×2.
    2 b. —— ——, extremity of toe, upper view, ×3.
    3. —— *vittatus*, Gray, p. 100, lower view of foot, ×2.
    4. —— *polyophthalmus*, Gthr., p. 101, lower view of foot, ×3.

Fig. 5. *Diplodactylus steindachneri*, Blgr., p. 102.
   5 a. —— ——, lower view of foot, ×3.
   6. —— *tessellatus*, Gthr., p. 103.
   6 a. —— ——, lower view of foot, ×3.

Plate IX.

Fig. 1. *Œdura ocellata*, Blgr., p. 105.
   1 a. —— ——, lower view of foot, ×2.
   1 b. —— ——, chin.
   2. —— *marmorata*, Gray, p. 104, chin.

Plate X.

Fig. 1. *Œdura robusta*, Blgr., p. 106.
   2. —— *lesueurii*, D. & B., p. 107, lower view of fourth toe, ×4.
   3. *Calodactylus aureus*, Bedd., p. 108, upper view of head.
   3 a. —— ——, lower view of foot, ×2.

Plate XI.

Fig. 1. *Thecadactylus australis*, Gthr., p. 112.
   1 a. —— ——, lower view of foot.
   1 b. —— ——, anal region and base of tail.
   2. *Hemidactylus reticulatus*, Bedd., p. 118.
   3. —— *echinus*, O'Sh., p. 123.
   4. —— *fasciatus*, Gray, p. 124.
   4 a. —— ——, lower view of foot, ×1½.

Plate XII.

Fig. 1. *Hemidactylus stellatus*, Blgr., p. 130.
   2. —— *bowringii*, Gray, p. 139.
   3. —— *richardsonii*, Gray, p. 143.
   4. *Perochirus guentheri*, Blgr., p. 155.
   4 a. —— ——, lower view of foot, ×2.

Plate XIII.

Fig. 1. *Spathoscalabotes mutilatus*, Gthr., p. 157.
    2. *Microscalabotes cowanii*, Blgr., p. 158.
    2 *a*. —— ——, lower view of foot, × 4.
    3. *Lepidodactylus ceylonensis*, Blgr., p. 164.
    4. —— *aurantiacus*, Bedd., p. 164.
    5. —— *pulcher*, Blgr., p. 166.
    5 *a*. —— ——, lower view of foot, × 2.
    6. —— *cyclurus*, Gthr., p. 167.

Plate XIV.

Fig. 1. *Hoplodactylus maculatus*, Gray, p. 171.
    1 *a*. —— ——, var.
    1 *b*. —— ——, lower view of foot, × 1½.
    2. —— *anamallensis*, Gthr., p. 175.
    2 *a*. —— ——, lower view of foot, × 3.
    3. *Naultinus elegans*, Gray, p. 168, lower view of foot, × 1½.

Plate XV.

Fig. 1. *Hoplodactylus granulatus*, Gray, p. 174.
    1 *a*. —— ——, lower view of foot, × 1½.
    2. *Luperosaurus cumingii*, Gray, p. 181.
    2 *a*. —— ——, lower view of foot, × 1½.

Plate XVI.

Fig. 1. *Tarentola ephippiata*, O'Sh., p. 198.
    2. *Pachydactylus formosus*, Smith, p. 203.
    3. —— *oshaughnessyi*, Blgr., p. 204.
    4. —— *maculatus*, Gray, p. 206.
    5. —— *mentomarginatus*, Smith, p. 207.
    6. —— *mariquensis*, Smith, p. 207.

Plate XVII.

Phelsuma newtonii, Blgr., p. 213.   *a.* The throat.

Plate XVIII.

Fig. 1. *Phelsuma lineatum*, Gray, p. 216.
 2. *Sphærodactylus nigropunctatus*, Gray, p. 220, end of snout, upper view, ×4.
 3. ——— *glaucus*, Cope, p. 221, end of snout, upper view, ×4.
 4. ——— *oxyrrhinus*, Gosse, p. 222, upper view of head, ×2.
 5. ——— *argus*, Gosse, p. 223.
 6. ——— *richardsonii*, Gray, p. 227.

Plate XIX.

Cryptodelma orientalis, Gthr., p. 242.
 a. Upper view of head, ×1¼.
 b. Anal region, × 1½.

Plate XX.

Lateral view of heads of:—
Fig. 1. *Draco reticulatus*, Gthr., p. 257.
 2. ——— *guentheri*, Blgr., p. 257.
 3. ——— *everetti*, Blgr., p. 258.
 4. ——— *cornutus*, Gthr., p. 258.
 5. ——— *rostratus*, Gthr., p. 261.
 6. ——— *bimaculatus*, Gthr., p. 263.
 7. ——— *blanfordii*, Blgr., p. 267.
 8. ——— *quinquefasciatus*, Gray, p. 269.

Plate XXI.

Gonyocephalus interruptus, Blgr., p. 290.

Plate XXII.

Fig. 1. *Acanthosaura armata*, Gray, p. 301.
    2. —— *crucigera*, Blgr., p. 302.
    3. —— *lamnidentata*, Blgr., p. 302.

Plate XXIII.

Fig. 1. *Otocryptis beddomii*, Blgr., p. 272.
    2. *Acanthosaura minor*, Gray, p. 304.
    3. —— *major*, Jerd., p. 306.

Plate XXIV.

Fig. 1. *Japalura variegata*, Gray, p. 308.
    2. —— *planidorsata*, Jerd., p. 311.

Plate XXV.

Fig. 1. *Calotes emma*, Gray, p. 324.
    2. —— *liolepis*, Blgr., p. 326.
    3. —— *elliotti*, Gthr., p. 330.

Plate XXVI.

Calotes liocephalus, Gthr., p. 329.

Plate XXVII.

Chelosania brunnea, Gray, p. 331.

Plate XXVIII.

Fig. 1. *Agama brachyura*, Blgr., p. 350.
    2. —— *kirkii*, Blgr., p. 354.

Plate XXIX.

Fig. 1. *Amphibolurus cristatus*, Gray, p. 383.
    2. —— *caudicinctus*, Gthr., p. 384.

Plate XXX.

Fig. 1. *Amphibolurus reticulatus*, Gray, p. 386.
    2. —— *pulcherrimus*, Blgr., p. 388.
2 a. —— ——, lower view.
    3. —— *pallidus*, Blgr., p. 388.

Plate XXXI.

Fig. 1. *Tympanocryptis cephalus*, Gthr., p. 393.
    2. *Diporophora bennettii*, Gray, p. 395.
    3. *Physignathus temporalis*, Gthr., p. 397.
    4. —— *longirostris*, Blgr., p. 398.

Plate XXXII.

Uromastix loricatus, Blanf., p. 409.

Nephrurus asper.

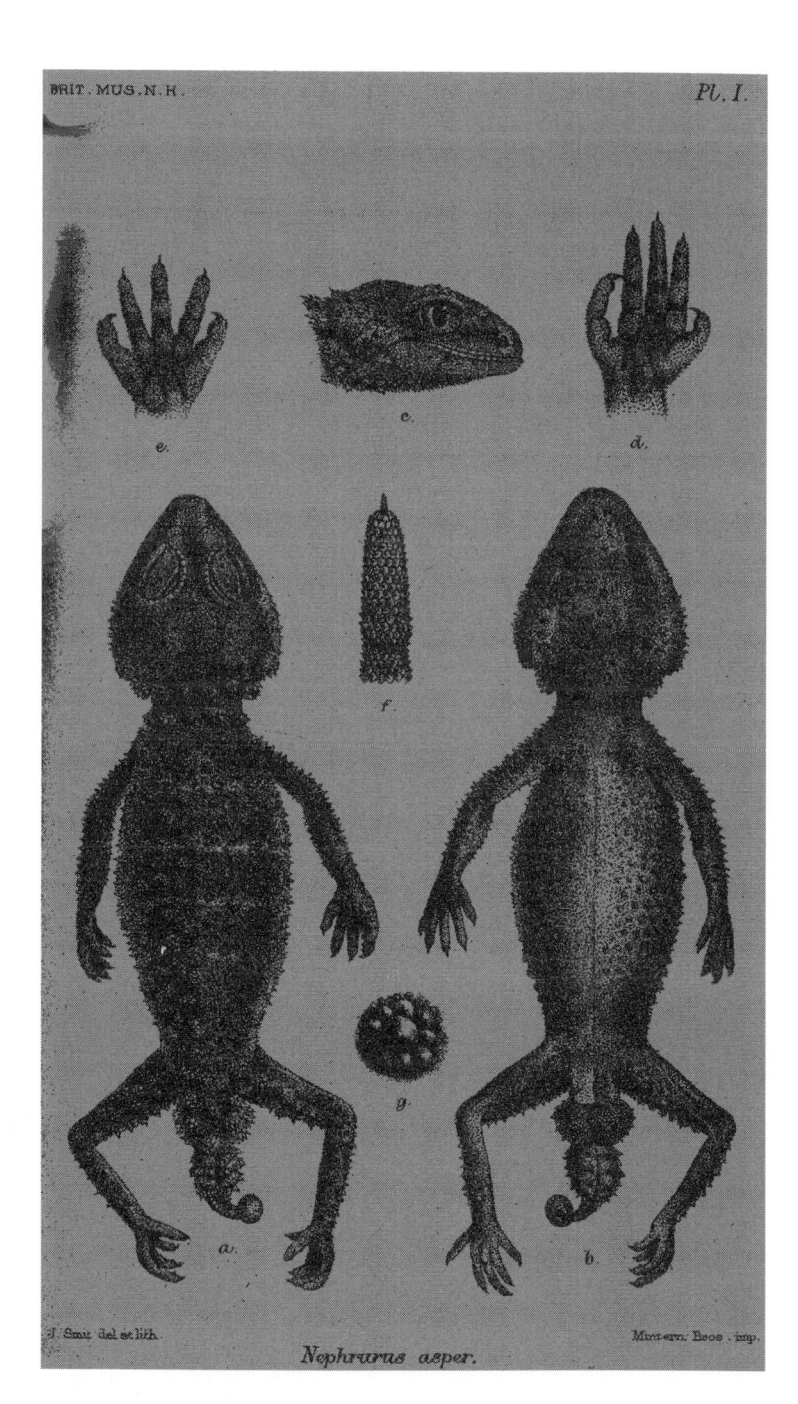

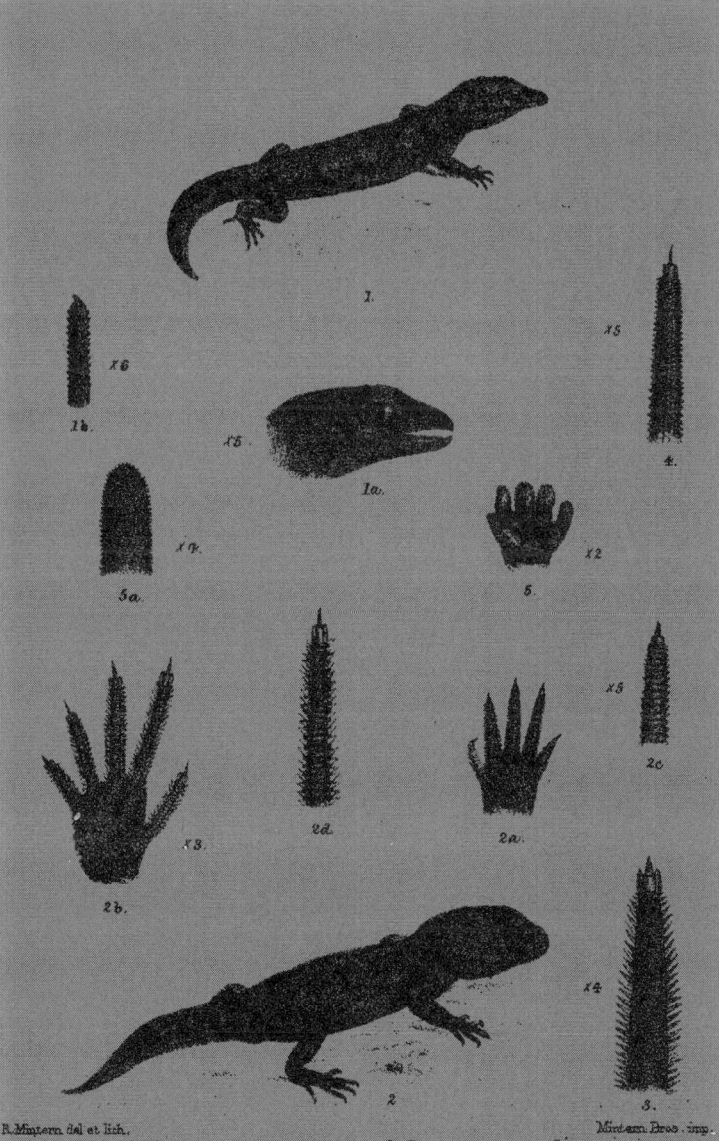

1. Rhynchœdura ornata.   2. Ptenopus garrulus.
3. Teratoscincus scincus.   4. Ceramodactylus doriæ.
5. Chondrodactylus angulifer.

BRIT. MUS. N. H.   Pl. III.

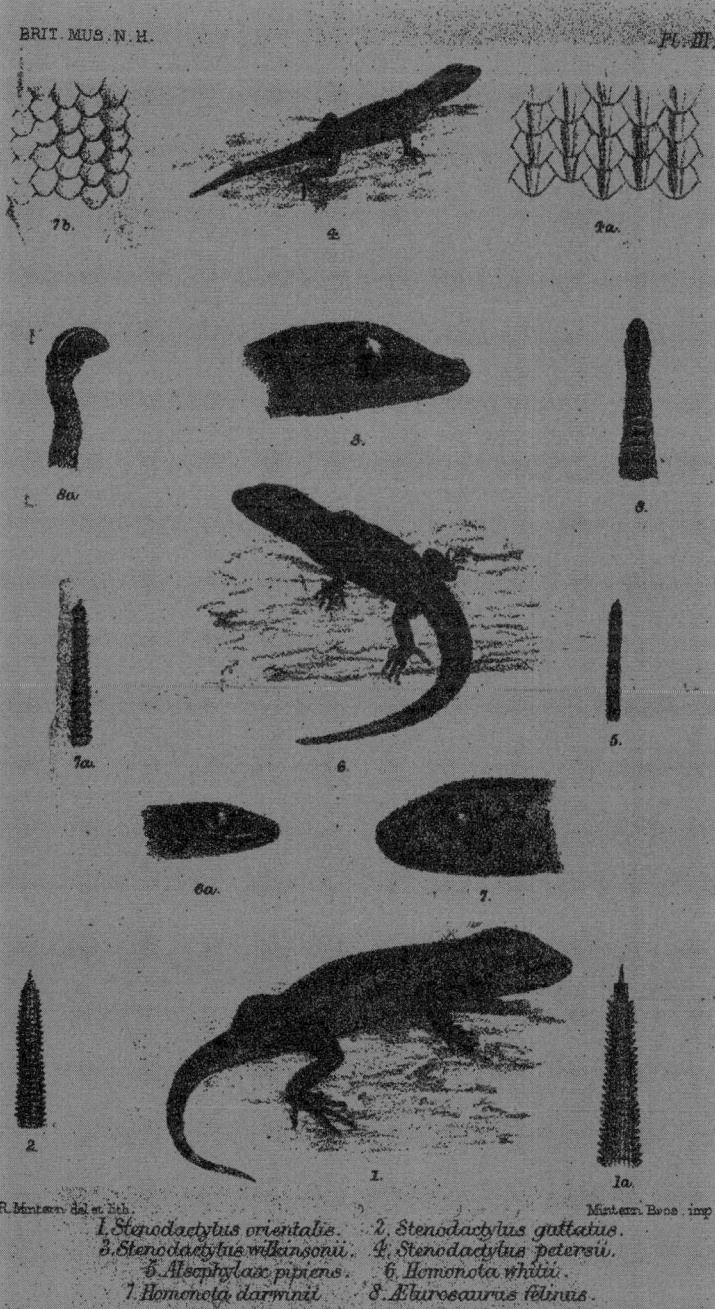

1. Stenodactylus orientalis. 2. Stenodactylus guttatus.
3. Stenodactylus wilkinsonii. 4. Stenodactylus petersii.
5. Alsophylax pipiens. 6. Homonota whitii.
7. Homonota darwinii. 8. Ælurosaurus felinus.

R. Mintern, del. et lith.
Mintern Brothers

1. Gymnodactylus nebulosus.   2. Gymnodactylus jeyporensis.
3. Gymnodactylus albofasciatus.

BRIT. MUS. N.H.   Pl. V.

B. Mintern del. et lith.   Mintern Bros. imp.

1. *Gonatodes ocellatus.*   2. *Gonatodes caudiscutatus.*
3. *Gonatodes humeralis.*   4. *Gonatodes kendallii.*

BRIT. MUS. N.H.   Pl. VI.

P. Mintern del et lith.   Mintern Bros. imp.

1. Gonatodes indicus.   2. Gonatodes wynadensis.
3. Gonatodes ornatus.   4. Gonatodes marmoratus.
5. Gonatodes gracilis.   6. Gonatodes littoralis.

1. *Phyllodactylus sancti-johannis*. 2. *Phyllodactylus macrodactylus*.
3. *Phyllodactylus guentheri*. 4. *Phyllodactylus affinis*.
5. *Phyllodactylus porphyreus*. 6. *Phyllodactylus marmoratus*.

1. *Ebenavia boettgeri.* 2. *Diplodactylus ciliaris.*
3. *Diplodactylus vittatus.* 4. *Diplodactylus polyophthalmus.*
5. *Diplodactylus steindachneri.* 6. *Diplodactylus tessellatus.*

1. Œdura ocellata. 2. Œdura marmorata.

1. Œdura robusta. 2. Œdura lesueurii.
3. Calodactylus aureus.

BRIT. MUS. N.H. Pl. XI.

R. Mintern del et lith. Mintern Bros. imp.
1. Thecadactylus australis. 2. Hemidactylus reticulatus.
3. Hemidactylus echinus. 4. Hemidactylus fasciatus.

1. *Hemidactylus stellatus.* 2. *Hemidactylus bowringii.*
3. *Hemidactylus richardsonii.* 4. *Peropus guentheri.*

Pl. XIII.

1. Spathoscalabotes mutilatus. 2. Microscalabotes cowanii.
3. Lepidodactylus ceylonensis. 4. Lepidodactylus aurantiacus.
5. Lepidodactylus pulcher. 6. Lepidodactylus cycturus.

1. *Hoplodactylus maculatus, & var.*   2. *Hoplodactylus anamallensis.*
3. *Naultinus elegans.*

1. Hoplodactylus granulatus. 2. Luperosaurus cumingii.

1. Tarentola ephippiata. 2. Pachydactylus formosus.
3. Pachydactylus oshaughnessyi. 4. Pachydactylus maculatus.
5. Pachydactylus mentomarginatus. 6. Pachydactylus mariquensis.

Pl. XVII.

Phelsuma newtonii.

BRIT. MUS. N. H.

R. Mintern del. et lith.   Mintern Bros. imp.

1. Phelsuma lineatum.   2. Sphærodactylus nigropunctatus.
3. Sphærodactylus glaucus.   4. Sphærodactylus oxyrrhinus.
5. Sphærodactylus argus.   6. Sphærodactylus richardsonii.

Cryptodelma orientalis

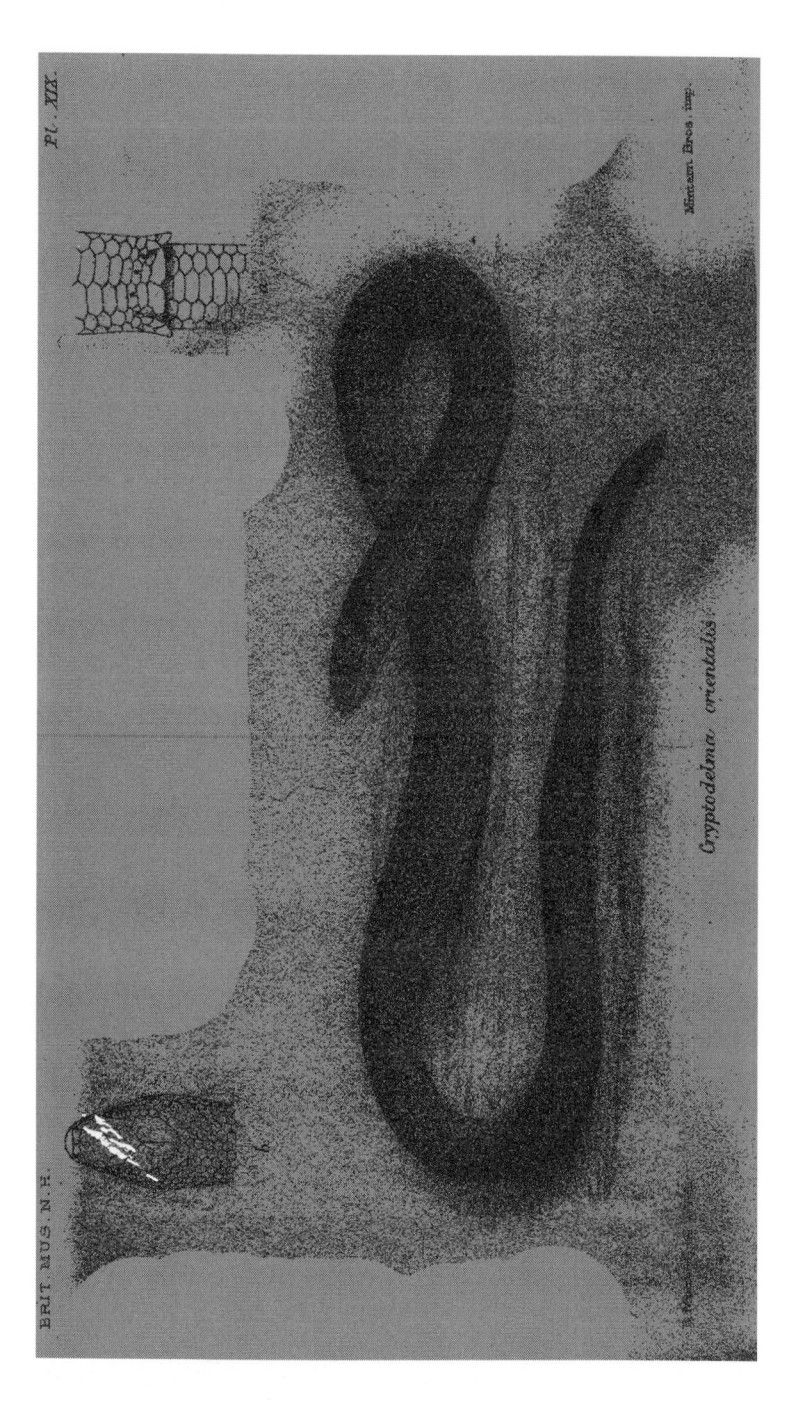

BRIT. MUS. N.H.   Pl. XX.

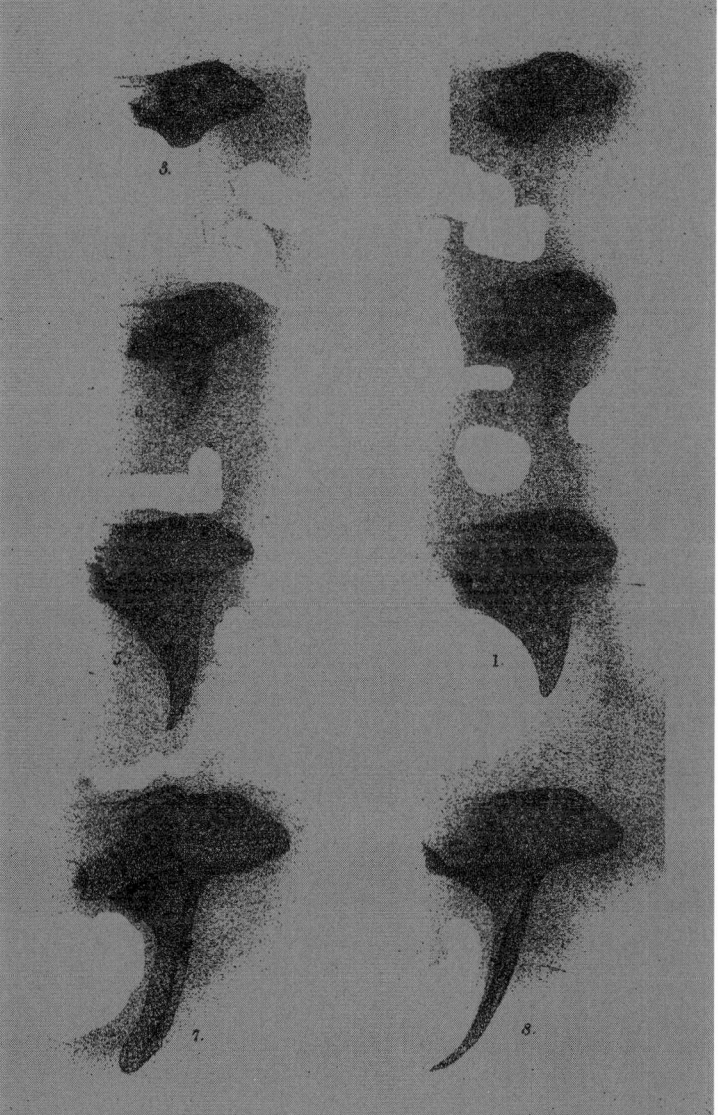

P. Smit del et lith.   Mintern Bros. imp.

1. Draco reticulatus.   2. Draco guentheri.   3. Draco everetti.
4. Draco cornutus.   5. Draco rostratus.   6. Draco bimaculatus.
7. Draco blanfordii.   8. Draco quinquefasciatus.

Gonyocephalus interruptus.

BRIT. MUS. N.H. Pl. XXII.

R. Mintern del et lith. Mintern Bros. imp.
1. Acanthosaura armata. 2. Acanthosaura crucigera.
3. Acanthosaura lamnidentata.

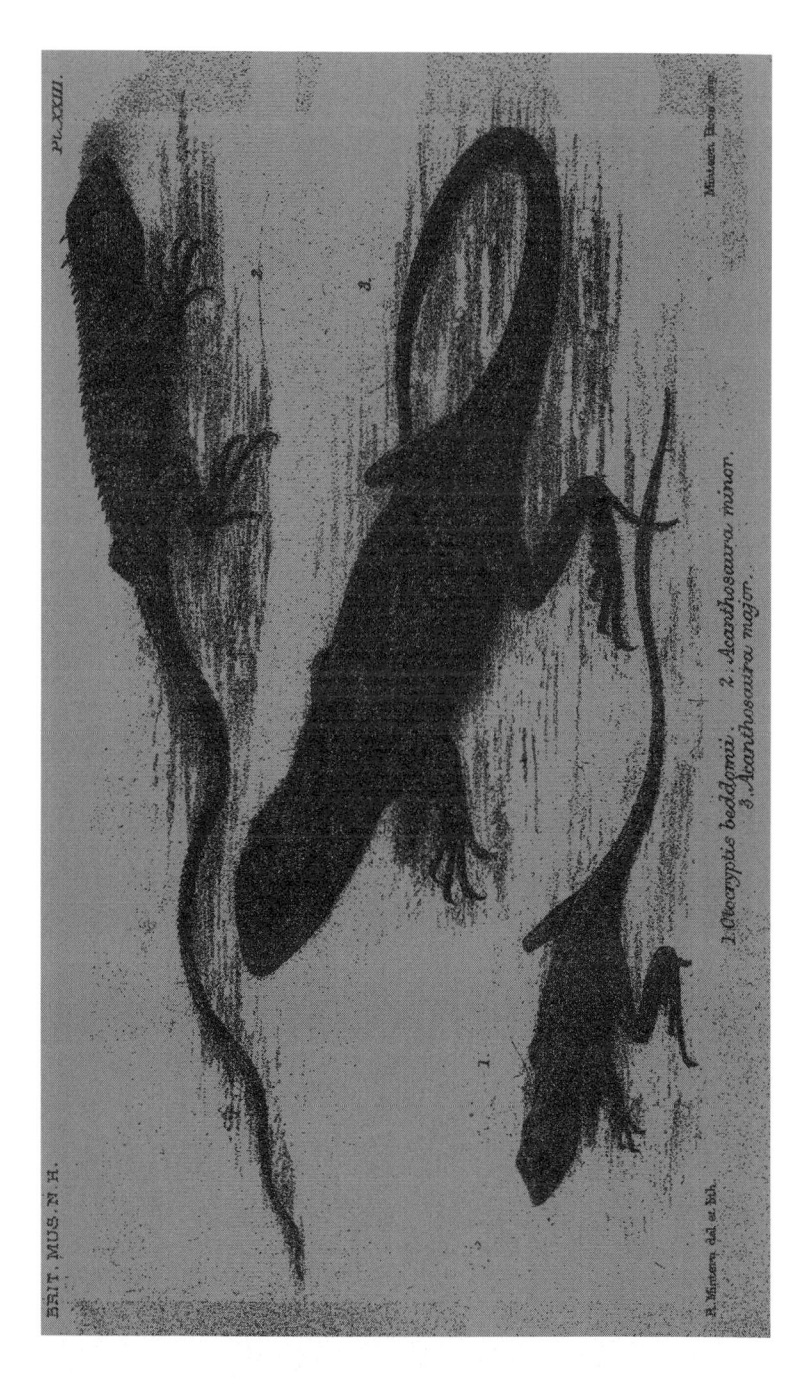

1. Otocryptis boddamii. 2. Acanthosaura minor.
3. Acanthosaura major.

1. Japalura variegata. 2. Japalura planidorsata.

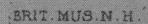

1. Calotes emma.   2. Calotes liolepis.
3. Calotes elliotti.

Calotes liocephalus.

Pl. XXVII.

BRIT. MUS. N. H.

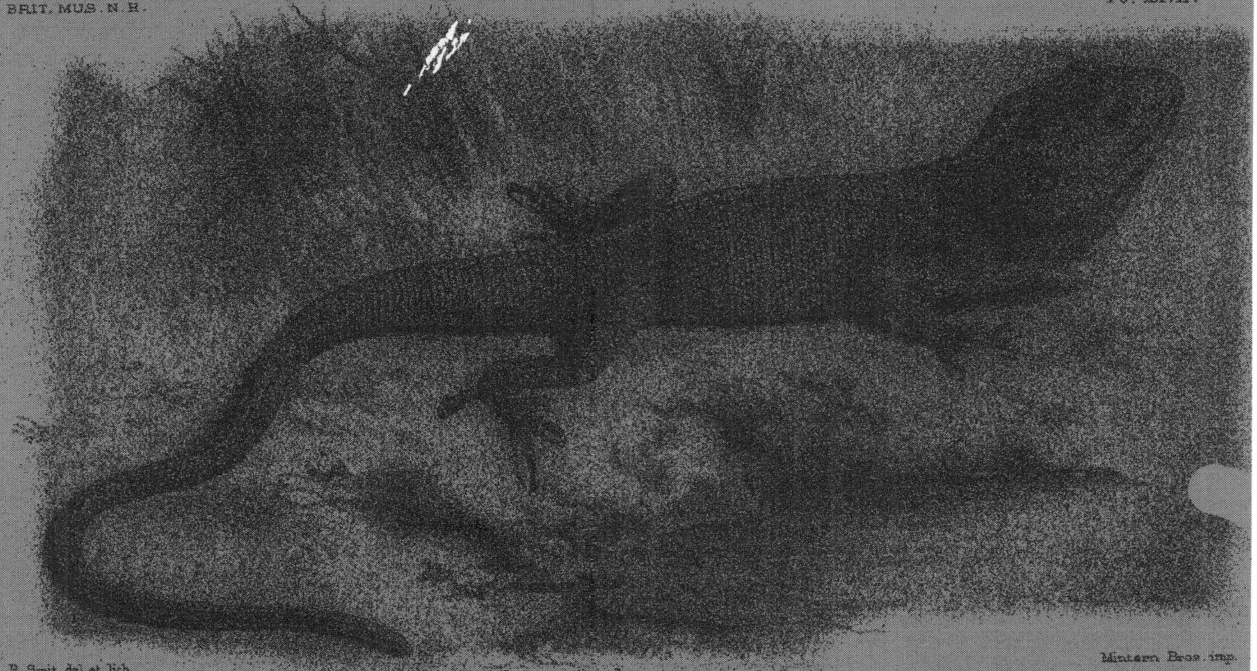

P. Smit del et lith.

Mintern Bros. imp.

Chelosania brunnea.

1. Agama brachyura. 2. Agama kirki.

1. *Amphibolurus cristatus*. 2. *Amphibolurus caudicinctus*.

1. *Amphibolurus reticulatus.* 2. *Amphibolurus pulcherrimus.*
3. *Amphibolurus pallidus.*

BRIT. MUS. N.H.   Pl. XXXI.

......t del et lith.

1. *Tympanocryptis cephalus*.   2. *Diporophora be..*
3. *Physignathus temporalis*.   4. *Physignathus longirostris*.